Günter Harder

Lectures on Algebraic Geometry II

Aspects of Mathematics

The texts published in this series are intended for graduate students and all mathematicians who wish to broaden their research horizons or who simply want to get a better idea of what is going on in a given field. They are introductions to areas close to modern research at a high level and prepare the reader for a better understanding of research papers. Many of the books can also be used to supplement graduate course programs.

Volumes of the series are listed on page 366

www.viewegteubner.de

Günter Harder

Lectures on Algebraic Geometry II

Basic Concepts, Coherent Cohomology, Curves and their Jacobians

VIEWEG+
TEUBNER

Bibliographic information published by the Deutsche Nationalbibliothek
The Deutsche Nationalbibliothek lists this publication in the Deutsche Nationalbibliografie;
detailed bibliographic data are available in the Internet at http://dnb.d-nb.de.

Prof. Dr. Günter Harder
Max-Planck-Institute for Mathematics
Vivatsgasse 7
53111 Bonn
Germany

harder@mpim-bonn.mpg.de

Mathematics Subject Classification
14-01, 14A01, 14F01, 14H01, 14K01

1st Edition 2011

Editorial Office: Ulrike Schmickler-Hirzebruch | Barbara Gerlach

Vieweg+Teubner Verlag is a brand of Springer Fachmedien.
Springer Fachmedien is part of Springer Science+Business Media.
www.viewegteubner.de

Cover design: KünkelLopka Medienentwicklung, Heidelberg
Printed on acid-free paper
Printed in Germany

ISBN 978-3-8348-0432-7

Preface

This is now, at last, the second volume of my "Lectures on Algebraic Geometry". When working on this second volume, I always had a saying by Peter Gabriel on my mind:

"Der Weg zur Hölle ist mit zweiten Bänden gepflastert!"

(The path to hell is paved with (never written?) second volumes.) Very often I felt like Sisyphos in Homer's Oddyssea. Sisyphos tries to push a rock over the ridge and just before he reaches top the rock rolls down again. Only at this very moment, when I am writing this preface, I am gaining some confidence that this second volume finally may come to life.

It is still valid what I said in the preface to the first volume, I plan to write a book on **Cohomology of arithmetic groups**. Actually there exists a very preliminary version of this "Volume III" on my home page at the Bonn university. The present book is also meant to provide background for "VolumeIII".
"Volume III" will be different in nature, we do not give an introduction into a field which is well established and already treated in other text books. It will rather be a description of a research area which is still developing, it will contain some new results, and it will put old results into a new perspective. I will formulate open questions and formulate problems, which are important on one hand but which are also tractable.
The first group of fundamental results in this book is proved in Chapter 8 when I discuss the finiteness results for the cohomology of coherent sheaves and the semi-continuity theorems. Here I use the theorems on sheaf cohomology which are proved in the first volume. I put a lot of emphasis on the relevance of the semi-continuity theorems for the construction of moduli spaces.

Moduli spaces are a central theme in this book. We discuss the moduli space of elliptic curves which are equipped in a nowhere vanishing differential form in Chapter 9. This moduli space and its generalization to moduli of abelian varieties will play a prominent role in "VolumeIII". On the other hand the representability of the modified Picard functor for curves (Chap. 10) is one of the main (and most difficult) results in this book.
At several places I give informal outlooks into further developments. In the last part of Chapter 9 I discuss the general version of the Grothendieck-Riemann-Roch theorem. Here I have to ask the reader to accept some concepts and results, for instance the existence of the Chow-ring, the theory of Chern classes and finally the Grothendieck-Riemann-Roch theorem itself.
The final goal of this book is to bring the reader to the foothills of the mountain range of étale cohomology. I give the definition of étale cohomology groups and "compute" these cohomology groups for curves. These first basic results on étale cohomology depend on the results proved in 10.2 and 10.3. Once we have some acquaintance with étale cohomology we can look to the giant peaks in the distance, for example the Weil conjectures and the modularity of elliptic curves. But we also see some peaks that so far nobody climbed, so for instance the Hodge and the Tate conjectures. We can define the L-functions attached to the cohomology of smooth projective algebraic varieties or even motives over number

fields. Then can formulate certain aspects of the Langlands program, these things will be discussed in "Volume III". There are excellent books which guide the reader into étale cohomology . (See [De1], [Mi], [F-K] [K-W].)

Again I want to thank my former student Dr. J. Schlippe, who went through this manuscript many times and found many misprint and suggested many improvements. I also thank J. Putzka who "translated" the original Plain-Tex file into Latex and made it consistent with the demands of the publisher. But he also made many substantial suggestions concerning the exposition and corrected some errors.

Günter Harder Bonn, February 2011

Contents

Introduction

This second volume starts where the first volume ends. In the first volume we did a lot of topology and also some analysis, in the last chapter we introduced compact Riemann surfaces. These are by definition compact complex manifolds of dimension one. But finally it turned out that they can be understood as purely algebraic objects; this is discussed in Vol. I, 5.1.7. In 5.1.14 we attach a locally ringed space to such a surface, and this locally ringed space is a scheme. This process of algebraization of analytic objects is continued in Vol. I, 5.3.

Hence we develop the language schemes in the first chapter of the second volume, and consequently this is Chapter 6 of the series. We discuss the basic abstract notions in the theory of schemes. Here the exposition has a higher level of abstractness and generality. In this chapter we also discuss the very abstract notions of descend. These notions play an important role in the last chapter. The reader may skip this part in first reading.

Chapter 7 is an introduction to commutative algebra and its implications in geometry. Here we are not very systematic and do not discuss all aspects in full generality. We only discuss very basic notions, we prove some of the easier theorems, and for the more difficult theorems we refer to the literature. As a byproduct the reader gets an introduction to algebraic number theory. We prove some of the fundamental theorems in algebraic number theory and formulate Dirichlet's theorem on units, the finiteness of the class number, and the unramified case of Artin's reciprocity law.

Chapter 8 is an introduction into projective algebraic geometry. After explaining the basic notions, we treat the fundamental finiteness theorems for the cohomology of coherent sheaves. After that we discuss the semi-continuity theorems, which fundamental in the construction of moduli spaces.

In the first part of Chapter 9 we consider projective curves, these are smooth projective varieties of dimension one. The first theme is the theorem of Riemann-Roch, here we emphasize that the theorem of Riemann-Roch, as it is usually stated, and Serre-duality should be considered as a unity. In my view these two theorems together should be called the Riemann-Roch theorem for curves. Our approach is different from the usual one, for our treatment is very close to the approach in the paper of Dedekind-Weber [De-We].

We then proceed and discuss some applications of the Riemann-Roch theorem. One of these applications concerns moduli problems. We show how the results on semi-continuity provide a tool to construct moduli spaces (elliptic curves together with a non vanishing differential, thm. 9.6.2). But after that we make some efforts to discuss the subtleties behind the notion of moduli spaces if the objects, which we want to classify, have automorphisms. This leads to the distinction between fine and coarse moduli spaces. The discussion also makes it clear that we can not hope for a moduli space of elliptic curves (or of curves of genus g). It is possible to define a more general class of objects, these are the so called stacks. It has been proved by Deligne and Mumford that the moduli stack of curves of genus g exists.

Finally we discuss the general Riemann-Roch theorem of Grothendieck. Here we can not prove everything, we have to accept the existence of the Chow ring, the theory of Chern classes and the isomorphism between two different definitions of K-groups. We formulate the general Grothendieck- Riemann-Roch theorem (GRR).

We also discuss and prove a special of GRR for products of curves over fields. Here we hope that the reader gets a glimpse of the proof of the general GRR. We use this version of GRR to prove the Hodge-index theorem for this special case.
In the last section of this Chapter we discuss curves over finite fields. In the beginning this looks rather innocent, but in my view it is a first culmination point in this book. We explain the relationship between the Riemann-Roch theorem and the Zeta-function of the curve. If we take the analogy between number fields and function fields into account, then the Zeta-function can be defined in terms of the function field. If we we look closer into this analogy-here we recommend strongly to read Neukirch's exposition in [Neu], Chapter VII. Then we see that the Riemann-Roch theorem (in the above sense) is essentially the Poisson summation formula and that this formula is the the basic reason for the functional equation of the Zeta-function.
But we go one step further and we give a proof of the analogue of the Riemann hypothesis, which in the realm of algebraic geometry is called the "Weil conjecture". Here we reproduce the arguments of Mattuck-Tate in [Ma-Ta] and of Grothendieck in [Gr-RH] and we show how the Riemann hypothesis follows from the Hodge index theorem applied to the product of the curve by itself.

In the last Chapter we discuss the the Picard functor on curves, in other words we investigate line bundles, or better the totality of line bundles on a given curve C/k. The first theorem is the the representability of some slightly modified Picard functors. This is a hard piece of work.
We prove that $\mathrm{Pic}^0_{C/k}$ is an abelian variety defined over k, this means it is a connected projective variety together with the structure of a commutative group scheme. It is called the Jacobian of the curve.
Starting from there we develop the theory of abelian varieties, and we study the Picard scheme of abelian varieties, we investigate their endomorphism rings and the ℓ-adic representation. This exposition overlaps with the book [Mu1], but we start from the Jacobians as prototypes of abelian varieties, whereas Mumford stubbornly avoids to speak of Jacobians.
Finally we give an outlook to the étale cohomology of schemes. We explain the concepts and formulate some of the basic theorems. Especially we formulate Deligne's theorem, i.e. we give the formulation of the Weil conjectures for smooth projective varieties. We prove this theorem (in a certain sense) for abelian varieties and for curves.

We conclude by discussing a degenerating family of elliptic curves. The purpose of this example is twofold. Firstly: Understanding such degenerations is important for the compactification of moduli spaces (stacks) of curves or abelian varieties. We describe in this special case how the theory of Θ -functions can be used to analyze elliptic curves in the neighborhood of their locus of degeneration, and write down explicit equations. This gives us a tool to compactify the moduli space. For the general case of abelian varieties we refer to [Fa-Ch].
Secondly we use this example to illustrate the final step in Deligne's proof of the Weil conjecture. This gives me the opportunity to finish this book with an exceptionally beautiful proof.

6 Basic Concepts of the Theory of Schemes

6.1 Affine Schemes

We consider commutative rings $A,B,\dots$ with identity $(1_A,1_B,\dots)$, homomorphisms $\phi : A \to B$ are always assumed to send the identity of A into the identity of B. We always assume that the identity in a ring is different from zero. A ring A is called **integral** if it does not have zero divisors.

For any such ring A we have the group of invertible elements (units):

Definition 6.1.1. *The* **group of invertible elements (units)** *of a commutative ring with identity is defined by $A^\times = \{a \in A \mid \exists\, a' \in A$ such that $aa' = 1_A\}$. An Element in $A^\times$ is called* **unit**.

Definition 6.1.2. *A* **proper** *ideal $\mathfrak{a} \subset A$ is an ideal with $1_A \notin \mathfrak{a}$, prime ideals are always proper.*

For any ring and any $f \in A$ we use the standard notation (f) for the principal ideal Af. If we have a homomorphism $\phi : A \to B$, then we will say that B is an **A-algebra** .

6.1.1 Localization

If we have a subset $S \subset A$, which is closed under multiplication and contains the identity $1_A \in S$, we can define a quotient ring A_S and a map $\phi_S : A \longrightarrow A_S$ such that the elements of S become invertible.
To do this we consider pairs $(a,s) \in A \times S$ and introduce an equivalence relation

$$(a,s) \sim (a',s') \iff \exists\, s'' \in S \text{ such that } (as' - a's) \cdot s'' = 0. \tag{6.1}$$

We consider the quotient A_S of $A \times S$ by this relation, let $\pi_S : A \times S \longrightarrow A_S$ be the projection to this quotient. We define a ring structure on A_S by

$$\begin{aligned} \pi_S((a,s)) + \pi_S((a',s')) &= \pi_S((as' + a's,ss')) \\ \pi_S((a,s)) \cdot \pi_S((a',s')) &= \pi_S((aa',ss')). \end{aligned} \tag{6.2}$$

We have a homomorphism of rings

$$\begin{aligned} \phi_S : A &\longrightarrow A_S \\ a &\longmapsto \pi_S((a,1)). \end{aligned}$$

We will write the elements of A_S simply as

$$\pi_S((a,s)) = \frac{a}{s} = as^{-1}.$$

Of course

$$\frac{a}{s} = \frac{as'}{ss'} = \frac{as's''}{ss's''} \cdots$$

Lemma 6.1.3. *The quotient ring has a universal property: For any ring B we can consider*

$$\mathrm{Hom}_{Rings,\ S\ to\ units}(A,B) = \{\phi : A \longrightarrow B | \phi(S) \in B^\times \ \ for\ all\ \ s \in S\}$$

and this set of homomorphisms is equal to $\mathrm{Hom}_{\mathrm{Rings}}(A_S,B)$, *where the identification is given by the diagram*

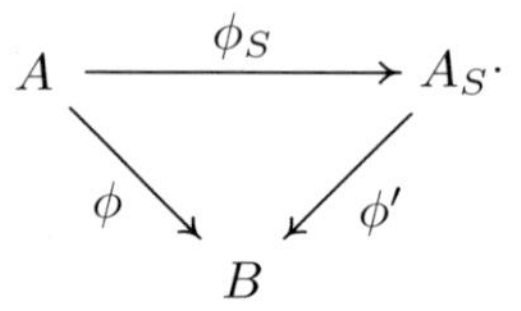

If $0 \in S$ then $A_S = \{0\}$. If $f \in A$ then we write $A_f = A_{\{f^n\}_{n=0,1\dots}}$.

6.1.2 The Spectrum of a Ring

Definition 6.1.4. *If A is a commutative ring with identity then we define the* **spectrum** *of A as* $\mathrm{Spec}(A) = \{\mathfrak{p} | \ \mathfrak{p}\ prime\ ideal\ in\ A\}$.

Lemma 6.1.5. *The spectrum of A is ordered. The ordering is given by the inclusion among prime ideals. The spectrum is functorial in A, if we have a homomorphism $\phi : A \longrightarrow B$ then it induces*

$$\begin{aligned} {}^t\phi : \mathrm{Spec}(B) &\longrightarrow \mathrm{Spec}(A) \\ {}^t\phi(\mathfrak{p}) &= \phi^{-1}(\mathfrak{p}) = \{f \mid \phi(f) \in \mathfrak{p}\} \end{aligned}$$

and ${}^t\phi$ respects the order relation.

Definition 6.1.6. *A* **maximal ideal** $\mathfrak{m} \subset A$ *is an ideal with $1_A \notin \mathfrak{m}$ and for any ideal $\mathfrak{m}'$ with $\mathfrak{m} \subseteq \mathfrak{m}' \subseteq A$ we have $\mathfrak{m} = \mathfrak{m}'$ or $\mathfrak{m}' = A$.*

The spectrum $\mathrm{Spec}(A)$ always contains the set of maximal ideals. We have a different characterization

Lemma 6.1.7. *An ideal $\mathfrak{m} \subset A$ with $1_A \notin \mathfrak{m}$ is maximal if and only if $A/\mathfrak{m}$ is a field. Maximal ideals are prime ideals.*

This is clear. The set of maximal ideals is denoted by $\mathrm{Specmax}(A) \subset \mathrm{Spec}(A)$.

Definition 6.1.8. *A* **chain** *of proper ideals is a totally ordered subset $\mathfrak{K}$ of the set of proper ideals, this means that for any pair $\mathfrak{a},\mathfrak{b} \in \mathfrak{K}$ we have $\mathfrak{a} \subseteq \mathfrak{b}$ or $\mathfrak{b} \subseteq \mathfrak{a}$.*

Zorn's lemma implies:

Proposition 6.1.9. *If A is a commutative ring with $1_A \neq 0$, then*

$$\mathrm{Specmax}(A) \neq \emptyset.$$

Proof: For any chain we can form $\bigcup_{\mathfrak{a}\in K} \mathfrak{a} = \mathfrak{a}^*$, this is an ideal with $1_A \notin \mathfrak{a}^*$ and $\mathfrak{a}^* \supset \mathfrak{a}$ for all $\mathfrak{a} \in \mathfrak{K}$. Hence we see that for any chain of proper ideals we can find a proper ideal which contains all elements of the chain. Now it is simply the assertion of Zorn's lemma that this implies the existence of a (proper) maximal ideal. □

This has as a consequence: We call an element $f \in A$ **nilpotent** if there exists an integer n such that $f^n = 0$. These elements form an ideal $\mathrm{Rad}\,(A)$, which is called the radical.

Lemma 6.1.10. *If $f \in A$ is not nilpotent, then $\mathrm{Spec}(A_f) \neq \emptyset$. Hence we get*

$$\mathrm{Rad}\,(A) = \textit{Ideal of nilpotent elements} = \bigcap_{\mathfrak{p}\in \mathrm{Spec}(A)} \mathfrak{p}.$$

Definition 6.1.11. *A commutative ring is called a* **reduced** *if it does not have non zero nilpotent elements.*

Definition 6.1.12. *A commutative ring is called a* **local ring** *if it has a unique maximal prime ideal.*

If $\mathfrak{p} \in \mathrm{Spec}(A)$ then the complement $S = A \setminus \mathfrak{p}$ is closed under multiplication. Then we write (abuse of notation)

$$A_{(A\setminus\mathfrak{p})} =: A_{\mathfrak{p}}.$$

Definition 6.1.13. *The ring $A_{\mathfrak{p}}$ is local and is called the* **local ring at $\mathfrak{p}$**. *The ideal*

$$\mathfrak{m}_{\mathfrak{p}} = \left\{ \tfrac{f}{g} \mid f \in \mathfrak{p}, g \notin \mathfrak{p} \right\}.$$

is the unique maximal ideal in this ring. The field $k(\mathfrak{p}) = A_{\mathfrak{p}}/\mathfrak{m}_{\mathfrak{p}}$ is called the **residue field at $\mathfrak{p}$.**

Lemma 6.1.14. *If we consider any multiplicatively closed $S \subset A$ and the localization $\phi : A \longrightarrow A_S$ then ${}^t\phi$ is an inclusion. We get*

$${}^t\phi : \mathrm{Spec}(A_S) \xrightarrow{\sim} \{\mathfrak{p} \in \mathrm{Spec}(A) | \mathfrak{p} \cap S = \emptyset\}.$$

If especially $S = \{f^n\}_{n=0,1...}$ then

$$\mathrm{Spec}(A_f) = \{\mathfrak{p} \in \mathrm{Spec}(A) | f \notin \mathfrak{p}\}.$$

If $\mathfrak{p} \in \mathrm{Spec}(A)$ then

$$\mathrm{Spec}(A_{\mathfrak{p}}) = \{\mathfrak{q} \in \mathrm{Spec}(A) \mid \mathfrak{q} \subset \mathfrak{p}\}.$$

The proof is left to the reader.

Remark 1 (Heuristical remarks).

1. The spectrum of a ring is a geometric object. At this point it is simply an ordered set, but soon we will put a topology onto this space (**The Zariski topology**). We already spoke of maximal ideals. If our ring A is integral, then the zero ideal (0) is also a prime ideal. It is the unique minimal element in $\mathrm{Spec}(A)$. This ideal (0) is called the **generic point** of $\mathrm{Spec}(A)$.

2. Intuitively we want to consider A as a ring of functions on $\mathrm{Spec}(A)$. This is not quite the case because these functions do not have a common domain of values. But it makes sense to say that $f \in A$ "vanishes" at $\mathfrak{p} \in \mathrm{Spec}(A)$: By this we mean that $f \in \mathfrak{p}$. Sometimes we will write $f(\mathfrak{p}) = 0$, (resp. $f(\mathfrak{p}) \neq 0$) for $f \in \mathfrak{p}$ (resp.$f \notin \mathfrak{p}$).

Example 1. *The ring $\mathbb{Z}$ and the polynomial ring $k[X]$ are principal ideal domains. This implies immediately that the maximal ideals are of the form $\mathfrak{p} = (p)$ resp. $\mathfrak{p} = (p(X))$ where p is a prime number (resp. $p(X) \in k[X]$ is a non-constant irreducible polynomial). Both rings contain one more prime ideal namely $\mathfrak{p} = (0)$ because they are integral. Hence*

$$\mathrm{Spec}(\mathbb{Z}) = (0) \cup \{(2),(3),(5),\dots\}$$
$$\mathrm{Spec}(k[X]) = (0) \cup \{(X),(X-1),\dots\}.$$

Of course not all irreducible polynomials are linear, but we cannot write down any other polynomial, which is irreducible regardless what the field k is.

Example 2. *Let us assume that k is a field. We consider the polynomial ring $A = k[X_1,X_2,\dots,X_n]$ in n variables. For any point $P = (a_1,a_2,\dots,a_n) \in k^n$ we get an evaluation homomorphism*

$$\phi_P : A \longrightarrow k$$
$$\phi_P : f \mapsto f(P),$$

whose kernel is the maximal ideal $\mathfrak{m}_P = (X_1 - a_1, X_2 - a_2, \dots, X_n - a_n)$: If our field k is algebraically closed then the Nullstellensatz of Hilbert (See 7.1.11 *) says that we get an identification*

$$\mathrm{Specmax}(k[X_1,X_2,\dots,X_n]) = k^n. \tag{6.3}$$

In other words the maximal ideals are exactly the ideals of the form $\mathfrak{m} = \mathfrak{m}_P$.

Exercise 1. Prove the Nullstellensatz in the case of a polynomial ring in one variable.

Exercise 2. Try to prove it in the case of a polynomial ring $A = k[X,Y]$ in two variables. Of course k is still algebraically closed.
We give a hint: Let $\mathfrak{m}$ be a maximal ideal. It cannot be the zero ideal.

Step 1: Assume $\mathfrak{m}$ contains an element of the form

$$F(X,Y) = Y^m + g_1(X)Y^{m-1} + \dots g_m(X)$$

where the g_i are polynomials in X. Now we get a diagram

$$\begin{array}{ccc} A & \longrightarrow & A/(F) = B\cdot \\ & \searrow \quad \swarrow & \\ & A/\mathfrak{m} & \end{array}$$

The ring B contains the polynomial ring $k[X] = B_0$ and over this ring it is generated by an element y, which satisfies the relation $y^m + g_1(X)y^{m-1} + \dots g_m(X) = 0$. The maximal ideal $\mathfrak{m} \subset A$ has as its image a maximal ideal $\overline{\mathfrak{m}}$ in B.

Prove that $\overline{\mathfrak{m}} \cap k[X]$ is a maximal prime ideal! In this case it suffices to show it is not zero. Hence $B_0/\overline{\mathfrak{m}} = k$ and since $B/\mathfrak{m}$ is a finite extension of k it follows that $B/\mathfrak{m} = k$.

Step 2: We know that $\mathfrak{m}$ contains a non zero polynomial

$$F(X,Y) = \sum_{\nu\mu} a_{\nu,\mu} X^\nu Y^\mu.$$

Write this as a polynomial in Y with coefficients polynomials in X. Now it will not be the case in general that the highest power of Y occurring in the polynomial has a constant coefficient as in Step 1. But if we make a substitution

$$\begin{aligned} X &\longrightarrow X + Y^m = X' \\ Y &\longrightarrow Y \end{aligned}$$

then $k[X',Y] = k[X,Y]$ and for $m >> 0$ the new polynomial will satisfy the assumption in step 1.

It is known that a polynomial ring $k[X_1,\dots,X_n]$ has unique factorization, we will discuss this fact in the Chapter VIII on commutative algebra. (see exercise 19 and Theorem 7.1.5) This implies that any non constant irreducible polynomial $f \in k[X_1,\dots,X_n]$ defines a prime ideal $\mathfrak{p} = (f)$. Therefore the ring $\mathrm{Spec}(k[X,Y])$ contains many more elements than just (0) and the maximal ones: Any irreducible polynomial

$$p(X,Y) = X + Y \text{ or } X^2 + Y^3 \text{ or} \dots$$

defines a prime ideal $\mathfrak{p} = (p(X,Y))$. If k is algebraically closed then the Nullstellensatz allows us to identify $\mathfrak{p} = (p(X,Y))$ with

$$V(\mathfrak{p}) = \{(a,b) \in k^2 | f(a,b) = 0 \text{ for all } f \in \mathfrak{p}\} \tag{6.4}$$

this is the set of common zeroes of the elements in $\mathfrak{p}$ or the set of zeroes of $p(X,Y)$ and also the set of maximal ideals containing $\mathfrak{p}$. Hence we get an injection into the power set

$$\begin{aligned} \mathrm{Spec}(k[X,Y]) &\longrightarrow \mathfrak{P}(\mathrm{Specmax}(k[X,Y])) &(6.5)\\ \mathfrak{p} &\longmapsto V(\mathfrak{p}). &(6.6) \end{aligned}$$

The maximal ideals correspond to the sets consisting of one element, the prime ideals $\mathfrak{p} = (p(X,Y))$ give hypersurfaces and $\mathfrak{p} = (0)$ gives us the entire plane.

Example 3. *Let k be arbitrary, we consider the $A = k[X,Y]/(XY) = k[x,y]$. The elements x,y satisfy $xy = 0$. Hence this ring has zero divisors.*
A prime ideal $\mathfrak{p}$ in A has to contain either x or y. On the other hand the principal ideals $\mathfrak{p} = (x)$ and $\mathfrak{q} = (y)$ are prime because after dividing by them we get polynomial rings in the other variable.
We see that

$$\mathrm{Spec}(k[x,y]) = \mathrm{Spec}(k[x]) \cup_{(0,0)} \mathrm{Spec}(k[y])$$

where the two spectra are identified at $(x,y) = (0,0)$.
This is an example of a **reducible** *spectrum. (See Def. 7.2.2)*

6.1.3 The Zariski Topology on $\mathrm{Spec}(A)$

We define a topology on the space X. To do so we have to define what open sets are. At first we declare the sets of the form

$$X_f = \mathrm{Spec}(A_f) \subset X$$

open. We saw that X_f was the set of prime ideals $\mathfrak{p}$, which do not contain f. In our remark 1 we said that this means f *does not vanish at* $\mathfrak{p}$. Hence our topology has the property that the sets, where a given $f \in A$ is not zero, i.e. does not vanish, are open sets.
This system of sets is closed under finite intersection because

$$X_{f_1} \cap \ldots \cap X_{f_s} = X_{f_1 \cdots f_s}.$$

These open sets are called **affine open sets** the reason is that they are again equal to a spectrum of a ring. This system of affine open sets forms a basis for the Zariski topology and this means that a set $U \subset X$ is open if and only if it is the union of the affine open sets, which are contained in U.
A subset $Y \subset X$ is closed if the complement $X \backslash Y$ is open. Of course this means that Y is the set of common zeroes of a collection of elements in A. Clearly the set of $f \in A$, which vanish on Y form an ideal $I(Y)$. If in turn we have given an ideal I then we may consider its set common zeroes $V(I)$. Clearly we always have $V(I(Z)) = Z$ but it is easy to see in examples that we may have a proper inclusion $I \subset I(V(Z))$. (What is $V(\{0\})$ and what is $I(X)$?) The topological space X will be called the **underlying space** .

The Zariski topology is not Hausdorff in general. It has other strange properties one has to get used to:

Exercise 3. If $\mathfrak{p} \in \mathrm{Spec}(A)$ then the closure of the set $\{\mathfrak{p}\}$ is given by

$$\overline{\{\mathfrak{p}\}} = \{\mathfrak{q} \in \mathrm{Spec}(A) | \mathfrak{q} \supset \mathfrak{p}\}.$$

A point $\mathfrak{p} \in \mathrm{Spec}(A)$ is closed if and only if $\mathfrak{p}$ is maximal.

If $\mathfrak{q}$ is in the closure of $\{\mathfrak{p}\}$ then we say that $\mathfrak{q}$ is a **specialization** of $\mathfrak{p}$.

Example 4. *In our rings* $\mathbb{Z}$ *and* $k[X]$ *the closed points are the principal ideals* (p) *resp.* $(p(x))$ *where* p *is a prime (resp.* $p(x)$ *is a non-constant irreducible polynomial). The generic point* (0) *is dense in* $\mathrm{Spec}(A)$ *in both cases. The open sets are the complements of finite sets of closed points and the empty set. Here it becomes quite clear that* $\mathrm{Spec}(\mathbb{Z})$ *and* $\mathrm{Spec}(k[X])$ *are not Hausdorff.*
For an integral ring A *the generic point* (0) *is always dense in the space* $\mathrm{Spec}(A)$*. Every prime ideal* $\mathfrak{p} \in \mathrm{Spec}(A)$ *is a specialization of the generic point.*

General Remark: We have put some further structure onto the set $\mathrm{Spec}(A)$: Now it is a topological space. But still this space does not yet contain a lot of information on the original ring A. If for instance A is a field, then $\mathrm{Spec}(A)$ is a single point, which will never be able to recover the field A.
This is different for finitely generated algebras $A = k[x_1,x_2,\ldots,x_n]$ over an algebraically closed field k. In the next section we will see that the Nullstellensatz (7.1.11) implies

$$\bigcap_{\mathfrak{m}\in \mathrm{Specmax}(A)} \mathfrak{m} = \mathrm{Rad}(A). \tag{6.7}$$

If our k-algebra is reduced, i.e. if $\mathrm{Rad}(A) = 0$, then this implies that we can view A as an algebra of k-valued functions on $\mathrm{Specmax}(A)$. Then we will see that $\mathrm{Specmax}(A)$ contains a lot of information on A. (See following exercise.)This discussion is resumed in the sections 6.2.6

Exercise 4. Let k be an algebraically closed field and let $A = k[x_1,x_2,\ldots,x_n]$, $B = k[y_1,y_2,\ldots,y_m]$ be two finitely generated reduced k-algebras. Let $\phi : A \longrightarrow B$ be a homomorphism, which is the identity on k (a k-algebra homomorphism). Show that this induces a map $\phi^* : \mathrm{Specmax}(B) \longrightarrow \mathrm{Specmax}(A)$. We assume in addition that A and B do not contain nilpotent elements. Observe that ϕ^* is the restriction of ${}^t\phi$ to $\mathrm{Specmax}(B)$ and hence it contains less information than ${}^t\phi$.

Prove:

(a) ϕ^* is injective if and only if ϕ is surjective.

(b) If ϕ^* is surjective, then ϕ is injective.

(c) The map ϕ^* determines ϕ.

We return to the general discussion. The following property of the topological space $\mathrm{Spec}(A)$ is very important and perhaps a little bit surprising at the first glance.

Proposition 6.1.15. *The space* $X = \mathrm{Spec}(A)$ *is quasicompact, this means that for any covering* $\bigcup_{i\in I} U_i$ *by open sets* U_i *we can find a finite subcovering, i. e. we can find a finite subset* $E \subset I$ *such that* $X = \bigcup_{i\in E} U_i$.

Proof: The U_i are open, hence we can cover each of them by open sets of the form $X_{f_{i,\nu}}$. Therefore it is clear that we may assume that the U_i themselves are of this form $U_i = X_{f_i}$. Now we consider the ideal generated by the f_i, it consists of the finite linear combinations

$$\mathfrak{a} = \left\{ \sum_{i \in I} g_i f_i \mid \text{ almost all } g_i = 0 \right\}.$$

This cannot be a proper ideal because otherwise we could find a maximal ideal $\mathfrak{m}$ containing $\mathfrak{a}$ (see Proposition 6.1.9). Then we have $f_i(\mathfrak{m}) = 0$ for all $i \in I$ and hence $\mathfrak{m} \notin \bigcup_{i \in I} X_{f_i}$. This implies that $\mathfrak{a} = A$ and hence the identity 1_A is in our ideal. We can find a finite linear combination $1_A = \sum_{i \in E} g_i f_i$ with $E \subset I$ finite. But then it is clear that $X = \bigcup_{i \in E} X_{f_i}$ because if there would be a $\mathfrak{p}$ not contained in this union then we would have $f_i(\mathfrak{p}) = 0$ for all $i \in E$ and hence $1_a(\mathfrak{p}) = 0$, which cannot be the case if our ring is not the zero ring. But for this last ring the spectrum is empty so the claim is also clear. □

Our next goal will be to put more structure on $X = \mathrm{Spec}(A)$. Since it is already a topological space we have the notion of a sheaf on this space. We will construct the sheaf of regular functions on $\mathcal{O}_X = \mathrm{Spec}(A)$ and then $(X,\mathcal{O}_X)$ will be a locally ringed space.(See Vol I, 3.2.)

6.1.4 The Structure Sheaf on Spec(A)

We want to introduce the structure of a locally ringed space on $X = \mathrm{Spec}(A)$. This means that we want to construct a sheaf of rings $\widetilde{A}$ on X, which plays the role of the sheaf of regular functions on X. It will turn out – but this will be a theorem – that the ring of regular functions on the total space is again A.
We make the following Ansatz: If we have an open set $X_f \subset X$ then the element $1/f \in A_f$ should be a regular function on the affine open set X_f. Hence we define $\widetilde{A}'(X_f) = A_f$. If we have $h = gf$ then $A_h = (A_f)_g$ then the map

$$\phi_{\{g^n\}_{n=0,1,\dots}} : A_f \longrightarrow A_h$$

gives us a restriction map $\widetilde{A}'(X_f) \to \widetilde{A}'(X_{gf}) = \widetilde{A}'(X_h)$ where $X_{gf} = X_h$. This obviously satisfies the transitivity relation for presheaves this means that the restriction from $\widetilde{A}'(X_f)$ to X_{fg} and then composed with the restriction to X_{fgl} is equal to the restriction from X_f to X_{fgl}. We will denote the restriction map $\phi_{\{g^n\}_{n=0,1,\dots}} : A_f \to A_h$ also by $F \mapsto F \mid X_h$. Hence our $\widetilde{A}'$ is something like a presheaf except that it is not yet defined on all open set but only on the affine open sets of the form X_f.
We now have a proposition, which says that $\widetilde{A}'$ satisfies axioms (Sh1), (Sh2) in Vol. I, Definition 3.1.2 provided we restrict them to these special open sets.

Proposition 6.1.16. *If we have an arbitrary covering $X = \bigcup_{i \in I} X_{f_i}$ then the sequence*

$$A = \widetilde{A}'(X) \xrightarrow{p_0} \prod_{i \in I} \widetilde{A}'(X_{f_i}) \underset{p_2}{\overset{p_1}{\rightrightarrows}} \prod_{(i,j) \in I \times I} \widetilde{A}'(X_{f_i f_j})$$

is exact, this means that the first arrow is injective and its image is exactly the set of elements, which become equal under p_1 and p_2.

This proposition is really central.

Since our space is quasicompact, we can find a finite subset $E \subset I$ such that already $X = \bigcup_{i \in E} X_{f_i} = X$. We assume that we proved exactness for this finite covering. We want to show that then we have exactness for the original covering. We get a map from the diagram above to the corresponding diagram for our finite covering:

$$\begin{array}{ccccc}
\widetilde{A}'(X) & \xrightarrow{p_0} & \prod_{i\in I} \widetilde{A}'(X_{f_i}) & \overset{p_1}{\underset{p_2}{\rightrightarrows}} & \prod_{(i,j)\in I\times I} \widetilde{A}'(X_{f_i f_j}) \\
\downarrow & & \downarrow & & \downarrow \\
\widetilde{A}'(X) & \xrightarrow{p_0^E} & \prod_{i\in E} \widetilde{A}'(X_{f_i}) & \overset{p_1^E}{\underset{p_2^E}{\rightrightarrows}} & \prod_{(i,j)\in E\times E} \widetilde{A}'(X_{f_i f_j}).
\end{array}$$

The injectivity of the first arrow is quite obvious, because the arrow p_0^E for the second diagram is the composite of the p_0 for the first diagram and the projection from the product over I to the product over E. Now let us take an element $(\dots, \frac{g_i}{f_i^{n_i}}, \dots)_{i\in I}$ in the first diagram with

$$p_1\left(\left(\dots, \frac{g_i}{f_i^{n_i}}, \dots\right)\right) = p_2\left(\left(\dots, \frac{g_i}{f_i^{n_i}}, \dots\right)\right).$$

If we project it to the second diagram then the image is also equalized by the corresponding two arrows. Hence by our assumption it comes from an element $g \in A$, this means that the image of g in A_{f_i} is equal to $g_i/f_i^{n_i}$ for all $i \in E$. We have to show that this implies that g actually maps to $(\dots, g_i/f_i^{n_i}, \dots)_{i\in I}$. Hence we have show that g maps to $g_i/f_i^{n_i}$ for all $i \in I$. Let us pick an $i \in I$. We know that

$$X_{f_i} = \bigcup_{e\in E} (X_{f_i} \cap X_{f_e}) = \bigcup_{e\in E} X_{f_i f_e}.$$

But then we have

$$g\,|X_{f_i f_e} = (g\,|X_{f_e})|\,X_{f_e f_i} = \left.\frac{g_e}{f_e^{n_e}}\right| X_{f_e f_i} = \left.\frac{g_i}{f_i^{n_i}}\right| X_{f_e f_i}$$

for all $e \in E$. Now we need a little remark. The open set X_{f_i} is again the spectrum of a ring. Hence everything we proved for X is also valid for X_{f_i}. Especially we can assume that $A_{f_i} \to \prod_{e\in E} A_{f_i f_e}$ is injective. We have seen that $g \mid X_{f_i f_e} = g_i/f_i^{n_i} \mid X_{f_e f_i}$ for all $e \in E$ hence we conclude $g \mid X_{f_i} = g_i/f_i^{n_i} \mid X_{f_i}$ for all $i \in I$. Hence the reduction to the case of a finite covering is complete and therefore, we assume the the index set I is finite.

If the homomorphism p_0 is not injective then we have have an element $f \in A$ and $f \mid X_{f_i} = 0$ for all $i \in I$. This means that we can find exponents n_i so that $f f_i^{n_i} = 0$ in the ring A. Since I is finite we can assume that all these exponents are equal to a fixed integer n.

In the proof for the quasicompactness of X we have seen that we can find $g_i \in A$ such that

$$\sum_i g_i f_i = 1.$$

Raising this to a suitable high power N we get a relation

$$\sum_i G_i f_i^n = 1$$

and hence $f = f1 = \sum_{i\in I} G_i f_i^n f = 0$ and proves injectivity.
Now let us assume we have an element $(\ldots, g_i/f_i^{n_i}, \ldots)_{i\in I}$, for which

$$\frac{g_i}{f_i^{n_i}}\bigg| X_{f_i,f_j} = \frac{g_i}{f_j^{n_j}}\bigg| X_{f_i,f_j},$$

for all pairs $(i,j) \in I \times I$. Again we may assume that all n_i are equal. Then the equality means that we can find an integer N so that

$$(g_i f_j^n - g_j f_i^n)(f_i f_j)^N = 0.$$

We are searching an element $g \in A$, which satisfies $g \mid X_{f_i} = \frac{g_i}{f_i^n}$ for all $i \in I$. This is certainly true if $f_i^n g = g_i$ for all $i \in I$. But again we can find $H_\nu \in A$ for $\nu \in I$ such that

$$\sum_{\nu\in I} H_\nu f_\nu^n = 1$$

and we see that $g = \sum_{\nu\in I} H_\nu g_\nu$ solves our problem. □

Still we have not yet defined our sheaf $\widetilde{A}$. For an arbitrary open set $U \subset X$ we choose a covering $U = \bigcup_{i\in I} X_{f_i}$ and define as in Vol. I 3.1.3

$$\widetilde{A}(U) = \left(\prod_{i\in I} A_{f_i} \underset{p_2}{\overset{p_1}{\rightrightarrows}} \prod_{(i,j)\in I\times I} A_{f_i f_j}\right)[p_1 = p_2]. \tag{6.8}$$

We have to verify that this does not depend on the covering and really defines a sheaf. We will not do this in detail, the proof is a little bit tedious. To prove the independence of the covering we first pass to a refinement of the covering: We have $\tau : J \to I$ and

$$X_{f_i} = \bigcup_{\nu\in\tau^{-1}(i)} X_{f_i h_\nu}. \tag{6.9}$$

We put $\widetilde{h}_\nu = f_i h_\nu$ (the index ν determines the index i) and $X = \bigcup_{\nu\in J} X_{\widetilde{h}_\nu}$.
We get a diagram

$$\begin{array}{ccccc} \widetilde{A}_I(U) & \longrightarrow & \prod_{i\in I} A_{f_i} & \rightrightarrows & \prod_{(i,j)\in I\times I} A_{f_i f_j} \\ & & \downarrow & & \downarrow \\ \widetilde{A}_J(U) & \longrightarrow & \prod_{\nu\in J} A_{\widetilde{h}_\nu} & \rightrightarrows & \prod_{(\nu,\mu)\in J\times J} A_{\widetilde{h}_\nu \widetilde{h}_\mu}. \end{array} \tag{6.10}$$

In the two horizontal diagrams $\widetilde{A}_I(U)$,$\widetilde{A}_J(U)$ are the subrings where the two horizontal arrows take the same value. If we apply the Proposition 6.1.16 to the vertical arrows, then we get an isomorphism between $\widetilde{A}_I(U)$ and $\widetilde{A}_J(U)$. So we see that a refinement of a covering gives the same result for $\widetilde{A}$. Then we may compare two coverings by passing to a common refinement.

The fact that $U \to \widetilde{A}(U)$ is actually a sheaf can be derived by similar arguments as those used in the sheafification process. The intuitive meaning of $\widetilde{A}(U)$ is clear: These are the *regular functions* on U and these are "functions", which can locally be written in the form $\frac{g_i}{f_i^{n_i}}$ in such a way that $\frac{g_i}{f_i^{n_i}}, \frac{g_j}{f_j^{n_j}}$ match on the intersection of their **domains of definition** , which of course is $X_{f_i f_j}$.

Our proposition 6.1.16 implies that $\widetilde{A}(X_{f_i}) = A_{f_i}$ and this means that a "regular function" on affine open sets X_{f_i} has always a kind of "global" description, which uses only denominators of the form $f_i^{n_i}$. Especially we have $\widetilde{A}(X) = A$.
The sheaf $\widetilde{A}$ on $\mathrm{Spec}(A)$ is a sheaf of local rings on $\mathrm{Spec}(A)$ and $(\mathrm{Spec}(A),\widetilde{A})$ is the **affine scheme attached to** A, it is a locally ringed space in the sense of Vol. I, 3.2. Later we will suppress the second entry and we will simple write $\mathrm{Spec}(A)$ for this scheme. Hence the notation $\mathrm{Spec}(A)$ may become a little bit ambiguous because it may denote the locally ringed space or the topological space. It will be clear from the context what is meant.

6.1.5 Quasicoherent Sheaves

Our considerations can be generalized. If we have an A-module M a set $S \subset A$, which is closed under multiplication and contains 1_A, then we define

$$M_S = \{(m,s) \mid m \in M, s \in S\}/\sim \tag{6.11}$$

where the equivalence relation is

$$(m,s) \sim (m',s') \iff \exists s'' \in S \quad \text{such that} \quad (ms' - m's)s'' = 0.$$

It is quite clear that this defines an A_S-module M_S. Now we can construct a sheaf $\widetilde{M}$ of $\widetilde{A}$-modules just by defining

$$\widetilde{M}(X_f) = M_f$$

then verifying the proposition – just replace A by M everywhere – and then we put

$$\widetilde{M}(U) = \left(\prod_{i \in I} M_{f_i} \overset{\longrightarrow}{\underset{\longrightarrow}{}} \prod_{(i,j) \in I \times I} M_{f_i f_j} \right) [p_1 = p_2]. \tag{6.12}$$

The stalk of the sheaf $\widetilde{A}$ at a point $\mathfrak{p}$ is the local ring $A_{\mathfrak{p}}$, the stalk of $\widetilde{M}$ in $\mathfrak{p}$ is the $A_{\mathfrak{p}}$-module $M_{\mathfrak{p}} = M_{(A \setminus \mathfrak{p})}$.
It can happen that the stalk $M_{\mathfrak{p}}$ vanishes in some points. This is so if for any $m \in M$ we can find an $s \in A \setminus \mathfrak{p}$ such that $sm = 0$.

Definition 6.1.17. *The module M defines an ideal*

$$\mathrm{Ann}(M) = \{f \in A | fM = 0\}.$$

The ideal $\mathrm{Ann}(M)$ *is called* **annulator ideal**.

It is clear that $M_{\mathfrak{p}} \neq 0$ is equivalent to $\mathfrak{p} \supset \mathrm{Ann}(M)$. The set of these $\mathfrak{p}$ is called the **support** of $\widetilde{M}$ and is a closed subset in $\mathrm{Spec}(A)$.
We will say that $\widetilde{M}$ is the sheaf **obtained from** M.
It is not so that any sheaf $\mathcal{M}$ of $\widetilde{A}$-modules is automatically of the form $\widetilde{M}$ with some A-module M. On $\mathrm{Spec}(\mathbb{Z}_{(p)})$ we have the sheaf

$$\mathcal{M}(\{(0)\}) = \mathbb{Q}$$
$$\mathcal{M}(\mathrm{Spec}(\mathbb{Z}_{(p)})) = 0,$$

which is not of this form.

Definition 6.1.18. *The sheaves $\widetilde{M}$, which are obtained from an A-module M, are called the* **quasi-coherent** *sheaves on* $\mathrm{Spec}(A)$. *We can recover the A-module from the sheaf since $M = \widetilde{M}(X)$.*

It is clear that a sequence $0 \longrightarrow \widetilde{M_1} \longrightarrow \widetilde{M} \longrightarrow \widetilde{M_2} \longrightarrow 0$ is exact if an only if if the sequence of modules $0 \longrightarrow M_1 \longrightarrow M \longrightarrow M_2 \longrightarrow 0$ is exact. (Later we will solve an exercise 9 where we show that localization is an exact functor and this has our assertion as a consequence.)
We get quasi-coherent sheaves of ideals on $X = \mathrm{Spec}(A)$ by starting from an ideal $I \subset A$, this is an A-module and the sheaf

$$\widetilde{I} \subset \widetilde{A}$$

is a quasi-coherent sheaf of ideals.

6.1.6 Schemes as Locally Ringed Spaces

In Vol I. 3.2 we introduced the notion of a locally ringed space.

Definition 6.1.19. *An* **affine scheme** *is a locally ringed space of the form* $(X,\mathcal{O}_X) = (\mathrm{Spec}(A),\widetilde{A})$.

I recall the definition of a morphism between two locally ringed spaces. A morphism is a pair

$$(f,h) : (X,\mathcal{O}_X) \longrightarrow (Y,\mathcal{O}_Y)$$

where f is a continuous map from X to Y and h is a map of sheaves of rings

$$h : f^*(\mathcal{O}_Y) \longrightarrow \mathcal{O}_X,$$

which induces local homomorphisms

$$h_x : \mathcal{O}_{Y,f(x)} = f^*(\mathcal{O}_Y)_x \longrightarrow \mathcal{O}_{X,x} \tag{6.13}$$

on the stalks. This means that the maximal ideal of $\mathcal{O}_{Y,f(x)}$ is mapped into the maximal ideal of $\mathcal{O}_{X,x}$. We call f the **spacial component** of the morphism.
The locally ringed spaces form a category we have an obvious way of composing two morphisms.

Remark 2 (Heuristical remark)**.** The difficulty is as always that the sections of the sheaves are not actual functions, they are elements in very abstract rings. In our previous examples ($\mathcal{C}_\infty$-manifolds, complex manifolds (see Vol. I,3.2)) a continous map $f : X \longrightarrow Y$ between the spaces gave us a map h_0 from the sheaves of continuous functions on Y to the continuous functions on X. Then we made requirements that this map should respect certain distinguished subsheaves of functions, which define a *so and so* structure on X and Y. If this was the case, we called f a *so and so* map. The map h_0 was determined by f in such a case. Especially it is clear in these examples that a germ f at a point $y \in Y$ with $f(y) = 0$ is mapped by h_0 to a germ at $x \in f^{-1}(y)$, which vanishes at x. This means that h_0 is automatically local.
Here the situation is different, the map is $h : f^*(\mathcal{O}_Y) \longrightarrow \mathcal{O}_X$ is an extra datum. But something is left from the notion of functions: We know what it means that a section $f \in \mathcal{O}_X(U)$ vanishes in a point $x \in U$ (see remark 1.2 on p. 4).
Now the requirement that h induces local homomorphisms in the stalks becomes clear: A germ in $\mathcal{O}_{Y,f(x)}$, which vanishes in $f(x)$ must be sent by h to a germ in $\mathcal{O}_{X,x}$, which vanishes at x. The reader should observe that a germ in $\mathcal{O}_{Y,f(x)}$, which does not vanish at $f(x)$, is a unit and hence it goes automatically to a germ in $\mathcal{O}_{X,x}$, which does not vanish in x.

The following theorem is fundamental.

Theorem 6.1.20. *Let* $(X,\mathcal{O}_X) = (\mathrm{Spec}(A),\widetilde{A})$ *and* $(Y,\mathcal{O}_Y) = (\mathrm{Spec}(B),\widetilde{B})$ *be affine schemes. A morphism*

$$(f,h) : (X,\mathcal{O}_X) \longrightarrow (Y,\mathcal{O}_Y)$$

defines a map $h_X : \mathcal{O}_Y(Y) \to \mathcal{O}_X(X)$ *i.e. a homomorphism* $h_X : B \to A$*. The map*

$$\mathrm{Hom}_{\text{affine schemes}}((X,\mathcal{O}_X),(Y,\mathcal{O}_Y)) \longrightarrow \mathrm{Hom}_{\mathrm{Rings}}(B,A)$$

given by $(f,h) \to h_X$ *is a bijection.*

We start by constructing a map in the other direction and then we show that the maps are inverse to each other.
Given $\phi : B \longrightarrow A$ we have defined a map ${}^t\phi : \mathrm{Spec}(A) \to \mathrm{Spec}(B)$ by ${}^t\phi(\mathfrak{p}) = \phi^{-1}(\mathfrak{p})$. If we have an element $b \in B$ we get an open set $Y_b = \{\mathfrak{q} | b \notin \mathfrak{q}\}$ in Y and it is clear that

$${}^t\phi^{-1}(Y_b) = X_{\phi(b)}.$$

Hence our map is continuous and we get maps

$$\phi_b : \mathcal{O}_Y(Y_b) = B_b \longrightarrow f_*(\mathcal{O}_X)(Y_b) = \mathcal{O}_X(X_{\phi(b)}) = A_{\phi(b)},$$

which by the adjointness formula is nothing else than a map

$$\widetilde{\phi} : f^*(\mathcal{O}_Y) \longrightarrow \mathcal{O}_X$$

and hence we constructed a morphism $({}^t\phi,\widetilde{\phi})$ between locally ringed spaces.

We have to show that these maps are inverse to each other. At first we start from $\phi : B \to A$, we get $({}^t\phi,\widetilde{\phi})$. From this we construct again a homomorphism from $B \to A$. According to our rules we have to evaluate $\widetilde{\phi}$ on the pair X,Y and get $\widetilde{\phi}_X : B \to A$, which is our original map.
Now we start from (f,h). The map h can be evaluated on X,Y and this gives us $h_X : B \to A$. We have to prove at first that the map ${}^th_X : X \to Y$ is equal to f. We have ${}^th_X(\mathfrak{p}) = h_X^{-1}(\mathfrak{p}) = \mathfrak{q}$.
Since we know that h induces a morphism between the sheaves we get a diagram

$$\begin{array}{ccc} B & \xrightarrow{h_X} & A \\ \downarrow & & \downarrow \\ B_{f(\mathfrak{p})} & \xrightarrow[h_{\mathfrak{p}}]{} & A_{\mathfrak{p}}. \end{array}$$

This implies that h_X has to map elements $b \in B\backslash f(\mathfrak{p})$ to elements $h_X(b) \in A\backslash\mathfrak{p}$ because b becomes invertible in $B_{f(\mathfrak{p})}$ and hence $h_{\mathfrak{p}}$ has to map it to a unit in $A_{\mathfrak{p}}$. This implies

$$B\backslash f(\mathfrak{p}) \subset B\backslash h_X^{-1}(\mathfrak{p})$$

and hence $h_X^{-1}(\mathfrak{p}) \subset f(\mathfrak{p})$. We also know that $h_{\mathfrak{p}}$ maps the maximal ideal $\mathfrak{m}_{f(\mathfrak{p})}$ into the maximal ideal $\mathfrak{m}_{\mathfrak{p}}$. Hence it maps the elements of $f(\mathfrak{p})$ into $\mathfrak{p}$ and this implies $f(\mathfrak{p}) \subset h_X^{-1}(\mathfrak{p})$ and we have the desired equality for $f(\mathfrak{p}) = h_X^{-1}(\mathfrak{p}) = {}^th_X(\mathfrak{p})$.
The rest is clear, the map $\widetilde{h}_X$, which we construct from h_X is obviously equal to h since these two coincide on the global sections. □

Closed Subschemes

We start from an ideal $I \subset A$. We have the projection map $\pi : A \to A/I$ and we have $\mathrm{Spec}(A/I) = \{\mathfrak{p} \mid \mathfrak{p} \supset I\} = V(I)$. If $i : V(I) \to \mathrm{Spec}(A)$ is the inclusion then we consider the map

$$(i,\widetilde{\pi}) : (V(I),\widetilde{A/I})) \longrightarrow (\mathrm{Spec}(A),\widetilde{A}) \tag{6.14}$$

as a **closed subscheme** of $(\mathrm{Spec}(A),\widetilde{A})$.

If the ideal is generated by elements $\{f_i\}_{i\in E}$ then we write $I = (\dots,f_i,\dots,f_j,\dots)$. Consequently principal ideals are written as (f). The underlying space of $(V(I),\widetilde{A/I}))$ is the set of points where all the f_i vanish.
If we have an open subset $X_f = \mathrm{Spec}(A_f) \subset X = \mathrm{Spec}(A)$ and if we have a closed subscheme $V \subset X_f$, which is defined by an ideal $J \subset A_f$ then the **closure** Y **of** V **in** X is the subscheme defined by the ideal $I \subset A$ which is the inverse image of J, i.e. it consists of all $h \in A$, which map under ϕ_{f^n} into J. (See 6.1.1)

Sections

If have an A-algebra B, in other words a morphism $\pi : X = \mathrm{Spec}(B) \longrightarrow Y = \mathrm{Spec}(A)$ then we define the set of **sections to** π , this is the set of morphisms $s : Y \longrightarrow X$, for which $\pi \circ s = \mathrm{Id}_Y$. We denote this set of sections by $X(Y)$.(This is of course a categorical notion.) In this special situation a section is nothing else than a $A-$ homomorphism from $\phi : B \longrightarrow A$. The subscheme of $\mathrm{Spec}(B)$ defined by the kernel of ϕ is canonically isomorphic to $Y = \mathrm{Spec}(A)$.
If we have an open subset $Y_f = \mathrm{Spec}(A_f) \subset Y$, then the inverse image X_f under π is an open subscheme of X and we can consider the sheaf of sections $X(Y_f)$. Now it is clear that we can formulate a proposition 6.1.16 for this functor $Y_f \longrightarrow X(Y_f)$:

If we have a covering $Y = \bigcup_{i\in E} Y_{f_i}$ then we get an exact sequence (See Vol. I.3.1.3)

$$X(Y) \xrightarrow{p_0} \prod_{i\in E} X(Y_{f_i}) \overset{p_1}{\underset{p_2}{\rightrightarrows}} \prod_{(i,j)\in E\times E}(X(Y_{f_i f_j}))$$

In other words it is a sheaf if we restrict it to the open sets of the form Y_f.
This is of course an immediate consequence of prop. 6.1.16

A remark

1 At this point the reader might wonder: We made a lot of effort to show that something seemingly simple, namely the category of commutative rings with identity, is anti equivalent to a certain category of locally ringed spaces. This category consists of rather complicated objects and the morphisms are also not so easy to define.

The reason why we do this will become clear: These concepts allow us to glue affine schemes together so that we can build larger objects, namely schemes. Locally these schemes look like affine schemes but globally they look different. In other words we embed the category of affine schemes into a larger category. In many respects this larger category contains new objects, which have better properties than the affine schemes. For instance we have a certain notion of compactness, which then is the source for finiteness theorems. (See section 8.3)

2 Finally we notice that we have to live with the following fact: Our theorem says that the homomorphism h_X between the rings determines the morphism between the affine schemes. But the map between the underlying spaces does not determine the morphism between the schemes. Especially we can have that the map between the spaces is a (topological) isomorphism but the morphism between the schemes is not.

If for instance $L \longrightarrow K$ is a homomorphism between to fields, then we have that $\mathrm{Spec}(K), \mathrm{Spec}(L)$ are plainly single points and the map between them has no chance to be anything else than a bijection. A second example is given by the following situation. We consider a field k and its ring of dual numbers over it, this is the ring $k[\epsilon] = k[X]/(X^2)$. We have the inclusion $k \hookrightarrow k[\epsilon]$ but again the underlying spaces are single points.

In a certain sense these to examples are typical for this fact, which is seemingly a pathology.

6.2 Schemes

6.2.1 The Definition of a Scheme

Definition 6.2.1. *A* **scheme** *is a locally ringed space* $(X,\mathcal{O}_X)$*, which is locally isomorphic to an affine scheme. In other words we can find a covering* $X = \bigcup_\nu U_\nu$ *by open sets such that* $(U_\nu,\mathcal{O}_X \mid U_\nu)$ *is affine.*

This implies of course that $(U_\nu,\mathcal{O}_X \mid U_\nu) = (\mathrm{Spec}(A_\nu),\widetilde{A}_\nu)$ where $A_\nu = \mathcal{O}_X(U_\nu)$ is the ring of regular functions on U_ν. But in contrast to the case of affine schemes the ring of regular functions on X may be to small too contain enough information to recover the scheme $(X,\mathcal{O}_X)$. This will be demonstrated in the Chapter 8 on projective schemes.
Very often we will suppress the second entry $\mathcal{O}_X$ in the notation, i.e. if we say that X is a scheme then X is not only the underlying space, the sheaf $\mathcal{O}_X$ is also given to us.

Example 5. *Let us consider the polynomial ring* $A = k[f,g]$*, let* $X = \mathrm{Spec}(A)$*. We remove the point* $(0,0)$ *from* X*, the resulting space* U *inherits a topology and the structure of a locally ringed space. It is clearly a scheme since we may cover it by*

$$U = X_f \cup X_g.$$

But it is easy to see that U *is not an affine scheme. It is obvious that any element* $h \in \mathcal{O}_U(U)$ *extends to an element in* A*, i.e.* $\mathcal{O}_U(U) = A$ *but* $U \neq \mathrm{Spec}(A)$*.*
In the theory of holomorphic functions in several variables this phenomenon is known as Lemma of Hartogs: If $n > 1$ *then a holomorphic function on* $\mathbb{C}^n \setminus \{0\}$ *extends to a holomorphic function on* $\mathbb{C}^n$*.*

This example shows that any open subset of an affine scheme is a scheme but not an affine scheme in general.
If X is a scheme and if $A = \mathcal{O}_X(X)$ is the ring of global sections, it certainly has an identity element different from zero. Then every point $x \in X$ yields a prime ideal $\mathfrak{p}_x \subset A$, we get a map between the underlying sets $i : X \longrightarrow \mathrm{Spec}(A)$ and this is the spacial component of a morphism- also called i - from X to $\mathrm{Spec}(A)$. It is clear that X is affine if and only if this morphism is an isomorphism.

The gluing

Let assume we have a family $(U_\nu = \mathrm{Spec}(A_\nu))_{\nu\in E}$ of affine schemes. Let us also assume that for any pair of indices $(\nu,\mu) \in E \times E$ we have an open subset $U_{\nu,\mu} \subset U_\nu$, these $U_{\nu,\mu}$ are schemes and they have their structure sheaves $\mathcal{O}_{U_{\nu,\mu}}$. Furthermore we assume that for any pair (ν,μ) we have isomorphisms of schemes $\phi_{\nu,\mu} : U_{\nu,\mu} \xrightarrow{\sim} U_{\nu,\mu}$ where $\phi_{\nu,\nu}$ is always the identity. Finally we assume that for any triple (ν,μ,κ) the morphism $\phi_{\nu,\mu}$ sends $U_{\nu,\mu} \cap U_{\nu,\kappa}$ to the subscheme $U_{\mu,\nu} \cap U_{\mu,\kappa}$ and that these morphisms yield a commutative diagram of isomorphisms

$$\begin{array}{ccc} U_{\nu,\mu} \cap U_{\nu,\kappa} & \longrightarrow & U_{\mu,\nu} \cap U_{\mu,\kappa}. \\ & \searrow \quad \swarrow & \\ & U_{\kappa,\nu} \cap U_{\kappa,\mu} & \end{array}$$

Then we can form the disjoint union of underlying sets $\bigsqcup_{\nu\in E} U_\nu$ and introduce the equivalence relation

$$x \sim y \text{ if } x \in U_\nu, y \in U_\mu \text{ and } \phi_{\nu,\mu}(x) = y.$$

We divide this disjoint union by the equivalence relation. On the quotient set X the structure sheaves $\mathcal{O}_\nu$ introduce the structure of a locally ringed space and hence we get a scheme $(X,\mathcal{O}_X)$. We say that this scheme is obtained by gluing from the data $(U_\nu = \mathrm{Spec}(A_\nu))_{\nu\in E}, U_{\nu,\mu}, \phi_{\nu,\mu})$.
We will see this kind of construction in a very concrete case when we construct the projective space in 8.1.1.

Another very simple situation where we can apply this construction is the case of the affine space $\mathbb{A}^n_S = S[X_1,\dots,X_n]$: We start from a scheme S and cover it by affine schemes $U'_\nu = \mathrm{Spec}(A_\nu)$. For any of these affine we consider the scheme $U_\nu = \mathrm{Spec}(A_\nu[X_1,\dots,X_n])$. We define $U_{\nu,\mu} = U_{\mu,\nu} = U_\nu \cap U_\mu$ and choose for the $\phi_{\nu,\mu}$ the identity. Then we get gluing data and the resulting scheme is $\mathbb{A}^n_S = S[X_1,\dots,X_n]$. We can give the scheme an additional structure. If we are over one of our affine subsets $U'_\nu = \mathrm{Spec}(A_\nu)$ then the set of sections is simply the set of A_ν-homomorphism from $A_\nu[X_1,\dots,X_n]$ to A_ν and this is equal to A^n_ν. Hence we can put the structure of a free A_ν- module on this set the addition and scalar multiplication are defined componentwise. Clearly this also allows us to put the structure of a sheaf of $\mathcal{O}_S$-modules on the sheaf of sections. (See the section on vector bundles further down.)

Again we have the notion of a quasi-coherent $\mathcal{O}_X$-module.

Definition 6.2.2. *A* **quasi-coherent $\mathcal{O}_X$-module** $\mathcal{M}$ *is a $\mathcal{O}_X$-module such that for all open subsets U the sections $\mathcal{M}(U)$ form an $\mathcal{O}_X(U)-$module and for affine open subsets U the restriction $\mathcal{M} \mid U$ is obtained from the $\mathcal{O}_X(U)$ module $\mathcal{M}(U)$(see 6.1.18).*

Closed subschemes again

At his point it is rather clear what a **closed subscheme** of a general scheme is. We know how to define a quasi-coherent sheaf of ideals $\mathcal{I} \subset \mathcal{O}_X$: It is a sheaf of ideals, i.e. for any open subset $U \subset X$ the sections $\mathcal{I}(U) \subset \mathcal{O}_X(U)$ form an ideal in $\mathcal{O}_X(U)$ and for an affine U the restriction $\mathcal{I} \mid U$ is the sheaf associated to $\mathcal{I}(U)$. On this open affine subset U we have the closed subscheme $((V(\mathcal{I}(U)),(\widetilde{A}(U)/\mathcal{I}(U)))),(\widetilde{A}(U)/\mathcal{I}(U))) \hookrightarrow (U,\widetilde{A}(U))$. We define $V(\mathcal{I})$ to be the union of all these subsets $(V(\mathcal{I}(U)),(\widetilde{A}(U)/\mathcal{I}(U)))$ and a quotient sheaf $\mathcal{O}_X/\mathcal{I}$ by its restriction to the affine pieces. This yields the closed subscheme $(V(\mathcal{I}),\mathcal{O}_X/\mathcal{I}) \hookrightarrow (X,\mathcal{O}_X)$.

Given a closed subset $Y \subset X$ we consider its open complement $U \subset X$, then U has an obvious structure of an open subscheme. We simply take the restriction of the structure sheaf to U. (See also Vol. I, 3.4.2) The analogous process is not so simple if we want to do the same thing with Y. To give a the structure of a closed sub scheme to Y we have to define the structure sheaf on it. But in general the sheaf of ideals defining Y is not unique. The only way is to choose the ideal $\widetilde{I}(Y)$ of all functions that vanish on Y, i.e. for any open set $V \subset Y$, which is of the form $V = U \cap Y$ with U open in X we define $I'_U(V) = \{f \in \mathcal{O}_X(U) | f(y) = 0 \text{ for all } y \in V\}$. We take the limit over all such subsets

U, and get a presheaf $V \mapsto I'(V)$. The sheafification of this presheaf is our sheaf $\widetilde{I}(Y)$ and this defines the structure of a reduced scheme on Y. The subset Y together with this structure sheaf on it is the scheme $Y_{\mathrm{red}} \subset X$.

Annihilators, supports and intersections

If $\mathcal{M}$ is quasi coherent $\mathcal{O}_X$ -module sheaf then we can consider the annihilator $\operatorname{Ann}(\mathcal{M}) \subset \mathcal{O}_X$. It is a sheaf of ideals and its local sections $f \in \operatorname{Ann}(\mathcal{M})(U)$ are those elements, which satisfy $f\mathcal{M}(U) = 0$. Then $\mathcal{M}$ is a sheaf of $\mathcal{O}_X/\operatorname{Ann}(\mathcal{M})$ modules. The **support** $\mathbf{Supp}(\mathcal{M})$ **of** $\mathcal{M}$ is the subscheme $Y = \operatorname{Spec}(\mathcal{O}_X/\operatorname{Ann}(\mathcal{M}))$ but $\mathcal{M}$ is not necessarily a sheaf over Y_{red}.
If for any open affine subscheme $U \subset X$ the $\mathcal{O}_X(U)-$module $\mathcal{M}(U)$ is finitely generated then we can easily see

$$\mathfrak{p} \supset \operatorname{Ann}(\mathcal{M}) \iff \mathfrak{p} \in \operatorname{Supp}(\mathcal{M}) \iff \mathcal{M}_{\mathfrak{p}} \neq (0) \tag{6.15}$$

If we have two subschemes $Y_1,Y_2 \subset X$, which are given by ideals $\mathcal{I}_1,\mathcal{I}_2 \subset \mathcal{O}_X$, then we define the **intersection** $Y_1 \cap Y_2$ to be the subscheme defined by the ideal $(\mathcal{I}_1,\mathcal{I}_2)$, which is the ideal generated by $\mathcal{I}_1,\mathcal{I}_2$.
If $U \subset X$ is an open sub scheme and if $V \subset U$ is a closed subscheme then it is clear what the closure Y of V in X is: We cover U by affines and apply the construction from p. 14

6.2.2 Functorial properties

We know what a morphism $f : X \longrightarrow Y$ between schemes is. If $\mathcal{M}$ is a quasi coherent sheaf on X, then it is clear that $f_*(\mathcal{M})$ is a quasi-coherent sheaf on Y. But if $\mathcal{N}$ is a quasi coherent sheaf on Y then $f^*(\mathcal{N})$ is not necessarily quasi-coherent on X, simply because we do not have the $\mathcal{O}_X$-module structure on it. Therefore we define the inverse image of a quasi coherent sheaf as

$$f^*_{\mathrm{qcoh}}(\mathcal{N}) = f^*(\mathcal{N}) \times_{\mathcal{O}_Y,h} \mathcal{O}_X.$$

If for instance $V \subset Y$ is an open affine subset and if $U \subset X$ is open affine and if $f : U \longrightarrow V$, then the morphism f gives us a homomorphism $h : \mathcal{O}_Y(V) \longrightarrow \mathcal{O}_X(U)$. If now $\mathcal{N}$ restricted to V is obtained from a $\mathcal{O}_Y(V)$ - module N, then $f^*_{\mathrm{qcoh}}(\mathcal{N})$ is obtained from $\mathcal{N} \otimes_{\mathcal{O}_Y(V),h} \mathcal{O}_X(U)$. (See 6.1.18). If $i : U \hookrightarrow X$ is an open embedding and if $\mathcal{M}$ is quasi-coherent on X then we have of course $i^*(\mathcal{M}) = i^*_{\mathrm{qcoh}}(\mathcal{M})$
We change the notation. If we work with quasi coherent sheaves, then the sheaf theoretic inverse image does not play such a role, therefore, from now on $f^*(\mathcal{N})$ will be the quasi coherent inverse image of the quasi coherent sheaf $\mathcal{N}$.
We have to pay a price: The functor $\mathcal{N} \longrightarrow f^*(\mathcal{N})$ is not exact anymore because the tensor product is not an exact functor. This does not apply to the case of an open embedding $i : U \hookrightarrow X$, in this case the (modified) functor i^* is exact on quasi-coherent sheaves.
If we have a quasi-coherent sheaf $\mathcal{M}$ on X and if $x \in X$ is a point, then we get an inclusion of schemes $i_x : \operatorname{Spec}(k(x)) \longrightarrow X$ and $(i_x)^*(\mathcal{M})$ is a $k(x)$ -vector space, it is the *evaluation* of $\mathcal{M}$ at x. More generally we may consider a closed subscheme $i : Y \hookrightarrow X$. Then we call $i^*(\mathcal{M})$ the evaluation of $\mathcal{M}$ at Y.

Affine morphisms

It is rather clear what an affine morphism $f : X \longrightarrow Y$ is. This is a morphism, for which we can find a covering $Y = \bigcup_i V_i$ by open affine sub schemes such that $f^{-1}(V_i) = U_i$ is affine for all i. In this case we also say that X is affine over Y, this does not imply that X is affine. But it is not difficult to see that X is affine if Y is affine. (See proof of Prop. 8.1.16).

Sections again

Let $f : X \longrightarrow Y$ be a morphism of schemes. For any open subset $V \subset Y$ we have the open subscheme $f^{-1}(V) = X_V$ of X and the restriction $f : X_V \longrightarrow V$. We can attach to any open subset $V \subset Y$ the set of sections $X(V)$ from V to X_V, this gives us a presheaf $V \longrightarrow Y(V)$ and in fact we have

Proposition 6.2.3. *For any morphism of schemes $f : X \longrightarrow Y$ the functor $V \longrightarrow X(V)$ from open subsets to sets is a sheaf.*

This is rather clear. A morphism of schemes has two components, the map between the underlying sets and then an morphism between the structure sheaves. For the first component it is obvious that they satisfy the two conditions (Sh1),(Sh2) (See Vol. I.3.1.3). For the second component we have a local problem, we can easily reduce to the affine case and there we apply prop. 6.1.16.
The construction of the functor $V \longrightarrow X_V$ is a special case of the fibered product, which will be discussed in section 6.2.5, later on we will denote this construction by $X \times_Y V$.

6.2.3 Construction of Quasi-coherent Sheaves

We have an important way of constructing quasi-coherent $\mathcal{O}_X-$modules on X. Let us assume we have a covering $\mathfrak{U} = \{U_\nu\}_{\nu \in N}$ of a scheme $(X,\mathcal{O}_X)$ by affine subschemes. Let us also assume that we have given an $\mathcal{O}_X(U_\nu)-$module M_ν for all $\nu \in N$. Each of them defines a quasi-coherent sheaf $\widetilde{M}_\nu$ on the corresponding subscheme U_ν. Now let us assume that for any pair (ν,μ) of indices we have an isomorphism

$$g_{\nu,\mu} : \widetilde{M}_\nu \mid U_\nu \cap U_\mu \xrightarrow{\sim} \widetilde{M}_\mu \mid U_\nu \cap U_\mu \tag{6.16}$$

such that this system of isomorphism satisfies

1. $g_{\nu,\nu} = \mathrm{Id}$ for all ν
2. $g_{\nu,\mu} \circ g_{\mu,\nu} = \mathrm{Id}$ for all pairs ν,μ
3. and for any three indices ν,μ,λ we have the relation $g_{\nu,\mu} \circ g_{\mu,\lambda} = g_{\nu,\lambda}$ on $U_\nu \cap U_\mu \cap U_\lambda$.

Then we can construct a sheaf $\widetilde{M} = (M_\nu,\mathfrak{U},g_{\nu,\mu})$ on X by the glueing process: For an open set $V \subset X$, which is contained in at least one of the U_ν we define $\widetilde{M}(V)$ to be the set of vectors $m = (\dots,m_\nu,\dots,m_\mu,\dots)$ where the indices run over the subset of indices λ, for which $U_\lambda \supset V$, where the $m_\nu \in \widetilde{M}_\nu(V)$ and where

$$g_{\nu,\mu}(m_\nu) = m_\mu \text{ for all pairs } \nu,\mu. \tag{6.17}$$

Of course any of the components determines all the others. Then for an arbitrary V we may cover it by the $V \cap U_\nu$ and define $\widetilde{M}(V)$ by the conditions (SH1), (SH2) for sheaves ((see Vol. I, 3.1.3.)
We will not discuss an example for this kind of construction, for this we refer to chapter 8 on projective spaces.

Vector bundles

We have the notion of a **vector bundle** in the the world of schemes. First of all we have the trivial vector bundle $\mathbb{A}^n_S/S$, for which we gave an explicit construction above. We observe that for any open subset $V \subset S$ the space of sections $V \longrightarrow \mathbb{A}^n_V$ is equal to $\mathcal{O}_S(V)^n$. (Check this for affine V first.) We notice that the scheme $\mathbb{A}^n_S/S$ has a particular subgroup in its automorphism group. These are the *linear automorphisms*, which respect the structure of the sheaf of sections as $\mathcal{O}_S$-module. This group of automorphisms is clearly the group $\mathrm{GL}_n(\mathcal{O}_S(S))$.
Now we say that a scheme $X \longrightarrow S$ is a vector bundle if it has the following properties:

1. For any open subset $V \subset S$ the set of sections $X(V)$ has the structure of an $\mathcal{O}(V)$-module and this structure is compatible with the restriction map to smaller open sets. In other words the sheaf of sections has the structure of an $\mathcal{O}_S$-module.

2. We can find a covering of S by affine open subsets V_i and isomorphisms

$$\psi_i : X|_{V_i} \xrightarrow{\sim} \mathbb{A}^n_{V_i},$$

which induce a $\mathcal{O}_S(V_i)$-linear isomorphism on the sheaves of sections.

We may formulate this slightly differently. We say that $\pi : X \longrightarrow S$ is a vector bundle if we can find a covering $S = \bigcup_i V_i$ and an isomorphism

$$\psi_i : X_{V_i} \xrightarrow{\sim} \mathbb{A}^n_{V_i}$$

such that on the intersections $V_i \cap V_j$ the isomorphism

$$\psi_i \circ \psi_j|_{U_i \cap U_j} : \mathbb{A}^n_{V_i \cap V_j} \xrightarrow{\sim} \mathbb{A}^n_{V_i \cap V_j}$$

is $\mathcal{O}_S(V_i \cap V_j)$-linear.

Vector Bundles Attached to Locally Free Modules

Of course we know what a locally free $\mathcal{O}_X-$ module module is (See I, 4.3.1 - 4.3.3). Given we any locally free $\mathcal{O}_X-$module $\widetilde{M}$ of finite constant rank d we can also go in the opposite direction, we can attach to it a vector bundle $V(\widetilde{M}) \longrightarrow X$ such that its sheaf of sections is equal to $\widetilde{M}$.
To do this we start from the dual module $\widetilde{M}^\vee$ consider the symmetric tensor algebra

$$\mathrm{Sym}^\bullet(\widetilde{M}^\vee) = \bigoplus_{n=0}^{n=\infty} \mathrm{Sym}^n(\widetilde{M}^\vee),$$

where $\mathrm{Sym}^n(\widetilde{M}^\vee)$ is the quotient of the n-fold tensor product $\widetilde{M}^\vee \otimes \widetilde{M}^\vee \otimes \cdots \otimes \widetilde{M}^\vee$ by the sub module generated by tensors $\phi_1 \cdots \otimes \phi_\nu \otimes \cdots \otimes \phi_\mu \otimes \phi_n - \phi_1 \cdots \otimes \phi_\mu \otimes \cdots \otimes \phi_\nu \otimes \phi_n$, i.e exactly two components are interchanged. We get of course the same sub module if take all differences of tensors $x_1 \otimes \cdots \otimes x_\nu \otimes \cdots \otimes x_n - x_{\sigma(1)} \otimes \cdots \otimes x_{\sigma(\nu)} \otimes \cdots \otimes x_{\sigma(n)}$, where σ runs over all permutations.
If we restrict the locally free module to an affine open subset $U \subset X$ then the elements $\Phi \in \mathrm{Sym}^n(\widetilde{M}(U))$ are the symmetric n-linear forms on $M(U)$. If we restrict to a smaller affine open subset, which is still called U, then we may assume that $\widetilde{M}(U)$ is a free $\mathcal{O}(U)$-module with basis $e_1,e_2,\ldots,e_d$ then we denote the dual basis by $X_1,\ldots,X_d$. Obviously the symmetric algebra $\mathrm{Sym}^\bullet(\widetilde{M}^\vee(U)) = \mathcal{O}(U)[X_1,\ldots,X_n]$. Recall that we want to construct the vector bundle $V(\widetilde{M})/X$ and locally we solve this problem by defining $V(\widetilde{M})_U = \mathrm{Spec}(\mathrm{Sym}^\bullet(\widetilde{M}^\vee(U))$.
Now we choose a covering $X = \bigcup_i U_i$ by affine open sub schemes such that the restriction $\widetilde{M}|U_i$ is free. Then clearly

$$\mathrm{Sym}^\bullet(\widetilde{M}^\vee(U_i))|_{U_i\cap U_j} = \mathrm{Sym}^\bullet(\widetilde{M}^\vee(U_j))|_{U_i\cap U_j}$$

and we can glue these pieces together to the scheme $V(\widetilde{M})$.
Finally we have to show that the sheaf of sections has the structure of an $\mathcal{O}_X$-module and that this sheaf of section is equal to $\widetilde{M}$. This is obvious: Over our affine subset U a section attaches an element a_i to X_i. Hence the tuple $(a_1,\ldots,a_n)$ is a linear form on $\widetilde{M}^\vee(U)$ and hence an element in $\widetilde{M}(U)$. The scheme $V(M)/X$ is called the **vector bundle** attached to $\widetilde{M}$.
Of course it is obvious that a vector bundle over S provides a locally free module, we simple take the sheaf of sections. Hence we may say that vector bundles and locally free sheaves a are essentially the same kind of objects.
A locally free sheaf of constant rank one is called a **line bundle** or a **invertible sheaf.** The tensor product of two line bundles is again a line bundle and the isomorphism classes of these line bundles form a group under this operation (See 9.4.).

6.2.4 Vector bundles and GL_n-torsors.

If we have a vector bundle $X \longrightarrow S$ of rank n then we can define a new object $P \longrightarrow S$, which is a scheme with an action of GL_n on it. To be more precise: For an open set $V \subset S$ we define

$$P(V) = \{(e_1,e_2,\ldots,e_n)|e_i \in X(V)\}$$

such that the e_1 form a set of generators of the space of sections $X(V_1)$ for any open subset $V_1 \subset V$.
Of course $P(V)$ may be empty, it is not empty if and only if the restriction of the bundle $X|V$ is trivial. If this restriction is trivial then we have an action of $\mathrm{GL}_n(V)$ on $P(V)$, and this action is simply transitive. We call an element $\underline{e} = (e_1,e_2,\ldots,e_n)$ a trivialization over V.
Now it is clear that we can define a scheme $P \longrightarrow S$ whose sections over V are the trivializations of $X \longrightarrow S$ over V, the group GL_n acts from the left on $P \longrightarrow S$ and if $P(V) \neq \emptyset$ then the action of $\mathrm{GL}_n(V)$ is simply transitive. Such an object is called a principal homogenous space under GL_n/S or a GL_n/S -torsor. We will come back to this in section 6.2.10

6.2.5 Schemes over a base scheme S.

The schemes form a category. It is very important to study relative situations, i.e. to study the category of schemes over a fixed base scheme S.

Definition 6.2.4. *A* **scheme over** S *is a scheme* X *together with a morphism*

$$\begin{array}{c} X \\ \pi \big\downarrow \\ S. \end{array}$$

Sometimes we write X/S*, the morphism* π *is called the* **structure morphism** *and* S *is called the* **base scheme**.

If our two schemes are affine, i.e. $S = \mathrm{Spec}(A), X = \mathrm{Spec}(B)$ the the morphism π is nothing else than a homomorphism $\varphi : A \longrightarrow B$, i.e. B is an A-algebra.
Any commutative ring is in a unique a $\mathbb{Z}$-algebra, there is exactly one homomorphism $\mathbb{Z} \longrightarrow A$ because we assume that 1 goes to the identity 1_A. Therefore it is clear that any affine scheme is in a canonical way a scheme over $\mathrm{Spec}(\mathbb{Z})$. But then it is also clear that any scheme X admits a unique morphism $X \longrightarrow \mathrm{Spec}(\mathbb{Z})$, this is the **absolute morphism.**

Some notions of finiteness

Again the schemes over S form a category . If we have two schemes $X/S, Y/S$ then the S-morphisms $\mathrm{Hom}_S(X,Y)$ are those morphisms φ from X to Y, which render the following diagram

$$\begin{array}{ccc} X & \xrightarrow{\quad\varphi\quad} & Y \\ & \searrow \quad \swarrow & \\ & S & \end{array} \tag{6.18}$$

commutative.
A morphism $f : X \longrightarrow S$ is called of **finite type** if we have finite coverings $X = \cup_{i \in E} U_i, Y = \cup_i V_i$ by open affine sub schemes such that $f : U_i \longrightarrow V_i$ and such that can write $\mathcal{O}_C(U_i)$ as a quotient of a polynomial ring

$$\mathcal{O}_C(U_i) = \mathcal{O}(V_i)[Y_1, Y_2, \ldots, Y_n]/I_i,$$

where the ideal I_i is finitely generated.
A scheme X is of *finite type* if the absolute morphism $X \longrightarrow \mathrm{Spec}(\mathbb{Z})$ is of finite type.

Perhaps this is a good place to introduce the notion of an **affine scheme of finite type over** S. This is a scheme X/S, which is given as a closed subscheme of some vector bundle $V(M)/S$. If $S = \mathrm{Spec}(A)$ is affine then we get an example of such an affine scheme of finite type over S if we consider $X = \mathrm{Spec}(A[X_1, X_2, \ldots, X_n]/I$ where the X_i are independent variables and I is an ideal. Locally in the base S all schemes, which are affine and of finite type over S are of this form. Under certain finiteness assumptions on S an affine scheme of finite type over S is the same as an affine scheme $X \longrightarrow S$ where the algebras $\mathcal{O}_X(f^{-1}(V_i))$ are of finite type over $\mathcal{O}(V_i)$.

A typical example of such a relative situation is the scheme $\mathbb{A}^n_S \longrightarrow S$, which was introduced further up or more generally the vector bundle $V(M)/S$, which we attach to a locally free quasi- coherent sheaf of $\widetilde{M}$ of finite rank.

Fibered products

Given two schemes X/S,Y/S we have the notion of the **fibered product** of these schemes over S. This fibered product is nothing else than the product in the category of schemes over S.
Hence the fibered product is an object Z/S together with two arrows p_1,p_2

$$\begin{array}{ccc} Z & \xrightarrow{p_1} & X \\ {\scriptstyle p_2}\downarrow & & \downarrow \\ Y & \longrightarrow & S \end{array} \tag{6.19}$$

such that for any scheme T/S we have

$$\mathrm{Hom}_S(T,Z) \xrightarrow{\sim} \mathrm{Hom}_S(T,X) \times \mathrm{Hom}_S(T,Y) \tag{6.20}$$

where the identification is given by

$$\Psi \longmapsto (p_1 \circ \Psi, p_2 \circ \Psi).$$

We can do this for any category (See Vol. I .1.3.1). The reader is advised to consider the construction of fibered product in the category of sets as an example.

Theorem 6.2.5. *In the category of schemes fibered products exist.*

This theorem will not be proved here in detail. We will prove it for affine schemes in this section and at the end of the proof I will give some indication how to do it in general. (See also [Ha], Chap. II, Thm. 3.3 or [EGA1] 3.2.6.) In the section on projective schemes we will prove that the product of projective schemes is again projective and hence the existence of products in that case will be a by-product of this result.
Now we discuss the above theorem in the case of affine schemes, the discussion will be very detailed and perhaps to verbose.
We consider the category of affine schemes. If $X = \mathrm{Spec}(A)$ and $S = \mathrm{Spec}(R)$ then $\pi : X \to S$ is the same thing as a homomorphism of rings $\varphi : R \to A$.
At this point two remarks are in order

1. The datum $\varphi : R \to A$ is the same as giving the additive group A the structure of an R-module, i.e. giving a composition

$$\cdot : R \times A \longrightarrow A,$$

which satisfies the usual rules, especially we want $1_R \cdot a = a$ and we have to require in addition $r \cdot (a_1 a_2) = (r \cdot a_1) a_2$. This is clear because starting from φ we put

$$r \cdot a = \varphi(r) \cdot a.$$

On the other hand if we have given the R-module structure on A then

$$\Psi(r) = r \cdot 1_A$$

gives us the ring homomorphism.

2. To simplify the notation we will drop the name of the morphism, this means we will only write $R \to A$ instead of $\varphi : R \to A$. In view of the first remark this means: If we see $R \to A$ then this allows us to write $r \cdot a$ for $r \in R$ and $a \in A$ and this satisfies the obvious rules.

If we have given R-rings A,B then $\mathrm{Hom}_R(A,B)$ are exactly those homomorphisms, which are linear with respect to R, i.e. $\varphi \in \mathrm{Hom}_R(A,B)$ means that φ satisfies $\varphi(r \cdot a) = r \cdot \varphi(b)$. Now the two dots have different meanings.
Of course we don't make any assumption that $R \to A$ should be injective. For instance $\mathbb{Z} \to \mathbb{Z}/p$ make $\mathbb{Z}/p\mathbb{Z}$ into a ring over $\mathbb{Z}$. Actually it is clear that any ring A is in a unique way a ring over $\mathbb{Z}$, we simply send $1 \mapsto 1_A$.
We come back to the construction of fibered products in the category of affine schemes. We describe the problem in the category of rings and therefore, we turn the arrows around. We have

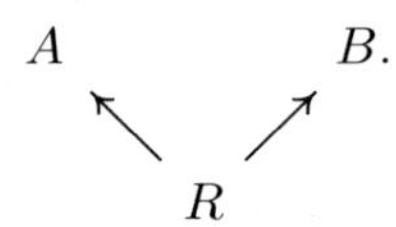

We are looking for a ring C over R together with two R-homomorphisms $\alpha : A \to C$, $\beta : B \to C$ such that for any other ring T over R we get: In the following diagram

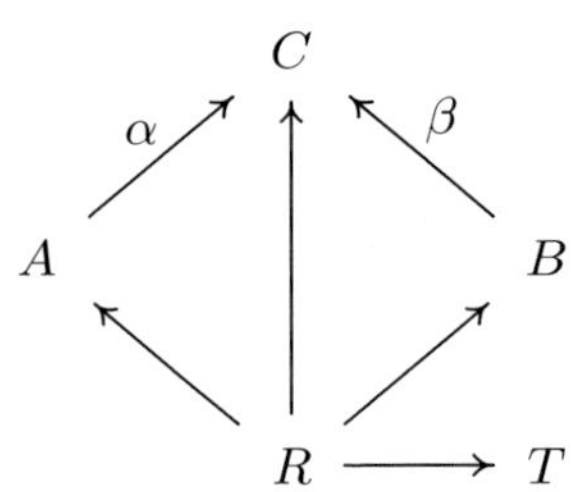

an R-homomorphism from C to T is the same thing as a pair of R-homomorphisms $f : A \to T$,$g : B \to T$.
How do we get such a C? Starting from f,g we get a map

$$\begin{aligned} A \times B &\longrightarrow T \\ (a,b) &\longmapsto f(a) \cdot g(b). \end{aligned}$$

This is an R-bilinear map from $A \times B$ to T. We have to verify

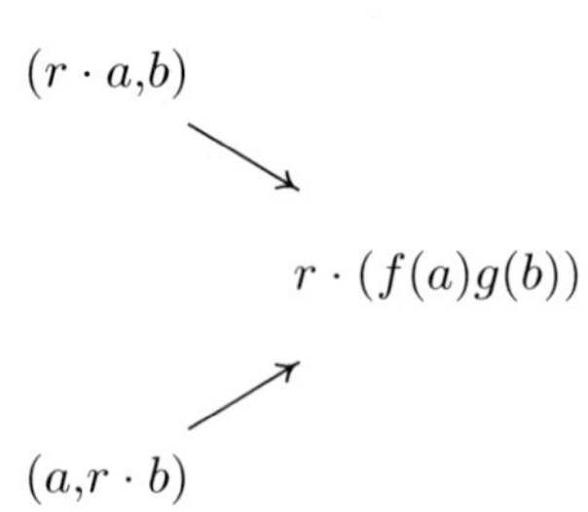

but

$$\begin{aligned}(r\cdot a,b) \longrightarrow f(r\cdot a)g(b) &= f((r\cdot a)g(b)\\ &= f(r\cdot 1_A)f(a)\cdot g(b) = (r\cdot f(1_A))f(a)g(b)\\ &= (r\cdot 1_C)f(a)g(b) = r\cdot (f(a)g(b))\end{aligned}$$

and the distributivity is clear.
But this tells us that the pair f,g provides us an R-linear map

$$f\otimes g : A\otimes_R B \longrightarrow T$$

where $A\otimes_R B$ is of course the tensor product of the two R-modules A,B.
We define a ring structure on $A\otimes_R B$: The elements of the tensor product are finite sums $a_1\otimes b_1 + a_2\otimes b_2 + \ldots + a_s\otimes b_s$ where we have the following rules

$$\begin{aligned}(r\cdot a)\otimes b - a\otimes r\cdot b &= 0\\ (a_1+a_2)\otimes b - a_1\otimes b - a_2\otimes b &= 0\\ a\otimes (b_1+b_2) - a\otimes b_1 - a\otimes b_2 &= 0.\end{aligned} \tag{6.21}$$

We introduce as multiplication

$$(a\otimes b)(a'\otimes b') = aa'\otimes bb',$$

we extend this by distributivity. Then we have to check that this is compatible with the rules above.
We put $C = A\otimes_R B$ with this ring structure, we have the homomorphism

$$r \longrightarrow r\cdot 1_B\otimes 1_b = 1_A\otimes r\cdot 1_B,$$

we have

$$\begin{aligned}\alpha : A &\longrightarrow A\otimes_R B & \qquad \beta : B &\longrightarrow A\otimes_R B\\ a &\longmapsto a\otimes 1_B & b &\longmapsto 1_A\otimes b\end{aligned}$$

Starting from f,g we already had the R-linear map from the R-module $A\otimes_R B$ to T. But the ring structure on $A\otimes_R B$ is made in such a way that $f\otimes g$ is a ring homomorphism. On the other hand, if $h : A\otimes_R B \longrightarrow T$ is a R-homomorphism then we may put $f = h\circ\alpha, g = h\circ\beta$ and it is clear that

$$\begin{aligned}h(a\otimes b) &= h((a\otimes 1_B)\cdot(1_A\otimes b))\\ &= h((a\otimes 1_B))\cdot h((1_A\otimes b)) = f(a)\cdot g(b)\\ &= (f\otimes g)(a\otimes b).\end{aligned} \tag{6.22}$$

After all this it should be clear that the diagram

$$\begin{array}{ccccc} & & \mathrm{Spec}(A\otimes_R B) & & \\ & \swarrow^{p_1} & & \searrow^{p_2} & \\ \mathrm{Spec}(A) & & & & \mathrm{Spec}(B)\\ & \searrow & & \swarrow & \\ & & \mathrm{Spec}(R) & & \end{array} \tag{6.23}$$

is a fibered product of $\mathrm{Spec}(A)$ and $\mathrm{Spec}(B)$ over the base $\mathrm{Spec}(R)$.
It is a standard terminology to say that B is a **finitely generated** A algebra if we can find $x_1,\ldots,x_r \in B$ such that any $b \in B$ can be written as a A-linear combination of monomials $x_1^{n_1}\ldots x_r^{n_r}$. We can also reformulate that and say: The A-algebra B is a quotient of the polynomial ring $A[X_1,\ldots,X_r]/I$ where $I \subset A[X_1,\ldots,X_r]$ is an ideal. (See the following examples 5 and 6)

Example 6. *If A and B are polynomial rings over R in finitely many variables, i.e. $A = R[X_1,\ldots,X_n]$ and $B = R[Y_1,\ldots,Y_m]$ then A,B are free R modules with a basis consisting of monomials $X_1^{\nu_1}\cdots X_n^{\nu_n}$, $Y_1^{\mu_1}\cdots Y_m^{\mu_m}$. Then $A\otimes_R B$ is free again and has as basis $X_1^{\nu_1}\cdots X_n^{\nu_n}\otimes Y_1^{\mu_1}\cdots Y_m^{\mu_m}$. But then it is obvious that $A\otimes_R B$ is actually isomorphic to the polynomial ring in $X_1,\ldots,X_n,Y_1,\ldots,Y_m$, i.e.*

$$R[X_1,\ldots,X_n]\otimes R[Y_1,\ldots,Y_m] = R[X_1,\ldots,X_n,Y_1,\ldots,Y_m].$$

The scheme $\mathrm{Spec}(R[X_1\ldots X_n])$ *is called* **the n-dimensional affine space over** $\mathrm{Spec}(R)$ *and if $S = \mathrm{Spec}(R)$ we write $\mathbb{A}_S^n$ for this scheme. Hence we get the truly exciting formula*

$$\mathbb{A}_S^n \times_S \mathbb{A}_S^m = \mathbb{A}_S^{n+m}.$$

Example 7. *If we have two R-algebras $R \longrightarrow A, R \longrightarrow B$ and we have given two ideals $I \subset A, J \subset B$ then these ideals define closed subschemes*

$$\begin{array}{ccccccc}
\mathrm{Spec}(A/I) & \hookrightarrow & \mathrm{Spec}(A) & & \mathrm{Spec}(B/J) & \hookrightarrow & \mathrm{Spec}(B) \\
 & \searrow & \searrow & & \swarrow & \swarrow & \\
 & & & \mathrm{Spec}(R) & & &
\end{array}$$

Hence we get a morphism from the fibered products

$$\begin{array}{ccc}
\mathrm{Spec}(A/I)\times_{\mathrm{Spec}(R)}\mathrm{Spec}(B/J) & \longrightarrow & \mathrm{Spec}(A)\times_{\mathrm{Spec}(R)}\mathrm{Spec}(B) \\
 & \searrow \quad \swarrow & \\
 & \mathrm{Spec}(R) &
\end{array}$$

and we claim that this is again a closed embedding. We leave it to the reader as an exercise to show that the arrow gives us an isomorphism of the fibered product of the subschemes to the subscheme defined by the ideal $(A\otimes_R B)(1_A\otimes_R J)+(A\otimes_R)(I\otimes_R B)\subset A\otimes_R B$.

Example 8. *If k is a field and K/k is a finite extension, then we have a map*

$$\begin{array}{c}
\mathrm{Spec}(K) \\
\downarrow \\
\mathrm{Spec}(k)
\end{array}$$

which on the underlying sets is just a map from a point to a point. But we have different rings of regular functions on these points, hence this morphism is not an isomorphism. If we take the fibered product we get

$$\operatorname{Spec}(K) \times_{\operatorname{Spec}(k)} \operatorname{Spec}(K) = \operatorname{Spec}(K \otimes_k K)$$

and $K \otimes_k K$ will not be a field in general. If for instance K/k is a separable normal extension then the Main theorem of Galois theory says

$$K \otimes_k K = \bigoplus_{\sigma \in \operatorname{Hom}_k(K,K)} K,$$

where the identification is given by $a \otimes b \mapsto (\dots,\sigma(a)b,\dots)_{\sigma \in \operatorname{Hom}_k(K,K)}$. Therefore

$$\operatorname{Spec}(K \otimes_k K) = \operatorname{Spec}\Big(\bigoplus_{\sigma \in \operatorname{Hom}_k(K,K)} K\Big) = \operatorname{Hom}_k(K,K)$$

as a set. Here we have an example where the underlying set of $X \times_S Y$ may differ from the set theoretic fibered product, which in our case is still a point.

Example 9. *As we explained earlier, we always have a canonical morphism* $\operatorname{Spec}(A) \to \operatorname{Spec}(\mathbb{Z})$ *and we may define the absolute product of two affine schemes as*

$$\operatorname{Spec}(A) \times \operatorname{Spec}(B) = \operatorname{Spec}(A \otimes_{\mathbb{Z}} B).$$

We can consider the situation that we have an R-algebra R_1 and two R_1 algebras, i.e. we have a diagram

$$\begin{array}{ccccc} & & & & A \\ & & & \nearrow & \\ R & \longrightarrow & R_1 & & \\ & & & \searrow & \\ & & & & B \end{array}$$

Then we have a morphism

$$\rho : A \otimes_R B \longrightarrow A \otimes_{R_1} B$$

In the ring on the right hand side we have the rule $r_1 a \otimes b = a \otimes r_1 b$ for $r_1 \in R$, which is not valid in the ring on the right hand side.
The map ρ is clearly surjective, hence we have

$$\operatorname{Spec}(A) \times_{\operatorname{Spec}(R_1)} \operatorname{Spec}(B) \hookrightarrow \operatorname{Spec}(A) \times_{\operatorname{Spec}(R)} \operatorname{Spec}(B).$$

We want to say a few words concerning the construction of fibered products for general schemes. First of all we have to cover the base scheme S by affines and to construct the fibered product over these affine subschemes and to glue the fibered products over the intersections. Over an affine base S we cover the schemes $X/S,Y/S$ by affines and constructs the fibered products for the pairs of affine covering sets. These will be glued together. For the details I refer to the references given above.
In the section on projective schemes I will discuss the construction of fibered products in a special case.

Base change

Let X/S be a scheme and $T \xrightarrow{f} S$ another scheme, i.e. we have a diagram

$$\begin{array}{ccc} X & & \\ \downarrow & & \\ S & \longleftarrow & T. \end{array} \tag{6.24}$$

Then we can form the fibered product $X \times_{S,f} T$ this is now a scheme over T. Most of the time we will drop the f in the subscript and write simply $X \times_S T$. This scheme is called the **base change** from X/S to T.

Very often the fibered product $X \times_S T$ is called the **pullback** of $X \longrightarrow S$ to T. We introduce the notation

$$X \times_S S' = f^*(X).$$

The same terminology is applied to an S morphism $h : X_1/S \longrightarrow X_2/S$, the induced morphism $h' : X_1 \times_S T \longrightarrow X_2 \times_S T$ is called the pullback of h and denoted by $f^*(h)$. If $\mathcal{M}$ is a quasi-coherent sheaf on X, the we also call $(\mathrm{Id} \times f)^*(\mathcal{M})$ the (quasi-coherent) pullback of $\mathcal{M}$ and we also denote it by $f^*(\mathcal{M})$.

6.2.6 Points, T-valued Points and Geometric Points

In the theory of schemes we have to be careful with the notion of a point. If we have a scheme $(X,\mathcal{O}_X)$ then the underlying space X is a set and the points of the scheme are the elements of this set. But we have seen that these points do not behave well under fibered products. (Example 8)

Definition 6.2.6. *Let X/S be a scheme and $T \xrightarrow{f} S$ another scheme, i.e. we have a diagram as above (6.24) then* **the T-valued points** *of X are simply the S-arrows from T to X, i.e.*

$$X_S(T) = \mathrm{Hom}_S(T,X). \tag{6.25}$$

If $S = \mathrm{Spec}(R)$ and $T = \mathrm{Spec}(B)$ then we denote the set of $T = \mathrm{Spec}(B)$-valued points also by $X(B)$ and speak of B-valued points.

Clearly the set of T valued points of X/S is equal to the set of T-valued points of the base change $X \times_S T \longrightarrow T$.
Therefore, we see that a scheme X/S defines a contravariant functor

$$F_X : \mathbf{Schemes}\,/S \longrightarrow \mathbf{Ens}\,, \tag{6.26}$$

where $F_X(T) = X_S(T)$.
It is the definition of the fibered product that the T-valued points behave well under fibered products. We have

$$(X \times_S Y)_S(T) = X_S(T) \times Y_S(T). \tag{6.27}$$

I want to discuss this concept in a couple of examples. Let k be a field and and $S = \mathrm{Spec}(k)$. We consider $A = k[X_1,\dots,X_n]/I = k[x_1,\dots,x_n]$ where I is an ideal in the polynomial ring, which is generated by polynomials $F_1(X_1,\dots,X_n),\dots F_r(X_1,\dots,X_n]$. We have the diagram

$$\begin{array}{ccc} \mathrm{Spec}(A) & & \\ \downarrow & & \\ \mathrm{Spec}(k) & \xleftarrow{\mathrm{id}} & \mathrm{Spec}(k)(=T). \end{array}$$

In this case the T-valued points are the k-homomorphisms $\varphi : k[x_1,\dots,x_n] \to k$. Such a φ is determined by its values $(\varphi(x_1),\dots,\varphi(x_n)) = (a_1,\dots,a_n) \in k^n$. But we have constraints on the n-tuples of values because we have relations among the x_i

$$F_1(x_1,\dots,x_n) = \dots = F_r(x_1,\dots,x_n) = 0 \tag{6.28}$$

and hence also $(a_1,\dots,a_n)$ has to satisfy

$$F_1(a_1,\dots,a_n) = \dots = F_r(a_1,\dots,a_n) = 0. \tag{6.29}$$

This means that the point $a = (a_1,\dots,a_n)$ has to be a solution of the polynomial equations $F_1 = \dots = F_r = 0$. If in turn $a = (a_1,\dots,a_n)$ solves the system of equations then we can look at the diagram

$$\begin{array}{ccc} k[X_1\dots X_n] & \longrightarrow & k[X_1\dots X_n]/I = A \\ & \Psi_a \searrow\!\nwarrow \quad \nearrow & \\ & k & \end{array}$$

where $\Psi_a(X_i) = a_i$. This homomorphism vanishes on the generators $(F_1,\dots,F_r)$ of the ideal I and hence it factors over A. Therefore we get:

Proposition 6.2.7. *For our A as above the k-valued points are given by*

$$\mathrm{Spec}(A)(k) = \{a \in k^n \mid F_1(a) = \dots = F_r(a) = 0\},$$

i.e. they are the solutions of the system of polynomial equations given by the $F_1,\dots,F_r = 0$, where we only allow solutions with coordinates in k.

We can embed our field k into an algebraic closure $\overline{k}$. Then this defines a morphism $\mathrm{Spec}(\overline{k}) \to \mathrm{Spec}(k)$ hence we have the diagram

$$\begin{array}{ccc} X & & \\ \downarrow & & \\ \mathrm{Spec}(k) & \longleftarrow & \mathrm{Spec}\left(\overline{k}\right). \end{array}$$

Definition 6.2.8. *The set $X\left(\overline{k}\right)$ is the set of solutions of the system of equations over $\overline{k}$. These are the so-called* **geometric points** *of the scheme $X \to \mathrm{Spec}(k)$.. The points in $X(k)$ are called k-***rational points** *or simply rational points.*

Of course it is much easier to find geometric points than k-valued points. We may consider for instance the polynomial $y^2 = x^3 - x - 1$ in $\mathbb{Q}[x,y]$ and want to find $\mathbb{Q}$ valued points. We may start from a value $a \in \mathbb{Q}$ for x, but now we need a good portion of luck if we want to find a point $(a,b) \in \mathbb{Q}$ satisfying the equation. We have to find a square root of $a^3 - a - 1$ in $\mathbb{Q}$, i.e. we have to choose an a such that this number is a square.
If we look for geometric points we do not have this problem. If k is not algebraically closed then we easily find a k algebra A/k, for which the set of k-valued points is empty.

Finding k-valued points on a scheme X/k is the classical problem of solving Diophantine equations. Diophantus solved the problem of finding all $\mathbb{Q}$ rational points on $\mathrm{Spec}(\mathbb{Q}[X,Y]/(X^2 + Y^2 - 1)$, this are the Pythagorean triples. Only recently it has been shown by A. Wiles [Wi] that for $n > 2$ the scheme $\mathrm{Spec}(\mathbb{Q}[X,Y]/(X^n + Y^n - 1)$ has only the trivial solutions were one of the variables goes to zero.
It is certainly the first basic problem of algebraic geometry to understand the "structure" of the set of geometric points of a scheme of finite type (see Def. 6.2.10 above) over a field k. Of course it is not clear what that means. If $k = \mathbb{Q}$ then in most cases this set is just a countable set. But for instance in the next chapter we will learn that under some assumption (irreducibility) it has a dimension and this dimension is an integer.
If $k = \mathbb{C}$ then all k-valued points are also geometric points. Then we have much more structure on the set of geometric points. For instance it is a topological space because we can realize affine pieces as subsets of $\mathbb{C}^n$. If we start from our equation $y^2 = x^3 - x - 1$ above and if we add a point at infinity, then the set of geometric points becomes a compact Riemann surface (See Vol. I, 5.1.7), actually it is even an elliptic curve. So we can ask for the cohomology groups of this space. This aspect was discussed already in Volume I.
In general we have the feeling, that the set of geometric points is a geometric object. For instance we may-in low dimensional cases-draw a picture of the $\mathbb{R}$-valued points if our scheme is defined over $\mathbb{R}$.

We discuss an example of such a drawing, but before doing this, we want to say a few words why we want to consider relative situations

$$\begin{array}{c} X \\ \pi \big\downarrow \\ S, \end{array}$$

where S is not just a point $\mathrm{Spec}(k)$. For any point $s \in S$ we can consider the stalk $\mathcal{O}_{S,s}$, this is a local ring and let $\mathfrak{m}_s$ be its maximal ideal. Then $k(s) := \mathcal{O}_{S,s}/\mathfrak{m}_s$ is a field and we have a morphism $S \leftarrow \mathrm{Spec}(k(s))$ and hence we get a base change $X \times_S \mathrm{Spec}(k(s))$ and this is a scheme over $\mathrm{Spec}(k(s))$.

Definition 6.2.9. *The scheme $X_s = X \times_s \mathrm{Spec}(k(s))$ over $\mathrm{Spec}(k(s))$ is called the* **fibre** *of X/S over s.*

Hence we get a family of schemes, which is parametrized by the points of S. We may even go one step further and embed $k(s)$ into an algebraically closed field $\bar{k}$. This gives us a morphism $\mathrm{Spec}(k(s)) \leftarrow \mathrm{Spec}(\bar{k})$ and the composition $S \leftarrow \mathrm{Spec}(\bar{k})$ is a geometric point of S. The base change to $S \times_S \mathrm{Spec}(\bar{k})$ is the **geometric fibre** over the geometric point.

A very simple example of such a situation is given if we consider the equation

$$y^2 = x^3 - (2 - \lambda)x^2 + x$$

over the scheme $S = \mathrm{Spec}(\mathbb{Q}[\lambda])$. Hence we get a family of curves, which is parameterized by the points on the affine line with coordinate function λ. In the pictures below we drew the set of real points of some members of the family, we see some interesting properties of these sets of real points and we see the dependence of these properties on λ.

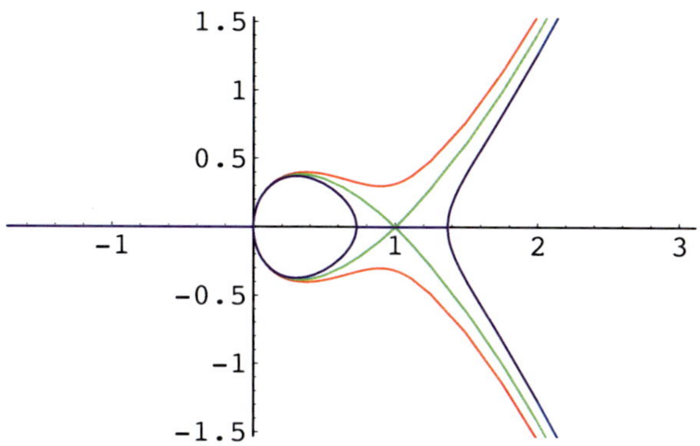

Figure 6.1 Pictures of real valued points

The above picture shows members of this family for the three values $\lambda = -1/10, 0, 1/10 =$ red,green,blue.

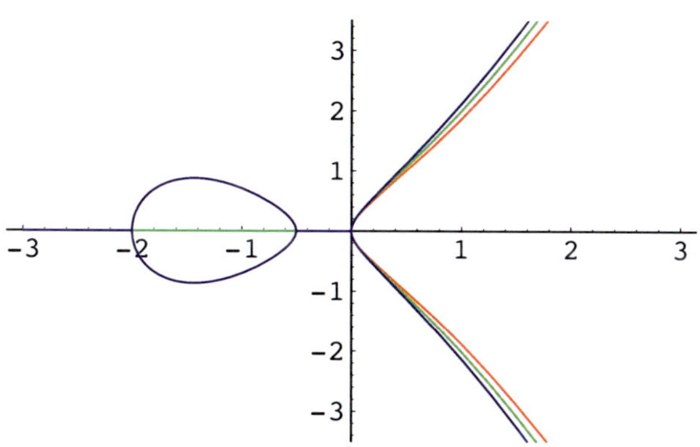

Figure 6.2 Pictures of real valued points

The second picture shows the same family for the values $\lambda = 7/2, 4, 9/2 =$ red,green,blue. These pictures tell us something: In the first picture we start with $\lambda < 0$ but moving to zero we get the blue curve. It has two connected components. If λ approaches zero the intersection points P_1,P_2 of the two components with the x-axis come closer and closer to each other. Eventually if $\lambda = 0$ we get the green curve, which has a singular point $(1,0)$. This is a so called double point, a more than natural terminolgy of course, the two points P_1,P_2 became one point. If now $\lambda > 0$ then we get the red curve, which has only one component. The double point splits again into two points but now the y coordinates of these two points are purely imaginary and they are not visible anymore.

An similar thing happens of λ approaches the value 4. If $\lambda > 4$ we get the blue curve. It has two components. If λ approaches 4, then the "circular" component becomes smaller and smaller, for $\lambda = 4$ the curve becomes green and the component shrinks to the point $(-1,0)$. This is again a double point, but the two branches have imaginary coordinates. So we see that the set of $\mathbb{R}$-valued points has interesting topological properties and these properties may vary if we move the scheme in a family.

Another interesting family is obtained if we take $S = \mathrm{Spec}(\mathbb{Z})$. In this case we may take $X = \mathrm{Spec}(\mathbb{Z}[X_1,X_2,\dots,X_n]]/I$, where $I = (F_1,\dots,F_r)$ is an ideal generated by the F_i and the F_i have coefficients in $\mathbb{Z}$. Now we can choose a prime p and put $T = \mathrm{Spec}(\mathbb{F}_p)$ We get a family of schemes $X \times_{\mathbb{Z}} \mathbb{F}_p$, which are parameterized by the primes.
In this case the set of $\mathbb{F}_p$-valued points $X(\mathbb{F}_p)$ is finite, this raises the question whether we can say something intelligent about the cardinality of this set. Since our scheme is over $\mathrm{Spec}(\mathbb{Z})$ we may also consider the topological space of $\mathbb{C}$-valued points. It is one of the great discoveries of the last century mathematics that the question of counting the number of points in $X(\mathbb{F}_p)$ is related to the topology of the space $X(\mathbb{C})$ of $\mathbb{C}$-valued points. We come back to this problem in 9.7.7 and section 10.4.2.

Closed Points and Geometric Points on varieties

Definition 6.2.10. *A scheme* $X = \mathrm{Spec}(k[X_1,X_2,\dots,X_n]/(F_1,\dots,F_r)) = \mathrm{Spec}(A)$ *is called an* **affine variety over** k. *A scheme* $X \longrightarrow \mathrm{Spec}(k)$ *is a* **variety over** k *if a has a finite covering by open affine varieties.*

We will use the terminology affine scheme of finite type over k (scheme of finite type over k) synonymously to affine variety (variety) over k.
Let us consider the case of an affine variety $\mathrm{Spec}(A)/k$. Then we have various kinds of points. We have the k *valued points*, we have *geometric* points but in general we still have many more points (see 6.4.)
If we have a geometric point $P : A \to \bar{k}$ then the kernel of P is a prime ideal $\mathfrak{m}_P$ of A. The quotient $A/\mathfrak{m}_P$ is a subring in $\bar{k}$, which contains k and hence it is a field. This implies that $\mathfrak{m}_P$ is maximal. Therefore we get a map $X\left(\bar{k}\right) \to \mathrm{Specmax}(A)$. If we have in turn a maximal ideal $\mathfrak{m}$ in A then it follows from the Nullstellensatz that $k(\mathfrak{m}) = A/\mathfrak{m}$ is a finite extension of k. We get a diagram

$$\begin{array}{ccc} A & \longrightarrow & A/\mathfrak{m} = k(\mathfrak{m}) \\ \nwarrow & & \nearrow \\ k & \longrightarrow & \bar{k} \end{array}$$

and it is clear that the k-homomorphisms $k(\mathfrak{m}) \to \bar{k}$ correspond one to one to the geometric points of X, which lie above $\mathfrak{m}$. Now Galois theory implies

Lemma 6.2.11.

(1) *The map* $X\left(\bar{k}\right) \to \mathrm{Specmax}(A)$ *is surjective with finite fibers. The points in the fibre are in one to one correspondence with the prime ideals in the* $\bar{k}$ *-algebra* $k(\mathfrak{m}) \otimes \bar{k}$. *(see p. 73)*

(2) *The cardinality of the fibre of* $\mathfrak{m}$ *of this map divides the degree* $[k(\mathfrak{m}) : k]$ *and it is equal to this degree if and only if* $k(\mathfrak{m})/k$ *is a separable extension.*

(3) *The Galois group of* $\bar{k}/k$ *acts transitively on the geometric points in the fibre over* $\mathfrak{m}$.

We know that any scheme X/S defines a functor F_X from the category of schemes over S to the category of sets. We also know that the functor recognizes X, this is the standard Yoneda lemma (See Vol. I. 1.3.). But we have to evaluate the functor on all schemes $T \longrightarrow S$.
We want to explain that for varieties X/k over a field k the value of the functor on $\bar{k}$, i-e. the set of geometric points, still contains a lot of information.

Definition 6.2.12. *We call an affine variety* $X = \mathrm{Spec}(A)/k$ *absolutely reduced if the algebra* $A \otimes_k \bar{k}$ *does not have non zero nilpotent elements. A scheme of finite type is absolutely reduced if it is covered by absolutely reduced affine schemes.*

We restrict our attention to affine varieties over k. An element $f \in A$ defines a $\bar{k}$-valued function on $X(\bar{k})$: By definition a $\phi \in X(\bar{k})$ is a homomorphism from A/k to $\bar{k}$ and we put $f(\phi) = \phi(f)$. If we assume that $A \otimes \bar{k}$ does not have nilpotent elements, then the Nullstellensatz implies that we get in inclusion

$$A \otimes \bar{k} \hookrightarrow \bar{k} \text{ valued functions on } X(\bar{k})$$

in other words we can view $A \otimes \bar{k}$ as a sub algebra of all $\bar{k}$ valued functions on $X(\bar{k})$. The set $X(\bar{k})$ is equal to the set of maximal ideals of this sub algebra, a point $a \in X(\bar{k})$ defines the maximal idea $\mathfrak{m}_a$ consisting of those functions, which vanish at this point. In the previous discussion our algebra always came with a set of generators and from this we saw that $X(\bar{k}) \subset \bar{k}^n$. Then we see that we can reconstruct the algebra from its set of geometric points. The algebra is simply the algebra of functions, which is generated by the coordinate functions x_i where $x_i((a_1, \dots, a_n)) = a_i$.
If we now have a second absolutely reduced affine variety of finite type $Y = \mathrm{Spec}(B)/k, B = k[Y_1, \dots, Y_m]/J$ then we may consider the regular maps from $X(\bar{k}) \longrightarrow Y(\bar{k})$. In a naive way we can say, that a map $f : X(\bar{k}) \longrightarrow Y(\bar{k})$ is regular if the coordinates b_j of $f((a_1, \dots, a_n))$ are given by evaluating polynomials $G_j(X_1, \dots, X_n)$ at $(a_1, \dots, a_n)$, i.e. $f((a_1, \dots, a_n)) = (G_1(a_1, \dots, a_n), \dots, G_m(a_1, \dots, a_n))$.
But it is much more elegant to say it this way:

If $X = \mathrm{Spec}(A), Y = \mathrm{Spec}(B)$ *are two absolutely reduced affine schemes over* k. *A map* ψ *between the sets of geometric points comes from morphism* $\widetilde{\psi} : X \to Y$ *if and only if the induced map* ${}^t\psi$ *between the* k*-valued functions maps the elements of* B *into elements of* A. *Hence we get from* ψ *a homomorphism of rings, which then in turn induces the map between the geometric points, from which we started. Note that* ${}^t(\psi_1 \circ \psi_2) = {}^t\psi_2 \circ {}^t\psi_1$

To summarize we can say that for absolutely reduced finitely generated k-algebras over an algebraically closed field k the set of geometric points together with the algebra A of regular k-valued functions contains all the information. This is of course tautological, because already the algebra A determines everything. But its realization as algebra of k-valued functions on the set of geometric points gives us some picture, which has some geometric flavor and is less abstract than the concept of a locally ringed space. Especially

in the case where a set of geometric points is given to us as a subset of $X(\bar{k}) = \Sigma \subset \bar{k}^n$ we can reconstruct the ring of regular functions, it is the polynomial ring $k[X_1,X_2,\dots,X_n]$ divided by the ideal of all polynomials, which vanish on Σ.
If our absolutely reduced scheme X/k is defined over a field k, which is not algebraically closed then we get an action of the Galois group $\mathrm{Gal}(\bar{k}/k)$ on the set of geometric points. If k is perfect, then we can reconstruct X/k from the scheme $X \times_k \bar{k}$ and the Galois action on the geometric points. This is discussed in the section on Galois descend (See 6.2.9 and see Exc. 6 below))

Exercise 5. Let us assume that k is a field of characteristic $p > 0$. We take $A = B = k[X]$. Then the set of geometric points is $\overline{k}$ and we have the bijective map $x \longrightarrow x^p$ on the set of geometric points. Show that this map comes from a morphism but its inverse does not.
This teaches us that a morphism between affine schemes of finite type over k, which induces a bijection between the sets of geometric points is not necessarily an isomorphism.

Exercise 6. We go back to the general situation that we have two reduced affine schemes X,Y of finite type over $\mathrm{Spec}(k)$. We assume now in addition that the field k has characteristic $p > 0$ and is **perfect** .This means that the map $x \mapsto x^p$ is surjective and hence bijective. The assumption on k also implies that the schemes are absolutely reduced. Let us assume that we have a morphism $\phi_{\overline{k}} : X \times_k \overline{k} \to Y \times_k \overline{k}$, which induces a map ${}^t\phi : X(\overline{k}) \longrightarrow Y(\overline{k})$. Now we can define an action of the Galois group $\mathrm{Gal}(\overline{k}/k)$ on the two set of geometric points. Show that $\phi_{\overline{k}}$ is defined over k, i.e. comes from a morphism $\phi : B \to A$ if and only if it commutes with the action of the Galois group.

We may also speak of integral solutions. If we have a scheme $A = \mathbb{Z}[X_1,\dots,X_n]/I$ then we may consider

$$\begin{array}{c} X = \mathrm{Spec}(A) \\ \downarrow \\ \mathrm{Spec}\,(\mathbb{Z}) \longleftarrow \mathrm{Spec}\,(\mathbb{Z})\,. \end{array}$$

and $X(\mathbb{Z}) = X(\mathrm{Spec}(\mathbb{Z}))$ are the integral solutions of the system of equations. For instance we can try to find $\mathbb{Z}$-valued points on $\mathrm{Spec}(\mathbb{Z}[x,y]/(y^2 - x^3 + x + 1))$, which is even harder than finding $\mathbb{Q}$-valued points.

6.2.7 Flat Morphisms

We now turn to something much more abstract, after some preparation we will discuss the notion of "descend", which is fundamental for the proof of representability results.
Let us consider a commutative ring A with identity and its category of A-modules. In this category we have the tensor product of two modules, which is again an A-module. (See Vol. I. 2.4.2)

Remark 3. Here we need the commutativity of A since we want

$$m \otimes (fg)n = (fg)m \otimes n = f(gm) \otimes n = gm \otimes fn = m \otimes g(fn) = m \otimes (gf)n.$$

If we fix a module M we can consider the functor $N \to N \otimes_A M$ of the category into itself. This is a right exact functor: If we have an exact sequence

$$0 \longrightarrow N' \longrightarrow N \longrightarrow N'' \longrightarrow 0$$

then the sequence

$$N' \otimes_A M \longrightarrow N \otimes_A M \longrightarrow N'' \otimes_A M \longrightarrow 0$$

is still exact. The right exactness is proved in [La], XVI, prop. 2.6. the proof is elementary if we use the construction by generators and relations. But the functor is not exact in general, it can happen that the first arrow is not injective anymore. We will give the examples, which we promised in Vol. I 2.4.2.

Exercise 7. To construct an example, which shows that the functor is not exact, let us consider an element $m \in M, m \neq 0$, which has a non-trivial annihilator in A, i.e. there is an $f \in A, f \neq 0$ and $fm = 0$. Consider the sequence of A modules

$$0 \longrightarrow Af \longrightarrow A \longrightarrow A/Af \longrightarrow 0.$$

Show: The element $f \otimes_A m \in Af \otimes M$ goes to zero in $A \otimes_A M = M$. Construct an example where you know that $f \otimes_A m \neq 0$.

The Concept of Flatness

Definition 6.2.13. *An A-module M is called* **flat** *if the functor $N \longrightarrow N \otimes_A M$ is exact.*

Example 10. *A simple example of a flat module is the free A-module over an arbitrary index set I: $A^I = \{(\dots, a_i, \dots)_{i \in I} \mid$ almost all $a_i = 0\}$. This is the direct sum $A^I = \bigoplus_{i \in I} A$ but the direct product $\prod_{i \in I} A$ is also flat.*

Definition 6.2.14. *An A-module M is called* **faithfully flat** *if it is flat and if in addition $N \otimes_A M = 0$ implies $N = 0$.*

The above examples of the direct sum and direct product are indeed faithfully flat. We will see flat modules, which are not faithfully flat in a minute.
Since an A-algebra $A \to B$ is also an A-module, we can speak of **flat** A-algebras. We may view $A \longrightarrow B$ also as a morphisms of affine schemes we can speak of **flat morphisms** $(\mathrm{Spec}(B), \widetilde{B}) \longrightarrow (\mathrm{Spec}(A), \widetilde{A})$ of affine schemes. We also observe that for an A-module N the module $B \otimes_A N$ has an obvious natural structure as a B-module, we simply put $b(b_1 \otimes_A n) = bb_1 \otimes_A n$.

Definition 6.2.15. *A morphism of affine schemes is called* **faithfully flat** *if the A-module B is faithfully flat.*

Exercise 8. Of course the polynomial ring $A[X_1, \dots, X_n]$ is faithfully flat over A because it is free as an A-module.

Exercise 9. An important case of a flat algebra is given by localization. It is not so difficult to check that for any subset $S \subset A$, which is closed under multiplication, the A-algebra $A \to A_S$ is flat. To see this one should also observe that $N \otimes_A A_S \simeq N_S$, where the isomorphism is simply given by $n \otimes \frac{f}{s} \mapsto \frac{fn}{s}$. Now it is clear that if $N' \subset N$ then $N'_S \subset N_S$.

Exercise 10. Let us assume that $S = \{f^n\}$ where $f \in A$ and not nilpotent. We have just seen that $A \to A_f$ is flat. But it will not be faithfully flat in general. To see this prove $A/fA \otimes_A A_f = 0$.

But on the other hand:

Exercise 11. If we have a covering $X = \mathrm{Spec}(A) = \bigcup_{i \in I} X_{f_i}$, where we may assume that I is finite (see Proposition 6.1.15), then $A \to \prod_{i \in I} A_{f_i}$ is faithfully flat. This corresponds to a map $X \leftarrow \bigsqcup X_{f_i}$ and this is a faithfully flat morphism of schemes. (See also Exercise 17)

Exercise 12. If B is the quotient of A by an ideal $A \to B = A/I$, then $\mathrm{Spec}(B) = \mathrm{Spec}(A/I) \longrightarrow \mathrm{Spec}(A)$ is a closed subscheme. In this case one can show that this is almost never flat.

Exercise 13. If A is a Dedekind ring (see 7.3.4), then an A-module N is flat if and only if it is torsion free, i.e. if $fn = 0$ with $f \in A, n \in N$ then f or n is zero.

If we have $A \to B \to C$ and an A-module N then we have a canonical isomorphism

$$\begin{array}{ccc} C \otimes_B (B \otimes_A N) & \xrightarrow{\sim} & C \otimes_A N \\ & \searrow^{\sim} & \downarrow \\ & & (C \otimes_B B) \otimes_A N \end{array}$$

We leave the verification to the reader.

Exercise 14. If we have $A \to B \to C$ and B is a flat A-algebra and C a flat B-algebra then C is also a flat A-algebra.

Exercise 15. If we have $A \longrightarrow B \longrightarrow C$ and if C is a flat A-algebra and if C is a faithfully flat B-algebra then B is a flat A-algebra.

The second assertion is important because it allows us to check flatness locally. The exercise implies

Exercise 16. We start from a homomorphism $A \longrightarrow B$ and we assume we have a family of elements $f_i \in B_i, i \in I$ such that the $X_{fi} \subset X = \mathrm{Spec}(B)$ form a covering of X. If $A \to B_{f_i}$ is flat for all i then B is a flat A-algebra.

The same principle can be applied if we want to check whether an A-module is flat: If $A \longrightarrow B$ is faithfully flat, then an A-module M is flat if and only if the B-module $M \otimes_A B$ is flat.

Definition 6.2.16. *A morphism between two schemes*

$$\begin{array}{c} X \\ \downarrow \pi \\ S \end{array}$$

is **flat** *if for any open affine subscheme $V \subset S$ and any open affine subscheme $U \subset \pi^{-1}(V)$ the $O_S(V)$-algebra $\mathcal{O}_X(U)$ is flat. It is* **faithfully flat** *if it is also surjective.*

To justify the last definition we need to solve the following

Exercise 17. Show that for a flat A-algebra B the two conditions are equivalent

1. B is faithfully flat
2. $\mathrm{Spec}(B) \to \mathrm{Spec}(A)$ is surjective.

Since the schemes over a given scheme S form a category any $X \longrightarrow S$ provides a contravariant functor $\mathcal{F}_X$ from the category of schemes over S to the category of sets $T \longrightarrow \mathrm{Hom}_S(T,X) = X(T) = \mathcal{F}_X(T)$. This functor $\mathcal{F}_X$ determines X/S up to a canonical isomorphism. A functor $\mathcal{F}$ from the category of schemes over S to the category of sets is representable if it is of the form $\mathcal{F}_X$.
The following fundamental fact gives us a constraint for the representability of a functor $\mathcal{F}$ from a category of schemes to the category of sets (see below proposition 6.2.18).

Theorem 6.2.17. *Let X/S be a scheme. Then for any faithfully flat morphism $S \leftarrow S'$ we get an exact sequence of sets*

$$X(S) \xrightarrow{p_0} X(S') \underset{p_2}{\overset{p_1}{\rightrightarrows}} X(S' \times_S S').$$

Recall that exactness means that p_0^* is injective and that the image of p_0^* is equal to the set of $x \in X(S')$, which satisfy $p_1^*(x) = p_2^*(x)$.
It is clear that it suffices to prove this for affine schemes. Then we have $S = \mathrm{Spec}(A)$, $X = \mathrm{Spec}(C)$ and $A \longrightarrow C$. If now $S' = \mathrm{Spec}(B) \longrightarrow S$ is faithfully flat this means that $A \longrightarrow B$ is faithfully flat. We have to show that the sequence

$$\mathrm{Hom}_A(C,A) \xrightarrow{p_0} \mathrm{Hom}_A(C,B) \underset{p_2}{\overset{p_1}{\rightrightarrows}} \mathrm{Hom}_A(C,B \otimes_A B)$$

is exact. Of course $A \longrightarrow B$ is injective, so the first arrow is indeed an injection. Now we have to show that an element $\phi : C \longrightarrow B$ in $\mathrm{Hom}_A(C,B)$, which is sent to the same homomorphisms under the two arrows

$$b \mapsto \begin{cases} b \otimes 1 \\ 1 \otimes b \end{cases}$$

is actually an element in $\mathrm{Hom}_A(C,A)$. But this is clear once we know that

$$A = \{b \in B \mid b \otimes 1 - 1 \otimes b = 0\}$$

We put $\delta(b) : b \mapsto b \otimes 1 - 1 \otimes b$, then last assertion says that the sequence

$$0 \longrightarrow A \longrightarrow B \xrightarrow{\delta} B \otimes_A B$$

is exact. Since $A \to B$ is faithfully flat this is equivalent to the exactness of the same sequence tensorized by B:

$$0 \longrightarrow B \longrightarrow B \otimes_A B \longrightarrow B \otimes_A B \otimes_A B$$

where the homomorphisms are $b \mapsto 1 \otimes b$ and $b_1 \otimes b_2 \mapsto (b_1 \otimes 1 - 1 \otimes b_1) \otimes b_2$. This is now a sequence of B-modules where B acts always on the last factor. We have an isomorphism

$$B \otimes_A B \otimes_A B \xrightarrow{\sim} B \otimes_A B \otimes_B B \otimes_A B,$$

which is given by $b_1 \otimes b_2 \otimes b_3 \mapsto b_1 \otimes 1 \otimes b_2 \otimes b_3$. (In the fourfold tensor product we are allowed to move the entry at place 2 to place 4 and backwards since we take the tensor product over B.) Hence we have to prove the exactness of

$$0 \longrightarrow B \longrightarrow B \otimes_A B \longrightarrow B \otimes_A B \otimes_B B \otimes_A B.$$

But now the inclusion $B \longrightarrow B \otimes_A B$ admits a splitting given by $b_1 \otimes b_2 \mapsto 1 \otimes b_1 b_2$ and hence we have $B \otimes_A B = B \oplus Y$ where Y is the kernel of that splitting. The elements of $y \in Y$ are mapped to

$$\delta(y) = 1 \otimes y + y \otimes 1 + \cdots \in B \otimes Y \oplus Y \otimes B \oplus Y \otimes Y$$

and therefore, it is clear that δ is injective on Y. □

Representability of functors

This theorem plays an enormous role since it gives us a necessary condition for the representability of a functor

$$\mathcal{F} : \mathbf{Schemes}\,/S \longrightarrow \mathbf{Ens}\,.$$

(See Vol. I.1.3.4). Recall that representability means that we can find a scheme X/S and an element – the **identity section** – $e_X \in \mathcal{F}(X)$ such that the map

$$\begin{aligned} \mathrm{Hom}_S(T,X) &\longrightarrow \mathcal{F}(T) \\ \varphi &\longmapsto \varphi^*(e_X) \end{aligned} \tag{6.30}$$

is a bijection for all $T \longrightarrow S$. Here φ^* is an abbreviation for $\mathcal{F}(\varphi)$ and e_X is called the identity section, because it is equal to $\varphi^*(\mathrm{Id}_X)$.

Certainly the reader has noticed that the exactness of the sequence in 6.2.17 is analogous to conditions the two conditions (Sh1),(Sh2), which have to be satisfied by sheaves on topological spaces (see Vol. I, 3.1.3). To make this analogy clear we start from a covering of a topological space $X = \bigcup_{i \in I} U_i$ by open sets. We consider the disjoint union of the open sets we get a continuous map $X \xleftarrow{p} X' = \bigsqcup U_i$. Then the fibered product is

$$X' \times_X X' = \bigsqcup_{(i,j)} U_i \cap U_j.$$

and we get the diagram

$$X \xleftarrow{\;p\;} X' \underset{p_2}{\overset{p_1}{\leftleftarrows}} X' \times_X X' \tag{6.31}$$

A presheaf $\mathcal{F}$ on X is a sheaf if and only if it satisfies the conditions (Sh1),(Sh2). But now it is clear that these two conditions together are equivalent to the exactness of the sequence

$$\mathcal{F}(X) \xrightarrow{p_0^*} \mathcal{F}(X') \underset{p_2^*}{\overset{p_1^*}{\rightrightarrows}} \mathcal{F}(X' \times_X X'), \tag{6.32}$$

A. Grothendieck introduced a much more general concept of topologies. Instead of considering coverings of spaces by open sets, he considers certain classes of morphisms $X' \longrightarrow X$, which are called coverings.
A very important example of such a **Grothendieck topology** is the the **flat topology** on a scheme X. This means that we replace coverings of the scheme X by Zariski open sets by faithfully flat morphisms $X \leftarrow X'$. In this more general context we can still define sheaves: These are functors

$$\mathcal{F} :(\textbf{Faithfully flat schemes}\, X' \longrightarrow X) \longrightarrow \textbf{Ens}\,, \tag{6.33}$$

which satisfy the extra condition that for all $X' \xrightarrow{p_0} S$ we get an exact sequence

$$\mathcal{F}(X) \xrightarrow{p_0^*} \mathcal{F}(S') \underset{p_2^*}{\overset{p_1^*}{\rightrightarrows}} \mathcal{F}(X' \times_X X'), \tag{6.34}$$

where again the p^* are called the restriction maps, these are the maps induced by the functor. We can summarize:

Proposition 6.2.18. *If we have a scheme X and a contravariant functor*

$$\begin{aligned} \mathcal{F} : \textbf{Schemes}\,/X &\longrightarrow \textbf{Ens} \\ T &\longmapsto \mathcal{F}(T) \end{aligned}$$

then a necessary condition for this functor to be representable is that its restriction to the faithfully flat topology is a sheaf.

We can say even more. For any such functor $\mathcal{F}$ and any scheme $T \xrightarrow{f} S$ we can define the restriction

$$\mathcal{F}_T : \textbf{Schemes}\,/T \longrightarrow \textbf{Ens}\,,$$

which is defined by the obvious definition

$$\mathcal{F}_T(T') = \mathcal{F}(T')$$

for any scheme $T' \longrightarrow T$, which then by the composition with f becomes a scheme over S. We have tautologically:

Proposition 6.2.19. *If our functor is representable by a scheme X/S, then for any scheme $T \to S$ the restriction $\mathcal{F}_T$ is represented by $X \times_S T$.*

Later on we will consider coverings, which satisfy some finiteness conditions. If $X' \longrightarrow X$ is faithfully flat and of finite type then we will call it a covering in the **ffft-topology**. Another topology will be the étale topology (see 7.5.14).

6.2.8 Theory of descend

In the last two chapters of this book we will discuss the representability of some functors, We will discuss the Picard functor or more precisely a modified Picard functor in detail. There we will encounter a problem of the following type:

Let S be a scheme and let $\mathcal{F}$ be a contravariant functor from schemes over S to **Ens**. Let us assume that $\mathcal{F}$ is a sheaf for the faithfully flat topology.

Let us also assume that we can find a faithfully flat scheme $S' \to S$ such that the restriction $\mathcal{F}' = \mathcal{F}_{S'}$ of our functor becomes representable over S', i.e. we have a scheme X'/S' and an element, the *identity section* $e_{X'} \in \mathcal{F}'(X')$, which represent the functor in the above sense. Under what conditions can we conclude the already $\mathcal{F}$ itself is representable?

This question has been analyzed by Grothendieck and we will describe the techniques, which allow us to construct – under certain conditions – an X/S, which represents $\mathcal{F}$.

We introduce some notation, we put $S' \times_S S' = S''$ and $S' \times_S S' \times_S S' = S'''$. We have the two projections

$$S' \underset{p_2}{\overset{p_1}{\longleftarrow\!\!\longleftarrow}} S'',$$

and we can take the pullback of X'/S' by these two arrows, i.e. we consider $X_1'' = X' \times_{S',p_1} S''$ and $X_2'' = X' \times_{S',p_2} S''$. These two schemes together with the restrictions $e_{X_1''}, e_{X_2''}$ of $e_{X'}$ represent the two restrictions of $\mathcal{F}'$ to S''. But these restrictions are also the restrictions of $\mathcal{F}$ to S'' via the composition of p_1 and p_2 with $p : S' \to S$. Since these two compositions are equal we see that $\mathcal{F}_1''(T) = \mathcal{F}_2''(T)$ for any object $T \to S''$. This means that the two restrictions of functors are the same, hence we have uniquely determined isomorphisms of schemes (inverse to each other)

$$X_1'' \underset{\varphi_{21}}{\overset{\varphi_{12}}{\rightleftarrows}} X_2'',$$

which send the restrictions $e_{X_1''}, e_{X_2''}$ of the section $e_{X'}$ into each other.
Now we go one step further and consider

$$S' \underset{p_2}{\overset{p_1}{\longleftarrow\!\!\longleftarrow}} S'' \;\overset{p_{12}}{\underset{p_{23}}{\overset{p_{13}}{\longleftarrow\!\!\longleftarrow\!\!\longleftarrow}}}\; S''' \tag{6.35}$$

and can consider the pullback

$$p_{ij}^*(\varphi_{12}) : (X' \times_{S',p_1} S'') \times_{S'',p_{ij}} S''' \longrightarrow (X' \times_{S',p_2} S'') \times_{S'',p_{ij}} S''' \tag{6.36}$$

and the same for $p_{ij}^*(\varphi_{21})$.

The composition $p_\nu \circ p_{ij}$ is always a projection to a factor (the α-th factor)

$$\pi_\alpha : S''' \longrightarrow S'.$$

and we have

$$\begin{array}{lll} p_1 \circ p_{12} = \pi_1 & p_1 \circ p_{13} = \pi_1 & p_1 \circ p_{23} = \pi_2 \\ p_2 \circ p_{12} = \pi_2 & p_2 \circ p_{13} = \pi_3 & p_2 \circ p_{23} = \pi_3 . \end{array}$$

This allows us to identify always two of the two step pullbacks to a one step pullback, we get

$$\begin{aligned} (X' \times_{S',p_1} S'') \times_{S'',p_{12}} S''' &= X' \times_{S',\pi_1} S''' = (X' \times_{S',p_1} S'') \times_{S'',p_{13}} S''', \\ (X' \times_{S',p_2} S'') \times_{S'',p_{12}} S''' &= X' \times_{S',\pi_2} S''' = (X' \times_{S',p_1} S'') \times_{S'',p_{23}} S''', \\ (X' \times_{S',p_2} S'') \times_{S'',p_{13}} S''' &= X' \times_{S',\pi_3} S''' = (X' \times_{S',p_2} S'') \times_{S'',p_{23}} S'''. \end{aligned}$$

Our φ_{ij} induce isomorphisms among these S''' schemes, for example

$$\begin{array}{ccc} X' \times_{S',\pi_1} S''' & \xrightarrow{p_{12}^*(\varphi_{12})} & X' \times_{S',\pi_2} S''' \\ & \searrow \quad \swarrow & \\ & S''' & \end{array}$$

and the assumption that X' represents the restriction $\mathcal{F}'$ of our functor $\mathcal{F}$ implies that the restrictions of the *identity sections* are mapped into each other. Hence we get the following 1-cocycle relation:

$$p_{13}^*(\varphi_{12})^{-1} \circ p_{23}^*(\varphi_{12}) \circ p_{12}^*(\varphi_{12}) = \mathrm{Id}\,, \tag{6.37}$$

where Id is the identity automorphism of $X' \times_{S',\pi_1} S'''$. Let us forget for a moment the functor $\mathcal{F}$.
Let us assume that we have a faithfully flat scheme $S' \to S$ and a scheme $f' : X' \to S'$.

Definition 6.2.20. *If we have an isomorphism $\varphi_{12} : X' \times_{S',p_1} S'' \to X' \times_{S',p_2} S''$, which satisfies the cocycle rule*

$$p_{13}^*(\varphi_{12})^{-1} \circ p_{23}^*(\varphi_{12}) \circ p_{12}^*(\varphi_{12}) = \mathrm{Id}$$

where we made the 3 identifications between the corresponding two step pullbacks, then we call this a **descent datum** *(after A. Grothendieck). Such a descent datum is called* **effective** *if we can find a scheme $X \xrightarrow{f} S$ and an isomorphism*

$$\begin{array}{ccc} X \times_S S' & \xrightarrow{h} & X' \\ & \searrow \quad \swarrow{\scriptstyle f'} & \\ & S' & \end{array}$$

such that: Firstly

$$X \times_S S'' = (X \times_S S') \times_{S',p_1} S'' = (X \times_S S') \times_{S',p_2} S'',$$

and secondly the diagram

$$\begin{array}{ccc} & & X' \times_{S',p_1} S'' \\ & \nearrow & \Big\downarrow \varphi_{12} \\ X \times_S S'' & & \\ & \searrow & \\ & & X' \times_{S',p_2} S'' \end{array}$$

is commutative. Then we say that $(X'/S',\varphi_{12})$ descends to S and (X,h) is the **realization** *of the descent datum.*

To be able to say this, we should convince ourselves that:

Proposition 6.2.21. *A realization of a descent datum is unique up to a canonical isomorphism.*

Proof: Let us assume we have two such realizations $(X/S,h)$ and $(X_1/S,h_1)$. Then we get from the definition an isomorphism

$$h_1^{-1} \circ h : X \times_S S' \longrightarrow X_1 \times_S S'$$

and from the compatibility with the descent datum we conclude that the two pullbacks

$$p_1^*(h_1^{-1} \circ h_2) : (X \times_S S') \times_{S',p_1} S'' \longrightarrow (X_1 \times_S S') \times_{S',p_1} S''$$

and

$$p_2^*(h_1^{-1} \circ h_2) : (X \times_S S') \times_{S',p_2} S'' \longrightarrow (X_1 \times_S S') \times_{S',p_2} S''$$

must be equal. Hence we get from Theorem 6.2.17 that $h_1^{-1} \circ h$ is the pullback of a uniquely determined isomorphism. □

Now we come back to our functor $\mathcal{F}$ whose restriction to $X \times_S S'$ was supposed to be representable by a scheme X'/S'. Then we constructed a descent datum for X'/S' from this information and we can summarize:

Proposition 6.2.22. *If our functor $\mathcal{F}$:* **Schemes** */S →* **Ens** *is a sheaf for the flat topology and if it is representable by a scheme X'/S' for some faithfully flat $S' \to S$, then it is representable by a scheme X/S if and only if the descent datum is effective.*

Here we can also discuss the descent for sheaves. If we have a flat morphism $S' \longrightarrow S$ and if we have a sheaf for the flat topology $\mathcal{F}'/S'$, we want to know, under which conditions we can find a sheaf $\mathcal{F}/S$, which restricts to $\mathcal{F}'$. We may also define descent data: We take the two restriction $\mathcal{F}'_1,\mathcal{F}'_2$ of $\mathcal{F}'$ by the two morphisms p_1,p_2 from $S' \times_S S'$ to S' and we assume that we have an isomorphism of sheaves $\varphi_{12} : \mathcal{F}'_1 \xrightarrow{\sim} \mathcal{F}'_2$.

We say that this isomorphism is a descent datum for the sheaf $\mathcal{F}'$ if its pullbacks via the different projections from $S' \times_S S' \times_S S'$ to $S' \times_S S'$ satisfy the 1-cocycle relation

$$p_{13}^*(\varphi_{12})^{-1} \circ p_{23}^*(\varphi_{12}) \circ p_{12}^*(\varphi_{12}) = \mathrm{Id}\,.$$

It is now easy to see that now the sheaf $\mathcal{F}'$ descends to a sheaf $\mathcal{F}$, if we have such a descent datum on it. The sheaf $\mathcal{F}$ is unique up to a canonical isomorphism.
We can now look at our previous discussion of descent data on schemes from a different point of view: If we have a scheme X'/S' then it defines a sheaf for the flat topology on schemes over S'. A descent datum defines also a descent datum for the sheaf. Then the sheaf descends to a sheaf over S on the flat topology over S. Now the descent datum is effective if and only if this resulting sheaf is representable.

Remark 4.

1. Of course it is clear that for a scheme X/S we have a descend datum on $X \times_S S'$, we can simply take the pullbacks of the identity.
2. Just to avoid possible misunderstandings: If we have a scheme X'/S', and if we can find a scheme X/S such that $X \times_S S' \xrightarrow{\sim} X'$, then this scheme X/S is not necessarily unique. Only if the isomorphism $X \times_S S' \xrightarrow{\sim} X'$ is compatible with the given descent datum, we have uniqueness (see 6.2.10.)

Effectiveness for affine descend data

We have a simple case where we have effectiveness of descend data. If all our schemes $S = \mathrm{Spec}(A)$, $S' = \mathrm{Spec}(A')$ and $X' = \mathrm{Spec}(B')$ are affine, then we have the diagram

$$\begin{array}{ccc} & & B' \\ & & \uparrow \\ A & \longrightarrow & A'. \end{array}$$

We have the two homomorphisms

$$A' \underset{i_2}{\overset{i_1}{\rightrightarrows}} A' \otimes_A A',$$

and we assume that we have an isomorphism of $A' \otimes_A A'$-algebras

$$B' \otimes_{A',i_1} A' \otimes_A A' \xrightarrow{\varphi} B' \otimes_{A',i_2} A' \otimes_A A'. \tag{6.38}$$

We consider the pullbacks

$$\varphi \otimes i_{\nu\mu} : (B' \otimes_{A',i_1} A' \otimes_A A') \otimes_{i_{\nu\mu}} A' \otimes A' \otimes A' \longrightarrow B' \otimes_{A',i_2} A' \otimes_A A'. \tag{6.39}$$

The $i_{\nu\mu}$ send an $a' \otimes a''$ to a threefold tensor with a $1_{A'}$ at the right place. Now we say that φ is a descent datum (we simply have to translate) if

$$(\varphi \otimes i_{12}) \circ (\varphi \otimes i_{23}) \circ (\varphi \otimes i_{13})^{-1} = \mathrm{Id}\,. \tag{6.40}$$

We want to show that such a descent datum is always effective. Let $1_{A'}$ be the identity of A'. We consider the algebra $B \subset B'$ consisting of elements

$$B = \{b \in B' \mid \varphi(b \otimes i_1(1_{A'})) = \varphi(b \otimes i_2(1_{A'}))\}.$$

This is an A-algebra and one checks easily that

$$B \otimes_A A' \simeq B'$$

6.2.9 Galois descend

In this book we only need a very special case for applying this method of descend. This case is also the historic origin of the method.
Let us assume that we have a field k and a finite separable normal extension L/k. These are our two schemes $S = \mathrm{Spec}(k)$, $S' = \mathrm{Spec}(L)$. Let us assume that we have an affine scheme $X'/\mathrm{Spec}(L)$, it is given as an affine L-algebra A', i.e. $X' = \mathrm{Spec}(A')$. Let us even assume for the moment that X is given as a subscheme of an affine space: We consider X' as a closed subscheme of A^n_L/L, this means that

$$X' = \mathrm{Spec}(L[X_1 \cdots X_n]/I_L]$$

where $I_L \subset L(X_1 \cdots X_n)$ is the defining ideal. The Galois group $\mathrm{Gal}(L/k)$ acts on L/k, we denote this action by $(\sigma,a) \mapsto \sigma(a)$, we also use the convention to write $\sigma(a) = a^\sigma$. Then one has to be aware that $(a^\tau)^\sigma = a^{\sigma\tau}$. It is clear that it acts on $L[X_1 \cdots X_n]$ via the action on the coefficients and we can define the conjugate of X' as

$$(X')^\sigma = \mathrm{Spec}(L[X_1 \cdots X_n]/I_L^\sigma \tag{6.41}$$

where of course

$$I_L^\sigma = \left\{\Sigma a^\sigma_{\nu_1 \cdots \nu_n} X_1^{\nu_1} \cdots X_n^{\nu_n} \mid \Sigma a_{\nu_1 \cdots \nu_n} X_1^{\nu_1} \cdots X_n^{\nu_n} \in I_L\right\}.$$

If we consider the set of geometric points $X'\left(\overline{k}\right)$, then we see

$$(X')^\sigma\left(\overline{k}\right) = \left(X'\left(\overline{k}\right)\right)^{\widetilde{\sigma}} \tag{6.42}$$

where $\widetilde{\sigma} \in \mathrm{Gal}(\overline{k}/k)$ maps to σ.
If we now want that $X'/\mathrm{Spec}(L)$ is obtained by a base extension of an affine scheme over k, then necessarily the two affine L-schemes X' and $(X')^\sigma$ must be isomorphic for any $\sigma \in \mathrm{Gal}(L/k)$.

What does it mean to have an isomorphism? This means that we should have an L-algebra isomorphism

$$\begin{array}{ccc} L[X_1 \cdots X_n]/I_L & \xrightarrow{\varphi_\sigma} & L[X_1 \cdots X_n]/I_L^\sigma \\ \| & & \| \\ A & \xrightarrow[\varphi_\sigma]{} & A^\sigma. \end{array}$$

We see that the element in the Galois group defines an isomorphism

$$\sigma : h \longmapsto h^\sigma \tag{6.43}$$

between the two rings A and A^σ. We have to be careful and observe that this isomorphism is not L-linear, but only σ-linear, this means that

$$\sigma(\lambda h) = \lambda^\sigma h^\sigma \tag{6.44}$$

for all $\lambda \in L$.
This allows us to get rid of the assumption that X is given as a subscheme of an affine space. We define A^σ as the L-algebra, which as a ring is equal to A but where the scalar $\lambda \in L$ acts by

$$\lambda *_\sigma h = \lambda^\sigma h \tag{6.45}$$

on the right hand side λ^σ act by the original L-algebra structure. This allows us to define the conjugate scheme X'^σ without reference to the embedding.
Now we reformulate the concept of descent datum for this special case. We say that a **Galois descent datum** is a family L-algebra of isomorphisms

$$\varphi_\sigma : A \longrightarrow A^\sigma,$$

which satisfies a compatibility condition. To formulate this condition we observe that for any τ and any σ the ring isomorphism $\varphi_\sigma : A \longrightarrow A$ provides also an isomorphism $\varphi^\sigma_\tau : A^\tau \longrightarrow (A^\sigma)^\tau = A^{\tau\sigma}$: We easily check this

$$\varphi^\tau_\sigma(\lambda^\tau h) = \varphi_\sigma(\lambda^\tau h) = (\lambda^\tau)^\sigma \varphi_\sigma(h) = \lambda^{\sigma\tau}\varphi_\sigma(h) = \lambda^{\sigma\tau}\varphi^\tau_\sigma(h)$$

Then the conditions for a Galois descend datum are

(1) $\varphi_1 = \mathrm{Id}$.

(2) For any pair σ,τ of elements in the Galois group the diagram

$$\begin{array}{ccc} A & \xrightarrow{\varphi_\tau} & A^\tau \\ & {\scriptstyle\varphi_{\sigma\tau}}\searrow \quad \swarrow{\scriptstyle\varphi^\tau_\sigma} & \\ & A^{\sigma\tau} & \end{array}$$

commutes, we have the cocycle condition

$$\varphi_{\sigma\tau} = \varphi^\tau_\sigma \circ \varphi_\tau$$

Proposition 6.2.23. **(i)** *If we have such a Galois-descent datum, then the k-algebra*

$$A_0 = \{a \in A \mid \varphi_\sigma(a) = a\}$$

defines an affine scheme over k (of course) and

$$\begin{aligned} A_0 \otimes_k L &\longrightarrow A \\ h \otimes \lambda &\longmapsto \lambda h \end{aligned}$$

is an L-algebra isomorphism.

(ii) *Such a Galois-descent datum is nothing else than a descent datum in the previous sense.*

The first assertion is a consequence of the main theorem in Galois theory. If $h \in A$ and if $\lambda \in L$, then we can form

$$\sum_{\sigma \in \mathrm{Gal}(L/k)} \lambda^\sigma \varphi_\sigma(h).$$

If $\tau \in \mathrm{Gal}(L/k)$, then

$$\begin{aligned}\varphi_\tau\left(\sum_{\sigma \in \mathrm{Gal}(L/k)} \lambda^\sigma \varphi_\sigma(h)\right) &= \sum_{\sigma \in \mathrm{Gal}(L/k)} (\lambda^\sigma)^\tau \varphi_\tau \varphi_\sigma(h) \\ &= \sum_{\sigma \in \mathrm{Gal}(L/k)} \lambda^{\tau\sigma} \varphi_\tau^\sigma \varphi_\sigma(h) = \sum_{\sigma \in \mathrm{Gal}(L/k)} \lambda^{\tau\sigma} \varphi_{\tau\sigma}(h)\end{aligned} \tag{6.46}$$

and hence we see that this element lies in A_0. But we know that we can find $d = [L : k]$ elements $\lambda_1 \cdots \lambda_d \in L$ such that the determinant of the matrix $\left(\lambda_i^\sigma\right)_{i=1,\dots,d,\sigma\in \mathrm{Gal}(L/k)}$ is non-zero. (Linear independence of the elements in the Galois group.) Hence we can write h as a unique linear combination of the elements

$$\sum_{\sigma \in \mathrm{Gal}(L/k)} \lambda_i^\sigma \varphi_\sigma(h),$$

and hence the assertion is clear.
We come to assertion (ii): Again we apply the main theorem of Galois theory, which can be summarized to

$$L \otimes_k L \xrightarrow{\sim} \bigoplus_{\sigma \in \mathrm{Gal}(L/k)} L$$

where

$$a \otimes_k b \longmapsto (\dots, a\sigma(b), \dots)_{\sigma \in \mathrm{Gal}(L/k)}.$$

If we look at the two arrows

$$i_1 : L \longrightarrow L \otimes L,\ i_2 : L \longrightarrow L \otimes L$$

then

$$\begin{aligned} i_1(\lambda) &= (\cdots, \lambda, \cdots)_{\sigma \in \mathrm{Gal}(L/k)} \\ i_2(\lambda) &= (\cdots, \lambda^\sigma, \cdots)_{\sigma \in \mathrm{Gal}(L/k)} \end{aligned}$$

and

$$k = \{\lambda \mid i_1(\lambda) = i_2(\lambda)\}$$

Now we skip some details and say that it is quite clear that for an L-algebra A the datum of a descent datum

$$\varphi : A \otimes_{L,i_1} L \otimes L \longrightarrow A \otimes_{L,i_2} (L \otimes_k L)$$

is exactly the same as the datum of a Galois-descent datum, we simply have to use the above description of $L \otimes_k L$. □

We have shown again :

Proposition 6.2.24. *A Galois-descent datum for affine schemes is always effective.*

It is quite clear at this point that we need the cocycle condition (2), i.e. the condition $\varphi_{\sigma\tau} = \varphi_\sigma^\tau \circ \varphi_\tau$. If we do not have this condition then we may try to modify the family φ_σ so that the cocycle condition holds for the modified family. But sometimes this is not possible and we end up with an obstruction in a second Galois-cohomology set. (See 9.6.2.)

A geometric interpretation

If we have a scheme of finite type X/k then we have an action of the Galois group $\mathrm{Gal}(\bar{k}/k)$ on the set $X(\bar{k})$ of geometric points. If $k_s \subset \bar{k}$ is the separable closure of k then $\mathrm{Gal}(\bar{k}/k)$ also acts on $X(k_s)$. We know that $X(k) \subset X(k_s)$ is exactly the set of fixed points under this action and an analogous statement holds for any finite separable extension $L \subset k_s$.
So we may ask the following question: Assume we have an absolutely reduced scheme $X'/\bar{k}$ of finite type and assume we have an action of the Galois group $\mathrm{Gal}(\bar{k}/k)$ on its set of geometric points. Does there exist a scheme of finite type X/k such that $X \times_k \bar{k} \xrightarrow{\sim} X'/\bar{k}$ such that the induced isomorphism between the sets of geometric points is invariant under the action of the Galois group?
Of course this action must satisfy some conditions. Let us assume our scheme is affine of finite type $X'/\bar{k} = \mathrm{Spec}(A'/\bar{k})$. We also assume that it is absolutely reduced. We get an action of the Galois group $\mathrm{Gal}(\bar{k}/k)$ on the algebra of all $\bar{k}$ valued functions by

$$\widetilde{\sigma}(f)(x) = \sigma(f(\sigma^{-1}x)).$$

Of course we demand that this induces an action of the Galois group on A', i.e. $\widetilde{\sigma}(f) \in A'$ if $f \in A'$.
Let us assume we have a finite, separable, normal extension L/k, and our scheme $\mathrm{Spec}(A')$ is obtained from an affine scheme $\mathrm{Spec}(A_L/L)$ by base change, i.e. $A' = A_L \otimes_L \bar{k}$ and our Galois action above restricted to $\mathrm{Gal}(\bar{k}/L)$ is the action on the geometric points of $\mathrm{Spec}(A_L/L)$.
We made the assumption that the action of $\mathrm{Gal}(\bar{k}/L)$ on $X'(\bar{k})$ extends to an action of $\mathrm{Gal}(\bar{k}/k)$. We denote this action by $(\sigma,x) \mapsto \sigma x$. Then we get an action of $\mathrm{Gal}(\bar{k}/k)$ on the $\bar{k}$ valued functions on $X'(\bar{k})$, and we have to demand that for $f \in A_L$ we must have $\widetilde{\sigma}(f) \in A_L$. Then a straightforward computation shows that $f \longrightarrow \widetilde{\sigma}(f)$ is in fact an isomorphism $\phi_\sigma : A_L \longrightarrow (A_L)^\sigma$. And this computation also shows immediately that the function $\sigma \longrightarrow \phi_\sigma$ satisfies the cocycle condition.

Therefore we see:
Let L/k be a finite separable extension. For an absolutely reduced affine scheme $X_L/L =$ $\mathrm{Spec}(A_L/L)$ a descend datum to k is the same as an extension of the Galois action of $\mathrm{Gal}(\bar{k}/L)$ on $X_L(\bar{k})$ to $\mathrm{Gal}(\bar{k}/k)$ where this extension satisfies $\sigma(A_L) = A_L$ for all $\sigma \in \mathrm{Gal}(L/k)$. (Note that $\mathrm{Gal}(\bar{k}/L)$ acts trivially on A_L)

It is also clear that in case of effectiveness of our descent datum this action of $\mathrm{Gal}(\overline{k}/k)$ on $X'(\overline{k})$ is exactly the action, which is provided by $h : X \times_k L \to X'$ and the action of $\mathrm{Gal}(\overline{k}/k)$ on $X(\overline{k})$.

Descend for general schemes of finite type

We want to say a word on Galois descend on absolutely reduced schemes of finite type X/k. It is quite clear that we can prove effectiveness for a descent datum $(X'/L, f_\sigma)_{\sigma \in \mathrm{Gal}(L/k)}$ if we know in addition that we can find a finite covering $X' = \{U'_i\}_{i \in E}$ by affine open subschemes, which is compatible with the descent datum, i.e.

$$f_\sigma : U'_i \longrightarrow (U'_i)^\sigma$$

for all σ and all indices $i \in E$. This is clear because then we apply that the datum on the affine pieces is effective, and we glue the descended affine pieces together. The detailed proof is pure routine.
There is a simple criterion that tells us that we can verify this assumption. If X/k is separated (see 8.1.4) and if we know that any finite set $\{x_1, \cdots, x_m\} \in X'(\overline{k})$ is contained in the set of geometric points of an affine open subset $U'/L \subset X'/L$, then our assumption is true. We simply start with any point $x \in X'(\overline{k})$ and look at its finitely many conjugates $\{\widetilde{\sigma}(x)\}_{\widetilde{\sigma} \in \mathrm{Gal}(\overline{k}/k)}$. We pick an open affine set $U'' \subset X'/L$, which contains all these points. It is clear that the conjugates $\widetilde{\sigma}(U''(\overline{k}))$ are also the sets of geometric points of an affine open subset, namely $f_\sigma^{-1}((U'')^\sigma)$, and we look at the intersection

$$U' = \bigcap_{\sigma \in \mathrm{Gal}(L/k)} \widetilde{\sigma}(U''). \tag{6.47}$$

Now we know that the intersection of finitely many open affine subschemes is again affine (see 8.1.4). It is affine and it contains our given point $x \in X'(\overline{k})$. Moreover it is clear that $f_\sigma : U' \to (U')^\sigma$, and we are finished.
In Chap. 8 we introduce the notion of a projective scheme X/k and it will be almost obvious that for any finite set $\{x_1, \cdots, x_m\} \in X(\overline{k})$ we can find an open affine subset containing these points, hence our above criterion for effectiveness applies in this case.

6.2.10 Forms of schemes

We briefly discuss another question. Let us consider two schemes $X/k, X'/k$ of finite type. Let us assume they are absolutely reduced. What can we say if these two schemes become isomorphic if base change them to $\overline{k}$ or the separable closure k_s.?
Let us assume that they become isomorphic over k_s. Then it is clear that we can find a finite normal extension L/k such that $X \times_k L \xrightarrow{\sim} X' \times_k L$. Now we analyze what this means, and the reader is asked to compare the reasoning to the considerations in Vol I 4.3.
First of all we introduce the functor the functor $\mathrm{Aut}(X)$: For any scheme $S \longrightarrow \mathrm{Spec}(k)$ we define $\mathrm{Aut}(X)(S)$ as the group of automorphisms of the scheme $X \times_k S \longrightarrow S$. We may also introduce the functor $\mathrm{Isom}(X, X')$. Our question is whether $\mathrm{Isom}(X, X')(k) \neq \emptyset$. We observe the following: If $S \longrightarrow \mathrm{Spec}(k)$ and we can find a $f \in \mathrm{Isom}(X, X')(S)$ then $\mathrm{Aut}(X)(S)$ acts on $\mathrm{Isom}(X, X')(S)$ and the action is simply transitive. Here one usually

says that Isom(X,X') is a **principal homogenous space** under $\mathrm{Aut}(X)(S)$ or a $\mathrm{Aut}(X)$ **-torsor.**

Now we come back to our problem. Our assumption says that we can find an isomorphism $f : X\times_k L \longrightarrow X'\times_k L$. Can we choose this isomorphism such that it is defined over k, i.e. that it is an extension of an isomorphism between X and X'? We have seen (See previous exercise 6) that f is defined over k if and only if $f = f^\sigma$ for all elements $\sigma \in \mathrm{Gal}(L/k)$. Hence we consider the map $\mathrm{Gal}(L/k) \longrightarrow \mathrm{Aut}(X \times_k L)$, which is given by

$$\sigma \mapsto g_\sigma = f^{-1} \circ f^\sigma.$$

and a straightforward computation shows that this map satisfies a cocycle condition

$$g_\tau \circ g_\sigma^\tau = g_{\tau\sigma} \text{ for all } \sigma,\tau \in \mathrm{Gal}(L/k).$$

The morphism f is defined over k if and only if this cocycle is trivial. But we can try to modify it: We can replace f by $f \circ h$ where h is an automorphism of $X \times_k L$. Then we change $\sigma \longrightarrow g_\sigma$ into $\sigma \longrightarrow g'_\sigma = h^{-1} \circ g_\sigma h^\sigma$. This defines an equivalence relation on the cocycles and we define

$$H^1(\mathrm{Gal}(L/k), \mathrm{Aut}(X \times_k L)) = \text{ set of cocycles divided by the equivalence relation}$$

Now it is obvious that X and X' are isomorphic over k if and only if the above cocycle defines the trivial class in $H^1(\mathrm{Gal}(L/k), \mathrm{Aut}(X \times_k L))$.

But we know more: If only X/k is given then we can consider a cohomology class $\xi \in H^1(\mathrm{Gal}(L/k), \mathrm{Aut}(X \times_k L))$ and represent it by by a cocycle $\sigma \longrightarrow g_\sigma$. We have the surjective homomorphism of the Galois group $\mathrm{Gal}(\bar{k}/k) \longrightarrow \mathrm{Gal}(L/k)$, we can interpret $\sigma \longrightarrow g_\sigma$ as a function (cocycle) on $\mathrm{Gal}(\bar{k}/k)$. We use this cocycle to define a modified action of the set $X(\bar{k})$:

$$\widetilde{\sigma}(a) = g_\sigma \sigma(a)$$

and an easy computation shows that it is exactly the cocycle condition that makes this an action. This action restricted to $\mathrm{Gal}(\bar{k}/L)$ is the old action but the extension to $\mathrm{Gal}(\bar{k}/k)$ is different.

Now we saw in the previous paragraph, that such an action defines a descend datum from L to k on $X \times_k L$. If this descend datum turns out to be effective, then we see that the cohomology class (better the representing cocycle) defines a scheme X'/k. It is almost tautological that the cohomology class in $H^1(\mathrm{Gal}(L/k), \mathrm{Aut}(X \times_k L)$ defined by X'/k is the given one.

Hence we get a fundamental principle

If we know that we have effectiveness of descend data, then the isomorphism classes of k-forms of X/k are given by the elements in

$$\varinjlim H^1(\mathrm{Gal}(L/k), \mathrm{Aut}(X \times_k L))$$

We want to give two simple examples.

Example 11. *Let k be any field with characteristic $\neq 2$, pick two elements $a,b \in k^\times$ such that $a,b,-ab \notin (k^\times)^2$. We consider the matrix algebra $M_2(k(\sqrt{a}))$ this are the (2,2) matrices with entries in $k(\sqrt{a})$. The Galois group $\mathrm{Gal}(k(\sqrt{a})/k)$ is of order 2, let σ be the non trivial element. It acts on $M_2(k(\sqrt{a}))$ by acting on the matrix coefficients. The fixed elements form the algebra $M_2(k)$. We identify the scalars with the diagonal matrices. Now we observe that the invertible elements of $M_2(k(\sqrt{a}))^\times = \mathrm{GL}_2(k(\sqrt{a}))$ act by conjugation on $M_2(k(\sqrt{a}))$, these conjugations induce automorphisms of the the algebra, and such a conjugation induces the identity if and only if the conjugating matrix is trivial. Hence we get a cocycle by sending*

$$\sigma \mapsto \begin{pmatrix} 0 & b \\ 1 & 0 \end{pmatrix}, e \mapsto \mathrm{Id}\,.$$

This cocycle defines a twisted action of the Galois group, we define

$$\widetilde{\sigma}(x) = \begin{pmatrix} 0 & b \\ 1 & 0 \end{pmatrix} x^\sigma \begin{pmatrix} 0 & 1 \\ b^{-1} & 0 \end{pmatrix}.$$

The elements fixed by this new action of the Galois group form a k-algebra D/k and we have seen that $D \otimes_k k(\sqrt{a})/k) \xrightarrow{\sim} M_2(k(\sqrt{a})/k)$. Hence D/k is a k-form of $M_2(k)$. Let α be one of the square roots of a in $k(\sqrt{a})$. Then we see easily that the matrices

$$u_a = \begin{pmatrix} \alpha & 0 \\ 0 & -\alpha \end{pmatrix} \quad u_b = \begin{pmatrix} 0 & b \\ 1 & 0 \end{pmatrix}$$

are elements in D, and the elements $1 = \mathrm{Id}\,, u_a, u_b, u_{ab} = u_a u_b$ form a basis of the k-vector space D. We have

$$u_a^2 = a, u_b^2 = b, u_{ab} = -ab.$$

This k-algebra D/k may have zero divisors or not. If it has zero divisors then it is isomorphic to $M_2(k)$ (exercise) and if not it is called a quaternion algebra. If $k = \mathbb{R}$ and $a = b = -1$ then we get the Hamilton quaternion algebra.
This k-algebra D/k is a k-form of the matrix algebra $M_2(k)$ if we extend the scalars to $k(\sqrt{a})$ then it becomes isomorphic to the matrix algebra.

Before we discuss the second example we want to say a few words about the general notion of an (affine) group scheme. Let S be any scheme. We consider an affine scheme G/S, which has the additional structure as an algebraic group scheme:
By this we mean that we have a morphism $m : G \times_S G \longrightarrow G$ such that this morphism defines a group structure on the set $G(T)$ of T valued points for any scheme $T \longrightarrow S$. To be more precise: For any $T \longrightarrow S$ a S-morphism $g : T \longrightarrow G \times_S G$ is nothing else than a pair of S-morphism $g_1, g_2 : T \longrightarrow G$ (Definition of the fibered product). Composing g with m we get a S-morphism $m \circ g : T \longrightarrow G$, which will be called $g_1 \cdot g_2$. We require that this defines a group structure on $G(T)$ for any $T \longrightarrow S$. It is clear that this group structure depends functorially on T, i.e. if we have another S-scheme T' and an S morphism $T' \longrightarrow T$ then the induced map $G(T) \longrightarrow G(T')$ is a group homomorphism.

We leave it to the reader to reformulate these requirements as properties of the S-morphism $m : G \times_S G \longrightarrow G$. One has to say what associativity, existence of the identity element and existence of the inverse mean. Of course we can extend this notion of group schemes to arbitrary schemes G/S. We will come back to this later (See 7.5.6).

As an example we may consider the group scheme $\mathrm{GL}_n/\operatorname{Spec}(\mathbb{Z})$. Its underlying affine scheme is

$$\operatorname{Spec}(\mathbb{Z}[X_{11},X_{12},\dots X_{1n},X_{21},\dots,X_{nn},Y]/(Y\det -1)$$

where det is the determinant of the matrix (X_{ij}). The structure of an algebraic group is given by matrix multiplication, this gives us the above group scheme.
We want to present an interesting case of a form of $\mathrm{GL}_n/\mathbb{R}$.

Example 12. *The group scheme $\mathrm{GL}_n/\mathbb{R}$ has an automorphism $\Theta : A \mapsto^t A^{-1}$ and $\Theta^2 = \mathrm{Id}$. Now we look at the extension $\mathbb{C}/\mathbb{R}$, the Galois group has as its non trivial element the complex conjugation, which we denote by σ. Again $\sigma \longrightarrow \Theta$ defines an cocycle and we can define a twisted action on $\mathrm{GL}_n(\mathbb{C})$ by*

$$\widetilde{\sigma}(A) = \Theta(A^\sigma) = \Theta(\bar{A}).$$

Then we get a $\mathbb{R}$-form of $\mathrm{GL}_n/\mathbb{R}$, which is called $U(n)/\mathbb{R}$ and whose real points are given by the matrices $A = \Theta(\bar{A})$ and if we unravel this we get

$$U(n)(\mathbb{R}) = \{A | A\,{}^t\bar{A} = \mathrm{Id}\}.$$

If we extend the scalars to $\mathbb{C}$ we get $U(n) \times_{\mathbb{R}} \mathbb{C} \xrightarrow{\sim} \mathrm{GL}_n/\mathbb{C}$.

6.2.11 An outlook to more general concepts

More generally we have the notion of G/S-torsors (or principal homogeneous G bundles) P/S for an arbitrary group scheme G/S. These are schemes P/S together with a left action $m : G \times P \longrightarrow P$, which satisfies two conditions.
a) Firstly we require that for any scheme $T \longrightarrow S$, for which $P(T) \neq \emptyset$ and $x_0 \in P(T)$ the map $G(T) \longrightarrow P(T)$ given by $g \mapsto gx_0$ is a bijection.
a) Secondly we require that the torsor is locally trivial in a certain sense (See Vol. I, 4.3.11. (Unfortunately we did not state the requirement of local triviality in 4.3.11, but it is clear from the context, because we speak of bundles.)) Here we have some flexibility. We may require that our torsor is locally trivial with respect to the Zariski topology, i.e. we have a covering by Zariski open sets $S = \bigcup_i U_i$ such that $P(U_i) \neq \emptyset$. If we put $S' = \bigsqcup_i U_i$ then this means that $P \times_S S'$ is trivial.
But in view of our examples 11,12 this is certainly not a good requirement if our base scheme is the spectrum of a non algebraically closed field. We need a more general notion of local triviality. We may require that our torsor is locally trivial for the flat topology. This means that for P/S we can find a faithfully flat morphism $S' \longrightarrow S$ such that $P(S') \neq \emptyset$.
Our considerations in 6.2.8 essentially mean that we do not loose any information if we pass from P/S to $P \times_S S'$ provided we keep track of the isomorphisms

$$(P \times_S S') \times_{S',p_1} (S' \times_S S') \xrightarrow{\sim} (P \times_S S') \times_{S',p_2} (S' \times_S S').$$

But the flat topology is very fine, it is necessary to introduce some coarser topologies. We may for instance require that $f : \widetilde{U} \longrightarrow U \subset S$ is of finite type, this gives the ffft-topology. We may require that $f : \widetilde{U} \longrightarrow U \subset S$ is étale (see 7.5.14, this yields the étale topology. For all these topology we have the notion of a covering $\{\widetilde{U}_i \longrightarrow S\}$ and this means that $\bigsqcup_i \widetilde{U}_i \longrightarrow S$ is faithfully flat.

If $S = \mathrm{Spec}(k)$ then an étale covering (see 7.5.14) is simply given by a finite separable extension $S' = \mathrm{Spec}(K)$, a finite flat covering is given by a finite extension K/k, which is not necessarily separable.
If we come back to our affine group schemes G/S and consider G torsors G/S which become trivial over a fixed covering $p_0 : S' \longrightarrow S$ (in some of our above topologies) then the isomorphisms classes of such torsors again are in one-to-one correspondence to a set of cohomology classes.
We define the set of cohomology classes first. We consider the sequence of groups

$$G(S) \xrightarrow{p_0^*} G(S') \underset{p_2^*}{\overset{p_1^*}{\rightrightarrows}} G(S' \times_S S') \xrightarrow[p_{23}^*]{\overset{p_{12}^*}{\longrightarrow}\; p_{13}^*} G(S' \times_S S' \times_S S')$$

We define the 1-cocycles to be the elements $g_{12} \in G(S' \times_S S')$, which satisfy

$$p_{12}^*(g_{12})p_{23}^*(g_{12})p_{13}^*(g_{12})^{-1} = 1$$

and on this set we define the equivalence relation

$$g_{12}' \sim g_{12} \text{ if and only if } \exists\, g' \in G(S') \text{ so that } g_{12}' = p_1^*(g')g_{12}p_2^*(g')^{-1}.$$

The set of 1-cocycles divided by this equivalence relation is the cohomology set

$$H^1(S'/S,G)$$

If now P/S is a G/S torsor, which becomes trivial over $S' \longrightarrow S$ (a morphisms in our given topology) then we find a section $x_0 \in P(S')$. We can take the pullbacks $p_1^*(x_0), p_2^*(x_0) \in P(S' \times_S S')$ and by definition we find an unique element g_{12} with $g_{12}p_1^*(x_0) = p_2^*(x_0)$. Now it is clear that g_{12} is a 1-cocycle, changing the section x_0 to another one yields an equivalent 1-cocycle. Hence P/S provides a class in $H^1(S'/S,G)$.
On the other hand it is easy to see that a 1-cocycle provides a descend datum on $G \times_S S'$, if we now assume that all these descend data are effective, then we get the canonical bijection

$$\begin{array}{c}\{\text{ Isomorphism classes of } G/S \text{ torsors } P/S, \text{ which become trivial over } S' \longrightarrow S\} \\ \xrightarrow{\sim} H^1(S'/S,G)\end{array} \tag{6.48}$$

It is now clear that the considerations in Volume I, 4.3. generalize to the situation here.
A morphism for our given topology $f_1 : S_1' \longrightarrow S$ is a refinement of $f : S' \longrightarrow S$ if we can find a morphism $g : S_1' \longrightarrow S'$ such that $f_1 = f \circ g$. Then we get obviously an injection from the set of isomorphism classes of G/S torsors, which become trivial under base change to S' to the set of those torsors, which become trivial over S_1' and hence an inclusion

$$H^1(S'/S,G) \hookrightarrow H^1(S_1'/S,G),$$

which does not depend on the choice of g. Then we can define the limit

$$H^1(S,G) = \lim_{S' \longrightarrow S} H^1(S'/S,G),$$

under our assumption on the effectiveness of descend data this is the set of isomorphism classes of G/S torsors, which are locally trivial in the given topology.
If $S = \mathrm{Spec}(k)$ is the spectrum of a field, then we can restrict our coverings $S' \longrightarrow S$ to finite k-algebras $k \longrightarrow L$. Then we define accordingly

$$H^1(k,G) = \lim_{L/k \text{ finite}} H^1(L/k,G).$$

If we restrict to finite separable algebras $k \longrightarrow L$ (i.e. L is a direct sum of separable extensions) then we can show easily that

$$H^1(k,G) = \lim_{L/k \text{ finite, separable, normal}} H^1(\mathrm{Gal}(L/k),G(L)).$$

The set $H^1(\mathrm{Gal}(L/k),G(L))$ is called the first **Galois cohomology (set)** of L/k (with coefficients in G) and the limit is called the first Galois cohomology of G/k. We considered these sets already above.
If our group scheme G/S is abelian then we can imitate the construction of the Čech complex (Vol. I, 4.5) and get a complex of abelian groups

$$0 \longrightarrow G(S) \longrightarrow G(S') \longrightarrow G(S' \times_S S') \longrightarrow G(S' \times_S S' \times_S S') \longrightarrow$$

and this allows us to define the cohomology groups $H^n(S'/S,G)$ in the usual way. Again we can take a limit over the ffft-coverings and define

$$H^n_{ffft}(S,G) = H^n(S,G) = \lim_{S' \longrightarrow S} H^n(S'/S,G).$$

7 Some Commutative Algebra

We want to collect some standard facts from commutative algebra. Here we will be rather sketchy because many good references are available. Some of the proofs are outlined in exercises.

7.1 Finite A-Algebras

Definition 7.1.1. *An A-module B is called* **finitely generated** *if there are elements* $b_1,b_2,\dots,b_r \in B$ *such that for all* $b \in B$ *we can find* $a_i \in A$ *such that* $b = a_1b_1+\dots+a_rb_r$ *with* $a_i \in A$.

If $\phi : A \longrightarrow B$ is a homomorphism of rings, then we say that an element $b \in B$ **is integral over** A if it is the zero of an monic polynomial, i.e. we can find a polynomial (with highest coefficient equal to 1)

$$P(X) = a_0 + a_1X \dots + X^n \in A[X] \tag{7.1}$$

such that

$$P(b) = \varphi(a_0) + \varphi(a_1)b + \dots + b^n = 0. \tag{7.2}$$

Definition 7.1.2. *A morphism* $\phi : A \to B$ *between two commutative rings is called* **finite** *if one of the following two equivalent conditions is satisfied*

1. *The A-module B is finitely generated.*

2. *The A-algebra B is finitely generated and all elements of B are finite over A.*

It is immediately clear that 2. implies 1. because we can use the polynomials to reduce the degree of the generating monomials. The proof that 1. implies 2. is amusing, we leave it as an exercise. (See also [Ei], Chap. I section 4 and [At-McD]). The following exercise gives a hint.

Exercise 18. 1. We have to show that any $b \in B$ is a zero of a monic polynomial in $A[X]$, i.e. it is integral over A. To see this we multiply the generators b_i by b and express the result again as A-linear combination of the b_i. This gives us an $r \times r$-matrix M with coefficients in A. If $\underline{b}$ is the column vector formed by the b_i we get a relation $b\underline{b} = M\underline{b}$ or $(M - b\,\mathrm{Id})\underline{b} = 0$. From this we have to conclude that $\det(M - b\,\mathrm{Id}) = 0$. This is clear if A is integral, but it suffices to know that the identity of B is contained in the module generated by the b_i, I refrain from giving a hint. Hence we see that b is a zero of the characteristic polynomial of the matrix M, but this polynomial equation has highest coefficient 1 and the other coefficients are in A.

2. This argument generalizes: Let us consider any A-algebra $A \to B$, but assume that B is integral. Show that an element $b \in B$ is integral over A if we can find a finitely generated A-submodule $Y \subset B, Y \neq 0$, which is invariant under multiplication by b, i.e. $bY \subset Y$.

If we have a morphism $\phi : A \longrightarrow B$, then the **integral closure** of A in B consists of all those elements in B, which are integral over A. It is an easy consequence of the two exercises above that the integral closure is an A-sub algebra of B. (For two integral elements b_1,b_2 consider the finitely generated module $\{b_1^\nu b_2^\mu\}$.)

Definition 7.1.3. *An ring A is* **normal** *if it is integral an if it is equal to its integral closure in its quotient field K, i.e. if any element $x \in K$, which is integral over A is already in A. For any integral ring A the integral closure of A in its quotient field is called the* **normalization.**

Synonymously we use the terminology A **is integrally closed** for A is normal.

Definition 7.1.4. *An element a in an integral ring A is irreducible if it is not a unit and if in any multiplicative decomposition $a = bc$ one of the factors is a unit. An integral ring A is called factorial if any element $x \in A$ has a finite decomposition $x = x_1 \ldots x_n$ into irreducible elements, where the irreducible factors are unique up to units and permutations.*

Exercise 19. 1. Show that a factorial ring is normal.

2. Show that an integral ring is factorial if for any irreducible element $\pi \in A$ the principal ideal (π) is a prime ideal.

3. Show that for any factorial ring A the polynomial ring $A[X]$ is again factorial. (This is essentially due to Gauss)

   Hint: Let K be the field of fractions. Let $P(X) = a_0+a_1X+\cdots+a_nX^n \in A[X]$,$a_n \neq 0$. Assume that this polynomial splits in $K[X]$. Then we find a $c \in A$,$c \neq 0$ such that we can factorize

$$ca_0 + ca_1X + \cdots + ca_nX^n = (b_0 + b_1X + \cdots + b_rX^r)(c_0 + c_1X + \cdots + c_sX^s)$$

   into a product of two polynomials in$A[X]$ of smaller degree. Now use 2) to show that for any irreducible divisor π of c one of the factors must be zero $\mod (\pi)$, hence we can divide on both sides by π. This process stops. Therefore we see that a polynomial in $A[X]$, which becomes reducible in $K[X]$ is also reducible in $A[X]$. Then the rest is clear.

4. Show that the ring of integers $\mathbb{Z}$ and the ring $k[X]$ of polynomials over a field k are normal.

5. Let us assume that $A \longrightarrow B$ are both integral and that $K \to L$ is the corresponding extension of their quotient fields. Let us assume that L/K is a finite extension. Furthermore we assume that A normal. For $x \in L$ we have a unique monic polynomial $F(X) \in K[X]$ such that $F(x) = 0$. The multiplication by x induces a linear transformation L_x of the K-vector space L. It is well known that x is a zero of the characteristic polynomial $\det(X\,\mathrm{Id} - L_x)$ of L_x. Show that

$$x \text{ is integral over } A \iff F[X] \in A[X] \iff \det(XId - L_x) \in A[X]$$

6. Under the above assumptions we have $\mathrm{tr}_{L/K}(x) \in A$ for any element $x \in L$, which is integral over A.

7. Again under the assumptions of 3. we can say: For any $x \in L$ we can find a non zero element $a \in A$ such that ax becomes integral over A

8. If an A module M is locally free of rank of one then we can find a finite covering $X = \mathrm{Spec}(A) = \bigcup X_{f_i}, X_{f_i} = \mathrm{Spec}(A_{f_i})$ such that $M \otimes A_{f_i}$ is free of rank one, i.e. $M \otimes A_{f_i} = A_{f_i} s_i$, where $s_i \in M$.

   Show that this implies that M is isomorphic to an ideal $\mathfrak{a}$, which is locally principal.

   Show that for a factorial ring A any locally principal ideal $\mathfrak{a} \subset A$ is itself principal Hint: Show that either $\mathfrak{a} = A$ or we can find an irreducible element π, which divides all elements of $\mathfrak{a}$ and hence $\mathfrak{a} \subset \pi^{-1}\mathfrak{a} \subset A$. The ideal $\pi^{-1}\mathfrak{a}$ is strictly larger than $\mathfrak{a}$. We apply the same argument to $\pi^{-1}\mathfrak{a}$ and get get an ascending chain of locally principal ideals. This chain has to stop because a non zero element of $\mathfrak{a}$ has only finitely many irreducible divisors.

   This implies that any locally free module of rank one over a factorial ring is free (See [Ma], Thm. 20.7)

The item (3) in the exercise above implies the following theorem, which we will use several times (See for instance [Ja-Sch], Chap. IV, Satz 4.4.)

Theorem 7.1.5. *For any factorial ring A the polynomial ring $A[X_1, X_2, \ldots, X_n]$ is factorial.*

We have the following fundamental theorem for finite morphisms

Theorem 7.1.6. *Assume that the ring homomorphism $\phi : A \to B$ is finite and injective. Then the induced map ${}^t\phi : \mathrm{Spec}(B) \to \mathrm{Spec}(A)$ is surjective, has finite fibers and the elements in the fibers are incomparable with respect to the order on* Spec(B).

This means in other words: For any $\mathfrak{p} \in \mathrm{Spec}(A)$ we can find a $\mathfrak{q} \in \mathrm{Spec}(B)$ such that $A \cap \mathfrak{q} = \mathfrak{p}$. The number of such $\mathfrak{q}$ is finite, whenever we have two of them $\mathfrak{q}_1, \mathfrak{q}_2$ we have $\mathfrak{q}_1 \not\subset \mathfrak{q}_2$. (See [Ei], I. 4.4,prop. 4.15, cor. 4.18)
To prove the theorem we need another famous result from commutative algebra, namely the Lemma of Nakayama.

Lemma 7.1.7. *[Nakayama] Let A be a local ring with maximal ideal $\mathfrak{m}$ and let M be a finitely generated A-module. If*

$$M \otimes (A/\mathfrak{m}) = M/\mathfrak{m}M = 0$$

then $M = 0$.

To see this we use the same trick as above: Express a system of generators of M as a linear combination of these generators but now with coefficients in $\mathfrak{m}$. We find that 1_A is a zero of a characteristic polynomial of a matrix with coefficients in $\mathfrak{m}$, which is only possible for the 0×0 -matrix. □

Now we sketch the proof of the theorem 7.1.6. We pick a prime $\mathfrak{p} \in \mathrm{Spec}(A)$. The residue class ring $A/\mathfrak{p}$ is integral, we have $\mathrm{Spec}(A/\mathfrak{p}) \hookrightarrow \mathrm{Spec}(A)$ and the zero ideal (0) is mapped to $\mathfrak{p}$. We localize at (0) and we get the quotient field $(A/\mathfrak{p})_{(0)}$. Taking fibered products we get a diagram of affine schemes

$$\begin{array}{ccc} \mathrm{Spec}(A) & \longleftarrow & \mathrm{Spec}(B) \\ \uparrow & & \uparrow \\ \mathrm{Spec}(A/\mathfrak{p}) & \longleftarrow & \mathrm{Spec}(B \otimes_A (A/\mathfrak{p})) \\ \uparrow & & \uparrow \\ \mathrm{Spec}(A/\mathfrak{p})_{(0)} & \longleftarrow & \mathrm{Spec}(B \otimes_A (A/\mathfrak{p})_{(0)}) \end{array}$$

The vertical arrows are inclusions and it is clear that the prime ideals $\mathfrak{q} \in \mathrm{Spec}(A)$, for which $\mathfrak{q} \cap A$ are exactly the elements in $\mathrm{Spec}(B \otimes_A (A/\mathfrak{p})_{(0)})$. To prove the surjectivity we have to show that this scheme is not empty. This follows from the lemma of Nakayama because we can obtain $\mathrm{Spec}(A/\mathfrak{p})_{(0)}$ also as the residue field $A_\mathfrak{p}/\mathfrak{m}_\mathfrak{p}$ of the local ring $A_\mathfrak{p}$. We have $B \otimes_A A_\mathfrak{p} \neq 0$ (only the zero divisors of $S = A \setminus \mathfrak{p}$ go to zero in this tensor product) and hence we get by Nakayama that $B \otimes_A A_\mathfrak{p}/\mathfrak{m}_\mathfrak{p} \neq 0$ and this implies $\mathrm{Spec}(B \otimes_A A_\mathfrak{p}/\mathfrak{m}_p) \neq \emptyset$. Now we have that $B \otimes_A (A/\mathfrak{p})_{(0)}$ is a finite dimensional vector space over the field $(A/\mathfrak{p})_{(0)}$, it is a finite $(A/\mathfrak{p})_{(0)}$-algebra. This implies that any prime ideal $\mathfrak{q} \in \mathrm{Spec}(B \otimes_A (A/\mathfrak{p})_{(0)})$ is maximal because the residue ring is automatically a field. Then it is also clear that $\mathrm{Spec}(B \otimes_A (A/\mathfrak{p})_{(0)})$ must be finite. The map

$$B \otimes_A (A/\mathfrak{p})_{(0)} \longrightarrow \prod_{\mathfrak{q}} B \otimes_A (A/\mathfrak{p})_{(0)}/\mathfrak{q}$$

is easily seen to be surjective. Hence we have proved that the fibers are finite and non empty and we have seen that the prime ideals in the fibers are incomparable.

7.1.1 Rings With Finiteness Conditions

In this section formulate some finiteness for rings collect some facts about these rings. We will not give proofs because these facts are easily available in the literature. On the other hand it may be a good exercise if the reader tries to find the proofs her(him)self.

Definition 7.1.8. *A commutative ring A with identity is called* **noetherian** *if it satisfies one of the following equivalent four conditions*

1. *Any ideal $\mathfrak{a} \subset A$ is finitely generated.*
2. *Any submodule N of a finitely generated A-module M is finitely generated.*
3. *Any ascending chain $\mathfrak{a}_\nu \subseteq \mathfrak{a}_{\nu+1} \subseteq \dots \mathfrak{a}_n \subseteq \dots$ becomes stationary, i.e. there exists an n_0 such that $\mathfrak{a}_{n_0} = \mathfrak{a}_{n_0+1} \dots = \mathfrak{a}_{n_0+m}$ for all $m \geq 0$.*
4. *Any ascending chain of A-submodules $N_\nu \subseteq N_{\nu+1} \subseteq \dots$ of a finitely generated A-module M becomes stationary.*

Example 13. *The ring $\mathbb{Z}$ is noetherian and of course we know that fields are so too.*

Theorem 7.1.9 (Hilbertscher Basissatz). *If A is a noetherian ring then the polynomial ring $A[X]$ is also noetherian.*

This implies that polynomial rings $A[X_1,\dots,X_n]$ are noetherian.
The following lemma is extremely important. We leave the proof to the reader, a special case of it was discussed in exercise (2) Step 2 in 6.1.2.

Lemma 7.1.10 (Noetherscher Normalisierungssatz). *Let $A = k[x_1,x_2,\dots,x_n]$ be a finitely generated k-algebra (k a field). Then we can replace the system of generators by another system $y_1,\dots,y_r,y_{r+1},\dots,y_m$ such that $B = k[y_1,y_2,\dots,y_r]$ is a polynomial ring in the variables $y_1,\dots,y_r$ and the other variables (or the algebra A) are integral over B.*

For the proof we refer to the standard books (See for instance [Ei],II,13). The reader is invited to write down a proof using the idea from the exercise mentioned above.
The theorem above has several important applications. One of them is the

Theorem 7.1.11 (Nullstellensatz von Hilbert). *Let k be a field and let $A = k[x_1,x_2,\dots,x_r]$ be a finitely generated k-algebra. For any maximal ideal $\mathfrak{m} \in \mathrm{Spec}(A)$ the residue field $A/\mathfrak{m}$ is a finite field extension of k. Especially if k is algebraically closed then $A/\mathfrak{m} = k$.*

Here we give the argument. We divide by $\mathfrak{m}$, in other words we may assume that A itself is already a field. We proceed by induction on r. If $r = 0$ then the theorem is obvious. Let $r \geq 1$. We apply the lemma above and find that A is finite over a polynomial ring $B = k[y_1,\dots,y_s] \subset A$. If $s \geq 1$ then B has non trivial prime ideals, which by our theorem 7.1.6 extend to non zero prime ideals in A, which is not possible. Hence we see that $B = k$ and therefore, A is a finite field extension of k.
The Nullstellensatz implies:

Corollary 7.1.12. *For a finitely generated k-algebra A/k the intersection of the maximal prime ideals is equal to the radical.*

Proof: Let $f \in A$ be not nilpotent. Then A_f is still a finitely generated k-algebra and by the Normalisierungsatz it is finite over a polynomial k-algebra B. This polynomial algebra has maximal ideals and then we apply 7.1.6.

7.1.2 Dimension theory for finitely generated k-algebras

Definition 7.1.13. *For an integral k-algebra A we define the* **dimension** $\dim(A)$: *The number $\dim(A)+1$ is equal to the maximal length of a chain of prime ideals $(0) \subset \mathfrak{p}_1 \subset \dots \subset \mathfrak{p}_r$ where all the inclusions are proper. We say say that the dimension is ∞ if chains of arbitrary length exist.*

A maximal chain is a chain that cannot be refined. We also define:

Definition 7.1.14. *The* **height** $h(\mathfrak{p})$ *of a prime ideal $\mathfrak{p}$ is defined as the number, for which $h(\mathfrak{p})+1$ is the maximal length of a chain $(0) \subset \mathfrak{p}_1 \subset \dots \subset \mathfrak{p}$ ending with $\mathfrak{p}$.*

There exists a theory of dimensions for arbitrary noetherian rings (see for instance [At-McD],[Ei],II,Chap. 8), which is more general than what we are doing here.
Before we discuss the main theorem in this section I want to mention

Proposition 7.1.15. *If A,B are two finitely generated integral k-algebras, if $i : A \to B$ is an inclusion and if B is integral over A then for any chain $(0) \subset \mathfrak{p}_1 \subset \ldots \subset \mathfrak{p}_r$ of prime ideals in A there exists a chain $(0) \subset \mathfrak{p}'_1 \subset \ldots \subset \mathfrak{p}'_r$ of prime ideals such that $\mathfrak{p}_i = \mathfrak{p}'_i \cap A$. Especially we can say that the two rings have the same dimension.*

Proof: Obvious from Theorem 7.1.6 □

Theorem 7.1.16. *The dimension of an integral finitely generated k-algebra A is finite and any maximal chain has the same length. The dimension of the polynomial ring $k[X_1,X_2,\ldots,X_n]$ is n.*

Again we give the argument. We proceed again by induction on the number of generators. If we have two integral k-algebras $A \subset B$ and if B is finite over A then $\dim(A) = \dim(B)$. Using the Noether Normalisierung we can write our algebra A as a finite extension of an algebra with less generators as long as we have non trivial relations among the generators. Hence we are reduced to the case of a polynomial ring $A = k[X_1,\ldots,X_n]$. At this point the reader should be aware that it is not clear at this point that we can find minimal non zero prime ideals at all, it could be possible that we can always go to smaller and smaller non zero prime ideals. This would of course imply that the dimension is ∞. To see that this is not the case we pick an arbitrary non zero element $F(X_1,X_2,\ldots,X_n)$ in A. We consider prime ideals $\mathfrak{p}$ containing the principal ideal $(F(X_1,X_2,\ldots,X_n))$.
Before we continue, we will show a proposition, which is a special case of the Hauptidealsatz of Krull.

Proposition 7.1.17. *We can find minimal prime ideals $\mathfrak{p} \supset (F(X_1,X_2,\ldots,X_n))$ and each such ideal is also a minimal non zero prime ideal in $k[X_1,X_2,\ldots,X_n]$.*

Again we apply the trick used in the proof of Noether Normalisierungsatz, which tells us that after some change of variables we can assume that this polynomial is of the form

$$F(X_1,X_2,\ldots,X_n) = a_0(X_1,\ldots,X_{n-1}) + a_1(X_1,\ldots,X_{n-1})X_n + \ldots + X_n^m,$$

i.e. it is unitary in X_n over the ring $k[X_1,\ldots,X_{n-1}]$. We rewrite our polynomial ring by increasing the number of generators

$$A = k[X_1,\ldots,X_n,Y]/(F(X_1,X_2,\ldots,X_n) - Y).$$

This gives us an embedding

$$k[X_1,\ldots,X_{n-1},Y] \subset k[X_1,\ldots,X_n,Y]/(F(X_1,X_2,\ldots,X_n) - Y)$$

where now $k[X_1,\ldots,X_{n-1},Y]$ is again a polynomial ring. (This requires a little argument.) We exchanged the variable X_n against the polynomial F. Now it is very easy to see that the principal ideal $(Y) = Yk[X_1,\ldots,X_{n-1},Y]$ is a non zero prime ideal and it is minimal with this property. Our original algebra A is integral over $k[X_1,\ldots,X_{n-1},Y]$. Our theorem tells us that we can find a prime ideal $\mathfrak{p}$, for which we have

$$\mathfrak{p} \cap k[X_1,\ldots,X_{n-1},Y] = (Y)$$

and hence $\mathfrak{p} \supset F(X_1, \ldots, X_n)$ and $\mathfrak{p}$ is minimal with this property. It is even clear that $\mathfrak{p}$ is a minimal non zero prime ideal because a smaller one would induce a non zero prime ideal contained in (Y). □

We continue with the proof of 7.1.16. We see that chains of prime ideals, in which the non zero members contain $F(X_1, \ldots, X_n)$ have the same length as the induced chains in $k[X_1, \ldots, X_{n-1}, Y]$ where the non zero members contain (Y). But these last chains are chains in $k[X_1, \ldots, X_{n-1}]$ and it follows from the induction hypothesis that the maximal chains in this ring have length n. Hence we see that maximal chains in A have length $n+1$ and the theorem is proved.
Finally I want to comment briefly on the *Hauptidealsatz von Krull*. We proved it in the course of the proof of the previous theorem for a polynomial ring, but the general case follows easily again by Noether normalization. In our situation it says:

Theorem 7.1.18 (Hauptidealsatz von Krull). *For a finitely generated k-algebra A, which is integral and a non zero element $f \in A$ and a minimal prime ideal $\mathfrak{p} \supset (f)$ we have* $\dim A - 1 = \dim(A/\mathfrak{p})$.

For arbitrary noetherian rings the first assertion in theorem 7.1.16 is not true. (See [At-McD], chap. 11, Exercise 4) But for local noetherian rings we have the concept of dimension and an analogue of this theorem. (see [At-McD],p.122, [Ei],II, Thm. 10).
We need the integrality of our algebra because otherwise $\operatorname{Spec}(A)$ can have several irreducible components (see 7.2) and these components may have different dimensions. This would have the effect that we can find non refinable chains of prime ideals, which have different length.
On the other hand we know that $\operatorname{Spec}(A) = \operatorname{Spec}(A/\operatorname{Rad}(A))$ and this implies that what we need is the integrality of $A/\operatorname{Rad}(A)$ to define the dimension.

7.2 Minimal prime ideals and decomposition into irreducibles

For any noetherian ring A we can consider the set of minimal prime ideals. Of course for an integral ring A the set of minimal prime ideals consists of just one element $(0) \in \operatorname{Spec}(A)$, this is the generic point. In the general case it is not a priory clear that minimal prime ideals exist.
We know $\operatorname{Spec}(A) = \operatorname{Spec}(A/\operatorname{Rad}(A))$, hence if we want to say something about the ordered topological space $\operatorname{Spec}(A)$ we may very well restrict our attention to the case that A is reduced. We state a theorem, which is a weak form of a theorem proved by E. Lasker.

Theorem 7.2.1 (E. Lasker). *Let A be a reduced noetherian ring. Then the set of minimal prime ideals is finite. To any minimal prime ideal $\mathfrak{p}$ we can find an $f \in A \setminus \mathfrak{p}$ such that*

$$\mathfrak{p} = \operatorname{Ann}_A(f) = \{x \in A | xf = 0\}.$$

I want to indicate the steps of the proof and leave it to the reader to fill the gaps.

Exercise 20. We prove that there exist minimal prime ideals. This is clear if A is integral. If not, then we find $f,g \in A \setminus \{0\}$ such that $fg = 0$.

1a) Consider $\mathrm{Ann}_A(f) = \mathfrak{a}$ and prove: If $\mathfrak{a}$ is not prime then we can find an $x \in A$ such that $f_1 = xf \neq 0$ and such that $\mathfrak{a}_1 = \mathrm{Ann}_A(f_1)$ is strictly larger than $\mathfrak{a} = \mathrm{Ann}_A(f)$.

1b) Show that this implies that we can find an $y \in A$ such that $\mathrm{Ann}_A(fy) = \mathfrak{p}$ is a prime ideal and that this prime ideal is minimal. Hence we see that minimal primes exist.

Let us write $yf = f_\mathfrak{p}$. It is clear that $f_\mathfrak{p} \notin \mathfrak{p}$.

Exercise 21. Prove that any prime ideal $\mathfrak{q}$ contains a minimal prime ideal $\mathfrak{p} \subset \mathfrak{q}$ of the form $\mathrm{Ann}_A(f_\mathfrak{p})$ and hence all minimal prime ideals are of this form.

Let us assume we picked an $f_\mathfrak{p}$ for any minimal prime ideal.

Exercise 22. Prove that for two minimal prime ideals $\mathfrak{p} \neq \mathfrak{p}_1$ the product $f_\mathfrak{p} f_{\mathfrak{p}_1} = 0$.

Exercise 23. Consider the ideal generated by these $f_\mathfrak{p}$ and combine the fact that this ideal is finitely generated and Exercise 22 above to show that these $f_\mathfrak{p}$ form a finite set.

Exercise 24. Let A be an arbitrary noetherian ring, let $\mathfrak{p}_1, \dots, \mathfrak{p}_r$ be the set of minimal prime ideals. Let us also assume that the spaces $\mathrm{Spec}(A/\mathfrak{p}_i)$ are disjoint. Then there is a unique collection of elements $e_1, \dots, e_r$ such that

$$\begin{aligned} e_i \notin \mathfrak{p}_i \text{ and } e_i &\in \mathfrak{p}_j && \text{for all } j \neq i \\ e_i^2 &= e_i && \text{for all } i \\ e_i e_j &= 0 && \text{for all } i \neq j \\ \sum_{i=1}^{r} e_i &= 1_A \end{aligned}$$

(See [Ei], I. 7.3.) We give a hint for the solution. Our assumption that the spaces $\mathrm{Spec}(A/\mathfrak{p}_i)$ are disjoint implies that we can find e_i' such that $e_i' \equiv 1 \mod \mathfrak{p}_i$ and $e_i \in \mathfrak{p}_j$ for all $j \neq i$. These e_i' satisfy all the relations if we compute modulo the radical $\mathrm{Rad}(A)$. Now we can modify $e_i' \to e_i' + r_i = e_i$ such that we have the idempotency $e_i^2 = e_i$. (Use the next exercise to show that $\sum_i e_i'$ is a unit.) Then all the other requirements are also fulfilled.

Exercise 25. If we have any noetherian ring R and if we consider the homomorphism $R \longrightarrow R/\mathrm{Rad}(R)$ then the group $R^\times$ of units of R is the inverse image of the units in $R/\mathrm{Rad}(R)$

This decomposition of $1_A = e_1 + \ldots + e_r$ is called the decomposition into orthogonal idempotents. It gives a decomposition of the ring

$$A = \bigoplus_i A e_i$$

If our ring has no radical, the $f_{\mathfrak{p}_i}$ are equal to the e_i.

Definition 7.2.2. *An affine scheme $X = \mathrm{Spec}(B)$ is called* **irreducible**, *if it cannot be written in a non trivial way as a union of two closed subschemes $X = X_1 \cup X_2$. (Non trivial means both subschemes X_1 and X_2 are not equal to X.)*

If $A = B/\operatorname{Rad}(B)$ is not integral then we get a non trivial decomposition $\operatorname{Spec}(B) = \cup \operatorname{Spec}(A/\mathfrak{p}_i)$ and hence we see that $\operatorname{Spec}(B)$ is not irreducible in this case. If on the other hand we have a non trivial decomposition $\operatorname{Spec}(B) = X_1 \cup X_2$ then we can find elements f_1,f_2, which are non zero such that f_i vanishes on X_i. Then $f_1 f_2$ vanishes on $\operatorname{Spec}(B)$, hence it must be in the radical. (7.1.12). This implies that A has zero divisors. We proved

Lemma 7.2.3. *The spectrum* $\operatorname{Spec}(B)$ *of a noetherian ring* B *is irreducible if and only if* $B/\operatorname{Rad}(B)$ *is integral.*

It follows from Lasker‘s theorem (Theorem 7.2.1) that for a noetherian ring A we have a unique finite decomposition of $\operatorname{Spec}(A)$ into irreducible subschemes $\operatorname{Spec}(A/\mathfrak{p}_i)$. These irreducible subschemes are called the irreducible components. If our ring A has no radical then the elements $f_\mathfrak{p}$ have the property that they vanish on all components except the one, which they define. This is a geometric interpretation of the result in Exercise 22 above.

Associated prime ideals

We want briefly discuss an extension of the theorem 7.2.1. Let A be a noetherian ring and let M be a finitely generated A module. A prime ideal $\mathfrak{p}$ is *associated to M* if it is the annihilator of an element in M. Let $\operatorname{Ass}_A(M)$ be the set of associated primes. Then we have the following

Theorem 7.2.4. 1 *The set* $\operatorname{Ass}_A(M)$ *is finite, each prime in* $\operatorname{Ass}_A(M)$ *contains* $\operatorname{Ann}_A(M)$ *and the minimal prime ideals containing* $\operatorname{Ann}_A(M)$ *are in* $\operatorname{Ass}(M)$.

2 *The union of the associated primes is the set of zero divisors of M plus the zero element.*

3 *If S is a multiplicatively closed subset of A not containing* 0 *then*

$$\operatorname{Ass}_{A_S}(M_S) = \operatorname{Ass}(M)_S$$

For the proof of this theorem and further background we refer to [Ei],I. 3.1.

The restriction to the components

For our given noetherian ring A we consider the natural homomorphism

$$\pi : A \longrightarrow \prod A/\mathfrak{p}_i$$

where the $\mathfrak{p}_i$ run over the minimal prime ideals. The kernel is the radical of A. (see Exercise 19, 1b))

It is not necessarily surjective. If we have a pair of irreducible components, which have a non empty intersection then we can find a prime ideal $\mathfrak{p}$ of A, which contains the two different minimal prime ideals $\mathfrak{p}_i,\mathfrak{p}_j$. Then an element $f \in A, f \notin \mathfrak{p}$ has a non trivial image in $A/\mathfrak{p}_i$ and in $A/\mathfrak{p}_j$. If f_i is the image of f in $A/\mathfrak{p}_i$, then $f_i \notin \mathfrak{p}\mathfrak{p}_i$ and hence the element $(0,\dots,f_i,0,\dots,0) \in \prod A/\mathfrak{p}_i$, is not in the image of π.

A slightly different way of saying this: The spectrum $\mathrm{Spec}(\prod A/\mathfrak{p}_i)$ is the disjoint union of the irreducible connected components, this may be different from $\mathrm{Spec}(A)$.

Definition 7.2.5. *A noetherian ring is called* **local artinian** *ring if its unique maximal ideal is also minimal.*

Definition 7.2.6. *A non zero element $f \in A$ is called a* **non zero divisor** *if $fg = 0$ implies $g = 0$.*

The non zero divisors form a multiplicatively closed subset $S \subset A$. It is clear that $f \in A$ is a non zero divisor if and only if all its components under the projection map π are non zero or what amounts to the same if f is not contained in any of the minimal prime ideals.

Definition 7.2.7. *The* **total quotient ring** *of A is the localization $A_S = \mathrm{Quot}(A)$.*

The spectrum of A_S is simply the set of generic points of the irreducible components. If we pick a minimal prime ideal $\mathfrak{p}_i$ then the localization $A_{\mathfrak{p}_i}$ is a local artinian ring. We get

Proposition 7.2.8. *The ring A_S is the direct product of the finitely many local artinian rings $A_{\mathfrak{p}_i}$ and our map π_S is defined component wise as $\prod A_{\mathfrak{p}_i} \longrightarrow \mathrm{Quot}(A/\mathfrak{p}_i)$. This map is surjective and the kernel is $\mathrm{Rad}(A)_S$*

In principle this is the situation in exercise 24, since we localize at the generic points of the components, we make the irreducible components disjoint. Instead of taking this radical step we could have chosen non zero divisors $f_{ij} \subset A$, which are zero on $\mathrm{Spec}(A/\mathfrak{p}_i) \cap \mathrm{Spec}(A/\mathfrak{p}_j)$. If we take the product of all these f_{ij} we get a non zero divisor $F \in A$ and then $\mathrm{Spec}(A_F)$ has now the virtue that its irreducible components are disjoint. Hence we may apply exercise 24 and write

$$A_F \xrightarrow{\sim} \bigoplus A_F e_i.$$

Here we replaced the generic points by actual open sets, which looks a little bit more geometric.
An element $f \in \mathrm{Quot}(A)$ can be written as $f = g/h$ where h is a non zero divisor. Then we have of course $f \in A_h$ and f is a regular function on $\mathrm{Spec}(A_h)$(See p.11.) We look at all different ways to write f as a quotient, then we see that f is defined on the union of all the $\mathrm{Spec}(A_h)$. This open set is called the **domain of definition** for f. It is a dense subset of $\mathrm{Spec}(A)$.

Decomposition into irreducibles for noetherian schemes

Definition 7.2.9. *If we have an arbitrary scheme X we say that this scheme is a* **noetherian scheme** *if we have a finite open covering $X = \bigcup U_i$ by affine schemes $U_i = \mathrm{Spec}(A_i)$ where the rings A_i are noetherian rings.*

Again we can speak of irreducible schemes. If one of the covering sets U_i is not irreducible, then we write it non trivially as $U_i = V_i \cup W_i$. If we take the closures of these subschemes in X we clearly get two closed subschemes, which are not equal to X. Then we get a non trivial decomposition of $X = \overline{V}_i \cup \overline{W}_i \cup Z_i$ where Z_i is the complement of U_i.
Hence we see:

Proposition 7.2.10. *A noetherian scheme $X = \bigcup U_i$ is irreducible if and only if all the U_i are irreducible.*

From this we can deduce very easily that X itself has always a unique finite decomposition into irreducible subschemes. This is usually proved directly without reference to affine schemes.

If we have an affine scheme of finite type $X = \mathrm{Spec}(k[X_1, \dots ,X_n]/I) = \mathrm{Spec}(A)$ then it is automatically noetherian. Hence our previous considerations apply.

If X/k is an irreducible and reduced scheme of finite time and if $U \subset X$ is an open affine subscheme, which is not empty, then $\mathcal{O}_X(U)$ is integral and the quotient field K of this affine k-algebra is independent of U. This quotient field will be denoted by $k(X)$ and it is called the **field of meromorphic functions** on X/k or simply the function field of X/k.

If we apply the Noether -normalization than we can write $\mathcal{O}_X(U)$ as a finite extension of a polynomial algebra $k[T_1,T_2, \dots ,T_r]$ and then we have seen that

$$\dim(\mathcal{O}_X(U)) = \dim(k[T_1,T_2, \dots ,T_r]) = r$$

and this number r is also by definition equal to the transcendence degree of the function field $k(X)/k$. Hence we define for any irreducible scheme of finite type

$$\dim(X/k) = \mathrm{trdeg}(k(X)/k).$$

Local dimension

It is clear that it does not make sense to speak of the dimension of the dimension of a finite type scheme X/k, which is not irreducible. But if we have point $x \in X$ and if this point lies on exactly one of the irreducible components, then we can speak of the local dimension of X/k at x. It is simply the dimension of this irreducible component. The point x lies in exactly one irreducible component if and only if the local ring $\mathcal{O}_{X,x}$ is integral and in this case the local dimension at x is also the dimension of the local ring $\mathcal{O}_{X,x}$.

7.2.1 Affine schemes over k and change of scalars

In this section we want to consider k-algebras A/k. We want to change the terminology a little bit.

Definition 7.2.11. *A k-algebra is called an* **affine k-algebra** *if it is finitely generated over k as an algebra. This is synonymous to k-algebra of finite type.*

For these schemes over a field k we always have the option of base change to $\overline{k}$ or to another field extension, i.e. we can consider the algebra $A \times_k L$ especially $A \otimes_k \overline{k}$. We want to see what happens to irreducibility, reducedness and dimensions under such a base change. We also want to discuss the behavior under the formation of products.

Definition 7.2.12. *An affine algebra A/k is called* **absolutely irreducible** *if $A \otimes_k \overline{k}$ is irreducible.*

We know already what it means that A/k is absolutely reduced, it means that $A \otimes_k \bar{k}$ has no non trivial nilpotent elements (see Def.6.2.12.)
Again it can happen that an irreducible (resp. reduced) k-algebra A/k becomes reducible (non reduced) if extend the ground field. To see this we can start from a finite, non trivial extension K/k. Then we know that $K \otimes_k \bar{k}$ is not a field. If K/k is separable then $K \otimes_k \bar{k}$ is a direct sum of copies of $\bar{k}$, if K/k is not separable then $K \otimes_k \bar{k}$ contains non trivial nilpotent elements.
In the following exercise we will show: If an integral k-algebra A becomes reducible or non reduced after extension of scalars, then we can find a finite extension of K/k, which is contained in the field of fractions of A, and which is "responsible" for that.

Exercise 26. Let A/k be an integral affine k- algebra. If it is not absolutely irreducible or not absolutely reduced then we can find a non zero $f \in A$ such that the localization A_f contains a non trivial finite extension of k.

This is a little bit tricky, we outline a strategy for a solution. First reduce the problem to a base change $A \otimes_k L$ where $L = k[X]/(p(X))$ with an irreducible polynomial $p(X) \in k[X]$. Investigate, under which conditions the algebra $A \otimes_k L$ can become reducible or non reduced. This means that we can find two polynomials $g_1(X)$,$g_2(X) \in A[X]$ whose degrees are strictly less than the degree of $p(X)$ such that $p(X)$ divides the product $g_1(X)g_2(X)$. Look at all such pairs and pick one where the degree of say $g_1(X)$ is minimal. Use the arguments you learned in your first algebra course to show that there must be a non zero element $f \in A$ such that $g_1(X)$ divides $fp(X)$. This means that $g_1(X)/f \in A_f[X]$ divides $p(X)$. Now conclude that the coefficients of $g_1(X)/f$ must be algebraic over k. But not all of them are in k and we found the non trivial extension.

Definition 7.2.13. *We call an element $y \in A_f$ a* **pseudoconstant** *if it generates a finite extension of k, i.e. if it is algebraic over k. We call it a* **constant** *if it is actually in A.*

It looks a little bit bizarre that we can have pseudoconstants, which are not constants, but this may be the case.
To give an example we consider the $\mathbb{R}$-algebra $A = \mathbb{R}[x,y]/(x^2 + y^2)$. Then our scheme $\mathrm{Spec}(A)$ is irreducible but $A \otimes \mathbb{C}$ is not. The elements x/y,y/x are pseudoconstants but they are not constants. The geometric points are given by two lines intersecting in the origin. If we remove the origin, the value of say x/y is $\pm i$ on the two lines. In the origin the function x/y is just a little bit confused and does not know, which value to take.
If k is a non perfect field of characteristic 2 and if $a \in k$ is not a square then $A = k[x,y]/(x^2 + ay^2)$ is integral. Again we see that x/y,y/x are pseudoconstants but not constants. The terminus "constants" is a little bit misleading. The "constants" are only constant if we restrict them to the irreducible components. We claim:

Exercise 27. If A/k is integral and of finite type then the pseudoconstants generate a finite algebraic extension of K/k.

Hint: To see this we apply the Normalisierungslemma and write A as a finite extension of a polynomial ring B. Hence A is a finite B-module, let m be the number of elements in a system of generators of this B-module. It is easy to see that for any non zero $f \in A$ we can find an $F \in B$ such that $A_f \subset A_F$ and A_F is still integral over B_F. If our field k is infinite then we can find a k-rational point $a \in \mathrm{Spexmax}(B_F)$ let $\mathfrak{m}_a$ be the corresponding maximal ideal. We get an injection of the field of constants of A_F to $A_F/\mathfrak{m}_a A_F$. Now $A_F/\mathfrak{m}_a A_F$ is a vector space over $B/\mathfrak{m}_a B = k$, which is generated by the images of the generators of the B-module A. Hence the degree of this extension is bounded by m. Therefore the degree of the field of pseudoconstants is bounded. This proves the assertion if k is infinite. If k is finite then we can replace k by an extension k_1/k of prime degree where this prime is larger than m and such that we find a k_1-valued point a of B_F. Then we see that the tensor product $A_F \otimes_k k_1$ is still a field and injects into $A_F \otimes B_F/\mathfrak{m}_a$ and we have the same argument as before.

It is clear that a pseudoconstant has a dense domain of definition. Hence we see that we have an open dense subset $U \subset \mathrm{Spec}(A)$, which is the common domain of all pseudoconstants. We also see that all the pseudoconstants lie already in a suitable sub algebra A_f where f is a non zero divisor. (See 7.2.8).
We drop the assumption that A/k is integral and consider an arbitrary finitely generated k-algebra. We can consider its total quotient $\mathrm{Quot}(A)/k$ and define the sub algebra of pseudoconstants $L/k \subset \mathrm{Quot}(A)/k$ as being the ring of elements, which are finite over k. It is obvious that this k-algebra L/k is exactly the inverse image of the pseudoconstants under the map

$$\pi : \mathrm{Quot}(A) \longrightarrow \bigoplus_{\mathfrak{p}_i \text{ minimal}} \mathrm{Quot}(A/\mathfrak{p}_i),$$

where now on the right hand side we have a sum of fields. On the left hand side we may have nilpotent elements.
We conclude- using exercise 27- that L/k is finite dimensional, because the radical is of finite dimension over k. It is also clear that the map

$$\mathrm{Spec}(\mathrm{Quot}(A)) \longrightarrow \mathrm{Spec}(L)$$

is bijective, the left hand side is just the set of generic points of the irreducible components of $\mathrm{Spec}(A)$.
From these exercises we get

Lemma 7.2.14. *An integral affine k-algebra is absolutely reduced and absolutely irreducible if and only the field of pseudoconstants is equal to k. The k-algebra L/k of pseudoconstants is preserved under change of scalars, i.e. $L \otimes \overline{k}$ is the $\overline{k}$-algebra of pseudoconstants of $A \otimes \overline{k}$. We can find a finite separable extension K/k such that the irreducible components of $A \otimes K$ will be absolutely irreducible.*

We can give this a slightly different formulation. We can consider the k-algebra homomorphism

$$L \otimes \overline{k} \longrightarrow \mathrm{Quot}(A) \otimes \overline{k}.$$

The lemma says that this induces bijections

$$\mathrm{Spec}(L \otimes \overline{k}) \longleftarrow \mathrm{Spec}(\mathrm{Quot}(A) \otimes \overline{k})$$
$$\mathrm{Rad}(L \otimes \overline{k}) \longrightarrow \mathrm{Rad}(\mathrm{Quot}(A) \otimes \overline{k})$$

The proof is based on an observation which will also play a role in the next section.

Let us consider an element $F \in A \otimes \overline{k}$. We can write this element in the form

$$F = \sum f_i \otimes \alpha_i$$

where the α_i are taken from a finite normal extension K/k. This extension has a maximal separable sub extension K_{sep} in it. We know that any element $\alpha \in K$ raised to a sufficiently high power of the characteristic p of our field k will fall into K_{sep}. Hence we get that

$$F^{p^r} = \sum f_i^{p^r} \otimes \alpha_i^{p^r} \in A \otimes K_{\text{sep}}$$

If we now form the norm of this power, this means we form the product over all conjugates by elements $\sigma \in \text{Gal}(K_{\text{sep}}/k)$, then its norm is

$$G = \prod_\sigma \sigma(F^{p^r}) = \prod_\sigma \left(\sum f_i^{p^r} \otimes \sigma\left(\alpha_i^{p^r}\right)\right) \in A.$$

This has consequences: If for instance F is a non zero divisor, then F^{p^r} is a non zero divisor, then all $\sigma(F^{p^r})$ are non zero divisors. We see that $FF_1 = G \in A \subset A \otimes \overline{k}$ is still a non zero divisor. Therefore $(A \otimes \overline{k})_F \subset A_G \otimes \overline{k}$ and this implies that

$$\text{Quot}(A) \otimes \overline{k} = \text{Quot}(A \otimes \overline{k}).$$

Now let us assume that $F \in \text{Quot}(A) \otimes K$ is a pseudoconstant. This is so if and only if $G = F^{p^r}$ is a pseudoconstant for any $r \geq 1$. As before we rewrite G in the form

$$G = \sum g_i \otimes \beta_i$$

where the $\beta_i \in K_{\text{sep}}$ form a basis over k. We have the trace map $\text{tr} : K_{\text{sep}} \longrightarrow k$ and consider the elements

$$\text{tr}(G\beta_j) = \sum_i g_i \,\text{tr}(\beta_i\beta_j).$$

These elements are pseudoconstants and since the determinant of the matrix $\text{tr}(\beta_i\beta_j)$ is non zero we get that the coefficients $g_i \in \text{Quot}(A)$ must be pseudoconstants. We have the decomposition of $\text{Quot}(A) \otimes \overline{k}$ into a direct sum of local artinian algebras and we have the decomposition of the identity into orthogonal idempotents $1_A = \sum e_i$. The e_i are clearly pseudoconstants in $\text{Quot}(A) \otimes \overline{k}$. We apply our previous argument and find that suitably high powers $e_i^{p^r}$ lie in $L \otimes \overline{k}$. Since the e_i are idempotents we know $e_i \in L \otimes \overline{k}$ and more precisely we see that the e_i are already in $L \otimes K$ where K/k is separable. □

Exercise 28. Let us assume that A/k is absolutely reduced. In this case we may interpret $A \otimes_k \bar{k}$ as a ring of $\bar{k}$-valued functions on the set of geometric points of $\text{Spec}(A)/k$. The union of the intersections of pairs of two different irreducible components is a closed subset Y of $X = \text{Spec}(A \times_k \bar{k})$. The complement of $U = X \setminus Y$ is a disjoint union of irreducible components U_i each of them is a non empty open subset of an irreducible component of X. We can find an element $f \in A$ which vanishes on Y, but which is non zero on any of the components. Show that

$$A_f \otimes_k \bar{k} = \bigoplus (A_f \otimes_k \bar{k}) e_i,$$

the e_i are pseudoconstants and each of them is identically equal to one on exactly one of the irreducible components U_i and identically zero on all the others. If and only if all the irreducible components of $\mathrm{Spec}(A \otimes_k \bar{k})$ are disjoint we have $e_i \in A$! Of course the e_i are not really constant on the the set of geometric points of $\mathrm{Spec}(A)$, but they are defined everywhere and constant on the irreducible components.

In our discussion above we gave an argument, which also can be used to show:

Exercise 29. If K/k is separable then $\mathrm{Rad}(A \otimes K) = \mathrm{Rad}(A) \otimes K$
Hint: The only thing that has to be proved is the equality concerning the radicals in the reformulation of the Lemma. For this we may assume that our field k is already separably closed and that A/k is integral. Now we repeat the argument in the exercise above. We reduce the problem to the case of an extension $K = k(\alpha)$ where $\alpha^p = a \in k$. Then this argument yields: If $A \otimes K$ has nilpotent elements then we can find a $\beta \in \mathrm{Quot}(A)$ such that $\beta^p = \alpha^p = a$. This shows the desired equality of radicals for this small extension. Then we proceed by looking at $(A \otimes K)/\mathrm{Rad}(A \otimes K)$ and apply the same argument.

Let us consider two affine $\overline{k}$-algebras $A_1/\overline{k}, A_2/\overline{k}$, then we have the following

Lemma 7.2.15. *We have*

$$\mathrm{Rad}(A_1 \otimes A_2) = \mathrm{Rad}(A_1) \otimes A_2 + A_1 \otimes \mathrm{Rad}(A_2)$$

and especially the tensor product $A_1 \otimes A_2$ is reduced if the factors are reduced. If the two algebras are irreducible, then tensor product $A_1 \otimes A_2$ is also irreducible.

Proof: Choose a basis $\{g_1, g_2, \ldots\}$ of A_2/k, where the first say t basis vectors form a basis of $\mathrm{Rad}(A_2)$. Let $h = \sum f_i \otimes g_i$ be an element in the radical. We evaluate this element at geometric points $x \in \mathrm{Specmax}(A_1)$ and then the element $\sum f_i(x) g_i \in A_2$ must be in the radical of A_2. This implies that for all such points x we have $f_i(x) = 0$ for all indices $i > t$. The Nullstellensatz gives us $f_i \in \mathrm{Rad}(A_1)$ for all these $i > t$ and this proves the first assertion.
The proof of the second assertion is similar. We can assume that both algebras are actually integral. We take the same basis and consider a pair of zero divisors

$$\left(\sum f_i \otimes g_i\right)\left(\sum f_i' \otimes g_i\right) = 0$$

If we evaluate at the geometric points x as above, then for each such x we must have that one of the factors vanishes (irreducibility of $\mathrm{Spec}(A_2)$). Hence we may consider the two closed subschemes of $\mathrm{Spec}(A_1)$, which are defined by the vanishing of the f_i or the f_i' respectively. Their union contains all geometric points, hence they cover $\mathrm{Spec}(A_1)$. Since $\mathrm{Spec}(A_1)$ is irreducible we get that one of them must be the whole $\mathrm{Spec}(A_1)$ and this proves that the corresponding factor in the product must be zero because the product is absolutely reduced. □

Proposition 7.2.16. *If A_1, A_2 are two irreducible $\overline{k}$-algebras then we have*

$$\dim(A_1 \otimes A_2) = \dim(A_1) + \dim(A_2).$$

Proof: This is rather clear. We apply Noether Normalisierungssatz to our algebras and write them as finite extensions of polynomial rings. Then the number of variables is equal to the dimension in either case (prop.7.1.15 and thm. 7.1.16). The tensor product is finite over the polynomial ring in the disjoint union of the variables. □

What is $\dim(Z_1 \cap Z_2)$***?***

Let us assume that X/k is an irreducible scheme of finite type, let $Z_1,Z_2 \subset X$ be irreducible subschemes. We consider the intersection $Z_1 \cap Z_2$ (See p. 18). This intersection can be arbitrarily complicated, in general it will not be neither irreducible nor reduced. Of course we can consider its decomposition into irreducibles, $Z_1 \cap Z_2 = \bigcup Y_i$ and we can ask for the dimension of the irreducible components. Even this may be difficult to answer. Instead of stating a theorem, I explain some reasoning how to attack this question.
Let us pick a closed point $P \in Z_1 \cap Z_2$ and let us assume that this point lies on exactly one irreducible component of the intersection. We look at affine neighborhoods $P \subset U$. We assume that we can find a neighborhood U such that $U \cap Z_2$ is defined by one equation, i.e. there is an $f \subset \mathcal{O}(U), f \neq 0, f(P) = 0$ such that $U \cap Z_2 = \mathrm{Spec}(\mathcal{O}(U)/(f)$. Then it follows from the Hauptidealsatz that $\dim(Z_2) = \dim(U) - 1 = \dim(X) - 1$. The subscheme $Z_1 \cup U \subset U$ is defined by an ideal $I_1 \subset \mathcal{O}(U)$, we have $Z_1 = \mathrm{Spec}(\mathcal{O}(U)/I_1)$. Then $Z_1 \cap Z_2 \cap U = \mathrm{Spec}(\mathcal{O}(U)/(I_1,f)$ Now we have two possibilities. It may be that the image of f in $\mathcal{O}(U)/I_1$ is zero, i.e. $(I_1,f) = I_1$. This means that $Z_1 \cap Z_2 = Z_2$ and hence $\dim(Z_1 \cap Z_2) = \dim(Z_2)$. If this is not the case then the Hauptidealsatz implies that for irreducible components $Y_i \subset Z_1 \cap Z_2$ we have $\dim(Y_i) = \dim(Z_1) - 1$. So the dimension stays the same or it drops by one.
The obvious induction yields the following technical proposition:

Proposition 7.2.17. *Let X/k be an irreducible scheme of finite type and of dimension d. Let Z_1,Z_2 be two irreducible sub-schemes of codimension a_1,a_2. Let Y be an irreducible component in the intersection, let $P \in Y$ be a closed point, which does not lie in another irreducible component. Now we assume that we can find an (affine) open neighborhood U of P such that the ideal I_2 defining Z_2 restricted to U is generated by a_2 elements. Then we have*

$$\dim(Y) \geq d - a_1 - a_2,$$

i.e. the codimension of Y is less or equal to $a_1 + a_2$.

This is obvious from the considerations above. □

The question we are discussing is local, hence we may assume that X/k is affine, we write $X = \mathrm{Spec}(A/k)$. We consider homomorphism $A \otimes_k A \longrightarrow A$, which corresponds to the embedding of the diagonal

$$\Delta : X \xrightarrow{\sim} \Delta_X \subset X \times_k X,$$

the diagonal is defined by the ideal I, which is generated by the elements $f \otimes 1 - 1 \otimes f$. Clearly we have $Z_1 \cap Z_2 = \Delta_X \cap (Z_1 \times_k Z_2)$.
Now we anticipate the notion of smoothness (See 7.5). If we have an irreducible component $Y \subset Z_1 \cap Z_2$ and a point $P \in Y$, which is not in any other irreducible component, and which is a smooth point on X then we can conclude

$$\dim_k(Y) \geq d - a_1 - a_2, \ \operatorname{codim}(Y) \leq a_1 + a_2 \tag{7.3}$$

This is an obvious consequence of the proposition above, if we take into account that in a smooth point the diagonal is defined by $d = \dim(X)$ equations (see prop. 7.5.16 2.)) and apply the above proposition 7.2.17

7.2.2 Local Irreducibility

Let A/k be an affine k-algebra. Now we assume that we have a geometric point $P \in \operatorname{Hom}_k(A,\overline{k})$. It induces a maximal ideal $\mathfrak{m}_P$ in $A \otimes_k \overline{k}$.

Definition 7.2.18. *We say that $A \otimes \overline{k}$ is* **locally integral** *at P if the local ring $(A \otimes_k \overline{k})_{\mathfrak{m}_P}$ is integral.*

The point P also induces a maximal ideal $\mathfrak{m}_P^0$ in A. Under these assumptions we have the following technical Lemma:

Lemma 7.2.19. *If $A \otimes \overline{k}$ is locally integral at P then the local ring $A_{\mathfrak{m}_P^0}$ is also integral, and $i : A_{\mathfrak{m}_P^0} \longrightarrow (A \otimes_k \overline{k})_{\mathfrak{m}_P}$ is an injection. We also have equality of dimensions $\dim(A_{\mathfrak{m}_P^0}) = \dim((A \otimes_k \overline{k})_{\mathfrak{m}_P})$. If $\mathfrak{p}^0$ is the unique minimal prime ideal in $A_{\mathfrak{m}_P^0}$, which is contained in $\mathfrak{m}_P^0$ then these dimensions are also equal to $\dim(A/\mathfrak{p}^0)$.*

Proof: The integrality of $A_{\mathfrak{m}_P^0}$ follows once we prove the injectivity of i. Let $x \in A_{\mathfrak{m}_P^0}$ be an element that maps to zero in $(A \otimes_k \overline{k})_{\mathfrak{m}_P}$. We can assume that $x \in A$ and via the inclusion $A \longrightarrow A \otimes_k \overline{k}$ we can view it as element in the ring $A \otimes_k \overline{k}$. We know that we can find an element $y_P \notin \mathfrak{m}_P$ such that $xy_P = 0$. We have a finite number of geometric points $P = P_1, P_2, \ldots, P_s$, which lie above $\mathfrak{m}_P^0$ and the Galois group of $\overline{k}/k$ permutes these points transitively (see 2.5.). We can find an element $z_P \notin \mathfrak{m}_P$ such that $z_P \in \mathfrak{m}_{P_i}$ for all $i = 2, \ldots, s$. If we take an element σ_i in the Galois group, which maps P to P_i then $\sigma(y_P z_P) = y_{P_i} z_{P_i} \notin \mathfrak{m}_{P_i}$ but it lies in all the other $\mathfrak{m}_{P_j}$ for $j \neq i$. We always have $xy_{P_i} z_{P_i} = 0$ and if we put $s = \sum y_{P_i} z_{P_i}$ then we have $s \notin \mathfrak{m}_{P_i}$ for all i and of course we still have $xs = 0$. We apply the argument from 2.2.2.1. to s and see that we can produce an element $t^{p^N} = \operatorname{Norm}(s^{p^N}) \in A$. Then we still have $xt^{p^N} = 0$ and $t^{p^N} \notin \mathfrak{m}_{P_i}$ for all i and hence $t^{p^N} \notin \mathfrak{m}_P^0$. But this tells us that the image of $x \in A$ maps to zero in $A_{\mathfrak{m}_P^0}$ hence we proved the injectivity of i.
Now we prove the assertions concerning the equality of dimensions. It is clear that we can find an $f \notin \mathfrak{m}_P^0$ such that A_f is integral and then we have $\dim(A_f) = \dim(A_{\mathfrak{m}_P^0})$, this follows from Theorem 7.1.16 and the fact that $\operatorname{Spec}(A_{\mathfrak{m}_P^0}) = \{\mathfrak{p} \in \operatorname{Spec}(A_f) \mid \mathfrak{p} \in \mathfrak{m}_P^0\}$. Now we have seen (Lemma 7.2.14) that we can find a finite extension K/k such that the irreducible components of $A_f \otimes K$ are absolutely irreducible. We consider the inclusion $A \longrightarrow A \otimes K$ it is finite. The algebra $A \otimes K$ may have zero divisors, its minimal ideals $\mathfrak{p}_i$ are transitively permuted by the Galois group and their intersection is the radical of $A \otimes K$. The composed homomorphism $A \longrightarrow A \otimes K \longrightarrow \prod A \otimes K/\mathfrak{p}_i$, (see 7.2) is injective and then it is clear that any of the projections $p_i : A \longrightarrow A \otimes K/\mathfrak{p}_i$ is still injective. But then proposition 7.1.15 implies $\dim(A) = \dim(A \otimes K/\mathfrak{p}_i)$ and hence $\dim(A_{\mathfrak{m}_P^0}) = \dim(A \otimes_k \overline{k})_{\mathfrak{m}_P}$ where now we choose i so that $\mathfrak{m}_P \supset \mathfrak{p}_i$. □

The property that A is locally integral at the geometric point P implies that P lies on exactly one irreducible component of $\mathrm{Spec}(A \otimes_k \bar{k})$ and this implies that it also lies on exactly one irreducible component of $\mathrm{Spec}(A)$. The converse is not true since the local ring at P may still have nilpotent elements. But for the definition of the dimension at P we only need that P lies on exactly one component. In this case we will also say that A is **locally irreducible at P.**

If we have an integral affine k-algebra A/k and a maximal ideal $\mathfrak{m}$, then we know that the dimension of A is equal to the dimension of the local ring $A_\mathfrak{m}$(see theorem 7.1.16). If our k-algebra is not integral and if $\mathfrak{m}$ is a maximal ideal, then we may still speak of $\dim(A_\mathfrak{m})$ provided this local ring is irreducible, which means that $\mathfrak{m}$ lies on exactly one irreducible component, i.e. there is exactly one minimal prime ideal $\mathfrak{p}_i \subset \mathfrak{m}$. If we pick an element $f \notin \mathfrak{m}$, which lies in all the other minimal prime ideals then $\mathrm{Spec}(A_f)$ is open and irreducible in $\mathrm{Spec}(A)$. We again have the equality of dimensions $\dim(A_f) = \dim(A_\mathfrak{m})$. If we have another maximal ideal $\mathfrak{m}_1 \supset \mathfrak{p}_i$, which does not lie in any other irreducible component then we can choose f so that $f \notin \mathfrak{m}_1$. This shows that the local dimension $\dim(A_\mathfrak{m})$ stays constant as long as $\mathfrak{m}$ moves within one component and avoids the set of points, which lie in several components.
If A/k is irreducible then we can speak of $\dim(A)$. But what happens if A/k is not absolutely irreducible? We claim that nothing happens. We have seen that we can find a finite separable extension K/k such that the irreducible components of $A \otimes K$ will be absolutely irreducible (Lemma 7.2.1). Then we see as in the proof Lemma 2.3.1. that $\dim(A) = \dim(A \otimes K)/\mathfrak{p}$ for any minimal prime ideal. We can summarize by saying that the dimension is stable under the extension of scalars.

The connected component of the identity of an affine group scheme G/k

If we have a scheme X/k of finite type, then it is clear that the set of those points, which lie in exactly one irreducible component, form an open subscheme $X_0/k \subset X/k$. The complement is a closed sub scheme $Y/k \subset X/k$. Let us now assume that we have an affine group scheme G/k, we briefly discussed this notion in 6.2.10. We can extend the scalars to $\bar{k}$ and $G \times_k \bar{k}$ is still a group scheme. It is rather obvious that in this case the irreducible components of $G \times_k \bar{k}$ are disjoint. (Because of the group structure the local rings in all geometric points must be isomorphic, hence if one of them is integral all of them must be integral.) Hence we see that this is also true for G/k. Let A/k be the affine algebra of G/k. Then we get the decomposition into irreducibles (See exercise 24)

$$A = \sum_i Ae_i$$

Now we now that $G(k)$ contains the identity element, this is a homomorphism $e : A/k \longrightarrow k$. Clearly we have exactly one e_i (call it e_0), which maps to one in k and the other ones map to zero.
We claim:
The algebra Ae_0 is absolutely irreducible and defines a sub group scheme of G/k This sub group scheme is called the connected component of the identity and denoted by $G^{(0)}/k$.

This is a rather easy consequence of our previous considerations. The main point is that in this case the pseudoconstants form a subfield of Ae_0 and since Ae_0 has a k-rational point the field of constants must be equal to k.
Actually we have a more general principle: If we have an irreducible scheme X/k and if the open sub scheme $X_0 \subset X$ has a k-rational point, then the field of pseudoconstants is k and X is absolutely irreducible. (See the example following the definition of pseudoconstants, this scheme has exactly one $\mathbb{R}$-rational point, but this point is not on X_0.)

7.3 Low Dimensional Rings

A noetherian ring is of dimension zero if every prime ideal is maximal (and minimal). In this case it is clear from Theorem 7.2.1 that $\mathrm{Spec}(A) = \{\mathfrak{p}_1, \dots, \mathfrak{p}_t\}$ is a finite set. Then the local rings $A_{\mathfrak{p}_i}$ are also of dimension zero and $A_{\mathfrak{p}_i}$ has only one prime ideal, which we call $\mathfrak{m}_{\mathfrak{p}_i}$ and hence $\mathfrak{m}_{\mathfrak{p}_i}$ is also the radical of this local ring. We get an isomorphism

$$A \longrightarrow \bigoplus_{i=1}^{t} A_{\mathfrak{p}_i} = \bigoplus_{i=1}^{t} Ae_{\mathfrak{p}_i}.$$

The $e_{\mathfrak{p}_i}$ are the idempotents (See Exercise 1 (5)).

Definition 7.3.1. *A ring is called* **artinian** *if any descending chain of ideals becomes stationary.*

The rings $A_{\mathfrak{p}_i}$ are local artinian and hence A is also artinian.

Finite k-Algebras

If k is a field, then a finite k-algebra A is a k-algebra, which is finite dimensional as a k-vector space. Then it is clear that this is a zero dimensional k-algebra and hence we apply step 5) in the proof of Theorem 7.2.1, we get $A \xrightarrow{\sim} \bigoplus Ae_\nu$, where the the Ae_ν are local finite (artinian) k-algebras. The k-algebra structure of Ae_ν is given by the injection $i_\nu : x \mapsto xe_\nu$.
Such a finite k-algebra A is called absolutely reduced or separable, if $A \otimes_k \overline{k}$ does not contain nilpotent elements. This is clearly equivalent to

$$A \otimes_k \overline{k} \xrightarrow{\sim} \bigoplus_{i=1}^{\dim A} \overline{k}. \tag{7.4}$$

We have a simple criterion for separability. To formulate this criterion, we define the trace $\mathrm{tr}_{A/k} : A \longrightarrow k$. To any element $x \in A$ we consider the linear endomorphism $L_x : y \mapsto xy$ and we put $\mathrm{tr}_{A/k}(x) = \mathrm{tr}(L_x)$. Then it is clear that:

Proposition 7.3.2. *The finite k-algebra A is separable if and only if the bilinear map $(x,y) \mapsto tr_{A/k}(xy)$ from $A \times A$ to k is non degenerate.*

To see that this is so one has to observe that degeneracy or non degeneracy are preserved, if we extend k to $\overline{k}$. For a nilpotent element $x \in A \otimes_k \overline{k}$ we have $\mathrm{tr}_{A\otimes_k \overline{k}/\overline{k}}(xy) = 0$ for all y. If we have a finite separable k-algebra A then we have

$$A \otimes_k \overline{k} = \oplus_{\mathrm{Hom}_k(A,\overline{k})} \overline{k}$$

where the isomorphism is given by $x \otimes a \mapsto \sum_{\sigma \in \mathrm{Hom}_k(A,\overline{k})} \sigma(x)a$. The linear map L_x is diagonal with eigenvalues $\sigma(x)$. Therefore we get the formula

$$\mathrm{tr}_{A/k}(x) = \sum_{\sigma \in \mathrm{Hom}_k(A,\overline{k})} \sigma(x) \tag{7.5}$$

for the trace. (This is the well know formula from an elementary course in algebra, which says that the trace of an element is the sum of its conjugates.)

One Dimensional Rings and Basic Results from Algebraic Number Theory

Now we consider integral rings A with $\dim(A) = 1$. This means that every non-zero prime ideal $\mathfrak{p}$ is already maximal. If we have any ideal $(0) \neq \mathfrak{a} \neq A$, then $\dim(A/\mathfrak{a}) = 0$ and $\mathrm{Spec}(A/\mathfrak{a}) \subset \mathrm{Spec}(A)$ is a finite subset by the previous results.
Hence we see that for a one dimensional ring A the open sets $U \subset \mathrm{Spec}(A)$ are the complements of a finite set of closed points (maximal prime ideals) and of course the empty set.

Definition 7.3.3. *If A is integral, of dimension one and local, then* $\mathrm{Spec}(A)$ *consists of two points* $\{\mathfrak{p},(0)\}$. *Such a ring is A is called a* **discrete valuation ring** *if $\mathfrak{p}$ is a principal ideal, i.e. we can find an element $\pi_\mathfrak{p}$ such that $\mathfrak{p} = A \cdot \pi_\mathfrak{p} = (\pi_\mathfrak{p})$. The element $\pi_\mathfrak{p}$ is called an* **uniformizing element**.

A uniformizing element $\pi_\mathfrak{p}$ is of course not unique in general, it can be multiplied by a unit and is still a uniformizing element. It is quite clear that any element $a \in A$ can be written as

$$a = \epsilon\, \pi_\mathfrak{p}^{\nu_\mathfrak{p}(a)} \tag{7.6}$$

where ϵ is a unit and where $\nu_\mathfrak{p}(a)$ is an integer. This exponent is called the order of a and can be considered as the order of vanishing of a at $\mathfrak{p}$.

The elements of the quotient field K are of the form

$$x = \frac{b}{c} = \frac{\epsilon\, \pi_\mathfrak{p}^{\nu_\mathfrak{p}(b)}}{\epsilon'\, \pi_\mathfrak{p}^{\nu_\mathfrak{p}(c)}} = \epsilon'' \cdot \pi_\mathfrak{p}^{\nu_\mathfrak{p}(a)-\nu_\mathfrak{p}(b)} = \epsilon'' \cdot \pi_\mathfrak{p}^{\mathrm{ord}_\mathfrak{p}(x)}. \tag{7.7}$$

We clearly have $\nu_\mathfrak{p}(x) \geq 0$ if and only if $x \in A$. We may say that x has a pole of order $-\nu_\mathfrak{p}(x)$ if $\nu_\mathfrak{p}(x) < 0$.
A very important class of one dimensional rings is provided by the Dedekind rings. We have the following

Definition 7.3.4. *A one-dimensional integral ring A is called a* **Dedekind ring** *if one of the following equivalent conditions is satisfied.*

1. *The ring is normal, i.e. integrally closed in its quotient field (See 7.1.3)*
2. *For every prime ideal $\mathfrak{p} \neq (0)$ the local ring $A_\mathfrak{p}$ is a discrete valuation ring.*

Proof: The inclusion 2. $\Rightarrow$ 1. is quite clear: We consider our element $x \in K$ and assume that it satisfies an equation as as in 7.1. We claim that for any $\mathfrak{p} \neq (0)$ we must have $x \in A_\mathfrak{p}$. Otherwise we could write $x = \epsilon \pi_\mathfrak{p}^{-r}$ with $r > 0$ and ϵ a unit in $A_\mathfrak{p}$. But then x can not satisfy the polynomial equation, because we can multiply the equation by $\pi_\mathfrak{p}^{rn}$ and then the first term is non zero $\mod \mathfrak{p}$ and the other terms are zero $\mod \mathfrak{p}$. But if $x \in A_\mathfrak{p}$ for all $\mathfrak{p}$ then this means that is regular at all points in $\mathrm{Spec}(A)$ and the assertion follows from proposition 6.1.16. The direction 1. $\Rightarrow$ 2. is a not so easy. Of course we may assume that A is already local. If $\mathfrak{p}$ is the maximal ideal then we consider the A-module $\mathfrak{p}^{-1}$ of all elements $x \in K$, which satisfy $x\mathfrak{p} \subset A$. We clearly have $\mathfrak{p}^{-1} \supset A$. The decisive point is to show that we can find an element $y \in \mathfrak{p}^{-1}$, which is not in A. To see this we pick a non zero element $b \in \mathfrak{p}$. The ring $A/(b)$ has dimension zero and therefore, the image of $\mathfrak{p}$ in this ring is equal to the radical. This implies that a suitable power $\mathfrak{p}^n \subset (b)$, we choose n minimal with this property. Then we know that we can find elements $p_1, \dots, p_{n-1} \in \mathfrak{p}$ such that the element $y = p_1 p_2 \dots p_{n-1}/b \notin A$. But if we multiply by any further element in $\mathfrak{p}$ then the result lies in A. Now we conclude $y\mathfrak{p} = A$ or $y\mathfrak{p} = \mathfrak{p}$. But the second case is impossible, because exercise 18. 2. implies that y is integral over A. Since A is integrally closed we get $y \in A$ this is a contradiction. The rest is clear: We can find a $\pi \in \mathfrak{p}$ such that $y\pi = 1$. Now it is clear that $\mathfrak{p} = (\pi)$ because if $p \in \mathfrak{p}$ then $yp = a \in A$ and this gives $p = \pi a$. □

This proposition is fundamental for the foundation of the theory of algebraic numbers.

If we have a Dedekind ring A and a non-zero ideal $(0) \neq \mathfrak{a} \subset A$, then the quotient $A/\mathfrak{a}$ has dimension zero and we just saw that

$$A/\mathfrak{a} = \prod_{\mathfrak{p} \supset \mathfrak{a}} (A/\mathfrak{a})_\mathfrak{p}.$$

If $\mathfrak{a}_\mathfrak{p}$ is the image of $\mathfrak{a}$ in the localization $A_\mathfrak{p}$ then $(A/\mathfrak{a})_\mathfrak{p} = A_\mathfrak{p}/\mathfrak{a}_\mathfrak{p}$. Now we know that $A_\mathfrak{p}$ is a discrete valuation ring hence we have $\mathfrak{a}_p = (\pi_p^{\nu_\mathfrak{p}(\mathfrak{a})})$ and $\nu_\mathfrak{p}(\mathfrak{a})$ is called the order of $\mathfrak{a}$ at $\mathfrak{p}$. It is not difficult to show that $A/\mathfrak{p}^{\nu_\mathfrak{p}(\mathfrak{a})} = A_\mathfrak{p}/(\pi_p^{\nu_\mathfrak{p}(\mathfrak{a})})$ and hence we get

$$A/\mathfrak{a} = \bigoplus_{\mathfrak{p} \supset \mathfrak{a}} A/\mathfrak{p}^{\nu_\mathfrak{p}(\mathfrak{a})}.$$

:

Exercise 30. a) Show this assertion implies $\mathfrak{a} = \prod_{\mathfrak{p} \supset \mathfrak{a}} \mathfrak{p}^{\nu_\mathfrak{p}(\mathfrak{a})}$.
Hint: What is in general the relation between the product $\mathfrak{a}\mathfrak{b}$ and the intersection of two ideals $\mathfrak{a}, \mathfrak{b}$ in an arbitrary ring A? Show that there is always an inclusion in one direction. Then verify that this inclusion becomes an equality if the two ideals generate the ring, or in other words if $\mathrm{Spec}(A/\mathfrak{a}) \cap \mathrm{Spec}(A/\mathfrak{b}) = \emptyset$.

b) Show: For any maximal prime ideal $\mathfrak{p}$ we can find an $x \in K$ (the field of fractions) such that $\mathrm{ord}_\mathfrak{p}(x) = -1$ and $\mathrm{ord}_\mathfrak{q}(x) \geq 0$ for all the other maximal ideals. Then $xp \in A$ for all $p \in \mathfrak{p}$.

Definition 7.3.5. *A* **fractional ideal** $\mathfrak{b}$ *of a Dedekind ring A is a finitely generated non zero A-submodule in the field of fractions K.*

For any fractional ideal $\mathfrak{b}$ we can find an $x \in K^*$ such that $x\mathfrak{b} \subset A$ becomes an integral (ordinary) ideal. We can multiply such fractional ideals and our previous results imply that:

Lemma 7.3.6. *The fractional ideals in a Dedekind ring form a group under multiplication.*

Definition 7.3.7. *The neutral element is obviously given by the ring A itself and exercise 30 b) above gives the inverse $\mathfrak{p}^{-1} = (1,x)$. This group is the free abelian group generated by the prime ideals. It is also called the group of* **divisors** Div(A)*. This group of divisors contains the subgroup of principal divisors $P(A)$, these are the ideals of the form (x) with $x \neq 0$. The quotient group*

$$\mathrm{Pic}(A) = \mathrm{Div}(A)/P(A)$$

is the so called **ideal class group** *of A. Sometimes it is also called the* **Picard group**.

The Picard group is an important invariant of the ring. By definition it is trivial if and only if A is a principal ideal domain.

If we have a Dedekind ring A with quotient field K and if L/K is an extension of finite degree, then we may consider the integral closure of A in L. This is the ring B consisting of those elements b, which satisfy an equation $b^n + a_1 b^{n-1} + \dots a_0 = 0$ with $a_i \in A$. We have seen in exercise 18 that they form an A-algebra.
We have the

Theorem 7.3.8 (Krull - Akizuki). *The integral closure of a Dedekind ring in a finite extension of its quotient field is again a Dedekind ring.*

This is not an easy theorem, we refer to the book of [Neu], prop. 12.8. The main problem is to show that B is again noetherian.

The following fundamental theorem is easier, we drop the assumption that A is a Dedekind ring, we only assume that it is integral, noetherian and normal (See 7.1.3).

Theorem 7.3.9.

1. *Let A be an integral ring, which is noetherian and normal. Let K be its quotient field and let L/K be a finite separable extension. Then the integral closure B of A in L is a finitely generated A- module. Hence B is clearly again an integral, normal and noetherian ring*

2. *If A is a normal integral ring, which is a finitely generated algebra over a field k, i. e. $A = k[x_1, \dots, x_n]$ and if L is any finite extension of the quotient field K of A, then the integral closure B of A in L is again a finitely generated algebra over k and hence noetherian and normal.*

For a proof see [Ei], II, 13.3, as an alternative the reader may fill the gaps in the following sketch of the proof.
To see that that first assertion is true we start from a basis $a_1, \cdots, a_n$ of the field L over K, which consists of integral elements over A. Write an element $b \in B$ as linear combination $b = a_1x_1 + a_2x_2 \dots a_nx_n$ with $x_i \in K$. We use the separability to invert this system of equations for the x_i. The traces $\mathrm{tr}_{L/K}(ba_\nu)$ are integral (use Exercise 18), and we find the relations

$$\mathrm{tr}_{L/K}(ba_\nu) = \sum \mathrm{tr}_{L/K}(a_i\, a_\nu)x_i.$$

Conclude that we can find an element $a \in A$, which does not depend on b such that $a_i a \in A$, hence $B \subset A\frac{a_1}{a} + \ldots + A\frac{a_n}{a}$ and therefore, is finitely generated.
To prove the second assertion we check that we may assume that L/K is normal (in the sense of field extensions). Then we have a maximal purely inseparable sub extension L_i/K. This is obtained by successive extraction of p-th roots. Hence we prove the assertion for extensions of the form $L = K[r^{1/p}]$ (not so easy) and proceed by induction and apply the first assertion at the end.
Without any further assumption on A or the extension L/K the assertion of the theorem above may become false.

We return to our assumption that A is a Dedekind ring. The theorem above has the following implication: Let us assume that we have a Dedekind ring A with quotient field K and a finite extension L/K and we assume that the assumptions of a) or b) are valid. Then we know that the integral closure B of A in L is a finitely generated A-module. Let us pick a maximal prime ideal $\mathfrak{p} \subset A$. We consider the $A/\mathfrak{p}$ algebra $B/\mathfrak{p}B$. First of all we claim that the dimension of $B/\mathfrak{p}B$ as an $A/\mathfrak{p}$-vector space is equal to the degree $[L : K] = \dim_K L$. This is almost obvious, we may assume that A is local and then B must be a free A-module of rank $[L : K]$ and this implies the claim. Now we have seen that

$$B/\mathfrak{p}B = \bigoplus_{\mathfrak{P} \supset \mathfrak{p}B} B/\mathfrak{P}^{\nu_{\mathfrak{P}}(\mathfrak{p}B)} e_{\mathfrak{P}} \tag{7.8}$$

where the $e_{\mathfrak{P}}$ are the idempotents. Then $B/\mathfrak{P}^{\nu_{\mathfrak{P}}(\mathfrak{p}B)}$ is a $A/\mathfrak{p}$ algebra.
For a $\mathfrak{P} \supset \mathfrak{p}$ we get a finite extension of residue fields $(B/\mathfrak{P})/(A/\mathfrak{p})$ and we denote its degree by $f_{\mathfrak{P}} = [B/\mathfrak{P} : A/\mathfrak{p}]$. Moreover we know that for any integer m the quotient $\mathfrak{P}^m/\mathfrak{P}^{m+1}$ is a $B/\mathfrak{P}$-vector space of dimension one and hence an $A/\mathfrak{p}$-vector space of dimension $f_{\mathfrak{P}}$. Hence we get that $B/\mathfrak{P}^{\nu_{\mathfrak{P}}}(\mathfrak{p}B)$ is an $A/\mathfrak{p}$-vector space of dimension $f_{\mathfrak{P}}\nu_{\mathfrak{P}}(\mathfrak{p}B)$. We call the numbers $\nu_{\mathfrak{P}}(\mathfrak{p}B) = \nu_{\mathfrak{P}}$ ramification indices. Counting the dimensions yields the formula

$$[L : K] = \sum_{\mathfrak{P} \supset \mathfrak{p}B} f_{\mathfrak{P}}\nu_{\mathfrak{P}}. \tag{7.9}$$

Definition 7.3.10. *The extension is called* **unramified at** $\mathfrak{p}$ *if all the* $\nu_{\mathfrak{P}} = 1$ *and if the extensions* $(B/\mathfrak{P})/(A/\mathfrak{p})$ *are separable.*

Since B is free over A we can define the trace $\mathrm{tr}_{B/A}$ in the same way as we did this in 2.4.1 and it is clear (we still assume that A is local):

Proposition 7.3.11. *The extension B/A is unramified if and only if the pairing*

$$B \times B \longrightarrow A, (x,y) \mapsto \mathrm{tr}_{B/A}(xy)$$

is non degenerate, i.e. if for $x \in B$ the trace $\mathrm{tr}_{B/A}(xy) \in \mathfrak{p}$ for all $y \in B$ then it follows that $x \in \mathfrak{p}B$.

Let us now assume that our field extension L/K is a normal extension, this means that it is normal and separable, and let us denote its Galois group by $\mathrm{Gal}(L/K)$. Let A,B be as above, let $\mathfrak{p}$ be a non zero prime ideal in A, we have the decomposition

$$B/\mathfrak{p}B = \bigoplus_{\mathfrak{P}\supset\mathfrak{p}B} B/\mathfrak{P}^{\nu_{\mathfrak{P}}(\mathfrak{p}B)} = \bigoplus_{\mathfrak{P}\supset\mathfrak{p}B} (B/\mathfrak{p}B)e_{\mathfrak{P}}.$$

The Galois group $\mathrm{Gal}(L/K)$ acts on B and permutes the prime ideals $\mathfrak{P} \supset \mathfrak{p}$ and the idempotents $e_{\mathfrak{P}}$. We will see in the next theorem that this action is transitive. This of course implies that all the ramification indices $\nu_{\mathfrak{P}}$ are equal, we denote this number by $\nu_{\mathfrak{p}}$, this is the ramification index of $\mathfrak{p}$.

Definition 7.3.12. *Let us denote by $D_{\mathfrak{P}} \subset \mathrm{Gal}(L/K)$ the stabilizer of $\mathfrak{P}$, this is the* **decomposition group** *of $\mathfrak{P}$.*

These decomposition groups for the different primes $\mathfrak{P}' \supset \mathfrak{p}$ are conjugate to each other. We get homomorphisms

$$D_{\mathfrak{P}} \longrightarrow \mathrm{Gal}((B/\mathfrak{P})/(A/\mathfrak{p})). \tag{7.10}$$

Definition 7.3.13. *The kernel of the homomorphism $D_{\mathfrak{P}} \to \mathrm{Gal}((B/\mathfrak{P})/(A/\mathfrak{p})$ is the* **inertia group $I_{\mathfrak{P}}$**.

For us the following result is basic for the theory of algebraic numbers.

Theorem 7.3.14. *Let K,L,A,B,$\mathfrak{p}$ as above. Then the action of the Galois group on the primes above $\mathfrak{P}$ is transitive. If the normal separable extension L/K is unramified at the prime $\mathfrak{p}$ then for any $\mathfrak{P} \supset \mathfrak{p}$ the homomorphism $D_{\mathfrak{P}} \to \mathrm{Gal}((B/\mathfrak{P})/(A/\mathfrak{p}))$ is an isomorphism.*

To see the transitivity we pick a pair $\mathfrak{P}$,$\mathfrak{P}'$ and assume that $\mathfrak{P}'$ is not in the orbit of $\mathfrak{P}$. We pick a uniformizing element $\Pi_{\mathfrak{P}} \in \mathfrak{P}$. Then our assumption implies that the conjugate elements $\sigma(\Pi_{\mathfrak{P}}) \notin \mathfrak{P}'$ for for all $\sigma \in \mathrm{Gal}(L/K)$. Hence the norm $a = \prod \sigma(\Pi_{\mathfrak{P}}) \notin \mathfrak{P}'$. But now $a \in A$ and since $\mathfrak{P} \cap A = \mathfrak{P}' \cap A = \mathfrak{p}$ we get a contradiction. If $g_{\mathfrak{p}}$ is the number of $\mathfrak{P}$ lying over $\mathfrak{p}$ we have $f_{\mathfrak{P}} g_{\mathfrak{p}} = [L : K] = \mathrm{Gal}(L/K)$, hence we see that we get the equality for the group orders: $\# D_{\mathfrak{P}} = \# \mathrm{Gal}((B/\mathfrak{P})/(A/\mathfrak{p}))$. Therefore it suffices to show that the homomorphism is surjective. To see this we consider the trace $\mathrm{tr}_{(B/\mathfrak{P})/(A/\mathfrak{p})}$, the trace of an element $x = xe_{\mathfrak{P}}$ is given by $\sum_{\sigma\in\mathrm{Gal}((B/\mathfrak{P})/(A/\mathfrak{p}))} \sigma(x)$. If we consider $x = xe_{\mathfrak{P}}$ as an element in $B/B\mathfrak{p}$, then we see that $\mathrm{tr}_{(B/\mathfrak{p})/(A/\mathfrak{p})}(x)e_{\mathfrak{P}} = \mathrm{tr}_{(B/\mathfrak{P})/(A/\mathfrak{p})}(x)$. If we lift x to an element $\widetilde{x} \in B$ so that $\widetilde{x} \mod \mathfrak{p} = x$ then we know

$$\mathrm{tr}_{(B/\mathfrak{p})/(A/\mathfrak{p})}(x) = \sum_{\tau\in\mathrm{Gal}(L/K)} \tau(\widetilde{x}) \mod \mathfrak{p}.$$

If we multiply this with $e_{\mathfrak{P}}$, then we obtain

$$\sum_{\sigma\in\mathrm{Gal}((B/\mathfrak{P})/(A/\mathfrak{p}))} \sigma(x) = \sum_{\tau\in D_{\mathfrak{P}}} \tau(\widetilde{x}) \mod \mathfrak{P}.$$

The left hand side is always an element in $(A/\mathfrak{p})$. But if the image $\overline{D_{\mathfrak{P}}}$ of $D_{\mathfrak{P}}$ is not the entire Galois group then we get for the right hand side

$$\sum_{\sigma \in \mathrm{Gal}((B/\mathfrak{P})/(A/\mathfrak{p}))} \sigma(x) = m_{\mathfrak{P}} \sum_{\tau \in \overline{D_{\mathfrak{P}}}} \tau(\widetilde{x}) \mod \mathfrak{P}$$

where $m_{\mathfrak{P}}$ is the index of $\overline{D_{\mathfrak{P}}}$ in $D_{\mathfrak{P}}$. Now we have two options: Either $m_{\mathfrak{P}}$ is zero in $A/\mathfrak{p}$, then the right hand side is identically zero and so is the left hand side. But this is impossible, because the extension $(B/\mathfrak{P})/(A/\mathfrak{p})$ is separable. Or we can find an x such that $m_{\mathfrak{P}} \sum_{\tau \in \overline{D_{\mathfrak{P}}}} \tau(\widetilde{x}) \mod \mathfrak{P} \notin A/\mathfrak{p}$ this is a contradiction because the left hand side is always in this field. □

Definition 7.3.15. *A finite extension K of $\mathbb{Q}$ is called an* **algebraic number field**.

Since the ring $\mathbb{Z}$ is a Dedekind ring we now know that its integral closure $\mathcal{O}_K$ in L is always a Dedekind ring. This ring is called the ring of **integers in K.** The study of these rings of integers is the subject of algebraic number theory. We briefly state a basic theorem of this theory. We need a little bit of notation. We consider the base extension $K \otimes_{\mathbb{Q}} \mathbb{R}$, this is a finite $\mathbb{R}$ algebra and hence a direct sum of copies of $\mathbb{R}$ and $\mathbb{C}$. Then

$$K \otimes_{\mathbb{Q}} \mathbb{R} = \mathbb{R}^{r_1} \oplus \mathbb{C}^{r_2},$$

this defines the numbers r_1 and r_2.

Theorem 7.3.16. *For any algebraic number field $K/\mathbb{Q}$ the ideal class group* $\mathrm{Pic}(\mathcal{O}_L)$ *is a finite abelian group.*
The group of units $\mathcal{O}_K^\times$ is a finitely generated group, it is the product $W \times E$, where W is the finite group of roots of unity and E is free of rank $r_1 + r_2 - 1$.

If in our situation above L/K is a finite normal extension of algebraic number fields and if this extension is unramified at a prime $\mathfrak{p}$ of $A = \mathcal{O}_K$, then the extensions $(B/\mathfrak{P})/(A/\mathfrak{p})$ are extensions of finite fields. Let $N(\mathfrak{p}) = \#(A/\mathfrak{p})$. Then we know that the Galois group of these extensions is the cyclic group generated by the Frobenius element $\Phi_{\mathfrak{P}} : x \mapsto x^{N(\mathfrak{p})}$. Hence we find a unique element, also called $\Phi_{\mathfrak{P}} \in D_{\mathfrak{P}} \subset \mathrm{Gal}(L/K)$, which maps to this generator. This elements of the Galois group are also called **Frobenius elements**. These Frobenii $\Phi_{\mathfrak{P}'}$ to the different $\mathfrak{P}' \supset \mathfrak{p}$ form a conjugacy class attached to $\mathfrak{p}$, it is the **Frobenius class**.
Since we are very close to it, we also state the simplest version of the main theorem of class field theory. We consider an algebraic number field L with its ring of integers $\mathcal{O}_L$. We consider finite normal extensions F/L, with the property that their Galois group $\mathrm{Gal}(F/L)$ is abelian, and which are unramified at all prime $\mathfrak{p}$ of $\mathcal{O}_L$. If we have two such extensions F_1,F_2 then we can form the tensor product $F_1 \otimes F_2$ and decompose this into a sum of fields

$$F_1 \otimes F_2 = \bigoplus_{\nu} F_\nu.$$

These extensions F_ν are again unramified and have an abelian Galois group.
Let us pick any such an extension. We construct a homomorphism from the group of fractional ideals to $\mathrm{Gal}(F/L)$: To do this we observe that the group of fractional ideals is the free abelian group generated by the prime ideals. To any prime ideal $\mathfrak{p}$ we pick a prime ideal $\mathfrak{P}$ and our homomorphism sends $\mathfrak{p}$ to the Frobenius element $\Phi_{\mathfrak{P}}$. Since our extension is abelian this extension does not depend on the choice of $\mathfrak{P}$. Now we can formulate the celebrated theorem, which has been proved by E. Artin in his paper [Art1]:

Theorem 7.3.17. *The above homomorphism is trivial on the principal ideals and hence it induces a homomorphism*

$$Art: \mathrm{Pic}(\mathcal{O}_L) \longrightarrow \mathrm{Gal}(F/L).$$

This homomorphism is surjective and there exists a maximal abelian, unramified extension H/L, for which this homomorphism becomes an isomorphism.

This maximal abelian, unramified extension is called the *Hilbert class field.*
Of course it is clear that for any normal ring A, which is also factorial, the Picard group $\mathrm{Pic}(A) = 0$. The opposite direction is also true if the ring is noetherian, see [Ei] , Cor. 11.7.

7.4 Flat morphisms

7.4.1 Finiteness Properties of Tor

In this section we prove two results concerning the structure of flat modules and properties of the functor $\mathrm{Tor}^A_\bullet$, which will become important later (See 8.4.1). For a more systematic treatment of these facts we refer to [Ei], I.6. The functor $\mathrm{Tor}^A_\bullet$ has a certain finiteness property.
If we have an A-module M, which is not flat, then this means that we can find another A-module N such that $\mathrm{Tor}^A_1(N,M) \neq 0$.

Proposition 7.4.1. *If M,N are A-modules such that $\mathrm{Tor}^A_1(N,M) \neq 0$ then we can find a finitely generated submodule $N' \subset N$ such that $\mathrm{Tor}^A_1(N',M) \neq 0$.*

Proof: We start from the beginning of a projective resolution of N,

$$0 \longrightarrow X \longrightarrow P_0 \longrightarrow N \longrightarrow 0$$

and our assumption implies that

$$X \otimes_A M \longrightarrow P_0 \otimes_A M$$

is not injective.
Hence we find an element $y = \sum x_\nu \otimes m_\nu \in X \otimes_A M$, which is not zero but, which goes to zero in $P_0 \otimes_A M$. We consider the element $\widetilde{y} = \sum(x_\nu,m_\nu)$ in the free abelian group generated by the elements in $P_0 \times M$. Since y goes to zero this element can be as a finite linear combination $\sum_j R_j$ where R_j is one of the relations, which we use to define the tensor product (see Vol. I,2.4.2). In these relations the first components occurring in the R_j together with the x_ν generate a finitely generated submodule $Y \subset P_0$. Then y goes already to zero in $Y \otimes_A M$. We get a sequence

$$0 \longrightarrow Y \cap X \longrightarrow Y \longrightarrow N' \longrightarrow 0$$

where $N' \subset N$ is finitely generated. Our element $y \in (Y \cap X) \otimes M$ non zero and maps to zero in $Y \otimes_A M$. Hence we see that

$$(Y \cap X) \otimes_A M \longrightarrow Y \otimes_A M$$

is not injective and this implies $\mathrm{Tor}_1^A(N',M) \neq 0$. □

Of course the result may be sharpened. We can put N' into an exact sequence $0 \longrightarrow N_1' \longrightarrow N' \longrightarrow N_2' \longrightarrow 0$, where the two outer modules a generated by fewer elements. If we look now at the exact sequence for Tor_1 it becomes clear that we can assume that N' is generated by one element. This has the consequence:

For a non flat A-module M we can find an ideal $\mathfrak{a} \subset A$ such that $\mathrm{Tor}_1^A(A/\mathfrak{a},M) \neq 0$.

Basically the same reasoning shows

If we have a flat A-module M, which is not faithfully flat, then we get by similar arguments that we can find an ideal $\mathfrak{a} \subset A, \mathfrak{a} \neq A$ such that $A/\mathfrak{a} \otimes M = 0$.
Hence we can say that the modules of the form $A/\mathfrak{a}$ recognize flat modules and among the flat modules they recognize those which are faithfully flat.

Proposition 7.4.2. *Let A be a noetherian ring and assume that M is a finitely generated A-module. For $\mathfrak{p} \in \mathrm{Spec}(A)$ the $A_\mathfrak{p}$ module $M_\mathfrak{p} = M \otimes A_\mathfrak{p}$ is free if and only if it is flat. The set of points $\mathfrak{p} \in \mathrm{Spec}(A)$ where $M_\mathfrak{p}$ is a free $A_\mathfrak{p}$-module is an open subset $U \subset \mathrm{Spec}(A)$.*

We pick a point $\mathfrak{p}$, then $M \otimes k(\mathfrak{p})$ is a vector space of finite dimension over $k(\mathfrak{p})$. If we lift the elements of a basis of this vector space to elements $m_1,\dots,m_s$ then these elements generate the $A_\mathfrak{p}$-module $M_\mathfrak{p}$ (Lemma of Nakayama). Hence we get an exact sequence

$$0 \longrightarrow R \xrightarrow{\ j\ } A_\mathfrak{p}^s \xrightarrow{\ h\ } M \longrightarrow 0.$$

Of course $M_\mathfrak{p}$ is free if and only if $R = 0$. We assume that $M_\mathfrak{p}$ is flat. Taking the tensor product with $k(\mathfrak{p})$ gives us an exact sequence

$$\mathrm{Tor}_1^A(M,k(\mathfrak{p})) \longrightarrow R \otimes k(\mathfrak{p}) \xrightarrow{\ j_\mathfrak{p}\ } A_\mathfrak{p}^s \otimes k(\mathfrak{p}) \xrightarrow{\ h_\mathfrak{p}\ } M \otimes k(\mathfrak{p}) \longrightarrow 0$$

where now the arrow $h_\mathfrak{p}$ in this sequence is an isomorphism by construction. Hence we see that the arrow $j_\mathfrak{p}$ is the zero map. Our assumption implies that $R \otimes k(\mathfrak{p}) = 0$. But since R is finitely generated (remember we assumed that A is noetherian) we get $R = 0$. Now the rest follows because it is clear that if $M_\mathfrak{p}$ is free then we can find an open neighborhood $U = \mathrm{Spec}(A_f)$ of $\mathfrak{p}$ such that $M_f = M \otimes_A A_f$ is free. □

We can easily extend the notion of flatness to quasi-coherent sheaves. If we have A-modules M,N then we compute the groups $\mathrm{Tor}_i^A(M,N)$ from a projective resolution of M:

$$\dots P_i \longrightarrow P_{i-1} \longrightarrow \dots P_0 \longrightarrow M \longrightarrow 0$$

and we get the A-modules $\mathrm{Tor}_i^A(M,N)$ as the homology groups of this complex. If $f \in A$ is a non nilpotent element then localization provides a projective resolution of M_f

$$\dots (P_i)_f \longrightarrow (P_{i-1})_f \longrightarrow \dots (P_0)_f \longrightarrow M_f \longrightarrow 0,$$

the complex stays exact because localization is exact and projective modules stay projective after localization. Hence we get

$$\mathrm{Tor}_i^A(M,N) \otimes A_f = \mathrm{Tor}_i^{A_f}(M_f,N_f).$$

If we consider the affine scheme $\mathrm{Spec}(A)$ and consider the associate sheaves $\widetilde{M},\widetilde{N}$ then we clearly get

$$\widetilde{\mathrm{Tor}_i^A(M,N)} = \mathrm{Tor}_i^{\mathrm{Spec}(A)}(\widetilde{M},\widetilde{N}).$$

This allows us to speak of flat quasi-coherent sheaves $\mathcal{M}$ on any scheme S and for any pair $\mathcal{M}, \mathcal{N}$ we can define

$$\mathrm{Tor}_i^S(\mathcal{M}, \mathcal{N}).$$

If we are dealing with non finitely generated A-modules, then we can not expect that a generalization of proposition 7.4.2 is true. This is of course clear: Given M we consider the points $\mathfrak{p}$ where $\mathrm{Tor}_1^A(M,N) = 0$ for all modules N. For a single module N it is of course clear that the set of points $\mathfrak{p}$ where $\mathrm{Tor}_1^A(M,N)_{\mathfrak{p}} = 0$ is open, but we have to check infinitely many modules N.

7.4.2 Construction of flat families

Let A be a noetherian ring, let $i : A \longrightarrow B$ be an A algebra, we assume that B is finitely generated over A, we write $B = A[x_1,x_2,\ldots,x_n]$. We introduce some notation. By $\underline{\nu} = (\nu_1,\ldots,\nu_n)$ we mean a multi index , i.e. $\nu_i \in \mathbb{N}$. We introduce the monomials

$$x^{\underline{\nu}} = x_1^{\nu_1} x_2^{\nu_2} \ldots x^{\nu_n}$$

and put $\deg(\underline{\nu}) = \sum \nu_i$. We fix an integer $N > 0$ and consider expressions

$$f = \sum_{\underline{\nu}:|\underline{\nu}|\leq N} c_{\underline{\nu}} x^{\underline{\nu}}$$

where the $c_{\underline{\nu}} \in A$ or more generally in an A algebra C.
Clearly we can view such an f as a C valued point in an affine scheme

$$T_A^N = \mathbb{A}_A^{k(N)} = A[\ldots,C_{\underline{\mu}},\ldots],$$

where the $C_{\underline{\mu}}$ are polynomial variables attached to the multi-indices $\underline{\mu}$ with $\deg(\underline{\mu}) \leq N$. For any A algebra C we have the special element $1 \in T_N(C)$, this is the element where $c_{\underline{0}} = 1$ and all other coefficients are zero.
The algebra $\widetilde{B} = B \otimes_A A[\ldots,C_{\underline{\mu}},\ldots]$ contains the "universal" element

$$F = \sum_{\underline{\nu},\, \deg(\underline{\nu})|\leq N} C_{\underline{\nu}} x^{\underline{\nu}}.$$

This element yields a principal ideal $(F) \subset \widetilde{B}$, it yields a quotient $A[\ldots,C_{\underline{\mu}},\ldots] \longrightarrow \widetilde{B}/(F)$ and hence an affine subscheme

$$\begin{array}{ccc} \mathrm{Spec}(\widetilde{B}/(F)) & \hookrightarrow & \mathrm{Spec}(\widetilde{B}) \\ & \searrow & \downarrow \\ & & T^N \end{array} \tag{7.11}$$

If we consider a point $\psi \in \mathrm{Spec}(A[\ldots,C_{\underline{\mu}},\ldots])$ with residue field $k(\psi)$, i.e. a homomorphism $\psi : A[\ldots,C_{\underline{\mu}},\ldots] \longrightarrow k(\psi),\psi : C_{\underline{\mu}} \longrightarrow c_\mu$ then we perform the base change and get a hyper-surface

$$\mathrm{Spec}(\widetilde{B} \otimes_{T^N,\psi} k(\psi))/(f) \subset \mathrm{Spec}(\widetilde{B} \otimes_{T^N,\psi} k(\psi)),$$

where $f = \sum_{\underline{\nu}\,\deg(\underline{\mu})\leq N} c_{\underline{\nu}} x^{\underline{\nu}}$ and the subscheme is defined by the equation $f = 0$. We call this the evaluation of the family at ψ.
We may view $\mathrm{Spec}(\widetilde{B})$ as a "constant" affine scheme over T^N and $\mathrm{Spec}(\widetilde{B}/(F))$ is a family of hyper-surfaces (of degree $\leq N$) inside this constant scheme.

We generalise our construction slightly. Let M be a B module of finite type we consider the $B \otimes A[\ldots C_{\underline{\mu}},\ldots]$ -module $\widetilde{M} = M \otimes_A A[\ldots C_{\underline{\mu}},\ldots]$. It is not difficult to verify that the annihilator of F in $\widetilde{M}$ is trivial, hence we get an exact sequence

$$0 \longrightarrow \widetilde{M} \xrightarrow{m_F} \widetilde{M} \longrightarrow \widetilde{M}/F\widetilde{M} \longrightarrow 0 \tag{7.12}$$

where m_F denotes the multiplication by F. Let $\psi \in \mathrm{Spec}(A[\ldots C_{\underline{\mu}},\ldots]),\psi : C_{\underline{\mu}} \mapsto c_\mu \in A$. The fibre over ψ is $\mathrm{Spec}(\widetilde{B} \otimes_{A[\ldots C_{\underline{\mu}},\ldots],\psi} A = \mathrm{Spec}(B)$ and if evaluate $\widetilde{M}/F\widetilde{M}$ at this fibre over ψ then we get the B-module M/fM where $f = 1 + \sum_{\underline{\nu}:\deg(\underline{\nu})\leq N} c_{\underline{\nu}} x^{\underline{\nu}}$.
We state an important theorem, the integer N is fixed.

Theorem 7.4.3. *Let $i : A \longrightarrow B = A[x_1,x_2,\ldots,x_n]$ as above, let M be finitely generated B-module, which is A-flat. The subset $U_M \subset \mathrm{Spec}(\widetilde{B})$ of points $\mathfrak{q} \in \mathrm{Spec}(\widetilde{B})$, where $(\widetilde{M}/F\widetilde{M})_{\mathfrak{q}}$ is A-flat, is a non empty open subset.*

This theorem is a special case of theorem of Thm. 24.3 in Chapter 8 of [Mat], the proof is not easy at all. We will not prove it here. Later in the chapter on projective schemes we will use a global version of this theorem and we will prove this global version using the finiteness theorems for coherent cohomology.

But we want to say a few words about the meaning of this theorem.
At first we remark that U_M is non empty: If we choose for f the above element $1 \in T^N(C)$ then $M/1M = 0$ and this module is flat.
Let us consider a very special case, namely $B = A[X_1,X_2,\ldots,X_n]$ where the capital letters letters signalise that B is the polynomial ring. Let us also assume that $M = B$. In this case $M \otimes A/\mathfrak{a} = (A/\mathfrak{a})[X_1,X_2,\ldots,X_n]$.
An element $f = \sum_{\underline{\nu}} c_{\underline{\nu}} x^{\underline{\nu}} \in T^N(A)$ yields an A-flat module B/fB if and only if the ideal generated by the coefficients $c_{\underline{\mu}}$ is equal to A, i.e. is not a proper ideal. The scheme $\mathrm{Spec}(A[\ldots C_{\underline{\mu}},\ldots]) = T^N \longrightarrow \mathrm{Spec}(A)$ has the structure of a vector bundle (See p. 20) and we have the zero section $\mathrm{Spec}(A) \longrightarrow T^N$, which is given by the homomorphism $C_{\underline{\mu}} \mapsto 0$ for all $\underline{\mu}$. This zero section defines a closed subscheme $\{0\} \subset T^N$. The condition on f can be reformulated: The element f yields an A-flat module M/fM if and only if $f \in (T^N \setminus \{0\})(A)$.

Now we assume again $M = B$ but $B = A[x_1,x_2,\ldots,x_n] = A[X_1,X_2,\ldots,X_n]/I$ where I is an ideal. Let us consider an A-valued point $\psi \in T^N(A)$ where $\psi = f = \sum_{\underline{\nu}\,\deg(\underline{\nu})|\leq N} c_{\underline{\nu}} x^{\underline{\nu}}$. We want to assume that the fibre over ψ is contained in U_M. What does this mean? Our element ψ yields a homomorphism $\mathrm{Id}\otimes\psi : \widetilde{B} \longrightarrow B$ and F is mapped to f under this homomorphism. (If we look at this in the category of schemes this means that we restrict $\mathrm{Spec}(\widetilde{B})$ to the fibre over ψ). We tensorize the sequence 7.4.2 via $\psi : A[\ldots,C_{\underline{\mu}},\ldots] \longrightarrow A$ by A and our assumption of flatness at ψ yields the exact sequence

$$0 \longrightarrow B \xrightarrow{m_f} B \longrightarrow B/fB \longrightarrow 0,$$

hence we see $\mathrm{Ann}_B(f) = 0$, i.e. f is not a zero divisor in B. But since flatness is preserved by base change our assumption surely implies that B/fB is flat over A.
Hence we have to show that for any ideal $\mathfrak{a}$ we have $\mathrm{Tor}_1^A(B/fB,A/\mathfrak{a}) = 0$. (See 7.4.1). We tensorize our exact sequence by $A/\mathfrak{a}$ and get

$$0 \longrightarrow \mathrm{Tor}_1^A(B/fB,A/\mathfrak{a}) \longrightarrow fB\otimes A/\mathfrak{a} \longrightarrow B \longrightarrow B/fB \longrightarrow 0$$

and hence we see that

$$\mathrm{Tor}_1^A(B/fB,A/\mathfrak{a}) \xrightarrow{\sim} \mathrm{Ann}_{B\otimes A/\mathfrak{a})}(f)$$

and M/fM is flat iff and only if for all ideals $\mathfrak{a}$ the element f is not a zero divisor in $B\otimes A/\mathfrak{a}$.
We consider the case that $\mathfrak{a}$ is prime, in other words that $\mathfrak{a}$ is a point $\mathfrak{p} \in \mathrm{Spec}(A)$. Let $k(\mathfrak{p})$ be its residue field. We consider the affine algebra $B\otimes k(\mathfrak{p})$ over $k(\mathfrak{p})$. This algebra may not be integral and $\mathrm{Spec}(B\otimes k(\mathfrak{p}))$ may have irreducible components, which correspond to minimal prime ideals. Hence f is not allowed to be a zero divisor in $B\otimes k(\mathfrak{p})$ Hence a least it has to avoid the minimal prime ideals (See 7.2), actually the precise condition is that this image is not in any of the finitely many associated ideals (See [Ei], I. 3) But this must be so if the point $\mathfrak{p}$ varies, so a requirement on f is that f avoids certain "forbidden" positions in the fibers $B\otimes k(\mathfrak{p})'$. So in a certain sense we can say that for a flat A-algebra B of finite type, the union of the associated ideals in the fibers $B\otimes k(\mathfrak{p})$ varies "continuously" with $\mathfrak{p}$ so that we can find an f that avoids them. It follows from the Hauptidealsatz that the dimension of the irreducible components of $\mathrm{Spec}(B\otimes k(\mathfrak{p}))$ drops by one if we intersect with the hyper-surface $f = 0$.

7.4.3 Dominant morphisms

If we have an arbitrary morphism $\pi : X \to Y$ between two schemes $X/k,Y/k$ of finite type then we would like to have some answers for the following questions:

1. What is the structure of the image?
2. What can we say about the dimension of the fibers?

To attack these questions we have the following reduction process.

1. We cover X,Y by affine schemes U_i,V_i such that $\pi : U_i \to V_i$, then we see that we can reduce these questions to the case of affine morphisms.

2. If now $\phi : A \to B, \operatorname{Spec}(B) \to \operatorname{Spec}(A)$ is a morphism of affine schemes then we see that we may assume that ϕ is injective, because it factors over $\operatorname{Spec}(A/\mathfrak{a})$, where $\mathfrak{a}$ is the kernel of ϕ.

3. Now we decompose both affine schemes into irreducible components and then we see that we may assume that the algebras are integral.

Let us assume that we have two integral affine k-algebras and a morphism

$$\begin{array}{ccc} A & \xrightarrow{\varphi} & B \\ & \nwarrow \quad \nearrow & \\ & k & \end{array}$$

Let us assume in addition that the morphism φ is injective, under these assumptions the morphism φ is called **dominant**. We get a morphism $X = \operatorname{Spec}(B) \to Y = \operatorname{Spec}(A)$. We can pass to the quotient fields K of A and L of B and we get an injection

$$\begin{array}{ccc} K & \xrightarrow{\varphi_1} & L \\ & \nwarrow \quad \nearrow & \\ & k & \end{array}$$

and now we get a surjective map $\operatorname{Spec}(L) \to \operatorname{Spec}(K)$ and the extension L/K is faithfully flat.
Now we are a little bit (too) optimistic and we try to prove that the image of $X = \operatorname{Spec}(B) \to Y = \operatorname{Spec}(A)$ is open. Assume that a given $\mathfrak{p} \in \operatorname{Spec}(A)$ is in the image of the map $\operatorname{Spec}(B) \longrightarrow \operatorname{Spec}(A)$. Following the proof of the theorem 7.1.6 we form the tensor product

$$A/\mathfrak{p} \otimes_A B \longrightarrow (A/\mathfrak{p})_{(0)} \otimes B$$

and our assumption means that $(A/\mathfrak{p})_{(0)} \otimes B \neq (0)$. Of course we are tempted to apply the previous proposition 7.4.2 this is still the case of we pass to a suitable open neighborhood of $\mathfrak{p}$. But this reasoning is wrong unless we assume that B is finite over A, i.e. the A-module B is finitely generated. To illustrate how subtle the situation is we recommend to solve the following exercise

Exercise 31. We consider the inclusion

$$A = k[U,V] \hookrightarrow B = k[U,V,W/(UW - V).$$

Determine the image of $\operatorname{Spec}(B) \longrightarrow \operatorname{Spec}(A)$. Show that the image is not open (Proposition 7.4.2 does not work) and the dimension of the fibers is not constant.

Of course our argument above shows

Proposition 7.4.4. *The morphism* $\operatorname{Spec}(B) \longrightarrow \operatorname{Spec}(A)$ *is surjective if* $A \longrightarrow B$ *is faithfully flat.*

We mentioned already that for a dominant morphism this morphism is faithfully flat in the generic fibre. The following proposition asserts that these two facts remain true after the restriction to suitable nonempty subsets $U \subset X, V \subset Y$, i.e. we get a faithfully flat morphism $U \longrightarrow V$.

Proposition 7.4.5. *Let $\varphi : A \to B$ be a dominant morphism between two affine k-algebras. Then we can find an $f \neq 0$ in A and a $g \neq 0$ in B such that $A_f \to B_{fg}$ is faithfully flat and then* $\mathrm{Spec}(B_{fg}) \to \mathrm{Spec}(A_f)$ *is surjective.*

Proof: We consider the field of fractions K of A. Then $B \otimes_A K$ becomes a finitely generated K-algebra, it is still integral. We apply the Normalisierungssatz 7.1.10 and this allows us to assume that $B \otimes_A K$ is of the form $K[x_1,x_2,\dots,x_n]$ where $K[x_1,x_2,\dots,x_r]$ is a polynomial ring and the remaining generators are integral over $K[x_1,x_2,\dots,x_r]$, i.e. we have equations

$$0 = a_\nu x_\nu^{m_\nu} + P_{\nu 1}(x_1,\dots,x_r)x_\nu^{m_\nu - 1} + \dots P_{\nu m_\nu}(x_1,\dots,x_r)$$

where $\nu = r+1,\dots,n$, the $a_\nu \in K$ and not zero. If H is the product of the a_ν, then we can find $f \in A, f \neq 0$ such that

$$fH \in A$$

and hence we see that

$$B_{fH} = B \otimes A_f \left[x_1,\dots,x_r,\frac{1}{fH}\right]$$

is integral over $A_f\left[x_1,\dots,x_r,\frac{1}{fH}\right]$. We look at the coefficients of fH, which are elements in A_f. At least one of these coefficients c is not zero. We replace f by fc, then one of the coefficients of $fH = g$ is a unit. Now we have

$$A_f \subset A_f\left[x_1,\dots,x_r,\frac{1}{g}\right] \subset B_{fg}$$

and B_{fg} is finite over $A_f\left[x_1,\dots,x_r,\frac{1}{g}\right]$.

We claim that under these conditions the A_f-algebra $A_f\left[x_1,\dots,x_r,\frac{1}{g}\right]$ is faithfully flat. First of all we observe that

$$A_f\left[x_1,\dots,x_r,\frac{1}{g}\right] = A_f\left[x_1,\dots,x_r,y\right]/(yg-1).$$

The ideal $(yg-1)$ is a free $A_f[x_1,\dots,x_r,y]$-module and hence we see that the A_f-modules $A_f[x_1,\dots,r,y]$ and $(yg-1)$ are free. Hence we have the sequence of A_f-modules

$$0 \longrightarrow (yg-1) \longrightarrow A_f[x_1,\dots,x_r,y] \longrightarrow A_f\left[x_1,\dots,x_r,\frac{1}{g}\right] \longrightarrow 0.$$

To check flatness it suffices to show that this sequence stays exact if we tensorize by quotients $A/\mathfrak{a}$. (See previous subsection)We get

$$(yg-1)\otimes A/\mathfrak{a} \longrightarrow A/\mathfrak{a}[x_1,\dots,x_r,y] \longrightarrow A/\mathfrak{a}\left[x_1 \dots x_r,\frac{1}{g}\right] \longrightarrow 0,$$

but now it is not difficult to see that the first arrow is still injective because $yg-1$ is not a zero divisor in $A/\mathfrak{a}[x_1,\ldots,x_r,y]$. To check that $A_f\left[x_1,\ldots,x_r,\frac{1}{g}\right]$ is faithfully flat, we need that $A/\mathfrak{a}\left[x_1,\ldots,x_r\frac{1}{g}\right]\neq 0$ for all ideals $\mathfrak{a}\in A,\mathfrak{a}\neq A$. This is obviously the case since we assumed that $g\not\equiv 0 \mod \mathfrak{a}$ for all ideals of this kind. Now we observe that the quotient field of B_{fg} is a finite extension of the quotient field of $A_f\left[x_1,\ldots,x_r,\frac{1}{g}\right]$.
Hence we see that B_{fg} is free in the generic point of $\mathrm{Spec}\left(A_f\left[x_1,\ldots,x_r,\frac{1}{g}\right]\right)$ and hence it is is flat over a non empty open subset of $\mathrm{Spec}\left(A_f\left[x_1,\ldots,x_r,\frac{1}{g}\right]\right)$ and then it is automatically locally free, hence faithfully flat.
This open set contains some $\mathrm{Spec}\left(A_f\left[x_1,\ldots,x_r,\frac{1}{g}\right]\right)_{g_1}$, but then we see that this again contains a subset of the form $\mathrm{Spec}\left(A_{f'}\left[x_1,\ldots,x_r,\frac{1}{g'}\right]\right)$. Hence we proved the assertion concerning faithful flatness.
The surjectivity follows from the faithfulness but can also easily be derived from the fact that B_{fg} is integral over $A_f\left[x_1,\ldots,x_r,\frac{1}{g}\right]$. □

Of course our considerations imply:

Proposition 7.4.6. *If $\varphi: A\to B$ be a dominant morphism between two affine k-algebras, and if r is the difference of the trancendence degrees of their fraction fields, then we can find a non empty open subsets $V=\mathrm{Spec}(A_f)\subset\mathrm{Spec}(A)$,$U\subset\mathrm{Spec}(B)$ such that $f_U: U\longrightarrow V$ is surjective and for any point $y\in V$ the dimensions of the irreducible components of the fiber $f_U^{-1}(y)$ are equal to r. Especially we have* $\dim(A)\leq\dim(B)$

Proof: This follows directly from the proof of the previous proposition. □

Now we can derive a general statement concerning the image of a morphism between arbitrary k-schemes of finite type. Actually what we would like to have is:

Hope: *Under some reasonable conditions the image of $f: X\longrightarrow Y$ is open in the Zariski closure of the image.*

Let us start from a dominant morphism $f: X\longrightarrow Y$ between affine schemes. Then we have seen that the image contains a non empty open subset $U\subset Y$. Let us consider the complement Y' of U, this is a closed subscheme, we get a morphism $X'=f^{-1}(Y')\longrightarrow Y'$. We consider two irreducible componets Y_1',X_1' of Y',X' respectively, such that $X_1'\longrightarrow Y_1'$ and we take open affine subsets V_1',U_1' such that $U_1'\longrightarrow V_1'$ and we get a homomorphism between the corresponding integral affine algebras $A'\longrightarrow B'$. But now there is no reason why this morphism is still dominant, we get a diagram

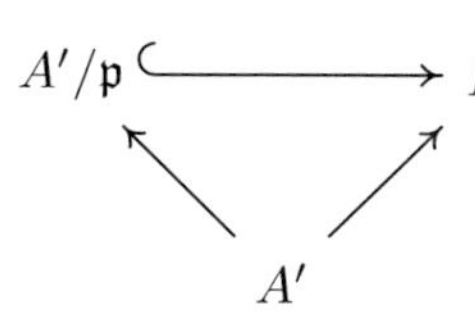

and this tells us that our morphism may factor through a proper closed irreducible subscheme V_1'' of V_1'. This morphism $U_1' \to V_1''$ is again dominant and therefore, its image contains a non empty open subset $\widetilde{V}_1'' \subset V_1''$. Again we can take its complement and proceed.
This argument tells us that the image under a morphism between schemes of finite type over a field k is always a so called **constructible** subset of Y. The family of constructible subsets is the smallest family of subsets, which contains the open subsets and is closed under finite intersections and the formation of complements. So open and closed subsets are constructible but also open subsets of closed subsets are constructible and so on.

Exercise 32. Construct an example of the form $k[U,V] \hookrightarrow k[U,V,Z]/(P(U,V,Z))$ where the image is the affine plane minus (one of the axes minus the point (0,0)).

If we analyse our reasoning in the previous section and the example in the exercise, then we see that taking the restriction from $Y = \mathrm{Spec}(A)$ to the complement of the open set U we may loose the dominance. Keeping the notations from previous section this means that the tensor product with the quotient field of A' $A'_{(0)} \otimes_{A'} B'$ may become zero. But now we look at the sequence

$$0 \longrightarrow A' \longrightarrow A'_{(0)} \longrightarrow A'_{(0)}/A' \longrightarrow 0$$

and take the tensor product with B'. We get

$$0 \longrightarrow B' \longrightarrow A'_{(0)} \otimes_{A'} B' \longrightarrow A'_{(0)}/A' \longrightarrow 0.$$

Now if B' is not zero (i.e. $X' \neq \emptyset$) then we see that B' is *not a flat* B' algebra if $A'_{(0)} \otimes_{A'} B' = 0$. This gives us the decisive hint how to formulate the assumption in the above assertion **Hope**

Proposition 7.4.7. *If $f : X \to Y$ is a flat morphism between to k-Schemes of finite type, then the image $f(X)$ is open in Y.*

Proof: Again we easily reduce this to the case $X = \mathrm{Spec}(A), Y = \mathrm{Spec}(B)$ and $\phi : A \to B$ is flat. It is a formal consequence of the concept of flatness that for any ideal $\mathfrak{a} \subset A$ the algebra homomorphism $A/\mathfrak{a} \to B \otimes_A A/\mathfrak{a}$ is flat. It is also clear that we can restrict the morphism to the inverse image of a connected component of Y. A component is given by a minimal prime ideal $\mathfrak{p}$. We remarked that $A/\mathfrak{p} \to B \otimes_A A/\mathfrak{p}$ is flat. Then we may have that $B \otimes_A A/\mathfrak{p} = 0$, then the inverse image of $\mathrm{Spec}(A/\mathfrak{p})$ is empty and nothing has to be proved. Otherwise we observed that $(A/\mathfrak{p})_{(0)} \otimes_{(A/\mathfrak{p})} B \neq 0$ and hence we see that B contains a minimal prime ideal $\mathfrak{q}$ such that ${}^t\phi(\mathfrak{q}) = \mathfrak{p}$. In other words $A/\mathfrak{p} \to B/\mathfrak{q}$ is injective. We are in the situation of proposition 7.4.5 and can conclude that $\mathrm{Spec}(B) \to \mathrm{Spec}(A/\mathfrak{p})$ has an open image. Now we proceed as indicated, we pass to the complement of the open image and use the fact that the restriction of the morphism to this complement is still flat. □

Birational morphisms

A dominant morphism $\phi : A \longrightarrow B$ between to affine k-algebras is called **birational** if ϕ induces an isomorphism between their quotient fields. It is very easy to see that this is the case if and only if we can find a non zero element $f \in A$ such that ϕ induces an isomorphism $\phi : A_f \xrightarrow{\sim} B_f$. Therefore we see that the morphism $\mathrm{Spec}(B) = X \longrightarrow \mathrm{Spec}(Y) = Y$ is birational if and only if we can find a non empty Zariski open subset $U \subset X$ and an open subset $V \subset Y$ such that ϕ induces an isomorphism $U \xrightarrow{\sim} V$.

The Artin-Rees Theorem

Let A be a noetherian local ring with maximal ideal $\mathfrak{m}$. We can provide it with the so called $\mathfrak{m}$-adic topology: The open neighborhoods of $m \in A$ are the sets $m + \mathfrak{m}^N$: We define the **completion**

$$i : A \longrightarrow \widehat{A} = \varprojlim A/\mathfrak{m}^N.$$

We may also consider the completion of an A-module M, we define $\widehat{M} = \varprojlim M/\mathfrak{m}^N M = M \otimes_A \widehat{A}$. The following two assertions are easy consequences of the theorem of Artin-Rees, which we formulate futher down.

Corollary 7.4.8. *If M is a finitely generated module over the noetherian local ring A then $M \to \widehat{M}$ is injective.*

This is of course equivalent to $\bigcap_{N \geq 1} \mathfrak{m}^N M = (0)$, in other words the $\mathfrak{m}$-adic topology is Hausdorff. The reader should verify that the module $\widehat{M}$ is complete with respect to the $\mathfrak{m}$-adic topology, i.e. any Cauchy sequence is convergent.
We consider finitely generated A-modules M,M' .... The main implication of the Artin-Rees theorem is:

Corollary 7.4.9. *In the category of finitely generated A-modules the functor*

$$M \longrightarrow \widehat{M} = M \otimes_A \widehat{A}$$

is faithfully flat, i.e. $M \longrightarrow M \otimes_A \widehat{A}$ is exact and $M \otimes_A \widehat{A} = 0$ implies $M = 0$:

The faithfulness is Nakayama's lemma. To prove flatness we have to show:
If M',M are finitely generated A-modules and if $M' \to M$ is injective then $\widehat{M'} \to \widehat{M}$ is again injective.
If we try to prove this we encounter the following problem: Consider $M' \bigcap \mathfrak{m}^N M$ for a very large N then this must be contained in $\mathfrak{m}^n M'$ where n is also large. We have to show that n goes to infinity if N goes to infinity.
The answer to our problem is the actual Artin-Rees theorem

Theorem 7.4.10 (Artin-Rees). *There exists an $r_0 > 0$ such that for all $r > r_0$ we have*

$$\mathfrak{m}^r M \cap M' = \mathfrak{m}^{r-r_0} (\mathfrak{m}^{r_0} M \cap M') \subset \mathfrak{m}^{r-r_0} M'.$$

For the proof we refer to the literature [Ei].I.5.
It is clear that it is the technical answer to our problem and it implies that $A \to \widehat{A}$ is faithfully flat on finitely generated modules.
We leave it as an exercise to verify that for any finitely generated module $\bigcap_{N \geq 1} \mathfrak{m}^N M = (0)$.

7.4.4 Formal Schemes and Infinitesimal Schemes

I want to discuss a situation, which is in some sense opposite to the one discussed above. Let k be a field, consider a polynomial ring $A = k[X_1,\dots,X_n]$ and we localize it at the maximal ideal $\mathfrak{m} = (X_1,X_2,\dots,X_n)$. This local ring $A_\mathfrak{m}$ is the ring of germs of regular functions at zero and hence we say that $\mathrm{Spec}(A_\mathfrak{m})$ is the germ of our affine space at zero. This ring is not of finite type over k anymore, but it is still a direct limit of finitely generated k-algebras. This local ring sits inside the ring $\widehat{A} = k[[X_1,\dots,X_n]]$ of power series in the variables $X_1,\dots,X_n$ as a subring: If $Q(X_1,\dots,X_n) \in A$ and if $a_0 = Q(0,\dots,0) \neq 0$ then we can write $Q = a_0 + R$ where R is a polynomial without constant term and

$$\frac{1}{Q(X_1,\dots,X_n)} = \frac{1}{a_0 + R} = \frac{1}{a_0}\left(1 - R/a_0 + (R/a_0)^2 + \dots\right).$$

It is easy to see that this power series ring can also be obtained as the projective limit $\varprojlim_N A_{\mathfrak{m}_P}/\mathfrak{m}_P^N$. It is also the completion of $A_\mathfrak{m}$ with respect to the $\mathfrak{m}$-adic topology. (See the previous subsection on the Artin Rees theorem)
The verifications of the following assertions are left to the reader.

1. $\mathrm{Hom}_k(A_\mathfrak{m},k) = \mathrm{Hom}_k(\widehat{A},k) = (0) \in k^n$. We lost all our geometric points except the origin. The geometric points are not dense and this does not contradict the Nullstellensatz because the local ring is not of finite type.
2. Let us assume that B is a k-algebra, which is local with maximal ideal $\mathfrak{n}$ and $B/\mathfrak{n} = k$. Then $\mathrm{Hom}_k(A_\mathfrak{m},B) \subset \mathfrak{n}^n$, where the identification is provided by $\phi \mapsto (\phi(X_1),\dots,\phi(X_n))$.
3. Can we construct rings B as above such that we still have $\mathrm{Hom}_k(\widehat{A},B) = (0)$.
4. If in addition the ideal $\mathfrak{n}$ in our k algebra B consists of nilpotent elements then we have $\mathrm{Hom}_k(A_\mathfrak{m},B) = \mathrm{Hom}_k(\widehat{A},B) = \mathfrak{n}^n$.

We have the inclusion of rings:

$$A \hookrightarrow A_\mathfrak{m} \hookrightarrow k[[X_1,\dots,X_n]],$$

which in turn induces morphism between schemes

$$\mathrm{Spec}(k[[X_1,\dots,X_n]]) \hookrightarrow \mathrm{Spec}(A_\mathfrak{m}) \hookrightarrow \mathrm{Spec}(A)$$

Definition 7.4.11. *We call an algebra B as above, for which the ideal $\mathfrak{n}$ consists of nilpotent elements an* **infinitesimal algebra**.

The spectrum of an infinitesimal algebra is a single point. But if we consider $\mathrm{Spec}(B)$-valued points of $\mathrm{Spec}(k[[X_1,\dots,X_n]])$ then we get n-tuples $(\epsilon_1,\epsilon_2,\dots,\epsilon_n) \in \mathfrak{n}^n$. These are considered to be points, which are infinitesimally small in the sense that the coordinates are not zero but if we raise them into a suitably high power they become zero. This leads us to the idea that $\mathrm{Spec}(k[[X_1,\dots,X_n]])$ should be considered as an infinitesimal neighborhood of the origin in $\mathrm{Spec}(k[X_1,\dots,X_n]) = \mathbb{A}^n$.
The scheme $\mathrm{Spec}(k[[X_1,\dots,X_n]])$ is also called the **formal completion** of the scheme $\mathrm{Spec}(k[X_1,\dots,X_n])$ at the origin.

7.5 Smooth Points

Let k be a field and X/k be a scheme of finite type (see 6.2.10). We consider the set of geometric points of X.

$$X(\overline{k}) = \mathrm{Hom}_{\mathrm{Spec}(k)}(X, \mathrm{Spec}(\overline{k}))$$

I want to introduce the notion of a **smooth** point $P \in X(\overline{k})$. To give a first idea of what it should be we give examples of a non smooth or **singular** point .

Example 14. *Consider the two algebras $A = k[x,y]/(x^2 - y^3)$ and $B = k[x,y]/(xy)$. The point $P = (0,0)$ is a geometric point on both of them and on both of them it is singular. If $k = \mathbb{R}$ and we draw a picture of the $\mathbb{R}$-valued points then we get an idea what a singular point is.*

The property of a point to be smooth will be local, this means that by definition a point will be smooth if and only if for all affine open neighborhoods $U \subset X, P \in U(\overline{k})$ the point is smooth on U. Hence we may assume that $X = \mathrm{Spec}(A)$, we could also pass to the germ of X at P.
The notion of a smooth point is also a "geometric" notion: If we embed $k \hookrightarrow \overline{k}$ and form the base extension $\overline{X} = X \times_{\mathrm{Spec}(k)} \overline{k} = \mathrm{Spec}(A \otimes_k \overline{k})$ then P is a geometric point of $\overline{X}$. The definition of a smooth point will be such that P is smooth as a geometric point on X if and only if it is smooth as a geometric point of $\overline{X}$. It will even turn out that the smoothness of the point is a property of the formal completion of our scheme at P (see above).
Our point P is now a homomorphism

$$\begin{array}{ccc} A \otimes_k \overline{k} & \xrightarrow{\quad P \quad} & \overline{k}. \\ & \nwarrow \qquad \nearrow & \\ & k & \end{array}$$

It defines a maximal ideal $\mathfrak{m}_P \subset A \otimes_k \overline{k}$. We have a decomposition of the $\overline{k}$-vector space

$$A \otimes_k \overline{k} = \overline{k} \oplus \mathfrak{m}_P,$$

which is given by $f = f(P) + (f - f(P))$ for $f \in A \otimes_k \overline{k}$.

Definition 7.5.1. *The point $P \in \mathrm{Hom}_k(A,\overline{k})$ is* **smooth** *if and only if the local ring $(A \otimes \overline{k})_{\mathfrak{m}_P}$ is integral and if the dimension of the $\overline{k}$-vector space $\mathfrak{m}_P/\mathfrak{m}_P^2$ is equal to the dimension of the local ring $(A \otimes_k \overline{k})_{\mathfrak{m}_P}$.*

The integrality of the local ring allows us to speak of its dimension (see 7.2.2) .
The maximal ideal $\mathfrak{m}_P$ induces also a maximal ideal $\mathfrak{m}_P^0$ in A itself. Recall (see 7.2.19) that we proved that the local ring $A_{\mathfrak{m}_P^0}$ is also integral, we have an injection $i : A_{\mathfrak{m}_P^0} \to (A \otimes_k \overline{k})_{\mathfrak{m}_P}$ and equality of dimensions

$$\dim(A_{\mathfrak{m}_P^0}) = \dim((A \otimes_k \overline{k})_{\mathfrak{m}_P})).$$

If $\mathfrak{p}^0$ is the unique minimal prime ideal in A, which is contained in $\mathfrak{m}_P^0$, then we have the equality of dimensions $\dim(A/\mathfrak{p}^0) = \dim(A_{\mathfrak{m}_P^0})$ (See 7.2.19). Clearly the closed point $\mathfrak{m}_P^0 \in \mathrm{Specmax}(A)$ is smooth if and only if one (and hence all) of the geometric points lying above this point is (are) smooth.

Now we want to discuss the implications of the smoothness of a point. For the following discussion we assume that $k = \overline{k}$ and hence $A \otimes_k \overline{k} = A$. We pick a point P, where A is locally irreducible.

We pick elements $t_1 \dots t_d \in \mathfrak{m}_P$, which provide a basis $\overline{t}_1 \dots \overline{t}_d$ in the quotient $\mathfrak{m}_P/\mathfrak{m}_P^2$. These elements generate the A-module $\mathfrak{m}_P$ by Nakayama's lemma. We consider the polynomial ring with the same number of generators $B = k[X_1,\dots,X_d]$ and we consider its maximal ideal $\mathfrak{m}_0 = (X_1,\dots,X_d)$. We get a k-algebra homomorphism $B_{\mathfrak{m}_0} \to A_{\mathfrak{m}_P}$ sending X_i to t_i. We compute modulo powers of the maximal ideals, i.e. we look at the rings $B_{\mathfrak{m}_0}/\mathfrak{m}_0^N$, $A_{\mathfrak{m}_P}/\mathfrak{m}_P^N$. Clearly our homomorphism induces maps $B_{\mathfrak{m}_0}/\mathfrak{m}_0^N \to A_{\mathfrak{m}_P}/\mathfrak{m}_P^N$. It is easy to see that these maps are surjective: The t_i generate the $A_{\mathfrak{m}_P}$-module $\mathfrak{m}_P$, hence any $f \in A_{\mathfrak{m}_P}$ can be written as $f = a + \sum g_i t_i$ with $a \in k$. Applying the same to the g_i we get $f = a + \sum b_i t_i + \textit{quadratic terms}$ with $b_i \in k$. Since this goes on forever the assertion becomes clear. We get projective systems

$$\begin{array}{ccccc} B_0/\mathfrak{m}_0^{N+1} & \longrightarrow & B_0/\mathfrak{m}_0^N & \longrightarrow & \cdots \\ \downarrow & & \downarrow & & \\ A_{\mathfrak{m}_P}/\mathfrak{m}_P^{N+1} & \longrightarrow & A_{\mathfrak{m}_P}/\mathfrak{m}_P^N & \longrightarrow & \cdots \end{array}$$

and if we consider the completions $\widehat{B} = \varprojlim B_0/\mathfrak{m}_0^N$ and $\widehat{A}_{\mathfrak{m}_P} = \varprojlim A_{\mathfrak{m}_P}/\mathfrak{m}_P^N$, the ring $\widehat{B}_{\mathfrak{m}_0}$ is isomorphic to the ring of formal power series $k[[X_1,\dots,X_d]]$. Any element in $\widehat{A}_{\mathfrak{m}_P}$ can be written as a power series $\sum_{\underline{\nu}} a_{\underline{\nu}} t^{\underline{\nu}}$, where $a_{\underline{\nu}} \in k$ and $\underline{\nu} = (\nu_1,\nu_2,\dots,\nu_d)$ is a multiindex. This yields a diagram

$$\begin{array}{ccc} B_{\mathfrak{m}_0} & \longrightarrow & A_{\mathfrak{m}_P} \\ \downarrow & & \downarrow \\ k[[X_1,\dots,X_d]] & \xrightarrow{i_P} & \widehat{A}_{\mathfrak{m}_P} \end{array}$$

The vertical maps in the diagram above are always injective by the Artin-Rees theorem. The homomorphism i_P is surjective by the above argument.

Theorem 7.5.2. *Let A/k be an affine algebra over the algebraically closed field k, let P be a geometric point such that $A_{\mathfrak{m}_P}$ is irreducible, i.e. P lies on exactly one irreducible component. Then we always have the inequality*

$$d = \dim_k \mathfrak{m}_P/\mathfrak{m}_P^2 \geq \dim(A_{\mathfrak{m}_P}) = d'.$$

We have equality if and only if the horizontal arrow in the bottom line i_P is an isomorphism. This is the case if and only if P is a smooth point.

This is obvious if we accept that the ring of formal power series is noetherian and if we accept the general dimension theory for noetherian rings. But since we did not explain this we give a different argument.

We prove the inequality. We proceed by induction on d'. Let $\mathfrak{a}$ be the kernel of i_P. For $d' = 0$ the maximal ideal $\mathfrak{m}_P$ is nilpotent and we have $d \geq 0$. Everything is clear. Now let $d' \geq 1$, then $\mathfrak{m}_P \neq \mathrm{Rad}(A_{\mathfrak{m}_P})$. We pick one of the generators, say t_1, which is not in the radical. We choose a minimal prime ideal $\mathfrak{m}_P \supset \mathfrak{p} \supset (t_1)$. The Hauptidealsatz tells us that $\dim((A/\mathfrak{p})_{\mathfrak{m}_P}) = d' - 1$. Now the number of generators of the image of $\mathfrak{m}_P$ in the local ring $(A/\mathfrak{p})_{\mathfrak{m}_P}$ has dropped by a number $r \geq 1$. Hence we get $d - r \geq d' - 1$ and this proves that always $d \geq d'$.
If $d = d'$ then $d - r \geq d - 1$. and we must have $r = 1$. We apply our construction above to $(A/\mathfrak{p})_{\mathfrak{m}_P}$ and get

$$k[[X_2, \ldots, X_d]] \longrightarrow \widehat{(A/\mathfrak{p})_{\mathfrak{m}_P}},$$

which is a bijection by the induction hypothesis. The kernel of

$$k[[X_1, X_2, \ldots, X_d]] \longrightarrow k[[X_2, \ldots, X_d]]$$

is the principal ideal (X_1) and hence $\mathfrak{a} \subset (X_1)$. Assume $\mathfrak{a} \neq (0)$. If $d = 1$ we get that $\mathfrak{a} = (X_1^m)$ for some m, but this is impossible because this implies $t_1^m = 0$ and the dimension of $A_{\mathfrak{m}_P}$ would be zero. If $d > 1$ then we could replace t_1 by any linear combination of the generators t_i and conclude that $\mathfrak{a}$ is contained in any principal ideal (f) where $f = a_1X_1 + a_2X_2 + \ldots a_dX_d +$ *higher order terms* with non zero linear term. But this tells us that any element $h \in \mathfrak{a}$ must be divisible by any such power series f, which is clearly impossible unless $h = 0$. Therefore we have proved that i_P is an isomorphism if $d = d'$, but then the local ring $(A/\mathfrak{p})_{\mathfrak{m}_P}$ must be integral and we proved that $d = d'$ implies that P is smooth.
If in turn i_P is an isomorphism, then we observe that X_1 is sent to $t_1 \in \mathfrak{m}_P$. Hence we see that (t_1) is a prime ideal and $k[[X_2, \ldots, X_d]] \to \widehat{(A/(t_1)_{\mathfrak{m}_P}}$ is an isomorphism. By induction this implies $d' - 1 = d - 1$ hence $d' = d$ and our point must be smooth. □

The theorem means that in an *infinitesimal neighborhood* of a smooth point a scheme X/k over an algebraically closed field k looks like the "infinitesimal neighborhood" of the origin of an affine space of dimension $\dim(X)$.

Now we assume that our k-algebra is given as a quotient of a polynomial ring, in other words we consider $X = \mathrm{Spec}(A)$ as embedded into an affine space. We write

$$0 \longrightarrow I \longrightarrow k[X_1, \ldots, X_n] \longrightarrow A \longrightarrow 0$$

where I is the defining ideal. Again we look at our smooth point $P : A \longrightarrow k$ and for simplicity we assume that $P(X_i) = 0$ for all i, i.e. our point P is the origin. We want to show that the smoothness of P has the consequence that the system of equations, which defines X locally at P as a subscheme has certain nice properties (Jacobi criterion). Let $\mathfrak{m}_0$ be the maximal ideal in A defined by P and $\mathfrak{M}_0 = (X_1, \ldots, X_n)$ the maximal ideal in the polynomial ring. We get an exact sequence (localization is flat)

$$0 \longrightarrow I_{\mathfrak{M}_0} \longrightarrow k[X_1, \ldots, X_n]_{\mathfrak{M}_0} \longrightarrow A_{\mathfrak{m}_0} \longrightarrow 0.$$

Let $d = \dim(A_{\mathfrak{m}_0})$, then we may assume that $X_1, \ldots, X_d$ map to a basis in $\mathfrak{m}_0/\mathfrak{m}_0^2$. We write x_i for the image of X_i in $A_{\mathfrak{m}_0}$. We get a diagram

$$\begin{array}{ccccccccc} 0 & \longrightarrow & I_{\mathfrak{M}_P} & \longrightarrow & k[X_1,\ldots,X_n]_{\mathfrak{M}_0} & \longrightarrow & A_{\mathfrak{m}_0} & \longrightarrow & 0 \\ & & \downarrow & & \downarrow & & \downarrow & & \\ 0 & \longrightarrow & \widehat{I}_{\mathfrak{M}_0} & \longrightarrow & k[[X_1,\ldots,X_n]] & \longrightarrow & k[[x_1,\ldots,x_d]] & \longrightarrow & 0 \end{array}$$

where $\widehat{I}_{\mathfrak{M}_0}$ is the completion of $I_{\mathfrak{M}_0}$ and the exactness on the left in the second sequence is again the Artin-Rees theorem. We also use our previous theorem, which gives us that $\widehat{A}_{\mathfrak{m}_P}$ is indeed the ring of power series in the given d variables.
Now we can find certain specific elements in $I_{\mathfrak{M}_0}$, which turn out to be generators of the ideal. We observe that for any $i = d+1,\ldots,n$ the element $x_i \in \mathfrak{m}_0$ must be up terms of higher order (i.e. elements in $\mathfrak{m}_0^2$) a linear combination of the x_ν with $\nu = 1,\ldots,d$. In other words, for any such i we get a relation in $A_{\mathfrak{m}_0}$:

$$x_i = L_i(x_1,\ldots x_d) + G_i(x_1,\ldots,x_n)$$

where L_i is linear in the $x_1,\ldots,x_d$ and where G_i is an expression, which contains only terms of degree at least one of the variables occurs with degree ≥ 2. If we lift this to the variable $X_1,\ldots,X_n$ this provides elements in $I_{\mathfrak{M}_0}$

$$F_i = X_i - L_1(X_1,\ldots,X_d) - G_i(X_1,\ldots,X_n) \in I_{\mathfrak{M}_0} \quad i = d+1,\ldots,n.$$

We claim:

Proposition 7.5.3. *The elements $F_{d+1},\ldots,F_n$ generate the ideal $I_{\mathfrak{M}_0}$.*

This is rather clear: If we look at the completion then we see that for any $i = d{+}1,\ldots,n$ we can express the image of x_i as power series in the variables $x_1,\ldots,x_d$. (Just substitute the expression for x_i into the G_i and you get an expression for the x_i in terms of a quadratic term in the $x_1,x_2,\ldots,x_d$ and an expression in all x_i where the terms are of degree ≥ 3, and so on.) Hence we can write the x_i for $i = d{+}1,\ldots,n$ as power series in the $x_1,\ldots,x_d$ and we get that the F_i viewed as elements in the ideal $\widehat{I}_{\mathfrak{M}_0}$ can be written as

$$F_i(X_1,\ldots,X_n) = X_i - P_i(X_1,X_2,\ldots,X_d) \quad i = d+1,\ldots,n,$$

where the $P_i(X_1,X_2,\ldots,X_d)$ are power series.
If we divide the power series ring $k[[X_1,\ldots,X_n]]$ by these F_i, then the resulting ring is the power series ring in the variables $X_1,\ldots,X_d$ and this is also the result if we divide by $\widehat{I}_{\mathfrak{M}_0}$. This makes it clear that these F_i generate the ideal $\widehat{I}_{\mathfrak{M}_0}$. Now $I_{\mathfrak{M}_0}/\mathfrak{M}_0 I_{\mathfrak{M}_0} = \widehat{I}_{\mathfrak{M}_0}/\mathfrak{M}_0\widehat{I}_{\mathfrak{M}_0}$ and the Nakayama lemma implies that the ideal $I_{\mathfrak{M}_0}$ is also generated by these F_i.
This has several consequences. It is clear that we can find an affine open neighborhood $U \subset \mathbb{A}_k^n$ of our point P such that the restriction $\widetilde{I}(U)$ of I to U contains the F_i and is generated by them. Hence:

1. In a suitable Zariski neighborhood of a smooth point P we can define a scheme $X/k \hookrightarrow \mathbb{A}_k^n$ by $n - \dim(X)$ equations, namely the above F_i.

2. If we form the Jacobi matrix of the relations F_i at our smooth point P, then the matrix $\left(\frac{\partial F_i}{\partial X_j}(P)\right)_{i=d+1,\ldots,n;j=1,\ldots n}$ has rank $n-d$.

This leads us to the famous Jacobi criterion, which gives a characterization of the smooth points of an embedded affine scheme $X \subset \mathbb{A}^n$.

The Jacobi Criterion

From now on we drop the assumption that k should be algebraically closed. We consider an affine scheme $A = k[X_1,\dots,X_n]/I$. A geometric point $P \in \mathrm{Hom}_k(A,\overline{k})$ defines a maximal ideal $\mathfrak{m}_P$ in A but it can happen that different P give the same maximal ideal. We have also the maximal ideal $\mathfrak{M}_P$ in the polynomial ring, which lies above $\mathfrak{m}_P$. We want to formulate a criterion for the smoothness of P in terms of the data over k, i.e. we do not extend the base to $\overline{k}$. This will imply for instance that all geometric point lying over a given maximal ideal $\mathfrak{m}$ will be smooth if one of them is so (see remark after 7.5.1).

Theorem 7.5.4 (Jacobi criterion). *Let I be an ideal in $k[X_1,X_2,\dots,X_n]$ and $A = k[X_1,X_2,\dots,X_n]/I$. Let P be a geometric point of* Spec(A), *let $\mathfrak{m}_P$ (resp $\mathfrak{M}_P$) be the corresponding maximal ideal in A (resp. $k[X_1,X_2,\dots,X_n]$). Let $F_1,\dots,F_r$ be a system of generators of $I_{\mathfrak{M}_P}$. We assume that the Jacobi matrix*

$$\left(\frac{\partial F_i}{\partial X_j}(P)\right)_{i=1,\dots,r;j=1,\dots n}$$

has rank r. Then P is smooth and $\dim(A_{\mathfrak{m}_P}) = n-r$.
If conversely P is smooth and $\dim(A_{\mathfrak{m}_P}) = n-r$ *then we can find generators $G_1,\dots,G_r$ of the ideal $I_{\mathfrak{M}_P}$ such that the Jacobi matrix built in these generators has rank r.*

Before we give the proof,we want to make a few comments. First of all we remark that the evaluation of an element $F \in k[X_1,\dots,X_n]$ at P has as result an element in $\overline{k}$. Hence we did not keep our promise not to extend the ground field. But our statement concerning the rank is equivalent to the assertion that a determinant of a suitable $r \times r$-matrix is non zero. This matrix has entries $\frac{\partial F_i}{\partial X_j} \in k[X_1,\dots,X_n]$ and hence the determinant is an element $F \in k[X_1,\dots,X_n]$. Now $F(P) \neq 0$ is equivalent to $F \notin \mathfrak{M}_P$, we restated our condition on the Jacobi matrix without extending the ground field. Then it is also clear that all the geometric points lying over a given maximal ideal are all smooth or none of them is.

Since we have a geometric definition of the smoothness of a point we have to extend the ground field to $\overline{k}$. We consider the ideal $I \otimes_k \overline{k} \subset \overline{k}[X_1,X_2,\dots,X_n]$. After making a substitution we may assume that $P = (0,\dots,0)$. Hence it is reasonable to denote the maximal ideals in $A \otimes_k \overline{k}$ resp. the polynomial ring by $\mathfrak{m}_0$ resp. $\mathfrak{M}_0$. We put $d = n-r$. If we renumber our variables we can assume that the partial Jacobi matrix

$$\left(\frac{\partial F_i}{\partial X_j}(P)\right)_{i=1,\dots,r;j=d+1,\dots,n}$$

has rank r. If we expand the polynomials at $P = (0,\dots,0)$ we get expressions

$$F_i(X_1,\dots,X_n) = \sum_{j=1}^{n} \frac{\partial F_i}{\partial X_j}(P)X_j + \textit{higher order terms}$$

for $i = 1,\dots,r$. Let us denote the homomorphism of the polynomial ring to our algebra A by Φ, the images of the X_i under Φ are again called x_i. Then we get in $(A \otimes_k \overline{k})_{\mathfrak{m}_P}$ the relations

$$0 = \sum_{j=1}^{n} \Phi(\frac{\partial F_i}{\partial X_j})x_j + \textit{higher order terms in the } x_i$$

If we ignore higher terms and evaluate at P then we get a system of linear equations for the $x_i \in \mathfrak{m}_0/\mathfrak{m}_0^2$ and our assumption on the partial Jacobi matrix implies that we can express the images x_i for $i = d+1, \dots, n$ as linear combinations of the x_i with $i = 1, \dots, d$. We get that the x_i with $i = 1, \dots, d$ generate $\mathfrak{m}_0/\mathfrak{m}_0^2$.
To prove the smoothness of P it suffices to show that $(A \otimes_k \overline{k})_{\mathfrak{m}_0}$ is integral, it has dimension d and that x_i with $i = 1, \dots, d$ form a basis of $\mathfrak{m}_0/\mathfrak{m}_0^2$.

To prove these facts we embed $(A \otimes_k \overline{k})_{\mathfrak{m}_0}$ into its $\mathfrak{m}_0$-adic completion $\widehat{(A \otimes_k \overline{k})}_{\mathfrak{m}_0}$. In this completion our defining equations just say that the x_i with $i = d+1, \dots, n$ can be expressed as power series in the x_i with $i = 1, \dots, d$ and hence it is clear that $\widehat{(A \otimes_k \overline{k})}_{\mathfrak{m}_0}$ is the power series ring in the variables x_i with $i = 1, \dots, d$. This implies integrality of $(A \otimes_k \overline{k})_{\mathfrak{m}_0}$ and it shows that $\dim(\mathfrak{m}_0/\mathfrak{m}_0^2) = d$.
As in our previous considerations we can construct an injective homomorphism from the localized polynomial ring

$$\overline{k}[X_1, \dots, X_d]_{\mathfrak{M}_0^0} \longrightarrow (A \otimes_k \overline{k})_{\mathfrak{m}_0}.$$

Since both sides are localizations of an affine k-algebra in closed points we can apply proposition 7.4.6 and get $\dim((A \otimes_k \overline{k})_{\mathfrak{m}_0}) \geq d$. Now it follows from our theorem 7.5.2 that $\dim(A_{\mathfrak{m}_0}) = d$. This proves the first half of the theorem.
To prove the converse we observe that we have seen that the ideal $I_{\mathfrak{M}_P} \otimes \overline{k}$ can be generated by $G_1, \dots, G_r \in \overline{k}[X_1, \dots, X_n]$ such that these have non vanishing Jacobi matrix at P. Let $F_1, \dots, F_t \in I$ be a system of generators. Then we can find an $H \in \overline{k}[X_1, \dots, X_n]$ with $H(P) \neq 0$ such that we can write

$$G_i H = \sum_j L_{ij} F_j$$

where $L_{ij} \in \overline{k}[X_1, \dots, X_n]$. Taking partial derivatives and evaluating at P yields a system of equations

$$H(P)\frac{\partial G_i}{\partial X_\nu}(P) = \sum_j L_{ij}(P)\frac{\partial F_j}{\partial X_\nu}(P).$$

This makes it clear that the Jacobi matrix built out of the F_j evaluated at P has rank r. But this means that the Jacobi matrix $\left(\frac{\partial F_j}{\partial X_\nu}\right)$ has rank r if we send it to the residue field $A/\mathfrak{m}_P$. Taking a sub matrix having the right number of columns gives us a subset of the set of generators say $F_1, \dots, F_r$, which satisfies the Jacobi criterion. This subset will necessarily generate $I_{\mathfrak{M}_P}$. □

The Jacobi criterion implies that the set of smooth geometric points of a scheme X/k of finite type is always open. To see this we consider the case of an affine k-algebra $A = k[X_1, \dots, X_n]/I$. We may assume that it is irreducible because smooth points lie on exactly one irreducible component. We pick generators $F_1, F_2, \dots, F_t$ of our ideal and consider the open set where the Jacobi matrix has rank equal to $n - \dim(A)$. This set can be empty. If for instance $A = k[X]/(X^2)$, then the only geometric point is $x = 0$, the dimension is zero, but the number d is one.

7.5.1 Generic Smoothness

Theorem 7.5.5 (Generic Smoothness). *Let A/k be a finitely generated k-algebra, assume that A/k is absolutely reduced. Then we can find a non empty open subset $U \subset \operatorname{Spec}(A)$ such that the morphism $\pi : U \to \operatorname{Spec}(k)$ is smooth.*

Proof: Since localization does not destroy reducedness we may assume that A is integral. We also may assume that $\overline{k} = k$. We write $A = k[x_1,x_2,\ldots,x_n] = k[X_1,X_2,\ldots X_n]/I$ and let $F_1,F_2,\ldots F_t$ be a set of generators of the ideal I. Let $\frac{\partial F_i}{\partial x_j}$ be the image of $\frac{\partial F_i}{\partial X_j} \in k[X_1,X_2,\ldots,X_n]$ modulo the ideal I. Let r be the rank of the Jacobi matrix $\left(\frac{\partial F_i}{\partial x_j}\right)$ considered as a matrix in the field of fractions of A and assume that

$$\left(\frac{\partial F_i}{\partial x_j}\right)_{i=1,\ldots,r;j=n-r+1,\ldots n}$$

has non zero determinant. Then this determinant is a unit in a suitable localization $\mathcal{O}_X(U) = \widetilde{A}(U)$. Let $d = n - r$. I claim that the sub algebra $k[x_1,\ldots,x_d]$ is a polynomial ring in d variables. If not, then the intersection $k[X_1,\ldots,X_d] \cap I \neq 0$. For all F in $k[X_1,\ldots,X_d] \cap I$ we have $\frac{\partial F}{\partial x_j} = 0$ for all $j = 1,\ldots,d$ because otherwise the rank of the Jacobian would be greater than r. This is equivalent to the assertion that for all $F \in k[X_1,\ldots,X_d] \cap I$ we have $\frac{\partial F}{\partial X_j} \in I$ for all $j = 1,\ldots,d$. Let us consider a non zero element $F \in k[X_1,\ldots,X_d] \cap I$ with lowest total degree, i.e. the sum of the exponents in the highest monomial occuring in F is minimal. Then $\frac{\partial F}{\partial X_j} \in I$ for all $j = 1,\ldots d$ and these polynomials have a strictly smaller total degree. Hence they are zero and F must be constant or of the form

$$F(X_1,\ldots,X_d) = \sum a_\nu X^{p\nu}$$

where p is the characteristic of k and $\nu = (\nu_1,\ldots,\nu_d)$ is a multiindex. Of course F cannot be constant. But then we find nilpotent elements in $\overline{k}[x_1,\ldots,x_d]$ because we can write

$$F(X_1,\ldots,X_d) = \sum a_\nu X^{p\nu} = \left(\sum a_\nu^{1/p} X^\nu\right)^p.$$

The element $(\sum a_\nu^{1/p} X^\nu)^p \notin k[X_1,\ldots,X_d] \cap I$ because it has smaller total degree than F. Its image in A is a non zero nilpotent element. This gives a contradiction to our assumption that A is absolutely reduced.
But once we know that $k[x_1,\ldots,x_d]$ is a polynomial ring we see that the dimension of A is greater or equal to d (2.5.3.) and the Jacobi criterion yields that the open subscheme where the above determinant is not zero must be smooth. □

The singular locus

Actually the proof gives us a little bit more: The algebra A/k is smooth in all those points where at least one of the $r \times r$-minors the $t \times n$- matrix $\left(\frac{\partial F_i}{\partial x_j}\right)_{i=1\ldots t,j=1\ldots n}$ is non zero. Or in other words: **The singular locus**, i.e. the set of non smooth points is a closed subscheme and it is the set of common zeroes of all the minors above.

We want to discuss a consequence of the smoothness of a geometric point for the original algebra A and the induced maximal ideal $\mathfrak{m}_P$. Let us assume we have an ideal $I \subset k[X_1,\dots,X_n]$, which is generated by $F_1,\dots,F_r$. Let X/k be the subscheme defined by the ideal, i.e. $X = \mathrm{Spec}(k[X_1,\dots,X_n]/I) = \mathrm{Spec}(A)$. We assume that these generators satisfy the Jacobi criterion at a point $P \in X(\overline{k})$, hence we know that this point is smooth. Let $\mathfrak{m}_P$ be the maximal ideal induced by P in our algebra A, we know that $A_{\mathfrak{m}_P}$ is integral. Again we put $d = n - r$. Now we assume for the coordinates $X_{d+1},\dots,X_n$ that

$$\det\left(\left(\frac{\partial F_j}{\partial X_\mu}(P)\right)_{j=1\dots,r,\mu=d+1,\dots,n}\right) \neq 0.$$

Let us denote by x_ν the image of the X_ν in A. Then we consider the sub algebra $B \subset A$ generated by the elements $x_1,\dots,x_d$. It will become clear in a minute that this algebra is indeed the polynomial algebra in these variables. We have the diagram

$$\begin{array}{ccc} k[x_1,\dots,x_d] & \longrightarrow & A \\ \downarrow & & \downarrow \\ \overline{k}[x_1,\dots,x_d] & \longrightarrow & A \otimes_k \overline{k}. \end{array}$$

The maximal ideal $\mathfrak{m}_P$ induces a maximal ideal (geometric point) $\mathfrak{n}_P$ in $k[x_1,\dots,x_d]$. Let $\overline{\mathfrak{n}_P}$ (resp. $\overline{\mathfrak{m}_P}$) be the maximal ideals induced by P in $\overline{k}[x_1,\dots,x_d]$ (resp. $A \otimes_k \overline{k}$). We have seen in the beginning of this section that after passing to the completions we get an isomorphism

$$\widehat{\overline{k}[x_1,\dots,x_d]}_{\overline{\mathfrak{n}_P}} \xrightarrow{\sim} (\widehat{A \otimes_k \overline{k}})_{\overline{\mathfrak{m}_P}}$$

and it follows from this that the sub algebra $B \subset A$ is the polynomial algebra.
We have the inclusion of local rings

$$B_{\mathfrak{n}_P} \hookrightarrow A_{\mathfrak{m}_P}.$$

We consider the residue field $B/\mathfrak{n}_P = k(\mathfrak{n}_P)$ and we claim

Lemma 7.5.6. *1. The tensor product*

$$A_{\mathfrak{m}_P} \otimes_B k(\mathfrak{n}_P) = A_{\mathfrak{m}_P}/A_{\mathfrak{m}_P} \cdot \mathfrak{n}_P$$

is a field, it is a finite separable extension of $k(\mathfrak{n}_P)$ *.*

2. *We have* $A_{\mathfrak{m}_P} \cdot \mathfrak{n}_P = \mathfrak{m}_P$ *and the residue field is* $k(\mathfrak{m}_P) = A_{\mathfrak{m}_P} \otimes_B k(\mathfrak{n}_P)$*.*

3. *If* A *is integral then the function field* $\mathrm{Quot}(A)$ *is a finite separable extension of* $L = k(x_1,\dots,x_d)$

Proof: The second assertion follows from the first. To prove the first assertion we put $K = k(\mathfrak{n}_P)$ and $\overline{A} = A_{\mathfrak{m}_P} \otimes_B k(\mathfrak{n}_P)$. We have that $\overline{A} = K[X_{d+1},\dots,X_n]/(\overline{F}_1,\dots,\overline{F}_r)$ where $\overline{F}_i$ is the image of F_i in $\overline{A}$. Our point P induces a maximal ideal $\overline{\mathfrak{m}_P} \subset \overline{A}$. Of course we have that the generators $\overline{F}_i, i = 1,\dots,r$, satisfy the Jacobi criterion at P and this means

$$\det\left(\left(\frac{\partial \overline{F}_j}{\partial \overline{X}_\mu}(P)\right)_{j=1\dots,r,\mu=d+1,\dots,n}\right) \neq 0.$$

Then it follows that all points P_i, which lie above P are smooth and the scheme $\overline{A} \otimes_K \overline{K}$ is of dimension zero and reduced. But then it follows from proposition 7.3.2 that $\overline{A}$ is a separable extension of K.
The third assertion follows from the same argument as the first, we simply replace K by L.

□

In our situation above we will call the elements $x_1, \dots, x_d$ a system of **local parameters** at the point P. We can interpret the inclusion $B \subset A$ as the morphism from X/k to $\mathbb{A}^r/k$, which is induced by the projection of the ambient affine space $\mathbb{A}^n/k$ to the first r coordinates. If we extend the ground field to $\overline{k}$ then this projection becomes an isomorphism between the infinitesimal neighborhoods of P in $X \times \overline{k}$ and its image in $\mathbb{A}^r/\overline{k}$. This is the algebraic version of the theorem of implicit functions. If we want to formulate what happens over k itself then this becomes a little bit less intuitive, we have to be aware that we may have non trivial residue fields.
Our notion of a system of local parameters is given in terms of an embedding of our scheme X/k into an affine space and with the help of the Jacobi criterion. In the next section we will give a much more elegant formulation using differentials. (See def. 7.5.13)

7.5.2 Relative Differentials

For any morphism

$$\pi : X \longrightarrow Y$$

we will define a quasi-coherent sheaf $\Omega^1_{X/Y}$ on X, which is called the **sheaf of relative differentials**. Since it is a quasi-coherent sheaf it is enough to define it in the case where $X = \mathrm{Spec}(B), Y = \mathrm{Spec}(A)$. Let $\phi : A \longrightarrow B$ be the homomorphism of rings corresponding to π.

We consider the fibered product of X/Y by itself

$$\begin{array}{ccc} B \otimes_A B & \qquad & X \times_Y X \\ \uparrow & & \downarrow \\ A & & Y \end{array}$$

and we have the diagonal $X \to X \times_Y X$ defined by $(\mathrm{Id}, \mathrm{Id})$. The diagonal corresponds to the multiplication

$$B \otimes_A B \xrightarrow{m} B$$

and is defined by the ideal I, which is the kernel of m, i.e. we have an exact sequence

$$0 \longrightarrow I \longrightarrow B \otimes_A B \longrightarrow B \longrightarrow 0.$$

Lemma 7.5.7. *The ideal I is generated as a $B \otimes_A B$-module by the elements $f \otimes 1 - 1 \otimes f$. The ideal I/I^2 is a B-module, which is generated by the elements $df = f \otimes 1 - 1 \otimes f$ mod I^2.*

Proof: If $\Sigma f_\nu \otimes g_\nu \in I$, i.e. $\Sigma f_\nu g_\nu = 0$, then

$$\Sigma f_\nu \otimes g_\nu = \Sigma(f_\nu \otimes 1)(1 \otimes g_\nu) = \Sigma(f_\nu \otimes 1 - 1 \otimes f_\nu)(1 \otimes g_\nu).$$

The B-module structure on I/I^2 is induced by the $B \otimes_A B$-module structure of I. □

Definition 7.5.8. *The B-module I/I^2 is denoted by $\Omega^1_{B/A}$ and is called the* **module of relative differentials**.

Proposition 7.5.9. *We have the product rule*

$$dfg = gdf + fdg$$

and $da = 0$ for $a \in A$.

Proof: This follows from

$$\begin{aligned} fg \otimes 1 - 1 \otimes gf &= fg \otimes 1 - f \otimes g + f \otimes g - 1 \otimes fg \\ &= (f \otimes 1)(g \otimes 1 - 1 \otimes g) + (1 \otimes g)(f \otimes 1 - 1 \otimes f) \end{aligned}$$

and $a \otimes 1 = 1 \otimes a$. □

We collect some facts where proofs can be found in the book by Matsumura [Ma], but since they are all not difficult to prove, the reader could try to find them her(-him)self.

Proposition 7.5.10. a) *The B-module $\Omega^1_{B/A}$ is a universal module of A-differentials: Whenever we have a B-module M and an A-linear map $d_1 : B \to M$ such that $d_1(fg) = fd_1g + gd_1f$ then we get a commutative diagram*

$$\begin{array}{ccc} \Omega^1_{B/A} & \xrightarrow{\ \varphi\ } & M \\ & {\scriptstyle d}\nwarrow \quad \nearrow{\scriptstyle d_1} & \\ & B & \end{array}$$

where φ is unique and B-linear. It is clear that $\varphi(f \otimes 1 - 1 \otimes f) = d_1(f)$ but it is not so clear why it is well defined. This can be checked by using **d)** *below and reducing to the case of a polynomial ring.*

a1) *If we have a homomorphism $f : B \longrightarrow C$, then C is an A-algebra and we can take $M = \Omega^1_{C/A}$, this is a B-module and we put $d_1(b) = d(f(b))$. Therefore we get a unique $\phi(f) : \Omega^1_{B/A} \longrightarrow \Omega^1_{C/A}$, which is the same as a C-module homomorphism $\phi(f) : \Omega^1_{B/A} \otimes_B C \longrightarrow \Omega^1_{C/A}$.*

b) *We start from our algebra homomorphism $A \to B$ and we pick an element $f \in B$. Then we have $A \to B \to B_f$ and hence we know what $\Omega^1_{B_f/A}$ is. It is easy to see that we have a canonical isomorphism $\Omega^1_{B_f/A} \xrightarrow{\sim} (\Omega^1_{B/A})_f$, which in principle comes from the rule*

$$d\frac{1}{f} = -\frac{df}{f^2}$$

Therefore we can view $\Omega^1_{B/A}$ *as a quasi-coherent sheaf on* $\mathrm{Spec}(B)$ *and then we can define* $\Omega^1_{X/Y}$ *for an arbitrary morphism of schemes* $\pi : X \to Y$.

b1) *If we have a commutative diagram of schemes*

$$\begin{array}{ccc} X & \xrightarrow{\quad f \quad} & Z \\ & \searrow \quad \swarrow & \\ & Y & \end{array}$$

then f *indices a canonical homomorphism*

$$\Delta_f : f^*(\Omega^1_{Z/Y}) \longrightarrow \Omega^1_{X/Y}.$$

c) *The module of relative differentials is compatible with base change. If we have a diagram*

$$\begin{array}{ccc} B & \longrightarrow & B' = B \otimes_A A \\ \uparrow & & \uparrow \\ A & \longrightarrow & A' \end{array}$$

then we have a canonical isomorphism $\Omega^1_{B/A} \otimes_B B' \xrightarrow{\sim} \Omega^1_{B'/A}$.

d) *Finally we consider the situation that our algebra* B *is a quotient of an* A*-algebra* C *by an ideal* J*, i.e. we have*

$$\begin{array}{ccccccccc} 0 & \longrightarrow & J & \longrightarrow & C & \xrightarrow{\pi} & B & \longrightarrow & 0. \\ & & & & \uparrow & \nearrow & & & \\ & & & & A & & & & \end{array}$$

It is clear that we have a surjective C*-module homomorphism* $\Omega^1_{C/A} \to \Omega^1_{B/A}$, *which sends* $g\,dh \in \Omega^1_{C/A}$ *to* $\pi(g)d\pi(h)$. *On the other hand we can map* J *to* $\Omega^1_{C/A}$ *by sending an* $f \in Y$ *to* df. *This yields a sequence of* B*-modules*

$$J/J^2 \longrightarrow \Omega^1_{C/A} \otimes_C B \longrightarrow \Omega^1_{B/A} \longrightarrow 0$$

this sequence is exact.

We prove the last assertion. We have to show that the elements in $\Omega^1_{C/A} \otimes_C B$, which go to zero come from J/J^2. To see this we start with the observation that $\Omega^1_{C/A} \otimes_C B = \Omega^1_{C/A}/J\Omega^1_{C/A}$. If we look at $C \otimes_A C \longrightarrow B \otimes_A B$ then we have seen that the kernel is the ideal $J \otimes_A C + C \otimes_A J$ (See 26). On the other hand we have the kernels of the multiplication maps

$$I_C = \ker(C \otimes_A C \longrightarrow C), I_B = \ker(B \otimes_A B \longrightarrow B).$$

We consider elements in I_C, which go to zero in $I_B/I_B^2 = \Omega^1_{B/A}$. But it is clear that $I_C^2 \longrightarrow I_B^2$ is surjective, hence we can represent an element in $\Omega^1_{C/A}$, which goes to zero in $\Omega^1_{B/A}$ by an element $h \in I_C \cap (J \otimes_A C + C \otimes_A J)$. Let us write this element

$$h := \sum j_\nu \otimes c_\nu + \sum c'_\mu \otimes j'_\mu$$

where the $j_\nu, j'_\mu \in J$. We have $\sum j_\nu \otimes c_\nu + \sum j'_\mu \otimes c'_\mu = 0$. The sum is equal to

$$\begin{aligned}
\sum j_\nu \otimes c_\nu + \sum j'_\mu \otimes c'_\mu &= \sum j_\nu \otimes c_\nu + \sum j'_\mu \otimes c'_\mu + \sum (c'_\mu \otimes j'_\mu - j'_\mu \otimes c'_\mu) \\
&= \sum_\nu (1 \otimes c_\nu)(j_\nu \otimes 1 - 1 \otimes j_\nu) \\
&\quad + \sum_\mu (1 \otimes c'_\mu)(j'_\mu \otimes 1 - 1 \otimes j'_\mu) \\
&\quad + \sum_\mu (1 \otimes j'_\mu)(c'_\mu \otimes 1 - 1 \otimes c'_\mu) \\
&\quad - \sum_\mu (1 \otimes c'_\mu)(j'_\mu \otimes 1 - 1 \otimes j'_\mu).
\end{aligned}$$

This yields an element in $\Omega^1_{C/A}$ namely

$$\sum_\nu c_\nu dj_v + \sum_\mu c'_\mu dj'_\mu + \sum j'_\mu dc'_\mu - \sum c'_\mu dj'_\mu,$$

which is clearly in the image of $J \mod J\Omega^1_{C/A}$

7.5.3 Examples

Example 15. *If $B = A[X_1 \dots X_n]$ then*

$$B \otimes_A B = A[X_1 \dots X_n, Y_1, \dots Y_n]$$

where $X_i = X_i \otimes 1$ and $Y_i = 1 \otimes X_i$. The ideal I is generated by the elements

$$X_i - Y_i = X_i \otimes 1 - 1 \otimes X_i$$

as a $B \otimes_A B$-module and I/I^2 is the free B-module generated by

$$dX_i = X_i - Y_i \mod I^2.$$

Example 16. *Assume that*

$$B = A[X_1,\dots,X_n]/\langle F_1,\dots,F_r\rangle$$

where $F_i = \sum_\nu a_{i\nu}X^\nu \in A[X_1,\dots,X_n]$. *Assume we have an* $f \in B$ *such that for all maximal ideals* $\mathfrak{m} \in \operatorname{Spec}(B_f)$ *the rank of*

$$\left(\frac{\partial F_i}{\partial X_j}\right)_{i=1,\dots,r,j=n-r+1,\dots,n} \mod \mathfrak{m}$$

is maximal, i.e. equal to r. *Then* $\Omega^1_{B_f/A}$ *is locally free of rank* $d = n-r$. *If at our maximal ideal* $\mathfrak{m}$ *the subdeterminant*

$$\det\left(\frac{\partial F_i}{\partial X_j}\right)_{i=1,\dots,r,j=n+1-r,\dots,n} \not\equiv 0 \mod \mathfrak{m}$$

then the differentials $dx_1,\dots,dx_d$ *(where the* x_i *are the images of the* X_i *in* B*) are free generators of* $\Omega^1_{B/A,\mathfrak{m}}$.

This is clear from our consideration above.

Example 17. *We look at our examples of non-smooth points and we will see how the fact that they are not smooth is reflected by the sheaf of relative differentials.*

(a) *If* $A = k[X,Y]/(XY) = k[x,y]$ *then the set of geometric points is the union of the* X *and the* Y*-axis in* $\mathbb{A}^2$. *The origin is not smooth. The* A*-module of differentials is generated by* dx *and* dy *and we have the relation*

$$x\,dy + y\,dx = 0.$$

In this example the origin is not smooth because already the condition that the local ring should be integral is violated. But we could easily modify the example by looking at

$$A_1 = k[X,Y]/(XY + X^5 + Y^7) = k[x,y].$$

Now the local ring at (O,O) *is integral and we have*

$$(y + 5x^4)dx + (x + 7y^6)dy = 0.$$

The dimension of A_1 *is one but the module of differentials is not free of rank one.*

This example is interesting for a different reason. If we look at the completions of our two rings then they become isomorphic

$$\widehat{A} \simeq \widehat{A}_1.$$

We leave this as an exercise to the reader, one has to show that one can construct power series

$$\widetilde{X} = X + P(X,Y)$$
$$\widetilde{Y} = Y + Q(X,Y)$$

such that the relation $XY + X^5 + Y^7 = 0$ *becomes* $\widetilde{X}\widetilde{Y} = 0$.

The reader should also prove that for $k = \mathbb{C}$ *and a small number* $\epsilon > 0$ *the complex space*

$$\{(x,y) \in \mathbb{C}^2 | xy + x^5 + y^7 = 0\} \bigcap B(0,\epsilon)$$

where $B(0,\epsilon) = \{(x,y) \in \mathbb{C}^2 ||x|^2 + |y|^2 < \epsilon\}$ *is indeed the union of two discs intersecting transversally at the origin.*

Hence we see that depending on the microscope, which we use to look at our singular point, the local ring may be integral (Zariski topology) or non-integral (analytic topology). The two branches of our complex space above in $B(0,\epsilon)$ *will come together very far out again.*

Such a singular point is called an **ordinary double point**.

(a1) *Let us consider an affine scheme* $T \longrightarrow \mathrm{Spec}(k)$, *which is reducible and whose irreducible components are two affine lines* $T_i = \mathrm{Spec}(k[X_i]), i = 1,2$. *Let us assume that their intersection is the origin, i.e. the point* $t_1 = t_2 = 0$. *Then we say that* T_1,T_2 *meet transversally if the completion of the local ring at the intersection point is isomorphic to* $k[[X_1,X_2]]/(X_1X_2)$.

(b) $A = k[X,Y]/(X^2 - Y^3)$.

Again we see that $\Omega^1_{A/k}$ *at the origin is generated by* dx *and* dy *with the relation* $2x\,dx - 3y^2\,dy = 0$, *the module of relative differentials is not free of rank 1.*

In our three examples the scheme is smooth outside the origin and this is also the open set where the sheaf of differentials is free of rank one. We will see that this is indeed a very general fact, which will allow us to define smooth morphisms in a very general context.
If we are in the situation of example 16 and if we consider a point $\mathrm{Spec}(k(y)) \hookrightarrow \mathrm{Spec}(A)$, i.e. $\Psi : A \to k(y)$, then we can consider the base change

$$\begin{array}{ccc} B & \longrightarrow & B \otimes_A k(y) \\ \uparrow & & \uparrow \\ A & \longrightarrow & k(y). \end{array}$$

This scheme is called the fibre over the point y, this is now a scheme over a field. It is clear that this scheme

$$\mathrm{Spec}(B \otimes_A k(y)) \longrightarrow \mathrm{Spec}(k(y))$$

is smooth of dimension d. Moreover it also follows that if we have given a system of local parameters at a closed point as above – say $x_1, \ldots ,x_d$ – then the differentials $dx_1, \ldots ,dx_d$ form a basis of $\Omega^1_{B\otimes_A k(y))/\mathrm{Spec}(k(y))}$ at this point.

Now we have enough insight to formulate our general definition.

Let us assume we have an arbitrary ring A and a finitely generated A-algebra

$$B = A[x_1, \ldots, x_n] = A[X_1, \ldots, X_n]/I.$$

Definition 7.5.11. *The morphism*

$$\begin{array}{c} \mathrm{Spec}(B) = X \\ \big\downarrow \pi \\ \mathrm{Spec}(A) = S \end{array}$$

is called **smooth** *if*

a) *it is flat*

b) *For all points $s \in S$ the fibre $X_s \longrightarrow \mathrm{Spec}(k(s))$ is smooth.*

Theorem 7.5.12. *The condition b) is equivalent to*

b1) *The fibers X_s are locally integral at each point and at any point $x \in X_s$ the sheaf $\Omega^1_{X/S}$ is locally free of rank $\dim(\mathcal{O}_{X_s,x})$.*

Before we come to the proof we want to make two comments.

1) We do not require that the fibers are integral since we never did that before. Hence it can happen that the fibers are disjoint unions of irreducible ones, which then may even have different dimensions.

2) Again it is clear that the notion of smoothness can be checked locally at the points. Hence it is clear that our definition extends to an arbitrary morphism $\pi : X \to S$ where X is of finite type over S.

To prove the theorem we start with the case that $S = \mathrm{Spec}(k)$ where k is a field. In this case the assumption a) is automatically fulfilled. We have seen in the discussion of example 16 that **b)** $\Rightarrow$ **b1)**.
If **b1)** is satisfied we write our algebra $B = k[x_1, \ldots, x_n]$ as a quotient

$$k[X_1, \ldots, X_n]/I \xrightarrow{\sim} B.$$

Let us assume that we have a geometric point P, which induces the maximal ideals m_p (resp. $\mathfrak{M}_p$) in B (resp. $C = k[X_1, \ldots, X_n]$). The module $\Omega^1_{B/k}$ is locally free at P and it is of course generated by $dx_1, \ldots, dx_n$. We can choose a basis locally at P let us assume it is of the form $dx_1, \ldots, dx_d$ where $d = \dim B_{m_p}$. As usual we put $r = n - d$. It is clear (see **(d)** above) that we can find $F_1, \ldots, F_r \in I$ such that $dF_1, \ldots, dF_r$ together with $dX_1, \ldots, dX_r$ provide a basis of $\Omega^1_{C/k}$ at P. But this implies that

$$\mathrm{Rank}\left(\frac{\partial F_i}{\partial X_\mu}\right) \mod \mathfrak{M}_p = r.$$

These $F_1, \ldots, F_r$ generate an ideal J and we get

$$\begin{array}{ccc} C_{\mathfrak{M}_p} = k[X_1,\dots,X_n]_{\mathfrak{M}_p} & \longrightarrow & B_{m_p} \\ & \searrow \quad \nearrow_{\Psi} & \\ & \widetilde{B}_{\widetilde{m}_p} & \end{array}$$

where $\widetilde{B}_{\widetilde{m}_p}$ is the quotient by the ideal J. The Jacobi criterion and $\dim \widetilde{B}_{\widetilde{m}_p} = r$ imply that $\widetilde{B}_{\widetilde{m}_p}$ is smooth at P . But then it is clear that Ψ must be an isomorphism and it follows that B is smooth at P.

Now we pass to the general case. From what we know it is clear that **b1)** $\Rightarrow$ **b)** because **b)** can be checked fibre by fibre. Now we assume **b)**, we pick a point $s \in S$ and a closed point $P \in X_s$. We want to show that $\Omega^1_{B/A}$ is locally free and its rank is $\dim(\mathcal{O}_{X_s,P})$. We may assume that $S = \mathrm{Spec}(A)$ is local and s is closed. Then we can write

$$A[X_1,\dots,X_n]/I = B = A[x_1,\dots,x_n].$$

If we pass to the special fibre we get an exact sequence

$$0 \longrightarrow I_0 \longrightarrow k(s)[X_1,\dots,X_n] \longrightarrow k(s)[x_1,\dots,x_n] = B \otimes_A k(y).$$

Now B is a flat A-algebra (we localized in between but this preserves flatness) and hence we get

$$I \otimes k(s) \xrightarrow{\sim} I_0.$$

Now we know that we can find $F_1^{(0)},\dots,F_r^{(0)} \in I_0$, which generate I_0, and which satisfy the Jacobi criterion at P. These elements can be lifted to elements $F_1,\dots,F_r$ in I and by Nakayama's lemma they generate I locally at P. Hence we see that locally at P we have

$$B \simeq A[X_1,\dots,X_n]/(F_1,\dots,F_r)$$

and we have seen in example 15 that this implies that $\Omega^1_{B/A}$ is locally free of rank d at P. □

We are now able to give a very intrinsic definition of a system of local parameters.

Definition 7.5.13. *If we have a smooth morphism $A \to B$, a point P and $f \in B$ with $f(P) \neq 0$ then we say that $x_1,\dots,x_d \in B_f$ is a* **system of local parameters** *at P, if the differentials $dx_1,\dots,dx_d$ form a basis of $\Omega^1_{B_f/A}$ in some neighborhood of P.*

We get a diagram

$$\begin{array}{ccc} A[X_1,\dots,X_d] & \longrightarrow & B_f \\ & \nwarrow \quad \nearrow & \\ & A & \end{array}$$

where we send the X_i to the x_i. We have seen in our previous proof that we can write

$$A[X_1,\dots,X_d,X_{d+1},\dots,X_n]/I = B_f \tag{7.13}$$

where $I = (F_1,\dots,F_r), r = 1,\dots,n-d$ and where

$$\det\left(\frac{\partial F_i}{\partial X_\mu}\right)(P)_{i=1,r,\mu=d+1,\dots,n} \neq 0. \tag{7.14}$$

If we localize further then we can choose our f so that this is true at all points in $\mathrm{Spec}(B_f)$. Then

$$i : A[X_1,\dots,X_r] \hookrightarrow B_f$$

is flat, smooth and $\Omega^1_{B_f/A[X_1,\dots,X_r]}$ is zero. This implies that the fibers are reduced zero dimensional. Then this inclusion i is an example of an étale morphism:

Definition 7.5.14. *A morphism $A \to B$ is called* **étale** *if it is smooth, of finite type and if the sheaf of relative differentials is zero.*

If we have a closed point $\mathfrak{n} \in \mathrm{Spec}(A)$ then $k(\mathfrak{n}) \longrightarrow B \otimes_A k(\mathfrak{n})$ is still smooth and the connected components are zero dimensional. Hence it follows that $k(\mathfrak{n}) \to B \otimes_B k(\mathfrak{n})$ is a finite separable algebra, i.e a finite sum of separable field extensions of $k(\mathfrak{n})$. Let us assume in addition that $A \to B$ is finite (or finite if we restrict to a neighborhood $\mathrm{Spec}(A_f)$ of $\mathfrak{n}$) . For our point $\mathfrak{n} \in \mathrm{Spec}(A)$ we can find an element $\bar{\theta} \in B \otimes_A k(\mathfrak{n})$ such that

$$B \otimes_A k(\mathfrak{n}) = A[\bar{\theta}] = A[X]/(\bar{F}(X)),$$

where $\bar{F}(X) = X^d + \bar{a}_1 X^{d-1} + \cdots + \bar{a}_d$ is a polynomial in $k(\mathfrak{n}))[X]$ of degree $d = \dim_{k(\mathfrak{n})}(B \otimes_A k(\mathfrak{n})$. (See [Ja-Sch], Chap. V , Satz 5.11). We know that $\bar{F}(X)$ and its derivative $\bar{F}'(X)$ are coprime. We can lift $\bar{\theta}$ to an element $\theta \in B$ and in a a suitable neighborhood $\mathrm{Spec}(A_f)$ of $\mathfrak{n}$ the elements $1,\theta,\dots,\theta^{d-1}$ will form a basis of B_f over A_f and we get an equation

$$\theta^d + a_1\theta^{d-1} + \cdots + a_d = 0.$$

Hence we see

Proposition 7.5.15. *A finite étale morphism $A \longrightarrow B$ is locally in the base of the form*

$$A_f \longrightarrow A_f[X]/(F(X)), F(X) = X^d + a_1X^{d-1} + \cdots + a_d \in A_f[X]$$

where F,F' are coprime.

We want to retain

Proposition 7.5.16. *1) If $j : A \longrightarrow B$ is smooth and if $x_1,\dots,x_d$ are local parameters on an open subset* $\mathrm{Spec}(B_f)$ *then j factorizes*

$$A \xrightarrow{i} A[X_1,\dots,X_d] \xrightarrow{h} B_f$$

where h is étale and $X_i \mapsto x_i$.
2) The diagonal $d : \Delta \longrightarrow \mathrm{Spec}(B) \times_{\mathrm{Spec}(A)} \mathrm{Spec}(B)$ is locally at a point P defined by the ideal, which is generated by $(x_1 \otimes 1 - 1 \otimes x_1,\dots,x_d \otimes 1 - 1 \otimes x_d)$ where $x_1,x_2,\dots,x_d$ are local parameters at P.

3) Let $Y = \mathrm{Spec}(C) \longrightarrow \mathrm{Spec}(A)$ *be a scheme and let* $f : Y \longrightarrow X = \mathrm{Spec}(B))$ *a* $\mathrm{Spec}(A)$*-morphism let* $h : B \longrightarrow C$ *be the corresponding homomorphisms of A-algebras. Let* $(Q,f(Q)) = P$ *a point of the graph of f. Let* $x_1,x_2,\dots,x_d \in B$ *be local parameters at* $f(Q)$*. Then the ideal* $\mathcal{I}_{\Gamma_f} \subset C \otimes_A B$ *defining the graph* $\Gamma_f \subset Y \times_{\mathrm{Spec}(A)} \mathrm{Spec}(B)$ *is locally at P generated by* $h(x_1) \otimes 1 - 1 \otimes x_1, \dots, h(x_d) \otimes 1 - 1 \otimes x_d$*.*
4) We apply this to the case where $f : Y \longrightarrow X = \mathrm{Spec}(B)$ *is an inclusion and therefore,* $C = B/\mathcal{I}$*. We assume that Y is also smooth over* $\mathrm{Spec}(A)$*. We refer to prop. 7.5.10 d), were here the roles of B,C are interchanged. We assert that that under our assumptions the homomorphism*

$$\mathcal{I}/\mathcal{I}^2 \longrightarrow \Omega^1_{B/A} \otimes_B C$$

is injective and $\mathcal{I}/\mathcal{I}^2$ *is locally free.*

The first assertion has been discussed above, the second follows easily from the definition of the module of relative differentials and the lemma of Nakayama. To prove the third one we consider the morphism

$$Y \times_{\mathrm{Spec}(A)} \mathrm{Spec}(B) \xrightarrow{f \times \mathrm{Id}} \mathrm{Spec}(B) \times_{\mathrm{Spec}(A)} \mathrm{Spec}(B)$$

and observe that Γ_f is the inverse of the diagonal. Therefore, 2) implies 3). To prove 4) we choose local parameters $x_1,x_2,\dots,x_d \in B$ at $f(Q)$. We can choose them in such a way that $h(x_1),\dots,h(x_r)$ are local parameters at Q and $x_{r+1},\dots,x_d \in \mathcal{I}$. The differentials $dx_1,dx_2,\dots,dx_n$ form (locally at Q) a basis of $\Omega^1_{B/A} \otimes C$. The homomorphism $\mathcal{I}/\mathcal{I}^2 \longrightarrow \Omega^1_{B/A} \otimes C$ may have a kernel $\mathcal{G}$, the image $\mathcal{B}$ of this homomorphism is locally free at Q. We want to show that the images of $x_{r+1},\dots,x_d$ in $\mathcal{I}/\mathcal{I}^2$ are free generators at Q and this means that the support of $\mathcal{G}$ does not contain Q. The image of Q in our base scheme $\mathrm{Spec}(A)$ is a prime ideal $\mathfrak{q}_0 \in \mathrm{Spec}(A)$. Since C is flat over A we get an exact sequence

$$0 \longrightarrow \mathcal{G} \otimes A/\mathfrak{q}_0 \longrightarrow \mathcal{I}/\mathcal{I}^2 \otimes A/\mathfrak{q}_0 \longrightarrow \mathcal{B} \otimes \mathcal{B}/\mathfrak{q}_0 \longrightarrow 0$$

If the support of $\mathcal{G}$ contains Q, then we have $\mathcal{G} \otimes A/\mathfrak{q}_0 \neq 0$. Then we can pick a geometric point, i.e. a $\phi : A/\mathfrak{q}_0 \longrightarrow \bar{k}$ and get the sequence

$$0 \longrightarrow \mathcal{G} \otimes A/\mathfrak{q}_0 \otimes \bar{k} \longrightarrow \mathcal{I}/\mathcal{I}^2 \otimes A/\mathfrak{q}_0 \otimes \bar{k} \longrightarrow \mathcal{B} \otimes A/\mathfrak{q}_0 \otimes \bar{k} \longrightarrow 0$$

and the term on the left is still not zero. All this holds in an open neighborhood of Q and we may replace Q by a $\bar{k}$ valued point Q_1 in the support of $\mathcal{G}$ and we still get the same sequence. We can assume that $x_1,\dots,x_r$ vanish at Q_1 and we pass to the completion at Q_1. Then we get from $0 \longrightarrow \mathcal{I} \longrightarrow B \longrightarrow C \longrightarrow 0$ the exact sequence

$$0 \longrightarrow \widehat{\mathcal{I}} \longrightarrow k[[x_1,x_2,\dots,x_r,x_{r+1},\dots,x_d]] \longrightarrow k[[x_1,x_2,\dots,x_r]]$$

and we conclude that $\widehat{\mathcal{I}}/(\widehat{\mathcal{I}})^2$ is the free $k[[x_1,x_2,\dots,x_r]]$ module generated by the images of images of $x_{r+1},\dots,x_d$. This implies that the completion $\widehat{\mathcal{G}} = 0$ and since completion is faithfully flat we get that $\mathcal{G}$ is zero at Q, in contrast to our assumption.

□

7.5.4 Normal schemes and smoothness in codimension one

We consider an integral, normal affine $k-$ algebra A/k. Then we know that the irreducible components of $\mathrm{Spec}(A \otimes_k \bar{k})$ are disjoint and

$$A \otimes_k \bar{k} = \bigoplus_i (A \otimes_k \bar{k}) e_i$$

where the e_i are constants (See exercise 28). If in addition the algebra A/k is absolutely reduced, then the $\bar{k}$-algebras $(A \otimes_k \bar{k})e_i$ are integral and we claim, that they are in fact still normal.

This is easy to see. First of all it is clear that under our assumption the field L/k of constants is separable. Since we have $L \subset A$ we can view A as an absolutely irreducible L-algebra, which is of course still normal. If we want to tensor by $\bar{k}$ we have to choose an embedding $\sigma : L \hookrightarrow \bar{k}$ from the k-algebra L/k to the k-algebra $\bar{k}$. Then our decomposition becomes

$$A \otimes_k \bar{k} = \bigoplus_{\sigma : \sigma L \hookrightarrow \bar{k}} \cdot A \otimes_{L,\sigma} \bar{k}.$$

(See example 8 on p. 26) Hence we have to show that for the absolutely irreducible L-algebra A and any σ the $\bar{k}$-algebra $A \otimes_{L,\sigma} \bar{k}$ is still irreducible. In other words it suffices to prove

Proposition 7.5.17. *Let A/k be absolutely irreducible. Then A/k is normal if and only if $A \otimes_k \bar{k}$ is normal.*

Of course $\mathrm{Quot}(A \otimes_k \bar{k}) = \mathrm{Quot}(A) \otimes_k \bar{k}$ and any element in $F \in \mathrm{Quot}(A \otimes_k \bar{k})$ is of the form

$$F = \sum F_i \otimes \omega_i,$$

where the $F_i \in \mathrm{Quot}(A)$ and the ω_i form a k- basis of some finite extension K/k. It suffices to show: If F is integral over $A \otimes_k K$, then the F_i are integral over A and hence in A. But this is clear because for any index ν the element $\omega_\nu F = \sum F_i \otimes \omega_\nu \omega_i = \sum_j (\sum_i F_i c_{ij}) \omega_j$ with some $c_{ij} \in k$, is integral. It is elementary linear algebra that we can write the F_i as linear combinations of the $\omega_\mu F$ with coefficients in K and this proves the claim. □

Now we can state

Theorem 7.5.18. *(Smoothness in codimension one) If A/k is an integral, normal and absolutely reduced k-algebra of finite type, then the singular locus is of codimension 2.*

This is a sharpening of the generic smoothness under the assumption of normality.

Our considerations above show, that we may assume that $k = \bar{k}$ and A/k is integral and normal. We consider the singular locus and we assume that it has an irreducible component of codimension one, This corresponds to a non zero prime ideal $\mathfrak{p} \subset A$ of height one, the local ring $A_\mathfrak{p}$ is of dimension one. It is still normal (easy exercise) and therefore, it is a discrete valuation ring. (See Def.7.3.4) Hence we see that we can find $f,g \in A$, $g \notin \mathfrak{p}$ such that in the localization A_g the ideal $\mathfrak{p}$ becomes principal, i.e. $\mathfrak{p} = (f/g)$. This tells us that we may assume that already $\mathfrak{p} = (f)$ is principal. Now we have that $\dim(A/(f)) = \dim(A) - 1$ and we know that $A/(f)$ is generically smooth. We can find a smooth geometric point $P \in \mathrm{Spec}(A/(f))$, and we show that this is also a smooth geometric point on $\mathrm{Spec}(A)$. This is almost obvious. Let $\mathfrak{m}_P \supset \mathfrak{p}$ the maximal ideal, then we know that $\dim((\mathfrak{m}_P/(f)/((\mathfrak{m}_P/(f))^2) = \dim(A) - 1$ (See Theorem 7.5.2), this implies $\dim((\mathfrak{m}_P/(\mathfrak{m}_P)^2) \leq \dim(A)$ and applying this theorem a second time we get the smoothness of P on A. But this shows that $\mathfrak{p}$ can not lie in the singular locus and we have a contradiction.

□

Regular local rings

At the end of section 7.1.2 we mentioned that integral, noetherian local rings A have a dimension, again it is defined as the length of a chain of prime ideals minus one. Let $\mathfrak{m} \subset A$ the maximal ideal, let $k(\mathfrak{m})$ be the residue field. Then our local ring is called a **regular** local ring if

$$\dim(A) = \dim_{k(\mathfrak{m})}(\mathfrak{m}/\mathfrak{m}^2).$$

If A/k is an absolutely reduced k algebra and if P is a geometric point, then we obtain a maximal ideal $\mathfrak{m}_P \subset A$ and a maximal ideal $\mathfrak{m}_0 \subset A \otimes_k \bar{k}$. We have seen in Thm. 7.5.2 that P is smooth if and only if $(A \otimes_k \bar{k})_{\mathfrak{m}_0}$ is regular. But we also have

Proposition 7.5.19. *If P is a smooth geometric point on an affine scheme* $\mathrm{Spec}(A)/k$ *then the local ring $A_{\mathfrak{m}_P}$ is regular.*

We sketch the argument. We encounter a difficulty if the extension of residue fields $k(\mathfrak{m}_P)/k$ is not separable. We write $A = k[X_1,X_2,\dots,X_n]/(F_1,\dots,F_r] = k[X_1,X_2,\dots,X_n]/I$ as in Thm. 7.5.4. (This is only valid locally at P) Then P is a k-homomorphism $A \longrightarrow \bar{k}$ and determined by the image $(a_{r+1},\dots,a_n) \in \bar{k}^d$ of the coordinate functions x_i mod $I, i = r+1,\dots,n$. Hence the maximal ideal $\mathfrak{m}_0 = (x_{r+1} - a_{r+1},\dots,x_n - a_n)$. If now the extension $k(a_{r+1},\dots,a_n)/k$ is separable, then for any $j = r+1,\dots,n$ we consider the set Σ_j of embeddings $\sigma : k(a_j)/k \hookrightarrow \bar{k}/k$. and see easily that the elements

$$(\prod_{\Sigma_{r+1}} (x_{r+1} - \sigma(a_{r+1})),\dots,\prod_{\Sigma_n}(x_n - \sigma(a_n)), j = r+1,\dots,n$$

are in $\mathfrak{m}_P$ and form a basis of the $k(\mathfrak{m})-$ vector space $\mathfrak{m}_P/\mathfrak{m}_P^2$. This proves the proposition if $k(\mathfrak{m}_P)/k$ is separable. Essentially the same argument-namely taking the product of conjugates of the generators- allows to assume that k itself is separably closed. Then we write again $\mathfrak{m}_0 = (x_{r+1} - a_{r+1},\dots,x_n - a_n)$ where $a_j \in \bar{k}$. We can forget those j, for which $a_j = 0$, i.e. we assume that all of them a non zero. Then the a_j generate a subgroup $< a_{r+1},\dots,a_n >$ in $\bar{k}^\times$ and from this we get a finite abelian p- group $< a_{r+1},\dots,a_n > \cdot k^\times/k^\times$. We apply the theorem of elementary divisors and after making some suitable substitutions we assume that $< a_{r+1},\dots,a_n > \cdot k^\times/k^\times$ is the direct product of the cyclic groups generated by the a_i, the cyclic groups are of order p^{n_j} and hence $a_j^{p^{n_j}} = b_i \in k$. It is clear that the elements

$$\{\dots,\prod_j a_j^{\nu_j},\dots\}_{0 \leq \nu_i < p^{n_j}}$$

form a basis of $k(a_{r+1},\dots,a_n)/k$. Now it follows from a simple calculation (here is a little gap to be filled) that the elements

$$\{(x_{r+1}-a_{r+1})^{p^{n_{r+1}}},\ldots,(x_j-a_j)^{p^{n_j}},\ldots,((x_n-a_n)^{p^{n_n}}\} =$$

$$\{x_{r+1}^{p^{n_{r+1}}}-a_{r+1}^{p^{n_{r+1}}},\ldots,x_j^{p^{n_j}}-a_j^{p^{n_j}},\ldots,x_n^{p^{n_n}}-a_n^{p^{n_n}}\}$$

form a basis for the $k(\mathfrak{m}_P)$-vector space $\mathfrak{m}_P/\mathfrak{m}_P^2$ and hence we have the proposition. The following theorem is deeper

Theorem 7.5.20. *A noetherian regular local ring is factorial.*

(See [Ei], Thm. 19.19)

7.5.5 Vector fields, derivations and infinitesimal automorphisms

We consider a smooth morphism of finite type $f : X \longrightarrow Y$. Then we define the *relative tangent sheaf* as

$$T_{X/Y} = \mathrm{Hom}_{\mathcal{O}_X}(\Omega^1_{X/Y},\mathcal{O}_X).$$

Since we assume that f is smooth we know that $\Omega^1_{X/Y}$ is locally free and hence $T_{X/Y}$ is also locally free.
If we have a point $x \in X$ and if $\mathcal{O}_{X,x}/\mathfrak{m}_x = k(x)$ is the residue field then $T_{X/Y}\otimes k(x)$ is the tangent space at x. We will sometimes also denote by $\pi : T_{X/Y} \longrightarrow X$ the associated vector bundle 6.2.3 then the tangent space at x is also equal to the fibre $\pi^{-1}(x)$.

We may define $T_{X/Y}$ by the same formula above, even if the morphism $f : X \longrightarrow Y$ not smooth. But hen it it is not so useful. Let us look at the example 17 (a). We describe the module $\Omega^1_{A/k}$ explicitly and we saw that it is locally free of rank 1 outside the origin and in the origin it is not locally free. But $\mathrm{Hom}_A(\Omega^1_{A/k},A)$ is in fact locally free at all points and hence in this case $T_{A/k}$ does not see the singularity.

Let us start from a diagram

$$\begin{array}{ccc} X & \xrightarrow{\quad f \quad} & Z \\ & {\scriptstyle g}\searrow \quad \swarrow{\scriptstyle h} & \\ & Y & \end{array} \tag{7.15}$$

then we stated in proposition 7.5.10 b1) that we get a homomorphism Δ_f between the sheaf of differentials. If now g and h are smooth then this yields a homomorphism

$$D_f : T_{X/Y} \longrightarrow f^*(T_{Z/Y}).$$

This means that in a point $x \in X$ we get a homomorphism

$$D_f : T_{X/Y,x} \longrightarrow T_{Z/Y,f(x)} \otimes_{\mathcal{O}_{f(x)}} \mathcal{O}_x \tag{7.16}$$

and this gives us a linear map between the tangent spaces

$$D_f(x) : (T_{X/Y,x}\otimes k(x)) \longrightarrow T_{Z/Y,f(x)}\otimes k(f(x))\otimes k(x) \tag{7.17}$$

This is now the algebraic version of our good old differential of a map between differentiable manifolds.
Let us assume in addition that f is an inclusion, we consider X as a sub scheme of Z. Then D_f is the inclusion of the tangent bundle $T_{X/Y}$ into the restriction $T_{Z/Y}|X$. The quotient bundle

$$N_{X/Y} = T_{Z/Y}|X/T_{X/Y}$$

is called the **normal bundle** of Y in X. Looking at prop. 7.5.16 4) and recalling the definition of the tangent bundle as dual the differentials we see

$$N_{X/Y} = (\mathcal{I}/\mathcal{I}^2)^\vee.$$

This is intuitively clear: If we have an $f \in \mathcal{I}$ and a section D in $T_{X/Y}$ then $D(f) = 0$ because f is identically zero on Y. This yields the pairing $N_{X/Y} \times \mathcal{I}/\mathcal{I}^2 \longrightarrow \mathcal{O}_Y$, which is non degenerate. The $\mathcal{O}_Y$ sheaf $\mathcal{I}/\mathcal{I}^2$ is locally free and it is called the **conormal bundle.**

.

We come back to the investigation of the properties of intersections of two schemes. (See p. 70). Again we assume that that Z_1,Z_2 are irreducible sub schemes of an irreducible scheme X/k of finite type. Let d_1,d_2 be their codimensions. Let $Y \subset Z_1 \cap Z_2$ be an irreducible component in the intersection.

Definition 7.5.21. *We say that Z_1 and Z_2 intersect transversally in Y if we can find an non empty open subset $U \subset Y$ such that any geometric point $P \in U(\bar{k})$ is smooth on Z_1,Z_2 and X and the tangent spaces $T_{Z_1,P}$,$T_{Z_2,P}$ intersect transversally and this means*

$$\dim_{\bar{k}}(T_{Z_1,P}) + \dim_{\bar{k}}(T_{Z_2,P}) = \dim_{\bar{k}}(T_{X,P}) + \dim_{\bar{k}}(T_{Y,P}).$$

This implies of course that P is a smooth point on Y. We can rephrase this by saying: Locally at P the ideals I_1,I_2 defining Y_1,Y_2 are generated by d_1,d_2 elements $f_1,\dots,f_{d_1}$ and $g_1,\dots,g_{d_2}$ and the sequence $\{f_1,\dots,f_{d_1},g_1,\dots,g_{d_2})$ yields local generators at P of the ideal defining Y. And proposition 7.2.17 implies

$$\operatorname{codim}(Y) = \operatorname{codim}(Z_1) + \operatorname{codim}(Z_2) \tag{7.18}$$

The global sections $H^0(X,T_{X/Y})$ are called *vector fields (along the fibers)*. We pass to a local situation $Y = \operatorname{Spec}(A)$,$X = \operatorname{Spec}(B)$ and $g : A \longrightarrow B$, then these vector fields are the derivations

$$\mathcal{D}_{A/B} = \{D \in \operatorname{Hom}_A(B,B) | D(b_1 b_2) = b_1 D(b_2) + b_2 D(b_1) \text{ for all } b_1,b_2 \in B\}.$$

This is clear: The sets $X_f = \operatorname{Spec}(B_f)$ form a basis of the Zariski topology and locally a section in $\Omega^1_{X/Y}$ is of the form $d(b/f) \in \Omega^1_{X/Y}(X_f)$ where $b \in B$. Now a derivation D has to yield a B_f-linear map $\widetilde{D}|_{X_f} : \Omega^1_{X/Y}(X_f) \longrightarrow B_f$ and we simply define $\widetilde{D}|_{X_f}(d(b/f) = (fdb - bdf)/f^2$. On the other hand if we have an element $\widetilde{D} \in \operatorname{Hom}_{\mathcal{O}_X}(\Omega^1_{X/Y},\mathcal{O}_X)$ then it especially yields a B-linear map $\widetilde{D}_X : \Omega^1_{X/Y}(X) \longrightarrow B = \mathcal{O}(X)$. The B-module $\Omega^1_{X/Y}(X)$ is generated by the differentials db and hence we define $D(b) = \widetilde{D}|_X(db)$. It is a simple calculation that the two assignments $D \longrightarrow \widetilde{D}$,$\widetilde{D} \longrightarrow D$ are well defined and provide the isomorphism

$$\mathcal{D}_{A/B} = T_{X/Y}(X)$$

where of course still $X = \mathrm{Spec}(B)$.
Let us keep the assumption that $X = \mathrm{Spec}(A)$,$Y = \mathrm{Spec}(B)$ are affine. We introduce the ring of dual numbers $B[\epsilon] = B[T]/(T^2)$, and consider diagrams

$$\begin{array}{ccc} X & & \\ P\uparrow\downarrow & \nwarrow t_P & \\ \mathrm{Spec}(B) & \underset{j}{\overset{\longleftarrow}{\longrightarrow}} & \mathrm{Spec}(B[\epsilon]). \end{array}$$

where P is a section, i.e. a Y-valued point of $X \longrightarrow Y$, where j is the morphism $\epsilon \mapsto 0$ and t_P is a $\mathrm{Spec}(B[\epsilon])$ valued point such that $P = t_P \circ j$. We say that t_P is in the $\epsilon-$ *cloud around* P.
We have seen that the section $P : Y \longrightarrow X$ (see 6.1.6) identifies Y to a closed sub scheme $i : P(Y) \hookrightarrow X$ and we can consider the restriction $i^*(T_{X/Y})$ (see 6.2.2) and hence we get a quasi-coherent sheaf $T_{X/Y,P} = (i \circ P)^*(T_{X/Y})$ on Y.
Now it is obvious

The set of t_P *satisfying* $P = t_P \circ j$, *i.e. the* t_P *in the* ϵ *-cloud around* P *is equal to* $H^0(Y, T_{X/Y,P})$. *If* $T \in H^0(Y, T_{X/Y,P})$ *then we denote the resulting point* t_P *by*

$$t_P = P + \epsilon T \tag{7.19}$$

If we drop the assumption that our schemes are affine, then we have to be a little bit careful. Our considerations are local in Y. It is clear how to define the scheme $\mathrm{Spec}(Y[\epsilon]$: We cover Y by affine schemes $V_i = \mathrm{Spec}(A_i)$ and then we glue the schemes $\mathrm{Spec}(A_i[\epsilon])$ together and get $\mathrm{Spec}(Y[\epsilon]$. But we have to make an assumption concerning the morphism $X \longrightarrow Y$. Without any further assumption we can not say that the section $P : Y \longrightarrow X$ defines a closed sub scheme $P(Y) \subset X$. For this to be true we need that the structural morphism $f : X \longrightarrow Y$ is separated, this is a global property of a morphism and will be discussed in the next chapter (see 8.1.4) (It means that the diagonal $\Delta_X \subset X \times_Y X$ is closed.) If now Y is not necessarily affine but f is separated then we can reformulate the above assertion into an assertion concerning the sheaf $T_{X/Y,P}$: We state the assertion only for the restriction to affine open subsets $V \subset Y$.
Therefore we can apply this to the following diagram: Let X/S be a separated scheme of finite type then we consider

$$\begin{array}{ccc} X \times_S X & & \\ \Delta\uparrow\downarrow p_1 & \nwarrow t_P & \\ X & \underset{j}{\overset{\longleftarrow}{\longrightarrow}} & \mathrm{Spec}(X[\epsilon]). \end{array}$$

and now locally on affine open subsets $U \subset X$ the $\epsilon-$ cloud around Δ is the space of sections $H^0(U, T_{X/S})$.

Automorphisms

For any scheme $X \longrightarrow S$ we can consider the functor of its automorphisms. This functor attaches to any scheme $T \longrightarrow S$ the group of automorphisms of the scheme $X \times_S T$. This functor $T \longrightarrow S \mapsto \mathrm{Aut}_F(X \times_S T/T) = \mathrm{Aut}(X)(T)$. Sometimes this functor is representable by a group scheme (See below.)
At this point we are interested in the kernel of the homomorphism

$$\mathrm{Aut}(X)(S[\epsilon]) \longrightarrow \mathrm{Aut}(X)(S)$$

and this group is called the group of **infinitesimal automorphisms.** It is clear from our considerations above that this group of infinitesimal automorphisms is equal to $H^0(X,T_{X/S})$: A global section $T \in H^0(X,T_{X/S})$ "displaces a point x into the infinitesimally close point $x + \epsilon T_x$".

7.5.6 Group schemes

We want to give a brief and informal outlook into the theory of group schemes over an arbitrary base. We consider a separated scheme $p : G \longrightarrow S$ of finite type. We assume that we have the structure of a group scheme on G/S (see also page 50.)This means that we have S-morphisms

$$m : G \times_S G \longrightarrow G,\mathrm{inv} : G \longrightarrow G,e : S \longrightarrow G,$$

which satisfy the following rules encoding the associativity, the existence of the identity and the inverse:

$$m \circ m \times_S \mathrm{Id} = m \circ \mathrm{Id} \times_S m, \qquad \textit{Associativity}$$

$$m \circ (e \times_S \mathrm{Id}) = \mathrm{Id}\, ,m \circ (\mathrm{Id} \times_S e) = \mathrm{Id}\, , \qquad \textit{Identityelement}$$

$$m \circ (\mathrm{Id} \times_S \mathrm{inv})) = e \circ p,m \circ (\mathrm{inv} \times_S \mathrm{Id})) = e \circ p. \qquad \textit{Inverse}$$

For any scheme $T \longrightarrow S$ the morphism m yields a composition $G(T) \times G(T) \longrightarrow G(T),(g_1,g_2) \mapsto g_1 \cdot g_2$ and this provides a group structure on $G(T)$.
It is that for any $S' \longrightarrow S$ the morphism $m' : G \times_S S' \times_{S'} G \times_S S' \longrightarrow G \times_S S'$ yields a group scheme structure on $G \times_S S'$.
A separated scheme $p : G \longrightarrow S$ together with the data m,inv,e is called a **group scheme over** S.. If in addition the structural morphism p is smooth, then it is called a **smooth group scheme** .

In the following we assume that our group scheme are smooth. Since we have the group structure the irreducible components of a fiber G_s all these irreducible components have the same dimension. This dimension is locally constant in s. (Theorem7.5.12) Hence we can speak of the dimension $\dim(G/S)$ if the base scheme S is connected. If S is not connected then let us assume that these dimensions are independent of s.
Our assumptions imply that $\Omega^1_{G/S},T_{G/S}$ are locally free of rank $\dim(G/S)$. We denote the restriction $e^*(T_{G/S})$ by $\mathfrak{g}$ or also by $\mathrm{Lie}(G/S)$. It is easy to see that for any affine open subset $U \subset S$ the elements in the ϵ-cloud $t_P = e + \epsilon T,T \in H^0(U,\mathrm{Lie}(G/S))$ form a subgroup and $(e+\epsilon T_1)\cdot(e+\epsilon T_2) = e+\epsilon(T_1+T_2)$. It follows immediately from the definition that the ϵ-cloud around e is also the kernel of the homomorphism $G(U[\epsilon]) \longrightarrow G(U)$. Hence we can say

$$H^0(U, \mathrm{Lie}(G/S)) = \ker(G(U[\epsilon]) \longrightarrow G(U)).$$

If we have a locally free sheaf $\mathcal{E}$ of finite rank r over our scheme S then we we can define the group scheme $\mathrm{GL}(\mathcal{E})/S$. For any $T \longrightarrow S$ the group $\mathrm{GL}(\mathcal{E})(T)$ is the group of $\mathcal{O}_T$-linear automorphisms of the bundle $\mathcal{E}_T$, which is the pullback of $\mathcal{E}$ to $X \times_S T$. Since $\mathcal{E}/S$ is locally trivial, we can find a covering by Zariski-open subsets $U \subset S$ such that $E/U \xrightarrow{\sim} \mathcal{O}_U^r$ and then $G \times_S U \xrightarrow{\sim} \mathrm{GL}_r/U$.
For this group scheme it is evident that

$$\mathrm{Lie}(\mathrm{GL}(\mathcal{E})/S) = \mathrm{End}_{\mathcal{O}_S}(\mathcal{E}).$$

A *representation* of a group scheme G/S is a S-homomorphism

$$\rho : G \longrightarrow \mathrm{GL}(\mathcal{E})$$

where $\mathcal{E}/S$ is a locally free $\mathcal{O}_S$ module. Then it is clear from our considerations above that we have a "derivative" of the representation (See 7.16)

$$d\rho : \mathfrak{g} = \mathrm{Lie}(G/S) \longrightarrow \mathrm{Lie}(\mathrm{GL}(\mathcal{E})/S) = \mathrm{End}_{\mathcal{O}_S}(\mathcal{E})$$

this is an $\mathcal{O}_S$-linear morphism of sheaves.
Every group scheme G/S has a very special representation, this is the the *Adjoint representation* . We observe that the group acts on itself by conjugation, this is the morphism

$$\mathbf{ad} : G \times_S G \longrightarrow G,$$

which on T valued points is given by

$$\mathbf{ad}(g_1,g_2) \mapsto g_1 g_2 (g_1)^{-1}.$$

This action clearly induces a representation

$$\mathrm{Ad} : G/S \longrightarrow \mathrm{GL}(\mathfrak{g})$$

and this is the adjoint representation. This adjoint representation has a derivative (see 7.16) and this is a morphism of locally free sheaves

$$D_{\mathrm{Ad}} = \mathrm{ad} : \mathfrak{g} \longrightarrow \mathrm{End}_{\mathcal{O}_S}(\mathfrak{g}).$$

If we assume that S is affine, i.e. $S = \mathrm{Spec}(A)$ then $\mathfrak{g}$ is simply a locally free A-module and ad is simply an A-module homomorphism from $\mathfrak{g}$ to $\mathrm{End}_{\mathcal{A}}(\mathfrak{g})$. We introduce the notation: For $T_1,T_2 \in \mathfrak{g}$ we put

$$[T_1,T_2] := \mathrm{ad}(T_1)(T_2).$$

Now we can state the famous and fundamental result

Theorem 7.5.22. *The map $(T_1,T_2) \mapsto [T_1,T_2]$ is bilinear and antisymmetric. It induces the structure of a Lie-algebra on $\mathfrak{g}$, i.e. we have the Jacobi identity*

$$[T_1,[T_2,T_3]] + [T_2,[T_3,T_1]] + [T_3,[T_1,T_2]] = 0.$$

We do not prove this here. In the case $G/S = \mathrm{GL}(\mathcal{E})$ it is easy to see that for $T_1,T_2 \in \mathrm{Lie}(\mathrm{GL}(\mathcal{E})) = \mathrm{End}(\mathcal{E})$ we have $[T_1,T_2] = T_1T_2 - T_2T_1$ and in this case the Jacobi Identity is a well known.

7.5.7 The groups schemes $\mathbb{G}_a$,$\mathbb{G}_m$ and μ_n

We want to introduce some simple affine group schemes. They can be defined over an arbitrary base scheme S, this means that theyc are obtained by base change from the two schemes $\mathbb{G}_m/\operatorname{Spec}(\mathbb{Z})$,$\mathbb{G}_a/\operatorname{Spec}(\mathbb{Z})$.
These two group schemes represent functors from the category of commutative rings to the category of groups: For any commutative ring $\mathbb{Z} \longrightarrow A$ we put

$$\mathbb{G}_m(A) = A^\times = \text{the multiplicative group of units of the ring } A,$$
$$\mathbb{G}_a(A) = A = \text{the additive group of the ring } A.$$

It is easy to see that these two functors are represented by the group schemes $\operatorname{Spec}(\mathbb{Z}[T,T^{-1}])$ and $\operatorname{Spec}(\mathbb{Z}[X])$, where the group structure on the affine algebras is given by the homomorphisms

$$m : T \mapsto T \otimes T, \mathrm{inv} : T \mapsto T^{-1}, e : T \mapsto 1$$

for the group scheme $\mathbb{G}_m$ and

$$m : X \mapsto X \otimes 1 + 1 \otimes X, \mathrm{inv} : X \mapsto -X, e : X \mapsto 0$$

for the group scheme $\mathbb{G}_a$.
If describe affine group schemes over $\operatorname{Spec}(\mathbb{Z})$ (or any affine base scheme) in terms of algebras then the homomorphism m is called the **comultiplication**. If k is an algebraically closed field an $S = \operatorname{Spec}(k)$ these are the are the only affine, connected, one dimensional group schemes over S.
On our group scheme $\mathbb{G}_m/\operatorname{Spec}(\mathbb{Z})$ we an endomorphism, which on $\mathbb{G}_m(A) = A^\times$ is given by $x \mapsto x^n$. This endomorphism has a kernel, this is the group scheme

$$\mu_n = \operatorname{Spec}(\mathbb{Z}[T,T^{-1}]/(T^n - 1))$$

where m,inv,e are given by the same formulae as for $\mathbb{G}_m$.
This group scheme is a finite group scheme over $\operatorname{Spec}(\mathbb{Z})$, but if $n > 1$ it it is not smooth anymore. If we pick a prime $p \mid n$ and perform the base change $\mathbb{G}_m \times_{\operatorname{Spec}(\mathbb{Z})} \operatorname{Spec}(\mathbb{F}_p)$ then we get the coordinate ring $\mathbb{F}_p[T,T^{-1}]/(T^n - 1)$ and this ring contains non zero nilpotent elements. Hence it cannot be smooth. The local ring at a smooth point does not contain nilpotent elements (see definition 7.5.1.) But it easy to see that its base change to $\operatorname{Spec}(\mathbb{Z}[\frac{1}{n}]$ is smooth.
Since we are at this point let us just discuss another interesting scheme. We start from the field $\mathbb{F}_p$ with p elements. We have $\mathbb{G}_a/\mathbb{F}_p = \mathbb{G}_a \times_{Spec(\mathbb{Z})} \operatorname{Spec}(\mathbb{F}_p)$. The affine algebra of this reduction $\bmod p$ is given by $\mathbb{F}_p[X]$ and the same m,inv,e as above. But for this scheme the map $X \mapsto X^p$ induces an endomorphism and the kernel of this endomorphism is a group scheme

$$\alpha_p/\mathbb{F}_p = \operatorname{Spec}(\mathbb{F}_p[X]/(X^p))$$

with still the same formulae for m,inv,e.
We conclude this section by stating a classical theorem

Theorem 7.5.23. (Hilbert's theorem 90) *Let V/k be a finite dimensional vector space over a field k. For any finite, normal separable extension L/k we have*

$$H^1(L/k,\mathrm{GL}(V)) = \{e\},$$

i.e it is trivial.

To see that this is true we observe that V/k is a scheme over k, we have an inclusion $\mathrm{GL}(V)/k \hookrightarrow \mathrm{Aut}(V)/k$ (the group on the right hand side is huge if $\dim_k(V) > 0$.) Now we know from section 6.2.10 that the image ξ' of a class $\xi \in H^1(L/k,\mathrm{GL}(V))$ in $H^1(L/k, \mathrm{Aut}(V))$ defines a form V'/k. But since this class is the image of ξ it is clear that V'/k is a vector space over k and has dimension $\dim_k(V)$. Therefore it must be isomorphic to V/k as a vector space and this shows that ξ must be trivial.
I want to point out that this argument is not the one given in most of the text books. In the text books, and also in Hilbert's original proof, a boundary is written down explicitly as a certain sum. This argument also occurs in this book at a different place.

7.5.8 Actions of group schemes

It is clear what it means that a group scheme G/S acts on a scheme X/S, this means that we have a morphism

$$\begin{array}{ccc} G\times_S X & \xrightarrow{\quad a \quad} & X \\ & \searrow \qquad \swarrow & \\ & S & \end{array}$$

such that the diagram

$$\begin{array}{ccccc} & & G\times_S X & & \\ & \overset{m\times_S \mathrm{Id}}{\nearrow} & & \overset{a}{\searrow} & \\ G\times_S G\times_S X & & & & X \\ & \underset{\mathrm{Id}\times_S a}{\searrow} & & \underset{a}{\nearrow} & \\ & & G\times_S X & & \end{array}$$

commutes and such that the composition $e\times_S \mathrm{Id} : X \longrightarrow G\times_S G$ with a is the identity. These axioms are equivalent with the requirement that we have a functorial action of $G(T)$ on $X(T)$ for all schemes $T \longrightarrow S$. To any such an action we can attach a functor from the category of schemes over S to the category of sets:

$$\widetilde{X/G} : \{T \longrightarrow S\} \longrightarrow X(T)/G(T). \tag{7.20}$$

It is one of the important issues of the theory of group schemes to discuss whether (or under which conditions) we have a reasonable quotient X/G for this action. The best we can hope for is that the functor $\widetilde{X/G}$ is representable, but this is only true for very specific cases. (See [M-F-K]).

We want to discuss very briefly actions of $\mathbb{G}_m/S$ on affine schemes $X = \mathrm{Spec}(A) \longrightarrow S$. This discussion is very informal and we leave the proofs of the statements as an exercise to the reader.
Let us assume that $S = \mathrm{Spec}(B)$. Then our action is given by a B-algebra homomorphism

$$a : A \longrightarrow A \otimes B[T,T^{-1}],$$

for $f \in A$ we write

$$f \mapsto \sum_{\nu \in \mathbb{Z}} a_\nu(f) \otimes T^\nu$$

where the sum is finite (depending on f) and where $a_\nu : A \longrightarrow A$ is B-linear. Our two axioms yield $a_\mu \circ a_\nu = \delta_{\nu,\mu} a_\nu$ and $f = \sum_\nu a_\nu(f)$. Therefore we see that

$$A = \bigoplus_{\nu \in \mathbb{Z}} A_\nu,$$

where $A_\nu = a_\nu(A)$. Since a is a homomorphism of B algebras it is clear that the A_ν are B-modules and that $A_\nu \cdot A_\mu \subset A_{\nu+\mu}$.
In other words: A $\mathbb{G}_m/\mathrm{Spec}(B)$-action on the affine scheme $B \longrightarrow A$ is simply a $\mathbb{Z}$-graduation of the B-algebra A.
If we now ask for a quotient $X/\mathbb{G}_m$ then it seems to be quite natural to define $X/\mathbb{G}_m = \mathrm{Spec}(A_0)$ where the projection $X \longrightarrow X/\mathbb{G}_m$ is given by the inclusion $A_0 \longrightarrow A$.This construction gives us a quotient, which has the following property: For any (affine) scheme $B \longrightarrow C$, which is endowed with the trivial $\mathbb{G}_m$ action we have

$$\mathrm{Hom}_{\mathbb{G}_m,S}(X, \mathrm{Spec}(C)) = \mathrm{Hom}(X/\mathbb{G}_m, \mathrm{Spec}(C)).$$

This means that our quotient is a categorical quotient in the sense of ([M-F-K],), but it may still have bad properties. Let us look a the case of the standard action of $\mathbb{G}_m$ on the affine line over a field k. Then $X = \mathrm{Spec}(k[U])$ and $a : U^\nu \mapsto U^\nu \otimes T^\nu$. The quotient is $Y = \mathrm{Spec}(k)$, and $Y(k)$ is simply one point. But if we consider the quotient $X(k)/\mathbb{G}_m(k) = k/k^\times$ then we get two points, namely the orbit of 1 and the orbit of zero. This is certainly not a very satisfactory situation.
The problem arises from the fact that the orbit of 1 under $\mathbb{G}_m$ is not closed. Hence we should formulate another assumption. We say that *all the geometric orbits are closed* if for all geometric points of $B \longrightarrow k$, where k is algebraically closed and all points $x \in X(k)$ the orbit Gx, i.e. the image of $G \times_B k$ under $G \times_B k \times_S \{x\} \longrightarrow X \times_B k$, is closed.
We formulate a criterion for an orbit to be closed. We assume that B is of finite type over $\mathbb{Z}$ and $A = B[x_1,x_2,\dots,x_n]$ is of finite type. Since we can write the x_i as sum of homogenous elements we may assume that the x_i themselves are homogenous of degree d_i.
Then a geometric point $x : B[x_1,x_2,\dots,x_n] \longrightarrow k$ yields a tuple $(a_1,a_2,\dots,a_n) \in k^n$ and an element $c \in \mathbb{G}_m(k) = k^\times$ acts by

$$(a_1,a_2,\dots,a_n) \mapsto (c^{d_1}a_1,c^{d_2}a_2,\dots,c^{d_n}a_n).$$

We claim that the orbit of x is closed if and only if one of the following two conditions holds

(i) There are two indices i,j, for which $a_i \neq 0, a_j \neq 0$ and $d_i > 0, d_j < 0$.

(ii) For all indices i with $a_i \neq 0$ we have $d_i = 0$

We leave it as an exercise to verify this criterion for the closedness of an orbit.
The following proposition is proved and explained in [M-F-K], Chap. I , § 2

Proposition 7.5.24. *Let $X \longrightarrow S$ be a scheme of finite type over S and let be an action of $\mathbb{G}_m/S$ on X/S. If all geometric orbits are closed then $X \longrightarrow X/\mathbb{G}_m$ is a geometric quotient and this means that for all geometric points $B \longrightarrow k$ the induced map*

$$X(k)/\mathbb{G}_m(k) \longrightarrow (X/\mathbb{G}_m)(k)$$

is a bijection.

It is important to notice that here k is algebraically closed, we are dealing with geometric points. If k is not algebraically closed and if we consider for instance the action of $\mathbb{G}_m/\operatorname{Spec}(k)$ on $X = \mathbb{G}_m/\operatorname{Spec}(k)$ itself, which is given by $(x,y) \mapsto x^2y$, then we clearly get $X/\mathbb{G}_m = \operatorname{Spec}(k)$ and $X/\mathbb{G}_m(k) = \{\mathrm{Id}\}$ is just one point. But in general $\widetilde{X/\mathbb{G}}(k) = X(k)/\mathbb{G}_m(k) = k^\times/(k^\times)^2$ will consist of more than one point. And we see that $X/\mathbb{G}_m$ does not represent $\widetilde{X/\mathbb{G}_m}$.
This kind of problem will play a role in the discussion of moduli problems (see 9.6.2 and 10.1.1). There exist concepts, which help us to deal with this difficulty. The starting point is to consider $X(k)/\mathbb{G}_m(k)$ not as a set but as a groupoid. A groupoid is a category where all morphisms between objects are isomorphisms. Hence the value of $\widetilde{X/\mathbb{G}_m}(T)$ will not be a set but a groupoid. In our situation the objects in our groupoid are the elements in $x,y \in X(T)$ (7.20 and the morphisms $\mathrm{Hom}_{\widetilde{X/G}}(x,y) = \{a \in \mathbb{G}_m(T) | ax = y\}$.
Now it does not make sense anymore to ask whether $\widetilde{X/G}$ is representable by a scheme. But we can ask whether it is representable by a "stack" $\widetilde{\widetilde{X/G}}$, where a stack is an object in a 2-category, it is a more general object than a scheme, it is some kind of quotient of a scheme by an equivalence relation. A 2-category $\mathcal{S}$ is a collection of objects $U,V\ldots$ where $\mathrm{Hom}_{\mathcal{S}}(U,V)$ is a category, or better a 1-category.
If for instance $X = pt = \operatorname{Spec}(k)$- here k is a field-, the we have the trivial action of $\mathbb{G}_m$ on pt and the stack $pt/\mathbb{G}_m$ is already a very sophisticated object.
If for instance $k = \mathbb{C}$, then we can consider the cohomology groups of the stack and it turns out that

$$H^\bullet(pt/\mathbb{G}_m,\mathbb{Z}) = \mathbb{Z}[x], x \text{ sits in degree } 2,$$

i.e. it is the cohomology of the infinite dimensional projective space $\mathbb{P}^\infty(\mathbb{C})$.
The stack $pt/\mathbb{G}_m$ is also what the topologists call the classifying space of $\mathbb{G}_m$.

8 Projective Schemes

8.1 Geometric Constructions

8.1.1 The Projective Space $\mathbb{P}^n_A$

Let A be a commutative ring with identity 1_A. Let $S = \mathrm{Spec}(A)$. We want to construct a scheme $\mathbb{P}^n_S$, which will be called the n-dimensional projective space over S.

To do this we consider the following $n+1$ affine schemes

$$U_i = \mathrm{Spec}(A[T_{i,0} \dots T_{i,n}]/(T_{i,i} - 1)). \tag{8.1}$$

The ring of regular functions of U_i is the quotient of the polynomial ring in $n+1$ independent variables variables $T_{i,0}$,$T_{i,1}, \dots$,$T_{i,i-1}$,$T_{i,i}$,$T_{i,i+1}, \dots$,$T_{i,n}$ divided by the relation $T_{i,i} = 1$. This means that all the U_i are copies of $\mathbb{A}^n_S$.
We denote by $t_{i,j}$ the images of the $T_{i,j}$ in the quotient ring $A[t_{i,0}, \dots ,t_{i,n}]$.This means that for a given i the $t_{i,0}$,$t_{i,1}, \dots t_{i,i-1}$,$t_{i,i+1}, \dots$,$t_{i,n}$ are independent polynomial variables and $t_{i,i} = 1$. Then $U_i = \mathrm{Spec}(A[t_{i,0},t_{i,1}, \dots t_{i,i-1},t_{i,i+1}, \dots ,t_{i,n}])$. For any index j we define the open subscheme of U_i:

$$\begin{aligned} U_{i,j} &= \mathrm{Spec}\left(A[T_{i,0}, \dots ,T_{i,n}]/(T_{i,i} - 1)\right)_{T_{i,j}} \\ &= \mathrm{Spec}(A[t_{i,0},t_{i,1}, \dots t_{i,i-1},t_{i,i+1}, \dots ,t_{i,n},t_{ij}^{-1}]. \end{aligned} \tag{8.2}$$

We have an isomorphism

$$f_{i,j} : U_{i,j} \longrightarrow U_{j,i},$$

which on the level of rings is given by

$$\phi_{i,j} : A[t_{j0},t_{j,1}, \dots t_{j,j-1},t_{j,j+1}, \dots ,t_{j,m},t_{ji}^{-1}] \longrightarrow A[t_{i,0},t_{i,1}, \dots t_{i,i-1},t_{i,i+1}, \dots ,t_{i,n},t_{ij}^{-1}] \tag{8.3}$$

$$\phi_{i,j}(t_{j\nu}) \mapsto t_{i\nu} \cdot t_{i,j}^{-1}.$$

We see that $\phi_{i,j}(t_{j,j}) = \phi_{i,j}(1) = t_{i,j} \cdot t_{i,j}^{-1} = 1$ and $\phi_{i,j}(t_{j,i}^{-1}) = t_{i,i} \cdot t_{i,j} = t_{i,j}$ as it must be. We allow $i = j$, in this case $\phi_{i,i}$ is the identity on U_i.
Given three indices i,j,k we get a commutative diagram

$$\begin{array}{ccccc} U_{i,j} \supset U_{i,j} \cap U_{i,k} & & \xrightarrow{\quad\phi_{i,j}\quad} & & U_{j,i} \cap U_{j,k} \subset U_{j,i}. \\ & \phi_{i,k} \searrow & & \swarrow \phi_{j,k} & \\ & & U_{k,i} \cap U_{k,j} & & \end{array}$$

of isomorphisms. This allows us to define an equivalence relation on the space $\bigsqcup_{i=0,\dots,n} U_i$, namely $u_i \sim u_j$ if and only if $u_i \in U_j, u_j \in U_{j,i}$ and $\phi_{i,j}(u_i) = u_j$. We divide by this equivalence relation and get the space

$$\bigsqcup U_i/_\sim = \mathbb{P}^n_A, \tag{8.4}$$

this will be the underlying space of $\mathbb{P}^n_A$. The projection map

$$\pi : \bigsqcup U_i \longrightarrow \mathbb{P}^n_A \tag{8.5}$$

provides a homeomorphism from U_i to an open subset in $\mathbb{P}^n_A$, we identify U_i with this open subset, i.e. we consider U_i as an open subset in $\mathbb{P}^n_A$. Now we define a sheaf $\mathcal{O}_{\mathbb{P}^n_A}$ on the space $\mathbb{P}^n_A$ simply by putting

$$\mathcal{O}_{\mathbb{P}^n_A}|U_i = A[T_{i,0}, \widetilde{\dots, T_{i,n}}]/(T_{i,i} - 1) \tag{8.6}$$

and we use the $\phi_{i,j}$ to glue $\mathcal{O}(U_i) \mid U_i \cap U_j$ with $\mathcal{O}(U_j)|U_i \cap U_j$.

This scheme $(\mathbb{P}^n_A, \mathcal{O}_{\mathbb{P}^n_A})$ is now the n-dimensional projective space over A. In accordance with an earlier convention (see p. 11) we will suppress the second variable and simply write $\mathbb{P}^n_A$.
After constructing $\mathbb{P}^n_A$ we can think of it in the following way: The scheme $\mathbb{P}^n_A$ admits an open covering

$$\mathbb{P}^n_A = \bigcup_{i=0,\dots,n} U_i$$

such that the U_i are affine spaces. The ring of regular functions on U_i is

$$\mathcal{O}_{\mathbb{P}^n_A}(U_i) = A\left[t_{i,0}, \dots, t_{i,n}\right].$$

where $t_{i,i} = 1$ and $t_{i,0}, \dots, t_{i,i-1}, t_{i,i+1} \dots, t_{i,n}$ are independent.
The intersection $U_i \cap U_j$ is affine and

$$\mathcal{O}_{\mathbb{P}^n_A}(U_i \cap U_j) = A\left[t_{i,0}, \dots, t_{i,n}, t_{i,j}^{-1}\right] = A\left[t_{j,0}, \dots, t_{j,m}, t_{j,i}^{-1}\right] \tag{8.7}$$

and

$$t_{i,\nu} = t_{j,\nu} \cdot t_{i,j}, t_{j,\mu} = t_{i,\mu} \cdot t_{j,i}. \tag{8.8}$$

Theorem 8.1.1. *The regular functions on $\mathbb{P}^n_A$ are the constants, this can be stated briefly as $\mathcal{O}_{\mathbb{P}^n_A}(\mathbb{P}^n_A) = A$.*

Proof: Let $f \in \mathcal{O}_{\mathbb{P}^n_A}(\mathbb{P}^n_A)$. We restrict f to the open sets U_i and get a polynomial

$$f \mid U_i = P_i(t_{i,0}, \dots, t_{i,n}) \in A[t_{i,0}, \dots, t_{i,n}].$$

For any pair of indices we have $P_i|U_i \cap U_j = P_j|U_i \cap U_j$ and this means that $\phi_{i,j}(P_j) = P_i$ and hence

$$P_i\left(t_{i,0}\dots t_{i,n}\right) = P_j\left(t_{i,0}\cdot t_{i,j}^{-1},\dots,t_{i,n}\cdot t_{i,j}^{-1}\right).$$

Now we write P_j as polynomial

$$P_j = \sum a_{\nu_0\dots\nu_n} t_{j,0}^{\nu_0}\cdots t_{j,j-1}^{\nu_{j-1}} t_{j,j+1}^{\nu_{j+1}}\cdots t_{j,m}^{\nu_n}. \tag{8.9}$$

We get

$$P_i\left(t_{i,0}\dots t_{i,n}\right) = \sum a_{\nu_0\dots\nu_n}\left(\frac{t_{i,0}}{t_{i,j}}\right)^{\nu_0}\cdots\left(\frac{t_{i,n}}{t_{i,j}}\right)^{\nu_n}, \tag{8.10}$$

this is a polynomial in $t_{i,0},\dots t_{i,i-1},t_{i,i+1}\dots t_{i,n}$. If $i\neq j$, then this is only possible if all exponents $\nu_0 = \dots\nu_{i-1},\nu_{i+1},\dots\nu_n = 0$ in other words, P_i has to be constant. This proves the theorem. □

We have a different way of looking at this argument:

Homogenous coordinates

We consider $n+1$ new variables $X_0,X_1,\dots,X_n$ and we endow the polynomial ring $A[X_0,X_1,\dots,X_n]$ with its standard graduation: To any monomial $X_0^{r_0}X_1^{r_1}\cdots X_n^{r_n}$ we attach the degree $d=\sum_i r_i$, then we define $A[X_0,X_1,\dots,X_n](d)$ to be the A-module generated by the monomials of degree d. This is the module of homogeneous polynomials (or forms) of degree d. We have the direct sum decomposition

$$\bigoplus_{d=0}^{\infty} A\left[X_0,X_1,\dots,X_n\right](d) = A\left[X_0,X_1,\dots,X_n\right].$$

We can localize this ring by inverting any of the variables and in any of these localizations we can consider the subring of elements of degree zero, i.e. we consider

$$A[X_0,X_1,\dots,X_n]_{X_i}^{(0)} = \left\{\frac{F}{X_i^d}\;\middle|\; F\in A[X_0,X_1,\dots,X_n](d)\right\}. \tag{8.11}$$

It is clear that the homomorphism sending $t_{i,j}\mapsto X_j/X_i$ induces an isomorphism

$$A[t_{i,0},\dots,t_{i,n}] \xrightarrow{\sim} A[X_0,X_1,\dots,X_n]_{X_i}^{(0)}$$

and the $\phi_{i,j}$ correspond to the obvious inclusion maps. We can always think that we have the relation $t_{i,j}=X_j/X_i$. Now our theorem above says that

$$\bigcap_{i=0\dots n} A[X_0,X_1,\dots,X_n]_{X_i}^{(0)} = A. \tag{8.12}$$

We want to describe the set of A-valued points in terms of homogenous coordinates. It is already little bit complicated to say what an A-valued point on $\mathbb{P}^n_A$ is. Such a point $x\in\mathbb{P}^n_A(A)$ is a section from $S=\mathrm{Spec}(A)$ to $\mathbb{P}^n_A$. We can find a covering $S=\cup V_\nu$ by open sets such that the section $x: V_\nu \longrightarrow \mathbb{P}^n_A\times V_\nu$ factors through one of the open subsets, say U_i. We may assume that these V_ν are affine, in other words $V_\nu=\mathrm{Spec}(A_{f_\nu})$. Then the restriction of x to A_{f_ν} is of the form

$$x_\nu = (a_{i,0}, \dots, 1, \dots, a_{i,n})$$

where $a_i \in A_{f_\nu}$, and where the entry with index i is one. The index i is not determined by ν and such a point may also lie in U_j. Then $a_{i,j}$ is a unit in A_{f_ν} and

$$x_\nu = \left(\frac{a_{i,0}}{a_{i,j}}, \dots, 1, \dots, \frac{1}{a_{i,j}}, \dots, \frac{a_{i,n}}{a_{i,j}}\right)$$

represents the same point. Of course the restriction of x_ν to $A_{f_\nu f_\mu}$ is equal the restriction of x_μ to the same set. If now x_ν factors through U_i and x_μ factors through U_j, then the restriction of these points to $A_{f_\nu f_\mu}$ factors through $U_i \cap U_j$ and if as above

$$x_\nu = (a_{i,0}, \dots, 1, \dots, a_{i,n}), x_\mu = (b_{j,0}, \dots, 1, \dots, b_{j,n})$$

then we have in $A_{f_\nu f_\mu}$ the necessary and sufficient relations

$$(a_{i,0}, \dots, 1, \dots, a_{i,n}) b_{j,i} = (b_{j,0}, \dots, 1, \dots, b_{j,n}) a_{ij}.$$

We should be aware that in general $\mathbb{P}^n(A) \neq \bigcup U_i(A)$.

The situation becomes much easier if we assume that A is a local ring. Then we can represent a point $x \in \mathbb{P}^n(A)$ by an element in

$$A_*^{n+1} = \Big\{(a_0, \dots, a_n) \mid \text{ at least one entry is a unit in } A\Big\},$$

and the equivalence class representing x is given by the elements

$$(a'_0, \dots, a'_n) = (a_0 b, \dots, a_n b) \quad \text{with} \quad b \in A^\times. \tag{8.13}$$

The vectors $(a_0, \dots, a_n)$ are the homogenous coordinates of the A-valued point x.
If we drop the assumption that A is local we also can represent a point by homogenous coordinates. Let us assume for simplicity that A is integral. Then we introduce the set

$$A_\times^{n+1} = \Big\{(a_0, \dots, a_n) \in A^{n+1} \mid \text{ not all entries are zero}\Big\}$$

where we require in addition that the ideal $\{a_0, \dots, a_n\}$ generated by the a_i is locally principal for the Zariski topology. We define any equivalence relation $\sim$: Two arrays $(a_0, \dots, a_n), (b_0, \dots, b_n)$ in A_*^{n+1} are equivalent if we can find an element $c \in \mathrm{Quot}(A)$ such that

$$(a_0, \dots, a_n) = c(b_0, \dots, b_n).$$

Then it is clear from our previous considerations that

$$A_*^{n+1}/\sim = \mathbb{P}_A^n(A).$$

With a little bit more effort we can find a formulation, which does not assume integrality of A.
The scheme $\mathrm{Spec}(A[X_0, X_1, \dots, X_n]) \longrightarrow S = \mathrm{Spec}(A)$ is of course equal to $\mathbb{A}_S^{n+1}$. We have the zero section $s : S \longrightarrow \mathbb{A}_S^{n+1}$, which ends all the coordinates to zero. The image of the zero section is a closed subscheme also denoted by $S \subset \mathbb{A}_S^{n+1}$, the complement is open and yields the scheme $\mathbb{A}_S^{n+1} \setminus S$. Our considerations above yield a diagram

$$\begin{array}{ccc} \mathbb{A}_S^{n+1}\setminus S & \xrightarrow{\ \pi\ } & \mathbb{P}_S^n \\ & \searrow \quad \swarrow & \\ & S & \end{array} \tag{8.14}$$

We have an action of the group scheme $\mathbb{G}_m$ on $\mathbb{A}_S^{n+1}\setminus S$, which on the S-valued points is given by the the the component wise multiplication: For $t \in \mathbb{G}_m(S),(a_0,\ldots,a_n)) \in \mathbb{A}_S^{n+1}\setminus S(S)$ we the action is given by $(t,(a_0,\ldots,a_n)) \mapsto (ta_0,\ldots,ta_n)$.
This action defines the structure of a $\mathbb{G}_m-$ torsor (See 6.2.4) on $\pi : \mathbb{A}_S^{n+1}\setminus S \longrightarrow \mathbb{P}_S^{n+1}$. The points in $\pi^{-1}(U_i)(A)$ are the elements $(a_0,\ldots,a_i,\ldots,a_n)$, for which $a_i \in A^\times$, over U_i the torsor is trivialised by choosing the section

$$s_i : (a_{i,0},\ldots,a_{i,i-1},a_{i,i+1},\ldots,a_{i,n}) \mapsto (a_{i,0},\ldots,a_{i,i-1},1,a_{i,i+1},\ldots,a_{i,n}). \tag{8.15}$$

Finally we remark, that the gluing argument allows us to replace the affine scheme $S = \mathrm{Spec}(A)$ by an arbitrary scheme, some of the formulations have to be modified appropriately.

8.1.2 Closed subschemes

Now we know of course what a closed subscheme of $\mathbb{P}_A^n$ is, see 6.2.2 page 17. We simply pick a quasi-coherent sheaf (see 6.2.2) of ideals $\mathcal{I}$ in $\mathcal{O}_{\mathbb{P}_A^n}$, then $\big(V(\mathfrak{I}),\mathcal{O}_{\mathbb{P}_A^n}\big)$ is a closed subscheme of $\mathbb{P}_A^n$. We get a commutative diagram

$$\begin{array}{ccc} \big(V(\mathfrak{I}),\mathcal{O}_{\mathbb{P}_A^n}/\mathfrak{I}\big) & \hookrightarrow & \mathbb{P}_A^n \\ & \searrow & \downarrow \\ & & \mathrm{Spec}(A), \end{array} \tag{8.16}$$

hence this subscheme is automatically a scheme over $\mathrm{Spec}(A)$.
Such a quasi-coherent sheaf of ideals $\mathcal{I}$ is simply a collection of ideals $I_i \subset \mathcal{O}_{\mathbb{P}^n}(U_i)$ such that I_i and I_j generate the same ideal in $\mathcal{O}_{\mathbb{P}^n}(U_i \cap U_j)$ (see 6.2.1).

We call an ideal $\widetilde{\mathcal{I}} \subset A[X_0,X_1,\ldots,X_n]$ homogeneous if for any $F \in \widetilde{\mathcal{I}}$ its homogeneous components F_d are also in the ideal.
We can get quasi-coherent sheaves of ideals from homogeneous ideals $\widetilde{I}$ in $A[X_0,X_1,\ldots,X_n]$. This is not difficult to see. To any homogeneous polynomial $F \in A[X_0,X_1,\ldots,X_n](d) \in \widetilde{I}$ of degree d we attach a collection of elements $\{f_i \in \mathcal{O}_{\mathbb{P}^n}(U_i)\}_{i=0,1\ldots n}$ simply by substituting $t_{i,\nu}$ for X_ν into the polynomial. These f_i generate an ideal I_i in $\mathcal{O}_{\mathbb{P}^n}(U_i)$. It is clear that the restrictions of f_i,f_j to $U_i \cap U_j$ satisfy $t_{i,j}^{-d} f_i|U_i \cap U_j = f_j|U_i \cap U_j$ and since $t_{i,j}$ is a unit in $\mathcal{O}_{\mathbb{P}^n}(U_i \cap U_j)$ it follows that I_i and I_j generate the same ideal in $\mathcal{O}_{\mathbb{P}^n}(U_i \cap U_j)$. If F runs over the homogeneous elements of $\widetilde{I}$ these F give us a quasi-coherent sheaf of ideals $\mathcal{I}$.
If in turn $Z \subset \mathbb{P}_A^n$ is a closed subset, we may consider the ideal $\widetilde{I}$ generated by homogeneous polynomials, which vanish on Z. It is clear that $V(\widetilde{I}) = Z$ (See 6.1.3).

Lemma 8.1.2. *For a homogeneous ideal $\widetilde{I} \subset A[X_0,X_1,\ldots,X_n]$ we have $V(\widetilde{I}) = \emptyset$ if and only if $\widetilde{I}(d) = A[X_0,X_1,\ldots,X_n](d)$ for d sufficiently large.*

Proof: Of course we know that $V(\widetilde{I}) = \emptyset$ if and only if $V(\widetilde{I}) \cap U_i = \emptyset$ for $i = 0 \ldots n$. This is equivalent to the assertion: For all i we can find $g_\nu \in \mathcal{O}_{\mathbb{P}^n}(U_i)$ and $f_\nu \in I_i$ such that

$$\sum g_\nu f_\nu = 1$$

(See prop 6.1.15 and its proof). Now we remember that we should think of $t_{i,j}$ as being X_j/X_i then we see that we can multiply this relation by a power of X_i and it yields a relation

$$\sum G_\nu F_\nu = X_i^{m_i}$$

where the G_ν,F_ν are homogeneous and $F_\nu \in I_i$. Hence we conclude that $V(\widetilde{I}) = \emptyset$ implies that for all i a suitable power of X_i is in the ideal, this proves one direction of the Lemma. The other direction is obvious. □

For our considerations above it was not necessary to assume that the base scheme $\mathrm{Spec}(A)$ is affine. All the constructions work over an arbitrary base scheme S.
We could have started with the ring $A = \mathbb{Z}$ and construct the scheme $\mathbb{P}^n_{\mathbb{Z}}$. If S is any scheme then we have the absolute morphism $S \longrightarrow \mathrm{Spec}(\mathbb{Z})$ and we could define $\mathbb{P}^n_S = \mathbb{P}^n_{\mathbb{Z}} \times S$.(See 6.2.5).
Of course it is again very easy to describe B-valued points of a closed subscheme $X \subset \mathbb{P}^n_A$ if this subscheme is given by a homogeneous ideal $I \subset A[X_0,\ldots,X_n]$. Then the B-valued points of X are given by

$$X(B) = \big\{(b_0,\ldots,b_n) \in B_*^{n+1} \mid f(b_0,\ldots,b_n) = 0 \,\forall\, f \in I \,\text{homogeneous}\big\}/\sim. \tag{8.17}$$

8.1.3 Projective Morphisms and Projective Schemes

Definition 8.1.3. *We call a morphism $\pi : X \to S$* **projective** *if we can find a commutative diagram*

$$\begin{array}{ccc} X & \xrightarrow{\ i\ } & \mathbb{P}^n_S \\ & \searrow & \downarrow \\ & & S, \end{array}$$

where i is a closed embedding, i.e. an isomorphism to a closed subscheme in $\mathbb{P}^n$. We also say that $\pi : X \longrightarrow S$ is a **projective scheme** *over S.*

Now we want to show that the fibered product of projective schemes exists and is again a projective scheme. We write $S = \mathrm{Spec}(A)$. We have the two schemes

$$\begin{array}{ccc} \mathbb{P}^n_S & & \mathbb{P}^m_S \\ & \searrow \quad \swarrow & \\ & S. & \end{array}$$

It will turn out that this fibered product of these two schemes exists and can be written as a closed subscheme of $\mathbb{P}_S^{nm+n+m}$.
To construct this closed subscheme we start from the usual covering by affine spaces. We change the numeration: We had $n+1$ open affine spaces to cover $\mathbb{P}_S^n$. Since we have $nm+n+m=(n+1)(m+1)-1$ hence we need $(n+1)(m+1)$ open affine spaces for the covering.
For $0 \leq i \leq n, 0 \leq j \leq m$ we write

$$U_{i,j} = \operatorname{Spec}\left(A[\ldots x_{i,j,\nu,\mu}\ldots]\right)$$

where of course $0 \leq \nu \leq n, 0 \leq \mu \leq m$, the $x_{i,j,\nu,\mu}$ are independent polynomial variables except that we have the relation $x_{i,j,i,j}=1$. We cover $\mathbb{P}_S^n$ and $\mathbb{P}_S^m$by

$$\begin{aligned} U_i &= \operatorname{Spec}\left(A[t_{i,0}\ldots t_{i,n}]\right) \\ V_j &= \operatorname{Spec}\left(A[t'_{j,0}\ldots t'_{j,m}]\right). \end{aligned}$$

We have already constructed the fibered product

$$U_i \times_S V_j = \operatorname{Spec}\left(A\left[t_{i,0},\ldots,t_{,in},t'_{j,0},\ldots,t'_{j,n}\right]\right).$$

We construct an morphism from $U_i \times V_j$ with a closed subscheme of $U_{i,j}$. To do this we construct an A-homomorphism of rings

$$A[\ldots,x_{i,j,\nu,\mu},\ldots] \longrightarrow A\left[t_{i,0},\ldots,t_{i,n},t'_{j,0},\ldots,t'_{j,m}\right]$$

and this homomorphism is given by

$$x_{i,j,\nu,\mu} \longrightarrow t_{i,\nu}\cdot t'_{j,\mu}.$$

We observe that $x_{i,j,i,j}$ maps to 1. We also observe that

$$\begin{aligned} x_{i,j,i,\mu} &\longrightarrow t'_{j,\mu} \\ x_{i,j,\nu,j} &\longrightarrow t_{i,\nu} \end{aligned}$$

and hence it is clear that the kernel of this homomorphism is the ideal $I_{i,j}$ generated by $x_{i,j,i,\mu}\cdot x_{i,j,\nu,j} - x_{i,j,\nu,\mu}$. If we divide by the ideal generated by these polynomials, then we get a polynomial ring generated by $x_{i,j,\nu,\mu}$ with $(\nu,\mu)\neq(i,j)$.
Hence this ideal defines a closed subscheme $V(I_{i,j})\subset U_{i,j}$, which is isomorphic to $U_i\times_S V_j$. This ideal defines a quasi-coherent sheaf $\widetilde{I}_{i,j}$ on $U_{i,j}$ and we will show that

$$\widetilde{I}_{i,j}|U_{i,j}\cap U_{i',j'} = \widetilde{I}_{i',j'}|U_{i,j}\cap U_{i',j'}. \tag{8.18}$$

To see this we consider the following diagram

$$\begin{array}{ccccc} U_i \times_S V_j & \xrightarrow{\sim} & V(I_{i,j}) \subset & & U_{i,j} \\ & & \cup & & \cup \\ & & V(I_{i,j})\cap V(I_{i',j'}) & \subset & U_{i,j}\cap U_{i',j'} \\ & & \cap & & \cap \\ U'_i \times_S V_j & \xrightarrow{\sim} & V(I_{i',j'}) \subset & & U_{i',j'}. \end{array}$$

We can construct the fibered product $(U_i \cap U_{i'}) \times_S (V_j \cap V_{j'})$ since both factors are affine and of course we can place it into the middle of the left column of the diagram above and we get open subschemes $(U_i \cap U_{i'}) \times_S (V_j \cap V_{j'}) \subset U_{i'} \times_S V_j$ and $(U_i \cap U_{i'}) \times_S (V_j \cap V_{j'}) \subset U_i \times_S V_j$. Of course we want a horizontal arrow from $(U_i \cap U_{i'}) \times_S (V_j \cap V_{j'})$ to $V(I_{i,j}) \cap V(I_{i',j'})$, which should be an isomorphism. To see that this works we rewrite the diagram in terms of rings, we drop the column in the middle

$$\begin{array}{ccc}
A[\dots t_{i\nu}, \dots, t'_{j\mu} \dots] & \longleftarrow & A[\dots x_{ij\nu\mu} \dots] \\
\cap & & \cap \\
A[\dots t_{i\nu}, \dots, t'_{j',\mu}, t_{ii'}^{-1}, t_{j,j'}^{-1}] & \longleftarrow & A[\dots, x_{i,j,\nu,\mu}, \dots, x_{i,j,i',j'}^{-1}] \\
\| & & \| \\
A[\dots t_{i',\nu}, \dots, t'_{j',\mu}, t_{i',i}^{-1}, t_{j',j}^{-1}] & \longleftarrow & A[\dots x_{i'j'\nu\mu} \dots, x_{i',j',i,j}^{-1}] \\
\cup & & \cup \\
A[\dots t_{i',\nu}, \dots, t'_{j',\mu} \dots] & \longleftarrow & A[\dots, x_{i',j',\nu,\mu}, \dots].
\end{array}$$

The two arrows in the middle are obtained from the arrow on the top and the bottom by the following rule: We send $x_{i,j,i',j'}^{-1}$ (resp. $x_{i',j',i,j}^{-1}$) to $t_{i,i'}^{-1} \cdot t_{j,j'}^{-1}$ (resp. $t_{i',i}^{-1} \cdot t_{j',j}^{-1}$). But using the rules for changing the coordinates yield that the two arrows in the middle are equal. The kernel of this arrow in the middle is clearly $(I_{i,j})_{x_{i,j,i',j'}} = (I_{i'j'})_{x_{i'j'ij}}$ and this proves $\widetilde{I}_{i,j} | U_{i,j} \cap U_{i',j'} = \widetilde{I}_{i',j'} | U_{i,j} \cap U_{i',j'}$. For later references we write our diagram again, but now we write the "completed diagram"

$$\begin{array}{ccccc}
U_i \times_S V_j & \xrightarrow{\sim} & V(I_{i,j}) \subset & & U_{i,j} \\
\cup & & \cup & & \cup \\
(U_i \cap U_{i'} \times_S (V_j \cap V_{j'}) & \xrightarrow{\sim} & V(I_{i,j}) \cap V(I_{i',j'}) & \subset & U_{i,j} \cap U_{i',j'} \\
\cap & & \cap & & \cap \\
U'_i \times_S V_j & \xrightarrow{\sim} & V(I_{i',j'}) \subset & & U_{i',j'}.
\end{array}$$

Hence the $I_{i,j}$ define a quasi coherent sheaf of ideals on $\mathbb{P}_S^{nm+n+m}$ and this defines a closed subscheme Y of $\mathbb{P}_S^{nm+n+m}$.
We can define projection maps

$$\begin{array}{ccc}
Y & \xrightarrow{p_1} & \mathbb{P}_S^n \\
& \searrow \quad \swarrow & \\
& S &
\end{array}
\qquad
\begin{array}{ccc}
Y & \xrightarrow{p_2} & \mathbb{P}_S^m, \\
& \searrow \quad \swarrow & \\
& S &
\end{array}$$

which on $U_{i,j} \cap Y = U_i \times_S V_j$ are the projections to the first and second factor. (It needs a little computation that these projections match on $U_{i,j} \cap U_{i',j'}$.)
Then it becomes clear that

$$\begin{array}{ccc}
& Y & \\
{}^{p_1}\swarrow & & \searrow^{p_2} \\
\mathbb{P}_S^n & & \mathbb{P}_S^m \\
\searrow & & \swarrow \\
& S &
\end{array}$$

is indeed a fibered product. If we have a scheme $T \to S$ and a pair of arrows

$$\begin{array}{ccc} T & \xrightarrow{f} & \mathbb{P}^n_S \\ & \searrow \quad \swarrow & \\ & S & \end{array} \qquad \begin{array}{ccc} T & \xrightarrow{g} & \mathbb{P}^m_S \\ & \searrow \quad \swarrow & \\ & S & \end{array}$$

then we find for any $t \in T$ an open neighborhood V such that $f(V) \subset U_i$,$g(V) \subset V_j$ for some i,j. Hence on V we get a map

$$\begin{array}{ccc} V & \longrightarrow & U_i \times_S V_j \simeq U_{i,j} \cap Y \\ & \searrow \quad \swarrow & \\ & S & \end{array}$$

and if we cover T by such $V's$ then the maps must match on the intersections. This follows from the "completed diagram" above.

We could also use the description of subschemes by homogeneous ideals. We introduce the ring $A[Z_{i,j}]$ where $i = 0,1,\ldots,n$,$j = 0,1,\ldots,n$ and in this ring we have the homogeneous ideal $\widetilde{\mathcal{P}}$ generated by the polynomials $Z_{i,j}Z_{ab} - Z_{aj}Z_{ib}$
for all quadruples of indices. Then the process of passing from homogeneous ideals to quasi-coherent sheaves of ideals gives us the ideal describing $\mathbb{P}^n_S \times_S \mathbb{P}^m_S$ as a subscheme of $\mathbb{P}^{(n+1)(m+1)-1}_S$.

The above embedding is the **Segre embedding**.

Locally Free Sheaves on $\mathbb{P}^n$

At this point we return to the construction of sheaves by the gluing process. We want to construct locally free sheaves and line bundles on $\mathbb{P}^n_A$. To do this we start from the collection of free modules of a fixed rank m on the open schemes U_i:

$$\mathcal{O}_{\mathbb{P}^n}(U_i)^m = A[t_{i,0},\ldots,t_{i,n}]^m = M_i.$$

They define sheaves $\widetilde{M_i}$ on the affine schemes U_i. Now we choose $\mathcal{O}_{\mathbb{P}^n}(U_i \cap U_j)$-linear isomorphisms

$$g_{i,j} : \widetilde{M_i}(U_i \cap U_j) \xrightarrow{\sim} \widetilde{M_j}(U_i \cap U_j),$$

this is nothing else than a collection of matrices $g_{i,j} \in Gl_m(\mathcal{O}_{\mathbb{P}^n}(U_i \cap U_j))$, which should satisfy the cocycle relation

$$\begin{aligned} g_{i,i} &= \mathrm{Id} \\ g_{i,j}g_{j,i} &= \mathrm{Id} \\ g_{i,j} \cdot g_{je} &= g_{i,e} \quad \text{on} \quad U_i \cap U_j \cap U_e. \end{aligned}$$

Then we get a locally free sheaf (vector bundle, see below) on $\mathbb{P}^n_A$ by glueing:

$$\widetilde{M} = (M_i, g_{i,j})_{i=0,\dots,n;j=0,\dots,n}.$$

We recall that for an open set a section $s \in \widetilde{M}(V)$ is a collection $s = (s_0, \dots, s_i, \dots, s_n)$ where $s_i \in \widetilde{M}_i(U_i \cap V)$ and where $g_{i,j}s_i = s_j$ on $V \cap U_i \cap U_j$. Especially we have $\widetilde{M}(U_i) = M_i$ because in this case the i-th component determines the others.
In the simple case $m = 1$ our matrix becomes a unit in $\mathcal{O}_{\mathbb{P}^n}(U_i \cap U_j)$ and we may choose for instance an integer r and define

$$g_{i,j} = t_{i,j}^{-r}.$$

This yields a locally free sheaf of rank one (or invertible sheaf), which is called $\mathcal{O}_{\mathbb{P}^n}(r)$ on $\mathbb{P}^n_A$. Again I point out that $\mathcal{O}_{\mathbb{P}^n}(r)(U_i) = \mathcal{O}_{\mathbb{P}^n}(U_i) = A[t_{i,0}, \dots, t_{i,n}]$.

Exercise 33. Let A be a factorial ring. Show that every line bundle $\mathcal{L}$ on $\mathbb{P}^n_A$ is isomorphic to $\mathcal{O}_{\mathbb{P}^n}(r)$ for some r. Hint: Exploit that for all i the ring $\mathcal{O}_{\mathbb{P}^n}(U_i)$ is a polynomial ring over A and hence factorial (See theorem 7.1.5). Then the restriction of $\mathcal{L}$ to U_i is free for all i, (see exercise 19.8) and the rest is clear.

Exercise 34. Compute $H^0(\mathbb{P}^n_A, \mathcal{O}_{\mathbb{P}^n}(r))$ and show that this is isomorphic to the A-module of homogeneous polynomials in $n+1$-variables of degree r. To be more precise: A section in $H^0(\mathbb{P}^n_A, \mathcal{O}_{\mathbb{P}^n}(r))$ is by definition a collection $s = (f_0, \dots, f_n)$ where $f_i \in A[t_{i,0}, \dots, t_{i,n}]$ which satisfies

$$t_{i,j}^{-r} \cdot f_i = f_j.$$

on $U_i \cap U_j$. A homogeneous polynomial

$$F(X_0, \dots, X_n) = \Sigma a_{\nu_0, \dots, \nu_n} X_0^{\nu_0} \dots X_n^{\nu_n}$$

with $\Sigma \nu_i = r$ provides such a section if we define

$$f_i(t_{i,0}, \dots, t_{i,n}) = F(t_{i,0}, \dots, 1, \dots, t_{i,n}).$$

Show that this gives us the isomorphism!

Exercise 35. Any homogeneous polynomial $F(X_0, \dots, X_n)$ defines a sheaf of ideals

$$\mathcal{O}_{\mathbb{P}^n}(F) \hookrightarrow \mathcal{O}_{\mathbb{P}^n}$$

which is defined by $\mathcal{O}_{\mathbb{P}^n}(F)(U_i) = \mathcal{O}_{\mathbb{P}^n}(U_i)f_i$, in other words: the restriction of the ideal to the open sets U_i is the principal ideal (f_i).
Show: Assume that for all $\mathfrak{p} \in \operatorname{Spec}(A)$ this polynomial is non zero in $A/\mathfrak{p}[X_0, \dots, X_n]$, then this is a locally free sheaf on $\mathbb{P}^n_A$, which is isomorphic to $\mathcal{O}_{\mathbb{P}^n}(-d)$ where $d =$ degree of F!
Hint: Write a section $\mathcal{O}_{\mathbb{P}^n}(F)(V)$ as a collection of elements $(\dots, h_i f_i, \dots)$ and show that we must have $t_{i,j}^d h_i = h_j$. Where do we use our assumption? Can it be replaced by a weaker assumption?

We refer to the exercise 35. The sheaf of ideals defines a closed subscheme $V(F)$, which is called a hypersurface of degree d. If we take for instance simply $F = X_0$ then this closed subscheme is the reduced scheme $H_0 = \mathbb{P}^n \setminus U_0$, it is isomorphic to $\mathbb{P}^{n-1}/\operatorname{Spec}(A)$. It is called the hyperplane at infinity (from the point of view of somebody who lives in U_0). We consider the other hyperplanes $H_i = V(X_i)$ as well. The exercise shows that $\mathcal{O}_{\mathbb{P}^n}(X_i) \xrightarrow{\sim} \mathcal{O}_{\mathbb{P}^n}(-1)$.

The tensor product of locally free sheaves is again a locally free sheaf and especially we have $\mathcal{O}_{\mathbb{P}^n}(r) \otimes \mathcal{O}_{\mathbb{P}^n}(s) \xrightarrow{\sim} \mathcal{O}_{\mathbb{P}^n}(r+s)$. The inclusion $\mathcal{O}_{\mathbb{P}^n}(X_0) \subset \mathcal{O}_{\mathbb{P}^n}$ combined with the isomorphism $\mathcal{O}_{\mathbb{P}^n}(X_0) \xrightarrow{\sim} \mathcal{O}_{\mathbb{P}^n}(-1)$ gives us an inclusion $\mathcal{O}_{\mathbb{P}^n}(-1) \hookrightarrow \mathcal{O}_{\mathbb{P}^n}$ and this induces after taking tensor products $\mathcal{O}_{\mathbb{P}^n}(r-1) \hookrightarrow \mathcal{O}_{\mathbb{P}^n}(r)$ and we get a chain of inclusions

$$\mathcal{O}_{\mathbb{P}^n} \hookrightarrow \mathcal{O}_{\mathbb{P}^n}(1) \hookrightarrow \mathcal{O}_{\mathbb{P}^n}(2) \hookrightarrow \ldots \hookrightarrow \mathcal{O}_{\mathbb{P}^n}(d) \ldots \tag{8.19}$$

where all the embeddings are obtained from $\mathcal{O}_{\mathbb{P}^n}(X_0) \hookrightarrow \mathcal{O}_{\mathbb{P}^n}$.
Of course it should be clear that we have many ways of mapping the sheaf $\mathcal{O}_{\mathbb{P}^n}$ into $\mathcal{O}_{\mathbb{P}^n}(d)$. To be more precise we can look at $\mathrm{Hom}_{\mathcal{O}_{\mathbb{P}^n}}(\mathcal{O}_{\mathbb{P}^n},\mathcal{O}_{\mathbb{P}^n}(d))$ and to give a homomorphism among sheaves we only need to know what happens to $1 \in H^0(\mathbb{P}^n,\mathcal{O}_{\mathbb{P}^n})$ because this section generates the stalk on each point. Hence we get

$$\mathrm{Hom}_{\mathcal{O}_{\mathbb{P}^n}}(\mathcal{O}_{\mathbb{P}^n},\mathcal{O}_{\mathbb{P}^n}(d)) = H^0(\mathbb{P}^n,\mathcal{O}_{\mathbb{P}^n}(d)) \tag{8.20}$$

and this is the A-module of homogeneous forms of degree d in $A[X_0,\ldots,X_n]$. Hence giving such an embedding $\mathcal{O}_{\mathbb{P}^n} \hookrightarrow \mathcal{O}_{\mathbb{P}^n}(d)$ amounts to pick a form of degree d. At this point we have chosen the form X_0^d.

$\mathcal{O}_{\mathbb{P}^n}(d)$ *as Sheaf of Meromorphic Functions*

If we look at $\mathcal{O}_{\mathbb{P}^n} \hookrightarrow \mathcal{O}_{\mathbb{P}^n}(d)$ and restrict this map to U_i then we get

$$\mathcal{O}_{\mathbb{P}^n}(U_i) \longrightarrow \mathcal{O}_{\mathbb{P}^n}(d)(U_i) = \mathcal{O}_{\mathbb{P}^n}(U_i)$$

and this map is given by

$$f_i \longmapsto t_{0,i}^d f_i. \tag{8.21}$$

We can embed $\mathcal{O}_{\mathbb{P}^n}(U_i)$ into the module $\frac{1}{t_{0,i}^d}\mathcal{O}_{\mathbb{P}^n}(U_i)$, which is the $\mathcal{O}_{\mathbb{P}^n}(U_i)$-module of meromorphic functions on U_i, which are regular on $U_i \cap U_0$ and have at most a "pole of order d" along the hyperplane $U_i \backslash U_i \cap U_0 = U_i \cap V(X_0)$. Then the map above gives an isomorphism

$$\frac{1}{t_{i_0}^d}\mathcal{O}_{\mathbb{P}^n}(U_i) \xrightarrow{\sim} \mathcal{O}_{\mathbb{P}^n}(d)(U_i). \tag{8.22}$$

The notion of a pole will also be discussed in section 9.1

(8.23)

Especially for $i = 0$ we have $t_{0,0} = 1$, which means that $V(X_0)$ does not meet U_0. We glue these modules $\frac{1}{t_{0,i}^d}\mathcal{O}_{\mathbb{P}^n}(U_i)$ over the intersections of two affine sets and hence we get the sheaf $\mathcal{O}_{\mathbb{P}^n}(dH_0)$ and call this the sheaf of "meromorphic functions on $\mathbb{P}^n$", which are regular on U_0 and have at most a pole of order d along the hyperplane at infinity. We have a diagram

$$\mathcal{O}_{\mathbb{P}^n} \hookrightarrow \mathcal{O}_{\mathbb{P}^n}(dH_0) \xrightarrow{\sim} \mathcal{O}_{\mathbb{P}^n}(d).$$

This also gives a slightly different view of the above chain of inclusions: The chains arise simply because we allow higher and higher orders of poles.
Of course we have also an interpretation for $r < 0$: In this case the sections are the germs of regular functions, which have a zero of order $\geq -r$ along the hyperplane.
Instead of looking at the hyperplane at infinity, we can choose an arbitrary homogenous polynomial F of degree d, again we make the assumption (*nonzero*) that

For all $\mathfrak{p} \in S = \mathrm{Spec}(A)$ the image of the polynomial is non zero in$A/\mathfrak{p}[X_0,X_1,\ldots,X_n]$.

Then we can define the sheaf $\mathcal{O}_{\mathbb{P}^n}(V(F))$ of functions h, which are regular on $\mathbb{P}^n \setminus V(F)$, and whose restriction to any $U_i \cap V(F)$ extends to a regular function on U_i after we multiply it by $f_i = F|U_i$. These are the functions, which are regular outside $V(F)$ and have at most a first order pole along (or at) $V(F)$.

The Relative Differentials and the Tangent Bundle of $\mathbb{P}^n_S$

The scheme

$$\begin{array}{c}\mathbb{P}^n_A \\ \downarrow \\ \mathrm{Spec}(A)\end{array}$$

is always smooth. Locally on one of the U_i the module of differentials is the free module generated by the $dt_{i,0},\ldots,dt_{i,i-1},dt_{i,i+1},\ldots,dt_{i,n}$, of course $dt_{i,i} = 0$.
We consider the n -th exterior power of this module

$$\Lambda^n \Omega_{\mathbb{P}^n/A} = \Omega^n_{\mathbb{P}^n/A}.$$

The restriction $\Omega^n_{\mathbb{P}^n/A} \mid U_i$ is the trivial line bundle generated by

$$dt_{i,0} \wedge \ldots \wedge dt_{i,i-1} \wedge dt_{i,i+1} \wedge \ldots \wedge dt_{i,n},$$

and on $U_i \cap U_j$ we have

$$t_{i,\nu} = t_{i,j} t_{j,\nu}, \tag{8.24}$$

and hence

$$dt_{i,\nu} = t_{i,j} dt_{j,\nu} + t_{j,\nu} \cdot dt_{i,j}. \tag{8.25}$$

Taking the highest exterior power we get

$$dt_{i,0} \wedge \ldots \wedge dt_{i,i-1} \wedge dt_{i,i+1} \wedge \ldots \wedge dt_{i,n} = \Bigg(t_{i,j} dt_{j,0} + t_{j,0} dt_{i,j} \Bigg) \wedge \ldots \wedge \Bigg(t_{i,j} dt_{j,n} + t_{j,n} dt_{i,j} \Bigg),$$

where on the right hand side we have to leave out the factor with index $\nu = i$. We assume $i \neq j$, the factor $\nu = j$ simplifies to $t_{j,j} dt_{i,j}$. Since $t_{i,j} = t_{j,i}^{-1}$ we have $dt_{i,j} = -t^2_{i,j} dt_{j,i}$. Hence we see that the right hand side is

$$(-1)^{i-j} t_{i,j}^{n+1}\ dt_{j,0} \wedge \ldots \wedge dt_{j,j-1} \wedge dt_{j,j+1} \ldots \wedge dt_{jn}.$$

Therefore we see that the line bundle $\Omega_{\mathbb{P}^n_A/A}$ is obtained from the cocycle

$$(-1)^{i-j} t_{i,j}^{n+1} = g_{i,j} \tag{8.26}$$

in the sense of the consideration on page 129. But we may of course change our generator of $\Omega^n_{\mathbb{P}^n_A/A}$ over U_i by the sign $(-1)^i$ and the cocycle modifies into $g_{i,j} = t_{i,j}^{n+1}$, which then implies

$$\Omega^n_{\mathbb{P}^n_A/A} \simeq \mathcal{O}_{\mathbb{P}^n_A/A}(-n-1). \tag{8.27}$$

We consider the dual sheaf of $\Omega^1_{\mathbb{P}^n/A}$, this is the sheaf of tangent vectors. We want to achieve a more geometric understanding of this bundle. We recall that we have the morphism

$$\pi : \mathbb{A}^{n+1}_S \setminus S \longrightarrow \mathbb{P}^n_S.$$

This morphism is a $\mathbb{G}_m$-torsor. It is trivialised over U_i by the sections s_i (See 8.15), we have $t_{i,j}s_i = s_j$ and hence the associated line bundle is $\mathcal{O}_{\mathbb{P}^n}(-1)$.
Then the tangent bundle of $\mathbb{A}^{n+1}_S \setminus S$ is trivial, we get an exact sequence of vector bundles on $\mathbb{A}^{n+1}_S \setminus S$

$$0 \longrightarrow T_{\mathbb{A}^{n+1}_S \setminus S/\mathbb{P}^n_S} \longrightarrow \mathbb{A}^{n+1}_{\mathbb{A}^{n+1}_S \setminus S} \longrightarrow f^*(T_{\mathbb{P}^n_S}) \longrightarrow 0 \tag{8.28}$$

This induces an exact sequence of bundles on $\mathbb{P}^n_S$: For any open subset $U \subset \mathbb{P}^n$ we consider a subspace of sections in the tangent bundle of $\mathbb{A}^{n+1}_S \setminus S$

$$\{f \in \mathbb{A}^{n+1}_{\mathbb{A}^{n+1}_S \setminus S}(\pi^{-1}(U)) | f \text{ is homogenous of degree } 1\} \tag{8.29}$$

i.e. $f(tx) = tf(x)$ for all $t \in \mathcal{O}_{\mathbb{A}^{n+1}_S \setminus S}(\pi^{-1}(U))^\times$. This space of sections is clearly equal to $\mathcal{O}_{\mathbb{P}^n_S}(U)(1)^{n+1}$. It follows from a simple calculation that the derivative $D_\pi(f) \in f^*(T_{\mathbb{P}_S})(\pi^{-1}(U))$ is actually constant, i.e. an element in $T_{\mathbb{P}_S}(U)$. Hence we get a surjective homomorphism $\mathcal{O}_{\mathbb{P}^n_S}(1)(U)^{n+1} \longrightarrow T_{\mathbb{P}_S}(U)$. The kernel $T_{(\mathbb{A}^{n+1}_S \setminus S)/\mathbb{P}^n_S} \longrightarrow \mathbb{A}^{n+1}_{\mathbb{A}^{n+1}_S \setminus S}$ in the sequence above is $\pi^*(\mathcal{O}_{\mathbb{P}^n_S}(-1))$. Eventually we get the exact sequence

$$0 \longrightarrow \mathcal{O}_{\mathbb{P}^n_S} \longrightarrow \mathcal{O}_{\mathbb{P}^n_S}(1)^{n+1} \longrightarrow T_{\mathbb{P}_S} \longrightarrow 0. \tag{8.30}$$

Further up we realised the sheaf $\mathcal{O}_{\mathbb{P}^n_S}(1)$ as the sheaf of meromorphic functions, which are regular on U_i and have at most a first order pole at the hyperplane H_i. This gives us an embedding $\mathcal{O}_{\mathbb{P}^n_S} \hookrightarrow \mathcal{O}_{\mathbb{P}^n_S}(H_i) = \mathcal{O}_{\mathbb{P}^n_S}(1)$. Hence we get an embedding

$$\mathcal{O}_{\mathbb{P}^n_S} \hookrightarrow \bigoplus_{i=0}^{i=n} \mathcal{O}_{\mathbb{P}^n_S}(H_i).$$

Now it follows from our formulae for the coordinate changes, that the tangent vector field $\frac{\partial}{\partial t_{i,\nu}}$ on U_i extends to a tangent vector field on $\mathbb{P}^n_S$. Hence for any $f \in \mathcal{O}_{\mathbb{P}^n_S}(H_i)(U_i)$, which is of the form $f = a_i + \sum_{\nu \neq i} a_\nu t_{i,\nu}$ we get a global tangent vector field $\partial f = \sum_{\nu \neq i} a_\nu \frac{\partial}{\partial t_{i,\nu}}$ and this yields another version of our sequence above

$$\mathcal{O}_{\mathbb{P}^n_S} \hookrightarrow \bigoplus_{i=0}^{i=n} \mathcal{O}_{\mathbb{P}^n_S}(H_i) \longrightarrow T_{\mathbb{P}_S} \longrightarrow 0.$$

Finally we observe that taking the $n+1$-th power we get the isomorphism

$$\Lambda^n(T_{\mathbb{P}_S}) \xrightarrow{\sim} \mathcal{O}_{\mathbb{P}^n_S}(n+1),$$

which dual to our formula above.

8.1.4 Seperated and Proper Morphisms

The property of a morphism to be projective is a global property in contrast to the property to be smooth or flat, which can be checked locally.
There are two other properties of morphism, which are global in nature namely a morphism $X \to S$ can be **separated** or it can be **proper** and I think here is the right place to discuss them.
If we have a scheme $\pi : X \to S$ then we can form the fibered product $X \times_S X/S$ and the identity $\mathrm{Id} : X \to X$ provides an element $(\mathrm{Id}, \mathrm{Id}) \in \mathrm{Hom}_S(X,X) \times \mathrm{Hom}_S(X,X)$. By the universal property this is nothing else than an element $\Delta_X \to X \times_S X$.

Definition 8.1.4. *The morphism π is called* **separated** *if Δ_X is a closed embedding.*

It is not too hard to see that this property is local in the base, hence if we discuss this notion we may assume that $S = \mathrm{Spec}(A)$. Then

Lemma 8.1.5. *A morphism $\pi : X \to \mathrm{Spec}(A) = S$ is separated if for any two affine open subsets $U,V \subset X$ the intersection $U \cap V$ is affine again and $\mathcal{O}_X(U \cap V)$ is generated by the restriction of $\mathcal{O}_X(U)|U \cap V$ and $\mathcal{O}_X(V)|U \cap V$.*

This is rather clear because - as we mentioned at the end of the section on fibered products- we can cover $X \times_S X$ by open affine subsets $U \times_S V$. The morphism Δ is a closed embedding if for any such pair $U \cap V \to U \times_S V$ is a closed embedding, But then $U \cap V$ is a closed subscheme of the affine scheme $U \times_S V$ and its ring of regular functions $\mathcal{O}_X(U \cap V)$ is a quotient of $\mathcal{O}_X(U) \otimes_A \mathcal{O}_X(V)$.
We should notice that a morphism $\pi : X \to \mathrm{Spec}(A) = S$ is separated if we can find some covering $\mathfrak{U} = \{U_i\}_{i \in I}$ by affine subschemes such that for any pair of indices i,j the affine schemes U_i,U_j satisfy the condition in the lemma above.
We have given an explicit construction of $\mathbb{P}^n_S \times_S \mathbb{P}^m_S$ as a projective subscheme of some $\mathbb{P}^N_S$. We apply this to the case $n = m$. In this case we can either verify that the system of affine sets $\{U_i\}_{i=0,\dots,n}$ satisfies the condition in the Lemma because the elements in $\mathcal{O}_{\mathbb{P}^n}(U_i)$, which we have to invert to get $\mathcal{O}_{\mathbb{P}^n}(U_i \cap U_j)$ lie in $\mathcal{O}_{\mathbb{P}^n}(U_j)$. Hence we see that that $\mathbb{P}^n \to S$ is separated. We could also argue that we can describe the diagonal as a closed subscheme of the product just by adding the polynomials $Z_{i,j} - Z_{j,i}$, to the ideal which describes the product as a subscheme of $\mathbb{P}_S^{(n+1)(m+1)-1}$.

Definition 8.1.6. *An open subscheme $Y \subset X/S$ of a projective scheme X/S is called* **quasi projective** *over S.*

It is now clear that quasi projective schemes Y/S are also separated.

Finally we want to introduce the notion of a proper morphism.

Definition 8.1.7. *A morphism $\pi : X \to S$ is called* **proper** *if it is separated and if it is universally closed. This means that for any base change $S' \to S$ and any closed subscheme $Z \subset X \times_S S'$ the image of Z under the projection $\pi \times_S S'$ is closed.*

Theorem 8.1.8. *A projective morphism $\pi : X \to S$ is always proper.*

We have just seen that it is separated. We recall the definition of a projective morphism. It means that we have a diagram

$$\begin{array}{ccc} X & \xhookrightarrow{\;i\;} & \mathbb{P}^n_S \\ & \searrow^{\pi} & \downarrow \\ & & S \end{array}$$

where i is a closed embedding. We have to show that for any base change $S' \to S$ and any closed subset $Y \subset X \times_S S'$ the image of π is closed. Since this closed subset is also a closed subset in $\mathbb{P}^n_{S'}$ it suffices to show that for any closed subset $Z \subset \mathbb{P}^n_S$ its image in S is closed.

The question is local in the base, we assume that $S = \mathrm{Spec}(A)$. Now we know that we can describe Z as the set of zeroes of a homogeneous ideal $\widetilde{I} \subset A[X_0,X_1,\dots,X_n]$ (see 1.1.2). Now we pick a point $s \in S$, which is not in the image of Z. We localize A at s, let $\mathfrak{m}_s$ be the maximal ideal of this local ring $A_{\mathfrak{m}_s}$. Now we have $Z_s = \emptyset$ and our Lemma 8.1.2 tells us that for a sufficiently large $d \gg 0$ we have

$$\widetilde{I}(d) \otimes A_{\mathfrak{m}_s}/\mathfrak{m}_s \simeq A_{\mathfrak{m}_s}/\mathfrak{m}_s[X_0,X_1,\dots,X_n](d)$$

or in other words

$$A_{\mathfrak{m}_s}[X_0,X_1,\dots,X_n](d)/\widetilde{I}(d) \otimes A_{\mathfrak{m}_s}/\mathfrak{m}_s = (0).$$

By the lemma of Nakayama it follows that

$$A_{\mathfrak{m}_s}[X_0,\dots,X_n](d)/\widetilde{I}(d) = 0.$$

We are basically through but since we passed to the localization at s, we need still a little finiteness argument. We know that any monomial $X_0^{r_0} \cdots X_n^{r_n}$ of degree d can be written in the form

$$X^{r_0} \dots X_n^{r_n} = \sum_u G_\mathfrak{u} \cdot F_\mathfrak{u}$$

where $F_\mathfrak{u} \in \widetilde{I}$ and $G_\mathfrak{u} \in A[X_0,\dots,X_n]$. The local ring $A_{\mathfrak{m}_s}$ is obtained from our original A by localization. But to write down the $G_\mathfrak{u}$ we need only finitely many denominators, which means that we can replace A by a localization A_f with some f with $f(s) \neq 0$. Hence we see that already

$$\widetilde{I}(d) \otimes A_f \simeq A_f[X_0,X_1 \dots X_n](d),$$

which proves that the image of Z does not intersect with the open neighborhood $\mathrm{Spec}(A_f)$ of s. This proves the theorem. □

We need some formal properties of proper morphisms.

Proposition 8.1.9. *Let $\pi : X \longrightarrow S$ be a proper morphism.*

a) *If $S' \longrightarrow S$ is any scheme over S, then $X \times_S S' \longrightarrow S'$ is also proper.*

b) *If $Y \longrightarrow S$ is another scheme and if $f : X \longrightarrow Y$ is an S-morphism, then f is proper.*

Proof: The first assertion is obvious, because for any scheme $S'' \longrightarrow S'$ we have $X \times_S S'' = (X \times_S S') \times_{S'} S''$. For the second assertion we notices that for any scheme $S' \longrightarrow Y$ the scheme $X \times_Y S'$ is a closed subscheme of $X \times_S S'$. □

8.1.5 The Valuative Criteria

I want to state criteria for separatedness and properness, which are extremely important, and which give a very intuitive idea of this notion. We will not give the proofs here, we refer to A. Grothendieck's book [Gr-EGA II]. The parts of that book concerning these valuative criteria is relatively self contained, so it can be read directly. For the central results proved in this book we do not need the concept of proper morphisms. At some points we have to work a little bit to circumvent the use of this notion.

Since we know that the question whether a morphism $\pi : X \to Y$ is separated (resp. proper) is local in the base, we consider the following situation. Let $Y = \mathrm{Spec}(A)$ where A is noetherian, let $\pi : X \to Y$ be of finite type, i.e. we can cover X by affines $U_\alpha = \mathrm{Spec}(B_\alpha)$ where B_α is a finitely generated A-algebra. Then we have:

Theorem 8.1.10. *Under our conditions above the morphism $\pi : X \to \mathrm{Spec}(A)$ is*

a) *separated if for any discrete valuation ring C with quotient field K and any morphism*

$$f : \mathrm{Spec}(C) \longrightarrow \mathrm{Spec}(A)$$

two $\mathrm{Spec}(C)$ valued points

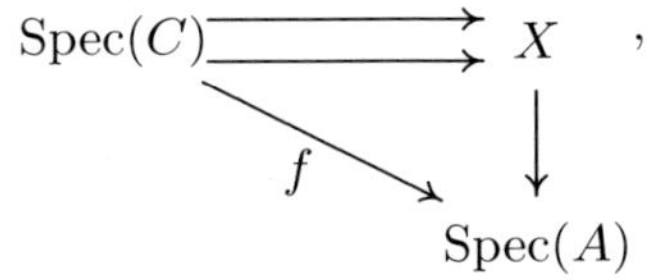

which become equal if we restrict them to $\mathrm{Spec}(K)$, are already equal,

b) *proper if for any such C and any f a $\mathrm{Spec}(K)$ valued point extends (uniquely) to a $\mathrm{Spec}(C)$ valued point.*

For the proof see [Gr-EGA II],§7, 7.2.3 and Remarks 7.2.4 and 7.3.8].

The Valuative Criterion for the Projective Space

The following property of a projective space is an algebraic substitute for the fact that the complex projective space $\mathbb{P}^n(\mathbb{C})$ is compact.

Let us start from a discrete valuation ring R. We want to study the $S = \mathrm{Spec}(R)$ valued points of $\mathbb{P}^n_S \to S$. Let π be a uniformizing element of R and let K be its quotient field. We have a diagram

$$\begin{array}{ccc} \mathbb{P}^n & & \\ \downarrow & & \\ \mathrm{Spec}(R) & \longleftarrow & \mathrm{Spec}(K) \end{array}$$

The valuative criterion asserts that the K valued points and the R valued points are the same. But this is almost obvious. We have seen on page 123

$$\mathbb{P}^n_S(\mathrm{Spec}(K)) = \mathbb{P}^n_S(K) = (K^{n+1} \setminus \{0\})/K^*.$$

But for any $x = (x_0,x_1 \dots ,x_n) \in K^{n+1} \setminus \{0\}$ we write $x_i = u_i\pi^{n_i}$ and pick the index i_0, for which n_{i_0} is minimal. Then $(\pi^{-n_{i_0}}x_0,\pi^{-n_{i_0}}x_1 \dots ,\pi^{-n_{i_0}}x_n)$ represents the same point, but now the coordinates are in R and one of the coordinates is a unit. Therefore this is a R-valued point and have have shown

Proposition 8.1.11. *If R is a discrete valuation ring with quotient field K then*

$$\mathbb{P}^n_{\mathrm{Spec}(R)}(R) = \mathbb{P}^n_{\mathrm{Spec}(R)}(K).$$

This is the valuative criterion for the projective space. This of course extends immediately to projective schemes over R.
This expresses in algebraic terms the *compactness* of the projective space. We should look at $\mathrm{Spec}(R)$ as a small disc, the generic point $(0) \in \mathrm{Spec}(R)$ corresponds to the disc minus the origin and the closed point (π) corresponds to the origin. The analogous object in function theory is a disc $D = \{z||z| < 1\}$ in the complex plane. Then we know from function theory that a meromorphic map $f : D \setminus \{0\} \to U_i(\mathbb{C}) \subset \mathbb{P}^n(\mathbb{C})$ extends in a unique way to a holomorphic map $f : D \to \mathbb{P}^n(\mathbb{C})$.

8.1.6 The Construction Proj(R)

We have a construction of projective schemes starting from a graded algebra. Let A be an arbitrary ring and let R be a graded A-algebra. This means that R is an A-algebra and we have a direct sum decomposition

$$R = R_0 \oplus R_1 \oplus \dots \oplus R_n \oplus \dots$$

such that the R_i are A-modules and $R_iR_j \subset R_{i+j}$. The identity element is in R_0 and the algebra homomorphism from A to R factors through R_0. We assume that $A \to R_0$ is surjective and we assume that R_1 is a finitely generated A-module, which generates the A-algebra.

Definition 8.1.12. *We define the set* $\mathrm{Proj}(R)$, *the* **projective spectrum**, *to be the set of homogeneous prime ideals of R, which do not contain R_1.*

If we have such a prime ideal $\mathfrak{p}$ we pick an $f \in R_1$ such that $f \notin \mathfrak{p}$ and we define

$$\operatorname{Proj}(R)_f = \{\mathfrak{q} \in \operatorname{Proj}(R) | f \notin \mathfrak{q}\}. \tag{8.31}$$

We form the ring

$$R_f^{(0)} = \left\{ \frac{g}{f^n} | g \in R_n \right\}, \tag{8.32}$$

i.e. the ring of elements of degree zero in the quotient ring R_f. It is very easy to see that

$$\operatorname{Proj}(R)_f = \operatorname{Spec}\left(R_f^{(0)}\right). \tag{8.33}$$

We use this to define a topology on $\operatorname{Proj}(R)$ and a structure of a ringed space. The open sets $V \subset \operatorname{Proj}(R)$ are those, for which $V \cap \operatorname{Proj}(R)_f$ is open for all f.
Then we define the sheaf of regular functions so that its restriction to the $\operatorname{Proj}(R)_f$ is simply the sheaf $\widetilde{R_g^{(0)}}$ on $\operatorname{Spec}\left(R_g^{(0)}\right)$. Now we have defined a scheme $(\operatorname{Proj}(R),\mathcal{O})$ for any such graded A-algebra. Usually we drop the $\mathcal{O}$ in the notation and $\operatorname{Proj}(R)$ will denote the scheme, i.e the underlying set plus the sheaf.
If we take for instance the polynomial graded algebra $A[X_0, \ldots, X_n]$ then

$$\operatorname{Proj}(A[X_0, \ldots, X_n],\mathcal{O}) = (\mathbb{P}^n, \mathcal{O}_{\mathbb{P}^n}).$$

If our A-algebra R is generated by elements $x_0, \ldots, x_n$ of degree one we have a homomorphism of graded A-algebras

$$\begin{array}{ccc} A[X_0, \ldots, X_n] & \longrightarrow & R_0[x_0, \ldots, x_n] \\ & \nwarrow \quad \nearrow & \\ & A & \end{array}$$

and we see that $\operatorname{Proj}(R)$ is a closed subscheme

$$\begin{array}{ccc} \operatorname{Proj}(R) & \longrightarrow & \mathbb{P}_A^n \\ & \searrow \quad \swarrow & \\ & \operatorname{Spec}(A). & \end{array}$$

The intersections $U_i \cap \operatorname{Proj}(R)$ are affine and clearly $U_i \cap \operatorname{Proj}(R) = \operatorname{Spec}(R_{x_i}^{(0)})$
This generalizes easily to the case where we replace $\operatorname{Spec}(A)$ by an arbitrary scheme S and where R is a sheaf of graded $\mathcal{O}_S$-algebras.

The assumption that R is generated by homogeneous elements in degree one is not essential. Let us assume that R is finitely generated by elements $x_0, x_1, \ldots, x_n$, which are of degree d_0, $d \geq 1$ respectibely. Then we can define

$$R_{x_i}^{(0)} = \left\{ \frac{f}{x_i} : \deg(f) = d_i m \right\} \subset R_{x_i}, \tag{8.34}$$

and we can consider Spec $\left(R_{x_i}^{(0)}\right)$. Now we define Proj$(R)$ as the space of homogeneous prime ideals $\mathfrak{p}$, which do not contain all the x_i. Then we have

$$\text{Proj}(R) = \bigcup_{i=1}^{n} \text{Spec}\left(R_{x_i}^{(0)}\right) \tag{8.35}$$

and we can proceed as before.

We get such graded algebras if we start from any A-module N and consider its symmetric graded algebra (See 6.2.3)

$$R = \text{Sym}^\bullet(N) = A \oplus N \oplus \text{Sym}^2(N) \oplus \dots .$$

If this module N is written as a quotient of a free A-module $M^\vee = AX_0 \oplus AX_1 \oplus \dots \oplus AX_n \longrightarrow N$, then we get a surjective homomorphism

$$A[X_0,X_1,\dots,X_n] \longrightarrow \text{Sym}^\bullet(N)$$

and hence a closed embedding $\text{Proj}(R) \hookrightarrow \mathbb{P}^n_S$.
Finally we may start from any scheme S and a locally free $\mathcal{O}_S$-module $\widetilde{M}$ of finite rank. Then we can use the standard gluing procedure to construct the scheme

$$\mathbb{P}(\widetilde{M}) = \text{Proj}(\text{Sym}^\bullet(\widetilde{M}^\vee))/S.$$

For any point $x \in S$ and the resulting local ring $\mathcal{O}_{S,x}$ the set of $\mathcal{O}_{S,x}$ -valued points of $\mathbb{P}(\widetilde{M})$ is the set of lines through the origin in $\widetilde{M} \otimes \mathcal{O}_{S,x}$. (See section on homogenous coordinates.)

A special case of a finiteness result.

The following result is a special case of a general theorem of Grothendieck, which will be stated later without a proof. For this special case we want to avoid the reference to the general result.

Definition 8.1.13. *A morphism $f : X \longrightarrow Y$ of schemes is called finite if it is affine, i.e. Y has a finite covering by affine open set V_i such that $U_i = f^{-1}(V_i)$ are affine and if in addition the restrictions $U_i \longrightarrow V_i$ are finite.*

We consider an affine scheme of finite type $\pi : Y \longrightarrow S$, the base is arbitrary. Then we have

Proposition 8.1.14. *If the morphism $\pi : Y \longrightarrow S$ is affine and proper, then the sheaf $\pi_*(\mathcal{O}_Y)$ is a locally finitely generated sheaf of $\mathcal{O}_S$-modules. Especially if $S =$ Spec(A) is affine and $Y =$ Spec$[A[X_1,X_2,\dots,X_n]/I =$ Spec$(A[x_1,\dots,x_n]) =$ Spec(B), then the A-algebra B is finite over A.*

Proof: The assertion is local in S, so we only have to prove the second assertion. The ideal I is generated by polynomials $f_\mu(X_1,X_2,\dots,X_n) = \sum_{\underline{\nu}} a_{\mu,\underline{\nu}} X^{\underline{\nu}}$, here $\underline{\nu}$ is a multiindex $\nu = (\nu_1,\dots,\nu_n)$ and $X^{\underline{\nu}} = X_1^{\nu_1} \dots X_n^{\nu_n}$. We put $\deg(\nu) = \sum \nu_i$. Let d_μ be the maximum $\deg(\nu)$, for which we have an $a_{\mu.\nu} \neq 0$.

We embed the affine variety into $\mathbb{P}^n/S$, to do this we make the equations homogenous: We introduce the new variable X_0 and define

$$F_\mu(X_0,X_1,X_2,\dots,X_n) = \sum_{\underline{\nu}} a_{\mu,\underline{\nu}} X^{\underline{\nu}} X_0^{d_\mu - \deg(\nu)},$$

let $\widetilde{I}$ be the homogenous ideal in $A[X_0,X_1,\dots,X_n]$. In the notation of 8.1.1 we have $X \subset U_0 \subset \mathbb{P}^n_S$. Since Y/S is proper we know that $Y \subset \mathbb{P}^n_S$ is a closed embedding (See prop. 8.1.9 b) and therefore, the projective scheme defined by the ideal $\widetilde{I}$ is equal to Y. Consequently this ideal defines the empty sub scheme in the complement H_0 of U_0. What does this mean? We restrict the ideal to this complement. To get this restriction we write

$$F_\mu(X_0,X_1,X_2,\dots,X_n) = X_0 G_\mu(X_0,X_1,\dots,X_n) + H_\mu(X_1,X_2,\dots,X_n)$$

where $H_\mu(X_1,X_2,\dots,X_n)$ is homogenous of degree d_μ, it collects the monomials, which do not contain X_0. The restriction of $\widetilde{I}$ to H_0 is now simply the ideal generated by the H_μ. These H_μ do not have a common zero on H_0 and therefore, for any index $i = 1,\dots,n$ we can find homogenous polynomials $R_{i,\mu}(X_1,\dots,X_n)$ and integers n_i such that

$$\sum_\mu R_{i,\mu}(X_1,\dots,X_n) H_\mu(X_1,X_2,\dots,X_n) = X_i^{n_i},$$

(See lemma 8.1.2) (We do not need that the indexing set of μ is finite, almost all of the $R_{i,\mu}(X_1,\dots,X_n)$ will be zero.)
We rewrite this for the original polynomials F_μ and get that the ideal $\widetilde{I}$ contains polynomials

$$\sum_\mu R_{i,\mu}(X_1,\dots,X_n) X_0 G_\mu(X_0,X_1,\dots,X_n) + X_i^{n_i}$$

for $i = 1,\dots,n$. Now we restrict these polynomials to U_0, this means we put $X_0 = 1$ and then we find that our ideal I contains polynomials

$$\sum_\mu R_{i,\mu}(X_1,\dots,X_n) G_\mu(1,X_1,\dots,X_n) + X_i^{n_i},$$

for $i = 1,\dots,n$ and where the total degree of the monomials in the $X_\nu, \nu = 1,\dots,n$ in $G_\mu(1,X_1,\dots,X_n)$ is less than n_i.
This implies that the $A-$module $A[x_1,\dots,x_n]$ is generated by monomials $x_1^{\nu_1}\dots x_n^{\nu_n}$ with $\nu_i < n_i$ and this is the finiteness. □

8.1.7 Ample and Very Ample Sheaves

Let S be a noetherian scheme. We want to discuss certain constructions, which allow us to show that a scheme $X \to S$ is projective or, which provide projective embeddings of this scheme. It will be clear that the results, which we are going to prove are local in the base S hence we always assume that $S = \operatorname{Spec}(A)$ and A is noetherian.

Let us go back briefly to the case of affine schemes. If we have an arbitrary scheme X and if we want to show that this scheme is affine we have only one chance: We consider the ring of global sections $B = \Gamma(X,\mathcal{O}_X)$ and we try to prove that $X \xrightarrow{\sim} \operatorname{Spec}(B)$.

This may be not so easy. Of course we always have a morphism $\pi : X \longrightarrow \mathrm{Spec}(B)$. It sends a point $x \in X$ to the prime ideal $\mathfrak{p}_x$ of functions, which vanish at x, we get an inclusion $h_x : B_{\mathfrak{p}_x} \hookrightarrow \mathcal{O}_{X,x}$ and this yields the morphism for the sheaves. (See 6.1.20). We have to show that both maps are isomorphisms. This makes it clear what kind of information we need if we want to be successful: We have find enough regular functions on X.
For instance we certainly need the information that the regular functions separate points:

Definition 8.1.15. *We say that the regular functions* **separate points** *if for any two points $x,y \in X$ such that y is not in the closure of $\{x\}$ (see 6.1.3) we can find an $f \in B$ such that $f(x) = 0$ and $f(y) \neq 0$.*

This would tell us that π is injective on the underlying space, but this is by far not good enough. For instance the example 32 tells us that π does not need to be surjective. Hence we have to assume the surjectivity of π or to make some assumptions, which allow to conclude that π is surjective.

We assume that $X \longrightarrow \mathrm{Spec}(A)$ is a separated scheme and it can be covered by finitely many affine open sub schemes. Let us consider a sub algebra $B \subset \Gamma(X,\mathcal{O}_X)$ and the resulting the diagram

$$\begin{array}{ccc} X & \xrightarrow{\ \pi\ } & \mathrm{Spec}(B) = Y \\ & \searrow \quad \swarrow & \\ & \mathrm{Spec}(A) = S & \end{array}$$

We formulate two strong assumptions
a) the morphism π is *closed*, this means that the image of any closed subset $Z \subset X$ is closed.
b) Any fibre $\pi^{-1}(y)$ is contained in an open affine subset of X.

The assumption b) is certainly true if B separates points because then $\pi^{-1}(y)$ is empty or a point. We claim:

Proposition 8.1.16. *Under the above assumptions the scheme X is affine.*

Proof: We pick a point $y \in Y$ and we choose an affine subset $U \subset X$, which contains the fibre $\pi^{-1}(y)$. We consider the complement of U in X and by our assumption we know that the image of this complement is a closed subset $Z \subset Y$, which of course does not contain y. Hence we find a regular function $g \in B$ such that $g(y) \neq 0$ and $g|Z = 0$. Then we see that $\pi^{-1}(Y_g) \subset U$. The element g is also a regular function on U and therefore, $\pi^{-1}(Y_g) = U_g$ is affine. We conclude that any point $y \in Y$ has an open neighborhood such that the inverse image of this neighborhood is affine. We can cover Y by affine subsets Y_{g_i} such that the $\pi^{-1}(Y_{g_i}) = X_i$ are affine. We get homomorphisms

$$\mathcal{O}_Y(Y_{g_i}) \longrightarrow \mathcal{O}_X(X_i).$$

The algebras $\mathcal{O}_Y(Y_{g_i})$ are localizations B_{g_i} and since $B \to \Gamma(X,\mathcal{O}_X)$ is injective, we get inclusions

$$B_{g_i} \longrightarrow \Gamma(X,\mathcal{O}_X)_{g_i} \longrightarrow \mathcal{O}_X(X_i).$$

We need that the last arrow is an isomorphism. This follows from a little lemma.

Lemma 8.1.17. *Let X be a separated scheme, which can be covered by a finite set of affine schemes. Let $g \in \Gamma(X,\mathcal{O}_X)$ and let $h \in \mathcal{O}_X(X_g)$. Then we can find an integer $n > 0$ such that $g^n h$ is the restriction of an element in $\Gamma(X,\mathcal{O}_X)$ to X_g.*

Proof: Let us write $X = \bigcup_{i\in E} U_i$ with U_i affine and E finite. We consider the restriction h_i of h to $X_g \cap U_i = U_{i,g}$. By definition we can write $h_i = \frac{f_i}{g^{n_i}}$ for some $f_i \in \Gamma(U_i,\mathcal{O}_X)$ and hence we can find an index n and functions $F_i \in \mathcal{O}_X(U_i)$ such that

$$g^n h = F_i \mid_{U_i \cap X_g} .$$

Now we compare F_i and F_j on $U_i \cap U_j$. We have

$$F_i \mid U_i \cap U_j \cap X_g = F_j \mid U_i \cap U_j \cap X_g.$$

Since we assumed that X is separated the intersections $U_i \cap U_j$ are affine (see definition 8.1.4 and the following lemma) and hence we can find an positive integer m such that

$$g^m F_i \mid U_i \cap U_j = g^m F_j \mid U_i \cap U_j$$

and then the $g^m F_i$ are restrictions of a function $F \in \Gamma(X,\mathcal{O}_X)$.
(If we want to avoid to assume that X is separated, we can assume instead that any open set has a finite covering by affines.)

□

Now it is clear that the morphism of schemes $X \longrightarrow \mathrm{Spec}(\Gamma(X,\mathcal{O}_X))$ is an isomorphism. It is obvious that $\Gamma(X,\mathcal{O}_X)$ separates points because already B separates the points in Y and the Lemma implies that $\Gamma(X_g,\mathcal{O}_X)$ separate the points on X_g. It is also clear that for any $x \in X$ the map $h : \Gamma(X,\mathcal{O}_X)_{\mathfrak{p}_x} \longrightarrow \mathcal{O}_{X,x}$ is an isomorphism, because we have the isomorphisms $\Gamma(X,\mathcal{O}_X)_{g_i} \longrightarrow \mathcal{O}_X(X_i)$.
It remains to prove the surjectivity. We go back to the beginning of the proof. If π is not surjective then we pick a point $y \in Y$, which is not in the image. We may take $U = \emptyset$ and choose our g as above. Then it is clear that g vanishes on X and hence it must be nilpotent. This contradicts our assumption that $g(y) \neq 0$. □

We have seen that the assumption a) is vital and it is not so clear how we can verify it in a given case. (See example 32).

Now we become a little bit more ambitious. If we have a scheme over $X \to \mathrm{Spec}(A)$ and we want to prove that it is projective. Then our strategy above cannot work since we do not have enough regular functions. But we have seen that the line bundles $\mathcal{O}_{\mathbb{P}^n_A}(r))$ on $\mathbb{P}^n_A$ have many sections if $r > 0$. Hence we replace the regular functions by the sections in the positive powers $\mathcal{L}^{\otimes n}$ of a suitable line bundle $\mathcal{L}$.
If we have a section $s \in H^0(X,\mathcal{L})$ then we can trivialize $\mathcal{L}$ on the open set X_s where s does not vanish. If now $f \in H^0(X,\mathcal{L}^{\otimes n})$ then f/s^n is regular functions on X_s. We see that a line bundle provides a tool to construct regular functions on certain open sets. For a particular class of line bundles - the so called *very ample* bundles - this method can be used to construct a projective embedding and this will be explained next.

First we go in the opposite direction. If we start from $X \to S$ and assume that we have an embedding

$$\begin{array}{ccc} X & \overset{i}{\hookrightarrow} & \mathbb{P}^n_A \\ & {\scriptstyle\pi}\searrow \quad \swarrow{\scriptstyle\pi_0} & \\ & \mathrm{Spec}(A) = S & \end{array}$$

This embedding provides the line bundle $\mathcal{O}_{\mathbb{P}^n_A}(1)$ on $\mathbb{P}^n_A$ and we may consider its pullback

$$\mathcal{L} = i^*(\mathcal{O}_{\mathbb{P}^n}(1)).$$

We explain how we can use the line bundle $\mathcal{L}$ to (re-)construct (the) an embedding from X into a projective space. We consider the direct image sheaf $R_1 = \pi_*(\mathcal{L})$. Since we assume that S is affine $R_1 = \pi_*(\mathcal{L})$ is an A-module.
We put $R_0 = A$ and form the graded A-algebra

$$R = R_0 \oplus R_1 \oplus R_2 \oplus \cdots = A \oplus (\bigoplus_{n \geq 1} \pi_*(\mathcal{L}^{\otimes n}))$$

and in this algebra we consider the sub algebra

$$R = R_0 \oplus R_1 \oplus R_2 \oplus \ldots,$$

which is generated by R_1.
We write $\mathbb{P}^n_A = \mathrm{Proj}(A[Y_0,Y_1,\ldots,Y_n]$ then $H^0(\mathbb{P}^n_A,\mathcal{O}_{\mathbb{P}^n_A}(1)) = \pi_{0,*}(\mathcal{O}_{\mathbb{P}^n_A}(1)) = AY_0 \oplus AY_1 \oplus \cdots \oplus AY_n$ (see p.123) and we have the restriction homomorphism $r : \pi_{0,*}(\mathcal{O}_{\mathbb{P}^n_A}(1)) \longrightarrow \pi_*(\mathcal{L})$. This homomorphism is not necessarily surjective, but if y_i is the image of Y_i then clearly $\phi_i : A[Y_0,Y_1,\ldots,Y_n]^{(0)}_{Y_i} \longrightarrow R^{(0)}_{y_i}$ is surjective and the kernel of ϕ_i is the ideal, which defines $X_i = X \cap U_i$ as a sub scheme of U_i. Hence $X_i = \mathrm{Spec}(R^{(0)}_{y_i})$ for all indices i, and it becomes clear that $X = \mathrm{Proj}(R)$, If we choose any submodule $R_1x_0 + Rx_1, \cdots + Rx_N$ in $\pi_*(\mathcal{L})$, which contains the image of r then this sub module generates a sub algebra $R' \subset A_0 \bigoplus_{n \geq 1} \pi_*(\mathcal{L}^{\otimes n})$ and this provides an embedding $X = \mathrm{Proj}(R') \hookrightarrow \mathbb{P}^N_A$.

We return to our original problem. We start from a proper scheme $X \longrightarrow S = \mathrm{Spec}(A)$. We consider a line bundle $\mathcal{L}$ on X and we want to investigate, under which condition the bundle $\mathcal{L}$ provides an embedding as above. We need a definition

Definition 8.1.18. *We say, that a line bundle $\mathcal{L}$ over a scheme $X \longrightarrow S$ has* **no base point**, *if for any point $s \in S$ we can find an affine neighborhood $V \subset S$ of s such that the sections*

$$\pi_*(\mathcal{L})(V) = \mathcal{L}\left(\pi^{-1}(V)\right)$$

generate the stalk of the line bundle at any point $x \in \pi^{-1}(s)$. This is the same as saying that for any x we can find a section $t \in H^0(\pi^{-1}(V),\mathcal{L})$, which does not vanish at x, i.e. it does not go to zero in $\mathcal{L} \otimes_{\mathcal{O}_x} k(x)$.

In the situation, which we considered above, our line bundle $\mathcal{L} = i^*(\mathcal{O}_{\mathbb{P}^n}(1))$ obviously has no base point.
Of course A is still noetherian, we assume that X is of finite type over $\mathrm{Spec}(A)$, let $\mathcal{L}$ be a line bundle without base point. Since under our given assumption the space X is quasi compact we find sections $t_0,t_1,\ldots,t_N \in \pi_*(\mathcal{L})$ such that the open sets X_{t_i} form a covering of X. Again we can consider the graded sub algebra $R' \subset A\bigoplus_{n\geq 1}\pi_*(\mathcal{L}^{\otimes n})$ generated by the t_i and if we write R' as a quotient

$$A[T_0,T_1,\ldots,T_N] \longrightarrow R', T_i \mapsto t_i$$

then we get a morphism, (which depends on the choice of the generators)

$$r_{\mathcal{L}} : X \longrightarrow \mathrm{Proj}(R') \hookrightarrow \mathbb{P}^N_A$$

and whose restriction

$$r_{\mathcal{L},i} : X_{t_i} \longrightarrow \mathrm{Spec}(R'^{,(0)}_{t_i}) \hookrightarrow U_i$$

is given by the homomorphism $T_n/T_i \mapsto t_n/t_i$.

Definition 8.1.19. *The line bundle $\mathcal{L}$ is called* **very ample** *if we can find sections $t_0,t_1,\ldots,t_N \in \pi_*(\mathcal{L})$ such that the X_{t_i} are affine and $r_{\mathcal{L},i} : X_{t_i} \xrightarrow{\sim} \mathrm{Spec}(R'^{,(0)}_{t_i})$. It is called* **ample** *if a suitable positive power $\mathcal{L}^{\otimes n}$ is very ample.*

Bundles of the form $\mathcal{L} = i^*(\mathcal{O}_{\mathbb{P}^n}(1))$ above are certainly ample. It is clear that for a very ample line bundle $\mathcal{L}$ the morphism $r_{\mathcal{L}}$ is a closed embedding (if we select enough t_i) and that for this embedding we get $\mathcal{L} = i^*(\mathcal{O}_{\mathbb{P}^n}(1))$.

We want to investigate, under which additional assumptions we get more precise information on $r_{\mathcal{L}}$, so far our arguments also apply to $\mathcal{L} = \mathcal{O}_X$ and in this case we only have shown that we get a morphism to S.
For any line bundle $\mathcal{L}$ on X the direct image $\pi_*(\mathcal{L})$ is a $\mathcal{O}_S$-module. In the following we assume that it is locally finitely generated (coherent, see 8.3.1). (This is not really an assumption, it follows from a general finiteness theorem of Grothendieck [Gr-EGA III]. This theorem will not be proved in full generality in this book, we will prove it only for projective morphisms (see theorem 8.3.2) whereas Grothendieck proves it for proper morphisms. A close look at the following arguments will show that we do not really need this finiteness result, it only makes the argument slightly more comfortable.)
We start from a line bundle $\mathcal{L}$ on X. Then $\pi_*(\mathcal{L})$ is coherent and hence by our assumption a finitely generated A-module $R_1 = H^0(X,\mathcal{L})$. We consider the graduated ring

$$R = A \oplus R_1 \oplus R_2 \oplus \ldots$$

where $R_i \subset H^0(X,\mathcal{L}^{\otimes i})$ is the submodule generated by the products of elements in R_1. Let us pick a set $t_0,t_1,\ldots,t_n$ of generators. As always we assume that R_1 has no base point and hence $X = \bigcup X_{t_i}$.
Now we formulate several assumptions:

a1) For any choice of $x \in X$ the fibers $r_{\mathcal{L}}^{-1}(r_{\mathcal{L}}(x)$ are finite and lie in an open affine subset of X.

a2) For all i the algebra $R_{t_i}^{(0)}$ seperates points on X_{t_i}.

(This assumption is equivalent to the assumption that $H^0(X,\mathcal{L})$ separates points.)

The strongest assumption is

a3) The algebra $R_{t_i}^{(0)}$ separates points and for any $x \in X_{t_i}, \bar{x} = r_{\mathcal{L}}(x)$ the homomorphism

$$R_{s,\bar{x}}^{(0)} \longrightarrow \mathcal{O}_{X,x}/\mathfrak{m}_{X,x}^2$$

is surjective.

Theorem 8.1.20. *Let $\pi : X \longrightarrow S$ be a proper morphism of finite type.*

(i) *Under the assumptions a1) or a2) the X_{t_i} are affine, and the morphism $r_{\mathcal{L}} : X \to \mathrm{Proj}(R)$ is finite.*

(ii) *Under the same assumption the morphism $X \longrightarrow S$ is projective.*

(iii) *Under the assumption a3), the morphism $r_{\mathcal{L}}$ is an isomorphism.*

Proof:
We apply our previous considerations for the affine case and we see that under the assumption a1) or under the assumption a2) the X_{t_i} must be affine. Hence we also know that $X_{t_i} = \mathrm{Spec}(\Gamma(X_{t_i},\mathcal{O}_X))$. By construction we have an inclusion $R_{t_i}^{(0)} \hookrightarrow \Gamma(X_{t_i},\mathcal{O}_X)$, the morphism $X_{t_i} \longrightarrow \mathrm{Spec}(R_{t_i}^{(0)})$ is proper. Hence proposition 8.1.14 implies that the algebras $\Gamma(X_{t_i},\mathcal{O}_{X_{t_i}})$ are finite over $R_{t_i}^{(0)}$. This proves (i).
We postpone the proof of (ii) and we assume a3). Then the algebra $R_{t_i}^{(0)}$ separates points, and we can conclude that $X_{t_i} \to \mathrm{Spec}\left(R_{t_i}^{(0)}\right)$ is always bijective. Now we are confronted with the problem we alluded to in the remark on p. 15. We have to show that the induced morphism between the sheaves is an isomorphism, this can be checked on the stalks. We have to show that for any point $x \in X_{t_i}$ and its image $\bar{x} \in \mathrm{Spec}(R_{t_i}^{(0)})$ the homomorphism $h_x : R_{s,\bar{x}}^{(0)} \to \mathcal{O}_{X_{s,x}}$ is an isomorphism. Let $\mathfrak{m}_{\bar{x}},\mathfrak{m}_x$ be the two maximal ideals. Since localization is flat we have the inclusion $R_{t_i,\mathfrak{m}_{\bar{x}}}^{(0)} \subset \Gamma(X_{t_i},\mathcal{O}_X) \otimes R_{t_i,\mathfrak{m}_{\bar{x}}}^{(0)}$ and this extension of rings is finite. We also know that there is only one maximal prime ideal in $\Gamma(X_{t_i},\mathcal{O}_X)$ lying over $\mathfrak{m}_{\bar{x}}$ this unique maximal ideal is $\mathfrak{m}_x$ and this implies that $\Gamma(X_{t_i},\mathcal{O}_X) \otimes R_{t_i,\mathfrak{m}_{\bar{x}}}^{(0)}$ is local and hence equal to $\mathcal{O}_{X_{t_i,x}}$. Putting everything together we conclude that $R_{t_i,\mathfrak{m}_{\bar{x}}}^{(0)} \subset \mathcal{O}_{X_{s,x}}$ is a finite extension of local rings. Now our assumption a3) implies that the homomorphism between the residue fields

$$R_{t_i,\mathfrak{m}_{\bar{x}}}^{(0)}/\mathfrak{m}_{\bar{x}} \xrightarrow{\sim} \mathcal{O}_{X_{t_i,x}}/\mathfrak{m}_x$$

is an isomorphism. The algebra $\mathcal{O}_{X_{t_i,x}}/\mathfrak{m}_{\bar{x}}$ is finite over $R_{t_i,\mathfrak{m}_{\bar{x}}}^{(0)}/\mathfrak{m}_{\bar{x}}$ and since it is local we know that $\mathfrak{m}_x^N \subset \mathfrak{m}_{\bar{x}}$ if N is sufficiently large. Now our assumption a3) also implies that $\mathfrak{m}_{\bar{x}} \longrightarrow \mathfrak{m}_x/(\mathfrak{m}_x)^2$ is surjective and this implies that $\mathfrak{m}_{\bar{x}} \longrightarrow \mathfrak{m}_x/(\mathfrak{m}_x)^N \longrightarrow \mathfrak{m}_x/(\mathfrak{m}_x\mathfrak{m}_{\bar{x}})$ is surjective. Hence we can conclude that

$$R_{t_i,\mathfrak{m}_{\bar{x}}}^{(0)} \longrightarrow \mathcal{O}_{X_{t_i,x}}/\mathfrak{m}_{\bar{x}}\mathcal{O}_{X_{t_i,x}}$$

is surjective. We noticed already that $\mathcal{O}_{X_{t_i x}}$ is finite over $R^{(0)}_{t_i,\mathfrak{m}_{\bar{x}}}$ and now the lemma of Nakayama implies that $R^{(0)}_{t_i,\mathfrak{m}_{\bar{x}}} \longrightarrow \mathcal{O}_{X_{t_i,x}}$ is surjective and hence an isomorphism. So we proved (iii) under the assumption a3).

We still have to prove (ii) under the assumption a1) or a2). This assertion follows from the following proposition

Proposition 8.1.21. *We consider a diagram of schemes of finite type*

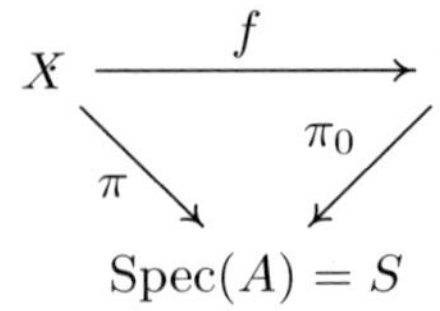

and we assume that f is finite, π is proper and π_0 is projective. Then π is also projective.

The proof of this proposition also finishes the proof of the theorem. We choose an embedding j of Y/S into some $\mathbb{P}^N_A$ and as before we put $\mathcal{L} = j^*(\mathcal{O}_{\mathbb{P}^N_A}(1))$. We consider the bundle $\mathcal{L}_1 = f^*(\mathcal{L})$ on X. We choose generators $t_0,t_1,\dots,t_N$ of $\pi_{0,*}(\mathcal{L})$ and define as usual R to be the graded sub algebra of $A \oplus \bigoplus_{n\geq 1} \pi_{0,*}(\mathcal{L}^{\otimes n})$ Then $Y_{t_i} = \mathrm{Spec}(R^{(0)}_{t_i})$, the elements t_i can also be seen as elements in $\pi_*\mathcal{L}_1$. Since f is finite the schemes $X_{t_i} = f^{-1}(Y_{t_i})$ are affine and $\mathcal{O}_X(X_{t_i})$ is finite over $\mathcal{O}_Y(Y_{t_i})$. Hence we can find finite sets of generators $h_{i,1},\dots,h_{i,\mu},\dots,h_{i,\nu_i}$, which generate the $\mathcal{O}_Y(Y_{t_i})$-module $\mathcal{O}_X(X_{t_i})$. Now a slight extension of the lemma 8.1.17 shows that we can find an integer $n > 0$ such that all $t_i^n h_{i,\mu}$ extend to sections in $\pi_*(\mathcal{L}^{\otimes n})$.
Hence we see: If start from the bundle $\mathcal{L}^{\otimes n} = \mathcal{L}_2$ and define accordingly the graded algebra $R = A \oplus \bigoplus_{n\geq 1} \pi_*(\mathcal{L}_2^{\otimes n})$ then the morphisms $X_{t_i} \longrightarrow R^{(0)}_{t_i}$ become isomorphisms and hence $r_{\mathcal{L}_2}$ provides an embedding.

□

8.2 Cohomology of Quasicoherent Sheaves

For any scheme X and any quasi coherent sheaf $\mathcal{F}$ on X we know how to define he cohomology groups $H^q(X,\mathcal{F})$. We recall that $\mathcal{F}$ is a sheaf on the underlying topological space, which here will also be denoted by X and this space is the first variable in $H^q(X,\mathcal{F})$. Furthermore $\mathcal{F}$ is a sheaf of abelian groups and hence the cohomology is defined in Volume I.
More generally we can consider morphisms $f : X \longrightarrow Y$ between schemes, then we have defined the direct image functor $\mathcal{F} \longrightarrow f_*(\mathcal{F})$ and its higher derived functors $R^q f_*(\mathcal{F})$. We have a first fundamental theorem:

Theorem 8.2.1. *Let $f : X \longrightarrow Y$ be an affine morphism between schemes. For any quasi coherent sheaf $\mathcal{F}$ the higher direct images vanish, i.e.*

$$R^q f_*(\mathcal{F}) = 0$$

for all $q > 0$.

Before proving this we want to indicate that this is highly plausible. First of all the assertion is local in the base. Hence we may assume that $X = \text{Spec}(A)$,$Y = \text{Spec}(B)$ are affine and then the morphism is given by a homomorphism $B \longrightarrow A$ of rings. At the end of section 6.1.5 we have seen that the category of quasi coherent sheaves on an affine scheme $\text{Spec}(C)$ is equivalent to the category of C modules. Now taking the direct image of a quasi coherent module $\widetilde{M}$ on X amounts to the same as considering the A-module M as a B-module and then taking the associated sheaf. Hence it is clear that for an exact sequence of quasi coherent sheaves on X the sequence of direct images is also exact. We find that there is *no need to have higher derived images* if we restrict to the category of quasi coherent sheaves.

But since the derived images are defined inside the category of *all sheaves*, something has to be proved. We need an acyclic resolution by injective sheaves and we do not know a priori whether these sheaves can be chosen to be quasi-coherent.

Proof: We consider two special cases of the theorem. In the first case $X = \text{Spec}(A)$ is an absolute scheme and we prove $H^q(X,\mathcal{F}) = 0$ for all $q > 0$. In the second case we pick an $f \in A$ and as morphism is an inclusion $i_f : X_f \hookrightarrow X$. Then we prove that $R^q i_{f,*}(i_f^*(\mathcal{F})) = 0$ for all $q > 0$ and any quasi coherent sheaf (See 6.2.2). We use an induction argument, which runs as follows: If we know the first case up to degree n, then we prove the second case up to degree n. Then we show that this in turn implies the first case up to $n+1$.

We start with the following observation. Let $n > 0$ be an integer, assume we proved that for any quasi coherent sheaf $\mathcal{F}$ on any affine scheme and any $0 < \nu \leq n$ the module $H^\nu(X,\mathcal{F}) = 0$. This is the first case up to the degree n. We claim that under this assumption $R^\nu i_{f,*}(i_f^*(\mathcal{F})) = 0$ for all $0 < \nu \leq n$. To see this we recall that the functor $\mathcal{F} \longrightarrow H^0(X_f,i_f^*(\mathcal{F}))$, which sends sheaves on X_f to abelian groups, is the composite of $i_{f,*}$, which sends sheaves on X_f to sheaves on X, and $\mathcal{G} \longrightarrow H^0(X,\mathcal{G})$ from sheaves on X to abelian groups. We have seen that this provides a spectral sequence, whose $E_2^{p,q}$-term is (see Vol. I, 4.6.3 example d))

$$H^p(X,R^q i_{f,*}(i_f^*(\mathcal{F}))) \Rightarrow H^\nu(X_f,i_f^*(\mathcal{F})) = 0 \text{ for } p+q=\nu.$$

Let us assume that we proved the vanishing of $R^q i_{f,*}(i_f^*(\mathcal{F}))$ for all indices $q < \mu_0 \leq n$. We consider the term $H^0(X,R^{\mu_0} i_{f,*}(i_f^*(\mathcal{F}))$. Looking at the differentials we see that we do not have any differentials going into this term. The outgoing differentials go to zero. Hence we see that we have a surjective homomorphism $H^{\mu_0}(X_f,i_f^*(\mathcal{F})) \longrightarrow H^0(X,R^{\mu_0} i_{f,*}(i_f^*(\mathcal{F})))$ and since the group on the left hand side is zero it follows that $H^0(X,R^{\mu_0} i_{f,*}(i_f^*(\mathcal{F}))) = 0$. But here we can pick any point $x \in X \setminus X_f$ and localise at this point by restricting to smaller affine schemes X_g with $g(x) \neq 0$. Then we always get $H^0(X_g,R^{\mu_0} i_{f,*}(\mathcal{F})) = 0$ and this proves that the stalk $R^{\mu_0} i_{f,*}(\mathcal{F})_x = 0$, hence $R^{\mu_0} i_{f,*}(\mathcal{F}) = 0$.

Now we assume $H^\nu(X,\mathcal{F}) = 0$, for all affine X, all quasi coherent sheaves $\mathcal{F}$ and all $0 < \nu < n+1$. We consider a quasi coherent sheaf $\mathcal{F}$ on X, let $M = H^0(X,\mathcal{F})$. We choose an $f \in H^0(X,\mathcal{O}_X)$ and compute the sheaf $i_{f,*}(i_f^*(\mathcal{F}))$. For any open subset $X_g \subset X$ we have $i_{f,*}(i_f^*(\mathcal{F}))(X_g) = i_f^*(\mathcal{F})(X_g \cap X_g) = \mathcal{F}(X_{fg})$. This tells us that $i_{f,*}(i_f^*(\mathcal{F})$ is quasi coherent and equal to the sheaf, which is obtained from the A-module M_f. We have the restriction map $\mathcal{F} \longrightarrow i_{f,*}(i_f^*(\mathcal{F}))$ on the level of sheaves and hence we get the restriction maps

$$H^n(X,\mathcal{F}) \xrightarrow{r_1} H^n(X,i_{f,*}(i_f^*(\mathcal{F})) \xrightarrow{r_2} H^n(X_f,i_f^*(\mathcal{F})).$$

We claim that r_2 is injective. This homomorphism is an edge-homomorphism and the kernel has a filtration, where the sub quotients in this filtration themselves are subquotients of the modules $H^{n-q-1}(X,R^q i_{f,*}(i_f^*(\mathcal{F}))$ with $0 < q < n+1$. But our induction hypothesis together with the above observation implies that $R^q i_{f,*}(i_f^*(\mathcal{F})) = 0$. So r_2 is injective.
We claim that that the cohomology groups are "effacable". This means that for any class $\xi \in H^q(X,\mathcal{F})$ we can find a finite covering $X = \bigcup X_{f_i}$ such that for all i the given class goes to zero if we restrict it to $H^q(X_{f_i},\mathcal{F})$. This is clear, we choose an injective resolution of $\mathcal{F}$ as in volume I. 4.2.1 and then we have $H^\bullet(X,\mathcal{F}) = H^\bullet(\mathcal{I}^\bullet(X))$. But the complex of sheaves $\mathcal{F} \longrightarrow \mathcal{I}^\bullet$ is exact, hence for any cycle $c \in \mathcal{I}^q(X)$ and any point $x \in X$ we find a neighbourhood $U = X_{f_\nu}$ of x such that the restriction of c to X_{f_ν} is a boundary. Hence the class represented by this cycle goes to zero after restriction to X_{f_ν}. We apply this to $q = n+1$ and see that for any class $c \in H^{n+1}(X,\mathcal{F})$ we can find a finite covering $X = \bigcup X_{f_\nu}$ such that c goes to zero under the map

$$H^{n+1}(X,\mathcal{F}) \longrightarrow \bigoplus_i H^{n+1}(X,i_{f_\nu,*}(i_{f_\nu}^*(\mathcal{F})).$$

We get an exact sequence

$$0 \longrightarrow \mathcal{F} \longrightarrow \bigoplus i_{f_\nu,*}(i_{f_\nu}^*(\mathcal{F})) \longrightarrow \mathcal{G} \longrightarrow 0$$

of quasi coherent sheaves and a long exact sequence in cohomology

$$H^n(X,\bigoplus i_{f_\nu,*}(i_{f_\nu}^*(\mathcal{F})) \longrightarrow H^n(X,\mathcal{G}) \longrightarrow H^{n+1}(X,\mathcal{F}) \longrightarrow H^{n+1}(X,\bigoplus i_{f_\nu,*}(i_{f_\nu}^*(\mathcal{F})).$$

Now we know that for $n \geq 1$ the cohomology $H^n(X,\mathcal{G}) = 0$ or that for $n = 0$ the homomorphism $H^0(X,\bigoplus i_{f_\nu,*}(i_{f_i}^*(\mathcal{F})) \longrightarrow H^0(X,\mathcal{G})$ is surjective. Since our class is in the image of the boundary map $H^n(X,\mathcal{G}) \longrightarrow H^{n+1}(X,\mathcal{F})$ we conclude that it has to be zero.
The rest is clear: If we have our affine morphism $f : X \longrightarrow Y$ and pick a point $y \in Y$ then the stalk of the sheaf $R^q f_*(\mathcal{F})$ in the point y is the limit of the cohomology groups $H^q(f^{-1}(V),\mathcal{F})$ where V runs over a system of neighbourhoods of y. (See Vol. I.4.4.2). If we take the V-s to be affine, then the $f^{-1}(V)$ are also affine and hence $H^q(f^{-1}(V),\mathcal{F}) = 0$. □

8.2.1 Čech cohomology

At this point we want to make a simple remark. When we defined the cohomology of a sheaf in Vol. I. 4.2.1 we said that the reader might be scared, because it seemed to be impossible to compute them. But then we developed some tools to compute the cohomology of some specific sheaves. Especially we considered the Čech resolution of sheaves, especially in the case of manifolds and local coefficient systems as sheaves. We had these coverings by convex sets (See Vol. I.4.8.2) they provided the acyclic (Čech) resolution of the the sheaf. The resolving sheaves were direct sums of sheaves, whose support was contractible and, which were constant on their support. At that point we applied a (difficult) theorem, namely that $H^q(X,\mathbb{Z}) = 0$ for a contractible space X.

Here the situation is quite similar, we want to compute the cohomology $H^n(X,\mathcal{F})$, where $\mathcal{F}$ is quasi coherent. We assume that X is separated. We cover our scheme by affine schemes and show that that the resulting Čech complex computes the cohomology.
More generally we consider a separated scheme $f : X \longrightarrow S$, let $\mathcal{F}$ be quasi coherent on X, we want to show that the sheaves $R^q f_*(\mathcal{F})$ on S are quasi coherent. For this we we assume that $S = \mathrm{Spec}(A)$ is already affine.
We have a covering $X = \bigcup_\alpha U_\alpha$, by affine subschemes. We make some finiteness assumption, the covering should be locally finite (See Vol. I.4.5.2, see also Vol. I.3.5.to justify the finiteness assumption.) Let I be the set of indices, the elements in I^{q+1} are denoted by $\underline{\alpha} = (\alpha_0,\alpha_a,\ldots,\alpha_q)$, then $d(\underline{\alpha}) = q$. As in Vol. I we put $U_{\underline{\alpha}} = U_{\alpha_0} \cap \ldots U_{\alpha_q}$ by $i_{\underline{\alpha}} : U_{\underline{\alpha}} \longrightarrow X$ we denote the inclusion and

$$\mathcal{F}^*_{\underline{\alpha}} = i_{\underline{\alpha},*} i^*_{\underline{\alpha}}(\mathcal{F}).$$

We consider the Čech resolution of $\mathcal{F}$ provided by this covering

$$0 \longrightarrow \mathcal{F} \longrightarrow \prod_{\alpha\in I} \mathcal{F}^*_\alpha \longrightarrow \prod_{\alpha,\beta} \mathcal{F}^*_{\alpha,\beta} \longrightarrow .$$

Our theorem 8.2.1 says that this resolution is acyclic, hence the complex of global sections computes the cohomology (See Vol. I. 4.6.6). Since $\mathcal{F}^*_{\underline{\alpha}}(X) = \mathcal{F}(U_{\underline{\alpha}})$ we see that

$$H^q(X,\mathcal{F}) = H^q(0 \longrightarrow \prod_{\alpha\in I} \mathcal{F}(U_\alpha) \longrightarrow \ldots \longrightarrow \prod_{\underline{\alpha}\in I^{q+1}} \mathcal{F}(U_{\underline{\alpha}}) \longrightarrow \ldots)$$

We get a presheaf of cohomology groups $V \longrightarrow H^q(X_V,\mathcal{F})$ where $V \subset S$ is open and $X_V = X \times_S V$ is the inverse image of V. We want to show that this presheaf is indeed a sheaf. To do this we consider affine subsets $S_f = \mathrm{Spec}(A_f) \subset S$. Then it is clear that $\mathcal{F}(U_{\underline{\alpha}} \cap X_f) = F(U_{\underline{\alpha}}) \otimes A_f$. Therefore we obtain

$$H^q(X_f,\mathcal{F}) = H^q(0 \longrightarrow \prod_{\alpha\in A} \mathcal{F}(U_\alpha)\otimes A_f \longrightarrow \ldots \longrightarrow \prod_{\underline{\alpha}\in I^{q+1}} \mathcal{F}(U_{\underline{\alpha}})\otimes A_f \longrightarrow \ldots).$$

We know that $A \longrightarrow A_f$ is flat therefore, the lemma below yields the equality

$$H^q(X_f,\mathcal{F}) = H^q(X_f,\mathcal{F}) \otimes A_f.$$

This proves

Proposition 8.2.2. *Let $f : X \longrightarrow S$ be a separated morphism, let $\mathcal{F}$ be a quasi coherent sheaf on X. Furthermore we assume that X has a locally finite covering by affine subschemes. Then we assert that for any open subscheme $S_\nu = \mathrm{Spec}(A_\nu) \subset S, X_\nu = f^{-1}(S_\nu)$, any degree q, the restriction of $R^q f_*(\mathcal{F})$ to S_ν is the quasi coherent sheaf obtained from $H^q(X_\nu,\mathcal{F})$. The sheaves $R^q f_*(\mathcal{F})$ are quasi coherent.*

The following lemma will be used later again.

Lemma 8.2.3. *Let A,A' be a rings and $A \longrightarrow A'$ a flat homomorphism. Let*

$$K^\bullet : 0 \longrightarrow K^0 \longrightarrow K^1 \longrightarrow \ldots$$

be a complex of A modules. Then

$$H^\bullet(K^\bullet \otimes_A A') = H^\bullet(K^\bullet) \otimes_A A'$$

Proof:
This is clear since in a given degree i we have

$$\begin{array}{ccccccccc}
0 & \longrightarrow & Z^{i-1} & \longrightarrow & K^{i-1} & \longrightarrow & B^i & \longrightarrow & 0 \\
 & & & & & & \downarrow & & \\
 & & & & & & Z^i & \longrightarrow & 0. \\
 & & & & & & \downarrow & & \\
 & & & & & & H^i(K^\bullet) & & \\
 & & & & & & \downarrow & & \\
 & & & & & & 0 & &
\end{array}$$

Now since Z^i is the kernel of $d : K^i \longrightarrow K^{i+1}$ we see that

$$Z^i \otimes_A A' = \ker\left(K^i \otimes_A A' \longrightarrow K^{i+1} \otimes_A A'\right) \tag{8.36}$$

and

$$B^i \otimes_A A' = \operatorname{Im}\left(K^{i+1} \otimes_A A' \longrightarrow K^i \otimes_A A'\right) \tag{8.37}$$

because $A \to A'$ is flat. Hence

$$H^i(K^\bullet \otimes A') = Z^i(K^\bullet \otimes_A A')/B^i(K^\bullet \otimes A') = Z^i \otimes_A A'/B^i \otimes_A A' \tag{8.38}$$

and the last quotient is equal to $H^i(K^\bullet \otimes_A A')$, which is clear if we tensorize the above vertical sequence by A' over A. □

8.2.2 The Künneth-formulae

Let S be any scheme and let $f : X \longrightarrow S$,$g : Y \longrightarrow S$ be two separated over S. Let $\mathcal{F}$ (resp. $\mathcal{G}$) be quasi coherent sheaves on X (resp. Y). We have the two projections $p_1 : X \times_S Y \longrightarrow X$,$p_2 : X \times_S Y \longrightarrow Y$. We consider the two inverse images $p_1^*(\mathcal{F})$,$p_2^*(\mathcal{G})$ on $X \times_S Y$(see 6.2.2), these are again quasi coherent and we define

$$\mathcal{F} \boxtimes \mathcal{G} = p_1^*(\mathcal{F}) \otimes p_2^*(\mathcal{G}).$$

As in Vol. I 4.6.7 we can construct a homomorphism

$$m : \bigoplus_{i+j=n} R^i f^*(\mathcal{F}) \otimes R^j g^*(\mathcal{G}) \longrightarrow R^n(f \times g)_*(\mathcal{F} \boxtimes \mathcal{G})$$

provided one of the sheaves is flat over S.

Theorem 8.2.4. *If the two sheaves are flat over S then m is an isomorphism.*

This is almost obvious. We may assume that $S = \mathrm{Spec}(A)$ is affine. Then we choose affine coverings $X = \bigcup_\alpha U_\alpha, Y = \bigcup_\beta V_\beta$. Then the products $U_\alpha \times V_\beta$ provide an affine covering of $X \times Y$ and the cohomology $H^\bullet(X \times_S Y, \mathcal{F} \boxtimes \mathcal{G})$ is computed from the resulting Čech complex. But with the above notations we have

$$\mathcal{F} \boxtimes \mathcal{G}(U_{\underline{\alpha}} \times_S V_{\underline{\beta}}) = \mathcal{F}(U_{\underline{\alpha}}) \otimes_A \mathcal{G}(V_{\underline{\beta}}),$$

this is an isomorphism of A-modules. Now the flatness implies that the cohomology of this Čech complex is equal to the tensor product of the two Čech complexes for $\mathcal{F}$ and $\mathcal{G}$. □

8.2.3 The cohomology of the sheaves $\mathcal{O}_{\mathbb{P}^n}(r)$

We investigate the cohomology of the sheaves $\mathcal{O}_{\mathbb{P}^n}(r)$ on $\mathbb{P}^n/S$ where $S = \mathrm{Spec}(A)$. Our main tool will be the exact sequence obtained from

$$\begin{array}{ccccccc} 0 & \longrightarrow \mathcal{O}_{\mathbb{P}^n}(X_0) & \longrightarrow \mathcal{O}_{\mathbb{P}^n} & \longrightarrow \mathcal{O}_{\mathbb{P}^{n-1}} & \longrightarrow 0. \\ & \| & & & \\ & \mathcal{O}_{\mathbb{P}^n}(-1) & & & \end{array} \tag{8.39}$$

(see exercise 35. Tensoring this with $\mathcal{O}_{\mathbb{P}^n}(r)$ we get

$$0 \longrightarrow \mathcal{O}_{\mathbb{P}^n}(r-1) \longrightarrow \mathcal{O}_{\mathbb{P}^n}(r) \longrightarrow \mathcal{O}_{\mathbb{P}^{(n-1)}}(r) \longrightarrow 0 \tag{8.40}$$

for all integers r. We get the chain of inclusions

$$\mathcal{O}_{\mathbb{P}^n} \hookrightarrow \mathcal{O}_{\mathbb{P}^n}(1) \hookrightarrow \mathcal{O}_{\mathbb{P}^n}(2) \hookrightarrow \ldots \mathcal{O}_{\mathbb{P}^n}(r) \ldots \hookrightarrow \mathcal{O}_{\mathbb{P}^n}(\infty) = i_{0,*}i_0^*(\mathcal{O}_{\mathbb{P}^n}).$$

Our theorem 8.2.1 implies that for $q > 0$ any class in $H^q(\mathbb{P}^n, \mathcal{O}_{\mathbb{P}^n}(r))$ vanishes if we send it to $H^q(\mathbb{P}^n, \mathcal{O}_{\mathbb{P}^n}(\infty))$. If we realise this class by a cochain in the Čech-complex we can bound this cochain by an element in $\prod_{\underline{\alpha} \in A^q} \mathcal{O}(\infty)(U_{\underline{\alpha}})$, but this cochain lies already in some $\prod_{\underline{\alpha} \in A^q} \mathcal{O}(r+s)(U_{\underline{\alpha}})$ and we conclude that any class in $H^q(\mathbb{P}^n, \mathcal{O}_{\mathbb{P}^n}(r))$ vanishes if we sent it to some $H^q(\mathbb{P}^n, \mathcal{O}_{\mathbb{P}^n}(r+s))$ with $s >> 0$. (We refer to this as the "limit argument").

We proceed by analyzing the information provided by the long exact cohomology sequence

$$\begin{array}{llll} 0 \longrightarrow H^0(\mathbb{P}^n, \mathcal{O}_{\mathbb{P}^n}(r-1)) & \longrightarrow H^0(\mathbb{P}^n, \mathcal{O}_{\mathbb{P}^n}(r)) & \longrightarrow H^0(\mathbb{P}^{n-1}, \mathcal{O}_{\mathbb{P}^{n-1}}(r)) & \xrightarrow{\delta} \\ H^1(\mathbb{P}^n, \mathcal{O}_{\mathbb{P}^n}(r-1)) & \longrightarrow H^1(\mathbb{P}^n, \mathcal{O}_{\mathbb{P}^n}(r)) & \longrightarrow H^1(\mathbb{P}^{n-1}, \mathcal{O}_{\mathbb{P}^{n-1}}(r)) & \xrightarrow{\delta} \end{array}$$

We know that $H^0(\mathbb{P}^n, \mathcal{O}_{\mathbb{P}^n}(r)) \to H^0(\mathbb{P}^{n-1}, \mathcal{O}_{\mathbb{P}^{n-1}}(r))$ is surjective unless we have $n = 1$ and $r < 0$. This follows from the Exercise 34, the map between the modules of homogeneous polynomials is given by putting $X_0 = 0$.

Hence we get for $n = 1$

$$\begin{array}{lllll} 0 \longrightarrow H^0(\mathbb{P}^1, \mathcal{O}_{\mathbb{P}^1}(r-1)) & \longrightarrow H^0(\mathbb{P}^1, \mathcal{O}_{\mathbb{P}^1}(r)) & \longrightarrow H^0(\mathbb{P}^0, \mathcal{O}_{\mathbb{P}^0}(r)) & \longrightarrow \\ H^1(\mathbb{P}^1, \mathcal{O}_{\mathbb{P}^1}(r-1)) & \longrightarrow H^1(\mathbb{P}^1, \mathcal{O}_{\mathbb{P}^1}(r)) & \longrightarrow 0 & \longrightarrow \\ H^2(\mathbb{P}^1, \mathcal{O}_{\mathbb{P}^1}(r-1)) & \longrightarrow H^2(\mathbb{P}^1, \mathcal{O}_{\mathbb{P}^1}(r)) & \longrightarrow 0 & \longrightarrow . \end{array}$$

If in addition $r \geq 0$ we see that for any $q \geq 1$ the map $H^q(\mathbb{P}^1,\mathcal{O}_{\mathbb{P}^1}(r-1)) \longrightarrow H^q(\mathbb{P}^1,\mathcal{O}_{\mathbb{P}^1}(r))$ is injective. Then the limit argument implies $H^q(\mathbb{P}^1,\mathcal{O}_{\mathbb{P}^1}(r)) = 0$ for $q \geq 2$ and all r and for $q = 1$ and $r \geq -1$. For $q = 1, r = -1$ our sequence becomes

$$\begin{array}{ccccccc} 0 \longrightarrow 0 \longrightarrow 0 \longrightarrow & H^0(\mathbb{P}^0,\mathcal{O}_{\mathbb{P}^0}(-1)) & \longrightarrow H^1(\mathbb{P}^1,\mathcal{O}_{\mathbb{P}^1}(-2)) \longrightarrow 0 \\ & \| & \\ & A & \end{array}$$

and hence

$$H^1(\mathbb{P}^1,\mathcal{O}_{\mathbb{P}^1}(-2)) \simeq A.$$

For $r \leq -2$ we get a short exact sequence

$$\begin{array}{ccccccc} & A & & & & & \\ & \| & & & & & \\ 0 \longrightarrow & H^0(\mathbb{P}^0,\mathcal{O}_{\mathbb{P}^0}(r)) & \xrightarrow{\delta} & H^1(\mathbb{P}^1,\mathcal{O}_{\mathbb{P}^1}(r-1)) & \longrightarrow & H^0(\mathbb{P}^1,\mathcal{O}_{\mathbb{P}^1}(r)) & \longrightarrow 0 \end{array}$$

and hence: For $r \leq -2$

$$H^1(\mathbb{P}^1,\mathcal{O}_{\mathbb{P}^1}(r)) = A^{(-r-1)}.$$

Now we state the theorem.

Theorem 8.2.5. *For $n \geq 1$ we have*

$$H^0(\mathbb{P}^n,\mathcal{O}_{\mathbb{P}^n}(r)) = \quad \textit{Module of homogeneous polynomials of degree } r \text{ in } A[X_0,\ldots,X_n].$$

Especially it is zero for $r < 0$.

$$H^i(\mathbb{P}^n,\mathcal{O}_{\mathbb{P}^n}(r)) = 0 \quad \textit{for} \quad 0 < i < n \quad \textit{or} \quad i > n.$$

$$H^n(\mathbb{P}^n,\mathcal{O}_{\mathbb{P}^n}(-r)) = \begin{cases} 0 & \textit{for} \quad r < n+1 \\ A & \textit{for} \quad r = n+1 \\ A^m & \textit{for} \quad r \geq n+1 \end{cases}$$

where $m =$ rank of $H^0(\mathbb{P}^n,\mathcal{O}_{\mathbb{P}^n}(-n-1+r))$.

We have proved this for $n = 1$. We get easily by induction that

$$H^i(\mathbb{P}^n,\mathcal{O}_{\mathbb{P}^n}(r-1)) \longrightarrow H^i(\mathbb{P}^n,\mathcal{O}_{\mathbb{P}^n}(r))$$

is injective for all $0 < i \leq n-1$ and $i > n$. Hence by the same argument as above we find $H^i(\mathbb{P}^n,\mathcal{O}_{\mathbb{P}^n}(r)) = 0$ for all $i \neq 0,n$. For $i = n$ we get

$$0 \longrightarrow H^{n-1}(\mathbb{P}^{n-1},\mathcal{O}_{\mathbb{P}^{n-1}}(r)) \longrightarrow H^n(\mathbb{P}^n,\mathcal{O}_{\mathbb{P}^n}(r-1)) \longrightarrow H^n(\mathbb{P}^n,\mathcal{O}_{\mathbb{P}^n}(r)) \longrightarrow 0.$$

For $r > -n$ the first term is zero hence we get an isomorphism between the second and third term. This implies again by the limit argument

$$H^n(\mathbb{P}^n,\mathcal{O}_{\mathbb{P}^n}(r)) = 0$$

for $r > -n$. For $r = -n$ we get an isomorphism

$$H^{n-1}(\mathbb{P}^{n-1}, \mathcal{O}_{\mathbb{P}^{n-1}}(-n)) \xrightarrow{\sim} H^n(\mathbb{P}^n, \mathcal{O}_{\mathbb{P}^n}(-n-1))$$

and for still smaller r we get

$$0 \longrightarrow H^{n-1}(\mathbb{P}^{n-1}, \mathcal{O}_{\mathbb{P}^{n-1}}(-n-s)) \longrightarrow H^n(\mathbb{P}^n, \mathcal{O}_{\mathbb{P}^n}(-n-1-s)) \longrightarrow$$
$$H^n(\mathbb{P}^n, \mathcal{O}_{\mathbb{P}^n}(-n-s)) \longrightarrow 0$$

and one checks easily that the recursion for the rank as a function of s is the same as for $H^0(\mathbb{P}^n, \mathcal{O}_{\mathbb{P}^n}(s))$. □

Remark: One make the point that the isomorphisms in the above theorem are not canonical, if we change X_0 into uX_0, where $u \in A^\times$ they will change by some power of u.

8.3 Cohomology of Coherent Sheaves

We consider the projective space $\pi : \mathbb{P}^n_S \longrightarrow S$ over an arbitrary locally noetherian base scheme S. Our results will be local in S, hence we assume that $S = \mathrm{Spec}(A)$, where A is noetherian.

Definition 8.3.1. *A quasi-coherent sheaf $\mathcal{F}$ on $\mathbb{P}^n$ is* **coherent** *if one of the following equivalent conditions is fulfilled:*

(1) *For all U_i the $\mathcal{O}_{\mathbb{P}^n}(U_i)$-module $\mathcal{F}(U_i)$ is finitely generated.*

(2) *For any affine open set $V \subset \mathbb{P}^n$ the $\mathcal{O}_{\mathbb{P}^n}(V)$-module $\mathcal{F}(V)$ is finitely generated.*

(3) *The stalks $\mathcal{F}_\mathfrak{p}$ are finitely generated $\mathcal{O}_{\mathbb{P}^n,\mathfrak{p}}$-modules for all points $\mathfrak{p} \in \mathbb{P}^n$.*

We leave it as an exercise to prove that these conditions are equivalent.
Of course we have this notion of coherence for any scheme X, which is locally noetherian, i.e. has a covering by affine schemes $\mathrm{Spec}(A_\nu)$, where the A_ν are noetherian. Then a quasi coherent sheaf $\mathcal{F}$ on X is coherent if its modules of sections over the $\mathrm{Spec}(A_\nu)$ are finitely generated.

Now we are ready for the celebrated coherence theorem.

Theorem 8.3.2 (Coherence Theorem)**.** *Let S be a locally noetherian scheme and let $\mathcal{F}$ be a coherent sheaf on $\mathbb{P}^n/S$. Then the sheaves $R^i\pi_*(\mathcal{F})$ are coherent and they are zero for $i > n$.*

Proof: The proof requires a series of steps. The theorem is local in the base, hence we may assume that $S = \mathrm{Spec}(A)$, where A is a noetherian ring. Our sheaf $\mathcal{F}$ has a support $\mathrm{Supp}(\mathcal{F})$ (see 6.1.5). This support is a union of irreducible components, these irreducible components have a dimension and let us denote by $\dim(\mathcal{F})$ the maximal dimension of an irreducible component.
Again we start from the exact sequence 8.39

$$0 \longrightarrow \mathcal{O}_{\mathbb{P}^n}(X_0) \longrightarrow \mathcal{O}_{\mathbb{P}^n} \longrightarrow \mathcal{O}_{\mathbb{P}^{n-1}} \longrightarrow 0$$

and the resulting exact sequence

$$0 \longrightarrow \mathcal{O}_{\mathbb{P}^n}(r-1) \longrightarrow \mathcal{O}_{\mathbb{P}^n}(r) \longrightarrow \mathcal{O}_{\mathbb{P}^{n-1}}(r) \longrightarrow 0.$$

We will consider the tensor products $\mathcal{F}(r) = \mathcal{F} \otimes \mathcal{O}_{\mathbb{P}^n}(r)$, but we have to be aware of a minor technical complication: Since the tensor product is not an exact functor tensoring $\mathcal{F}$ with this sequence yields only an exact sequence

$$0 \longrightarrow \mathcal{G} \longrightarrow \mathcal{O}_{\mathbb{P}^n}(X_0) \otimes \mathcal{F} \longrightarrow \mathcal{O}_{\mathbb{P}^n} \otimes \mathcal{F} \longrightarrow \mathcal{O}_{\mathbb{P}^{n-1}} \otimes \mathcal{F} \longrightarrow 0$$

where $\mathcal{G}$ is the kernel of the next arrow so the exactness is true by definition. Since $\mathcal{O}_{\mathbb{P}^n}(r)$ is locally free, we get for all $r \in \mathbb{Z}$ an exact sequence

$$0 \longrightarrow \mathcal{G}(r) \longrightarrow \mathcal{O}_{\mathbb{P}^n}(r-1) \otimes \mathcal{F} \longrightarrow \mathcal{O}_{\mathbb{P}^n}(r) \otimes \mathcal{F} \longrightarrow \mathcal{O}_{\mathbb{P}^{n-1}} \otimes \mathcal{F} \longrightarrow 0.$$

We need some information how the sheaf $\mathcal{G}$ looks like. Here we observe that locally we are in the following situation: We have an A-algebra B an element $f \in B$ and we consider the sequence of B modules

$$0 \longrightarrow fB \longrightarrow B \longrightarrow B/fB \longrightarrow 0.$$

This will be tensorized by a B-module N (the local $\mathcal{F}$) and we get

$$fB \otimes N \longrightarrow N \longrightarrow N/fN \longrightarrow 0$$

and the first arrow is $f \otimes n \mapsto fn$ and hence the kernel is exactly the annihilator of f in N. This means that we get

$$0 \longrightarrow \mathrm{Ann}_N(f) \stackrel{\cdot}{\longrightarrow} Bf \otimes N \longrightarrow N \longrightarrow N/fN \longrightarrow 0.$$

Now it is clear that $\mathrm{Ann}_N(f)$ is an A/fA-module, which is of course finitely generated. This implies for our sequence above that the kernel $\mathcal{G}$ is in fact a $\mathcal{O}_{\mathbb{P}^{n-1}}$-module sheaf, hence it is a coherent sheaf on the hyperplane at infinity, which is $\mathbb{P}^{n-1}$.

The next step in the proof of the coherence theorem will be the proof of

Theorem 8.3.3. *(Serre) Under the assumptions of the theorem above we can find an* $r_0 > 0$ *(depending on* $\mathcal{F}$*) such that*

1) *For all* r *the* A*-module* $H^0(\mathbb{P}^n,\mathcal{F}(r))$ *is a finitely generated. For* $r \geq r_0$ *the higher cohomology groups* $H^i(\mathbb{P}^n,\mathcal{F}(r)) = 0$ *for all* $i > 0$.

2) *For all* $r \geq r_0$ *the sheaf* $\mathcal{F}(r) = \mathcal{F} \otimes \mathcal{O}_{\mathbb{P}^n}(r)$ *is generated by a finite number of global sections. In other words we can find* $s_1, \ldots, s_N \in H^0(\mathbb{P}^n,\mathcal{F}(r))$ *such that for any point* $\mathfrak{q} \in \mathbb{P}^n$ *the stalk* $\mathcal{F}(r)_{\mathfrak{q}}$ *is generated by the restrictions of these sections to this stalk as an* $\mathcal{O}_{\mathbb{P}^n,\mathfrak{q}}$*-module.*

Proof: The proof of this second theorem will basically be obtained by induction on n. The case $n = 0$ is obvious. We break the four term exact sequences into two pieces

$$0 \longrightarrow \mathcal{G}(r) \longrightarrow \mathcal{O}_{\mathbb{P}^n}(r-1) \otimes \mathcal{F} \longrightarrow \mathcal{F}'(r-1) \longrightarrow 0$$
$$0 \longrightarrow \mathcal{F}'(r-1) \longrightarrow \mathcal{O}_{\mathbb{P}^n}(r) \otimes \mathcal{F} \longrightarrow \mathcal{O}_{\mathbb{P}^{n-1}}(r) \otimes \mathcal{F} \longrightarrow 0.$$

Since $\mathcal{G}$ and $\mathcal{O}_{\mathbb{P}^{n-1}}(r)\otimes\mathcal{F}$ are supported on $\mathbb{P}^{n-1}$ we can apply induction and find an $r_0>0$ such that

$$H^i(\mathbb{P}^n,\mathcal{G}(r))=H^i(\mathbb{P}^{n-1},\mathcal{G}(r))=0,H^i(\mathbb{P}^{n-1},\mathcal{O}_{\mathbb{P}^{n-1}}(r)\otimes\mathcal{F})=0$$

for all $r\geq r_0,i>0$. The first exact sequence yields

$$H^i(\mathbb{P}^n,\mathcal{F}'(r-1))=H^i(\mathbb{P}^n,\mathcal{O}_{\mathbb{P}^n}(r-1)\otimes\mathcal{F})$$

for $i>0,r\geq r_0$

We substitute this into the long exact cohomology sequence attached to the second short exact sequence of sheaves and get for $r>>0$

$$\begin{array}{lll}\longrightarrow H^0(\mathbb{P}^n,\mathcal{O}_{\mathbb{P}^n}(r)\otimes\mathcal{F}) & \longrightarrow H^0(\mathbb{P}^n,\mathcal{O}_{\mathbb{P}^{n-1}}(r)\otimes\mathcal{F}) & \longrightarrow \\ H^1(\mathbb{P}^n,\mathcal{O}_{\mathbb{P}^n}(r-1)\otimes\mathcal{F}) & \longrightarrow H^1(\mathbb{P}^n,\mathcal{O}_{\mathbb{P}^n}(r)\otimes\mathcal{F}) & \longrightarrow 0\end{array}$$

and for higher degrees $i\geq 2$ we get

$$H^i(\mathbb{P}^n,\mathcal{O}_{\mathbb{P}^n}(r-1)\otimes\mathcal{F})\xrightarrow{\sim}H^i(\mathbb{P}^n,\mathcal{O}_{\mathbb{P}^n}(r)\otimes\mathcal{F}).$$

In degree one we know by induction that $H^0(\mathbb{P}^n,\mathcal{O}_{\mathbb{P}^{n-1}}(r)\otimes\mathcal{F})$ is finitely generated, under the boundary operators the generators are mapped to finite set of classes in $H^1(\mathbb{P}^n,\mathcal{O}_{\mathbb{P}^n}(r-1)\otimes\mathcal{F})$. Our limit argument shows that these classes vanish, if we pass to a larger r. Hence we find a possibly larger r_0 such that we get $H^1(\mathbb{P}^n,\mathcal{O}_{\mathbb{P}^n}(r)\otimes\mathcal{F})=0$ if $r\geq r_0$. The same limit argument shows that $H^i(\mathbb{P}^n,\mathcal{O}_{\mathbb{P}^n}(\infty)\otimes\mathcal{F})=0$ implies that $H^i(\mathbb{P}^n,\mathcal{O}_{\mathbb{P}^n}(r)\otimes\mathcal{F})=0$ for $i\geq 2,r\geq r_0$.
Now we prove 2) We consider the restriction of $\mathcal{F}$ to the open sets U_i, we know that $\mathcal{F}|U_i=\widetilde{\mathcal{F}(U_i)}$. The $\mathcal{O}_{\mathbb{P}^n}(U_i)$-modules $\mathcal{F}(U_i)$ are finitely generated. We write $s_{i,1},\ldots,s_{i,\nu_i}$ for these generators. They also generate the stalks in the points inside U_i. We have the embedding $\mathcal{O}_{\mathbb{P}^n}\subset\mathcal{O}_{\mathbb{P}^n}(dH_i)$ and this induces a morphism among sheaves $\mathcal{F}\otimes\mathcal{O}_{\mathbb{P}^n}\longrightarrow\mathcal{F}\otimes\mathcal{O}_{\mathbb{P}^n}(dH_i)$. This morphism is an isomorphism on the stalks inside U_i. If we pass to the limit we see that $\mathcal{F}\otimes\mathcal{O}_{\mathbb{P}^n}(\infty H_i)(\mathbb{P}^n)=\mathcal{F}\otimes\mathcal{O}_{\mathbb{P}^n}(U_i)$ and hence we see that all the sections $s_{i\nu}$ extend to sections in the limit.
But then these extensions must already lie in some $\mathcal{F}\otimes\mathcal{O}_{\mathbb{P}^n}(rH_i)$ and since we have $\mathcal{O}_{\mathbb{P}^n}(rH_i)\xrightarrow{\sim}\mathcal{F}(r)=\mathcal{F}\otimes\mathcal{O}_{\mathbb{P}^n}(r)$ we see that at least the stalks inside U_i can be generated by global sections in $\mathcal{F}(r)$. Since this is so for any i the assertion 2) follows.
It follows from 2) that we get a morphism of sheaves $\mathcal{O}^N_{\mathbb{P}^n}\to\mathcal{F}(r)$ simply by sending $(0,\ldots,1,\ldots,0)\to s_i,\ldots$. This morphism is surjective and has a kernel $\mathcal{G}$, which is again coherent. We get an exact sequence, which we still can twist by $\mathcal{O}_{\mathbb{P}^n}(s)$ for an arbitrary integer s. Hence we obtain

$$0\longrightarrow\mathcal{G}(s)\longrightarrow(\mathcal{O}_{\mathbb{P}^n}(s))^N\longrightarrow\mathcal{F}(r+s)\longrightarrow 0,$$

which gives us for $s\gg 0$

$$0\longrightarrow H^0(\mathbb{P}^n,\mathcal{G}(s))\longrightarrow H^0(\mathbb{P}^n,\mathcal{O}^N_{\mathbb{P}^n}(s))\longrightarrow H^0(\mathbb{P}^n,\mathcal{F}(r+s))\longrightarrow H^1(\mathbb{P}^n,\mathcal{G}(s))=0$$

Then it follows from our computation of the cohomology of the sheaves $\mathcal{O}_{\mathbb{P}^n}(s)$ that $H^0(\mathbb{P}^n,\mathcal{F}(r+s))$ is finitely generated for $s >> 0$. But since for any $r_1 \leq r+s$ we have $H^0(\mathbb{P}^n,\mathcal{F}(r_1)) \subset H^0(\mathbb{P}^n,\mathcal{F}(r+s))$ this finishes the proof (inside the proof) . □

Now the coherence theorem can be proved easily by induction on n. But we want to prove a slightly stronger result. In section 8.4.1 we will prove a technical lemma (Lemma 8.4.7), which our given situation tells us the following:
For our sheaf $\mathcal{F}$ and a point $\mathfrak{p} \in \mathrm{Spec}(A)$ we can find an open neighbourhood $\mathrm{Spec}(A_g)$ of $\mathfrak{p}$, an $r > 0$ and a section $f \in H^0(\mathbb{P}^n_{A_g},\mathcal{O}_{\mathbb{P}^n_{A_g}}(r))$ such that the annihilator $\mathrm{Ann}_{\mathcal{F}\otimes A_g}(f) = 0$. The same holds true for the image $\bar{f} \in H^0(\mathbb{P}^n_{A_\mathfrak{p}/\mathfrak{p}},\mathcal{O}_{\mathbb{P}^n_{A_\mathfrak{p}/\mathfrak{p}}}(r))$, i.e. we have $\mathrm{Ann}_{\mathcal{F}\otimes A_\mathfrak{p}/\mathfrak{p}}(\bar{f}) = 0$. In other words f (resp. $\bar{f} = f \mod \mathfrak{p})$) are non zero divisors for the sheaves $\mathcal{F} \otimes A_g$ (resp. $\mathcal{F} \otimes A_\mathfrak{p}/\mathfrak{p}$.)
As we said, this is a very technical lemma. Its proof uses the above theorem 8.3.3, but this theorem we have already proved at this stage. The proof also uses another global argument, namely the we have to use the fact $\mathbb{P}^n_S \longrightarrow S$ is proper.
We pick such an f. Multiplication by f yields a homomorphism $m_f : \mathcal{O}_{\mathbb{P}^n_A} \xrightarrow{m_f} \mathcal{O}_{\mathbb{P}^n_A}(r)$ and we get an exact sequence

$$0 \longrightarrow \mathcal{O}_{\mathbb{P}^n_A} \xrightarrow{m_f} \mathcal{O}_{\mathbb{P}^n_A}(r) \longrightarrow \mathcal{O}_{\mathbb{P}^n_A}(r)/f\mathcal{O}_{\mathbb{P}^n_A} \longrightarrow 0,$$

and we get a corresponding sequence of sheaves on the fiber $\mathbb{P}^n \otimes A_\mathfrak{p}/\mathfrak{p}$. If we now tensorize the first sequence by $\mathcal{F}$, and restrict it to $\mathrm{Spec}(A_g)$ then we get an exact sequence

$$0 \longrightarrow \mathcal{F}_{A_g} \xrightarrow{m_f} \mathcal{F}_{A_g}(r) \longrightarrow \mathcal{F}_{A_g}(r)/f\mathcal{F}_{A_g} \longrightarrow 0,$$

here we use that f is a non zero divisor. On the fiber we get the analogous sequence

$$0 \longrightarrow (\mathcal{F} \otimes A_\mathfrak{p}/\mathfrak{p}) \xrightarrow{m_{\bar{f}}} (\mathcal{F} \otimes A_\mathfrak{p}/\mathfrak{p})(r) \longrightarrow \mathcal{F} \otimes A_\mathfrak{p}/\mathfrak{p}(r)/\bar{f}(\mathcal{F} \otimes A_\mathfrak{p}/\mathfrak{p}) \longrightarrow 0.$$

(The sheaf $\mathcal{G}$ disappears, but it did not do much harm to our argument) We can replace f by f^m and r_0 by $r = mr_0$, this means that we can make this degree arbitrarily large. Recall that we just proved that for degree r of f large enough that $\mathcal{F}_{A_g}(r)$ is acyclic.
We clearly have $\mathrm{Supp}(\mathcal{F} \otimes A_\mathfrak{p}/\mathfrak{p}(r)/\bar{f}(\mathcal{F} \otimes A_\mathfrak{p}/\mathfrak{p})) \subset \mathrm{Supp}(\mathcal{F} \otimes A_\mathfrak{p}/\mathfrak{p}(r))$. Since $\bar{f}$ is not a zero divisor we see that for any minimal prime ideal $\mathfrak{q} \in \mathrm{Supp}((\mathcal{F}(\mathfrak{p})(r))$, i.e any irreducible component of the support, the multiplication by f induces an isomorphism $m_{\bar{f}} : \mathcal{F} \otimes A_\mathfrak{p}/\mathfrak{p}_\mathfrak{q} \longrightarrow \mathcal{F} \otimes A_\mathfrak{p}/\mathfrak{p}(r)_\mathfrak{q}$. This implies that the irreducible components of $\mathrm{Supp}(\mathcal{F} \otimes A_\mathfrak{p}/\mathfrak{p}(r))$ are not contained in $\mathrm{Supp}(\mathcal{F} \otimes A_\mathfrak{p}/\mathfrak{p}(r)/\bar{f}(\mathcal{F} \otimes A_\mathfrak{p}/\mathfrak{p}))$. Hence we see that $d(\mathcal{F} \otimes A_\mathfrak{p}/\mathfrak{p}(r)/\bar{f}(\mathcal{F} \otimes A_\mathfrak{p}/\mathfrak{p})) < d(\mathcal{F} \otimes A_\mathfrak{p}/\mathfrak{p})$.
We apply the same reasoning to the quotient sheaf $\mathcal{F}_{A_g}(r)/f\mathcal{F}_{A_g}$ and get an acyclic resolution

$$0 \longrightarrow \mathcal{F} \longrightarrow \mathcal{F}^0(=\mathcal{F}_{A_g}(r)) \longrightarrow \mathcal{F}^1 \longrightarrow \dots \mathcal{F}^m \longrightarrow \dots \longrightarrow \mathcal{F}^k \longrightarrow .$$

Now we observe the after each step the number $d(\mathcal{F}^i \times A_\mathfrak{p}/\mathfrak{p}))$ drops and hence we see that for $k > d(\mathcal{F} \times A_\mathfrak{p}/\mathfrak{p})$ we get $\mathcal{F}^k \times A_\mathfrak{p}/\mathfrak{p} = 0$. This means that $\mathrm{Supp}(\mathcal{F}^k)$ has empty intersection with the fiber $\mathbb{P}^n_{A_\mathfrak{p}/\mathfrak{p}}$ and hence the image of this closed set in $\mathrm{Spec}(A_g)$ is closed (see Thm. 8.1.8). This means that we can localize further and find another $g_1 \notin \mathfrak{p}$ such that $\mathcal{F}^k \otimes A_{g_1} = 0$ for $k > d(\mathcal{F} \times A_\mathfrak{p}/\mathfrak{p})$.

Hence for a given coherent sheaf $\mathcal{F}$ we have constructed a finite acyclic resolution of $\mathcal{F}$ by coherent sheaves. This resolution is of course only local in the base Spec(A). Taking global sections over $\mathbb{P}^n$ we get a complex $H^0(\mathbb{P}^n,\mathcal{F}^\bullet)$ of A-modules. I refer to the *simple principle* in the section on homological algebra. This principle tells us that this complex computes the cohomology groups and

$$H^\bullet(\mathbb{P}^n,\mathcal{F}) = H^\bullet(H^0(\mathbb{P}^n,\mathcal{F}^\bullet))$$

and since the individual members of the complex are finitely generated A-modules the coherence theorem follows. The reasoning above shows that locally at $\mathfrak{p}$ we have $H^k(\mathbb{P}^n_{A_{g_1}},\mathcal{F}) = 0$ for $k > d(\mathcal{F}(\mathfrak{p}))$. Actually we proved the stronger statement:

Theorem 8.3.4. *In the derived category of coherent sheaves on $\mathbb{P}^n_A$ any coherent sheaf is - locally in the base* Spec(A) *-quasi-isomorphic to a finite complex of acyclic coherent sheaves.*

This finishes the proof of the coherence theorem. □

Of course we may replace Spec(A) by a more general base scheme S, we should assume that S has a finite covering by affine noetherian schemes. Then we consider a projective scheme $f : X \longrightarrow S$ and we get the same formulation, except that now the $\mathcal{F}^m$ are acyclic for the functor $\mathcal{G} \longrightarrow f_*(\mathcal{G})$.

The Hilbert polynomial

Our base scheme is a field k. We start from our sequence above (see 132)

$$0 \longrightarrow \mathcal{F} \longrightarrow \mathcal{F} \otimes \mathcal{O}_{\mathbb{P}^n}(V(f)) \longrightarrow \mathcal{F} \otimes (\mathcal{O}_{\mathbb{P}^n}(V(f))/\mathcal{O}_{\mathbb{P}^n}) \longrightarrow 0,$$

here we assume that f is of degree $r_0 > 0$. We tensorize this sequence by $\mathcal{O}_{\mathbb{P}^n}(r)$ and get an exact sequence

$$0 \longrightarrow \mathcal{F}(r) \longrightarrow \mathcal{F} \otimes \mathcal{O}_{\mathbb{P}^n}(V(f))(r) \longrightarrow \mathcal{F} \otimes (\mathcal{O}_{\mathbb{P}^n}(V(f))/\mathcal{O}_{\mathbb{P}^n})(r) \longrightarrow 0,$$

and now we observe that $\mathcal{F}\otimes\mathcal{O}_{\mathbb{P}^n}(V(f))(r) \xrightarrow{\sim} \mathcal{F}\otimes\mathcal{O}_{\mathbb{P}^n}(V(f))(r+r_0)$. For any coherent sheaf $\mathcal{G}$ on $\mathbb{P}^n_k$ we define its Euler characteristic

$$\chi(\mathbb{P}^n,\mathcal{G}) = \sum (-1)^i \dim H^i(\mathbb{P}^n,\mathcal{G}).$$

For the Euler characteristics of the sheaves in the exact sequence we get

$$\chi(\mathbb{P}^n,\mathcal{F}(r)) + \chi(\mathbb{P}^n,\mathcal{F}'(r)) = \chi(\mathbb{P}^n,\mathcal{F}(r+r_0)).$$

Now we make a little observation: If we have $d(\mathcal{F}') = 0$, i.e. its support is a finite number of points, then $\chi(\mathbb{P}^n,\mathcal{F}'(r))$ does not depend on r and it is equal to the number

$$\chi(\mathbb{P}^n,\mathcal{F}'(r)) = \sum_{\mathfrak{p}\in \mathrm{Supp}(\mathcal{F}')} [k(\mathfrak{p}) : k]\mathrm{lenght}(\mathcal{F}'_{\mathfrak{p}}) = \mathrm{length}(\mathcal{F}'),$$

where we observe that $\mathcal{F}'_{\mathfrak{p}}$ has a filtration by $k(\mathfrak{p})-$ vector spaces and the length is by definition the sum of the $k(\mathfrak{p})$ dimensions of these vector spaces.

Let us assume that $r_0 = 1$, this is possible if k is infinite. If k is finite, then we observe that we can extend the ground field without changing the Euler-characteristics. Hence we conclude that for $d(\mathcal{F}) = 1$ we have

$$\chi(\mathbb{P}^n, \mathcal{F}(r)) = c_1 r + c_0,$$

where $c_1 = \text{length}(\mathcal{F}')$. Then we get by the obvious induction argument

The function $r \longrightarrow \chi(\mathbb{P}^n, \mathcal{F}(r))$ is given by a polynomial

$$\chi(\mathbb{P}^n, \mathcal{F}(r)) = \frac{deg(\mathcal{F})}{d(\mathcal{F})!} r^{d(\mathcal{F})} + c_1 r^{d(\mathcal{F})-1} + \cdots + c_{d(\mathcal{F})}$$

where $\deg(\mathcal{F})$ *is equal to the length of the "last sheaf with zero dimensional support", which occurs at the end of the resolution, especially* $\deg(\mathcal{F}) \neq 0$

The polynomial is called the **Hilbert polynomial of** $\mathcal{F}$. If $Z/k \subset \mathbb{P}^n/k$ is a sub scheme defined by an ideal $\mathcal{I}$ then the degree of Z/k is defined by $d(Z) = d(\mathcal{O}_{\mathbb{P}^n_k}/\mathcal{I})$. It will become clear later that the degree $d(Z)$ is the number of points in the intersection $Z \cap H_1 \cap H_2 \cap \cdots \cap H_s$ where the H_i are hyperplanes in general position and s is the codimension of Z, this is the codimension of an irreducible component of maximal dimension.

8.3.1 The coherence theorem for proper morphisms

Of course we can consider arbitrary projective schemes $f : X \longrightarrow S$, where S is locally noetherian. Let $\mathcal{F}$ be a coherent sheaf on X. Then our two theorems above hold verbatim for this sheaf. This is clear because by assumption we have a diagram

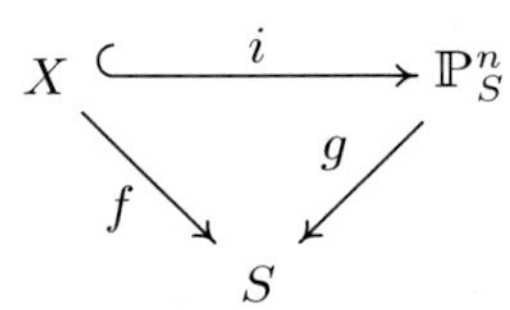

and if $\mathcal{F}$ is a coherent sheaf on X/S then $i_*(\mathcal{F})$ is coherent on $\mathbb{P}^n_S$ and we have

$$R^q f_*(\mathcal{F}) = R^q g_*(i_*(\mathcal{F})).$$

This gives us the theorem 8.3.2. The embedding provides the line bundle $\mathcal{L} = I^*(\mathcal{O}_{\mathbb{P}^n_S}(1)$ on X and we can use this bundle to twist coherent sheaves on X and to formulate theorem 8.3.3.

At this point we observe that $\mathcal{L}$ depends on the embedding, it is clear that we can replace it by any ample line bundle.

Now we want to explain A. Grothendieck's generalisation of the coherence theorem. We consider a base scheme S, which is locally noetherian. We consider a scheme $f : X \longrightarrow S$, which is of finite type and we assume that f is proper. Under this assumption it is still true that for any coherent sheaf $\mathcal{F}$ on X the sheaves $R^q f_*(\mathcal{F})$ are coherent sheaves on S. We will not prove this theorem here, but after the following digression we give some indications why it is true. Before stating the theorem and giving some hints why it is true we make a slight detour.

Digression: Blowing up and contracting

We want briefly explain some geometric constructions, which allow certain modifications of schemes. These modifications are of importance in the process of resolving singularities, an important subject, which is treated only marginally in this book.
We begin by discussing simple examples. Let us consider the affine space over a field k, i.e. $S = \mathrm{Spec}(k[Y_1,\ldots,Y_n])$. In the projective space $\mathbb{P}^{n-1}/S$ we define a closed subscheme $Z \hookrightarrow \mathbb{P}_S^{n-1}$ by the system of homogeneous (even linear) equations

$$Y_iX_j - Y_jX_i = 0.$$

(The $X_1,\ldots,X_n$ are the homogeneous variables for $\mathbb{P}^{n-1}$.) The inclusion composed with the projection to S a projective morphism provides

$$\pi : Z \longrightarrow S.$$

Now it is clear that for any point $s \in S$, the fibre consists of a single point if $(Y_1(s),\ldots,Y_n(s)) \neq 0$, because in this case the solutions are given by the line determined by $(Y_1(s),\ldots,Y_n(s))$. But if $(Y_1(s)\ldots,Y_n(s)) = 0$ then suddenly the system of equations degenerates to $0\,X_i - 0\,X_j = 0$ and the fibre is the full $\mathbb{P}^{n-1}/\,\mathrm{Spec}(k(s))$.

This process is called **blowing up a point** and is of considerable importance in algebraic geometry. The intuitive meaning is that the point is replaced by the projective space attached to its tangent space. To see how useful this is we give two examples.

Example 18. *Let us assume we have the meromorphic function*

$$f(X,Y) = \frac{X+Y}{X-Y}$$

in the function field of the affine plane $S = \mathrm{Spec}(k[X,Y])$ *where* k *is a field and* $\mathrm{char}(k) \neq 2$. *It is regular outside the diagonal* $X = Y$ *and provides a morphism*

$$f : \mathrm{Spec}(k[X,Y])\backslash\Delta \longrightarrow \mathrm{Spec}(k[T]) = \mathbb{A}_k^1.$$

We can extend this to a morphism

$$\mathrm{Spec}(k[X,Y])\backslash\{(0,0)\} \longrightarrow \mathbb{P}_k^1$$

if we send the points on the diagonal Δ *to infinity. But in the origin we do not know what to do.*
Composing this morphism with the morphism $\pi : Z \longrightarrow S$ *we get a morphism* $Z \setminus \pi^{-1}((0,0)) \longrightarrow \mathbb{P}_k^1$*.then a point in the fibre over* $(0,0)$ *tells us in "which direction" we approach* $(0,0)$ *and we can take a "limit". This allows us to extend* f *to a regular morphism* $\widetilde{f} : Z \to \mathbb{P}^1$.

Example 19. *Consider a plane curve* $C \subset \mathrm{Spec}(k[X,Y]) = S$, *for example something like*

$$f(X,Y) = XY + X^7 + Y^7 = 0,$$

then the origin $O = (0,0)$ is a geometric point and the Jacobi criterion shows that this point is singular. The curve has two tangents at $O = (0,0)$, these are the X and the Y-axis. Now we blow up the origin $O \in S$ then the inverse image of our curve has two irreducible components, one is the "proper transform" $\widetilde{C}$ and the other is the fibre of π over O. The curve $\widetilde{C} \subset Z$ meets the fibre $\pi^{-1}(O)$ in two points (the two tangents) and is smooth in these two points. We have resolved the singularity.

This process of blowing up a point is a special case of a much more general procedure. Let us consider any scheme X, we may assume that it is affine, i.e. $X = \mathrm{Spec}(A)$. Let $Y = \mathrm{Spec}(A/I)$ be a subscheme, which is defined by an ideal I. Now we form the graded A-algebra

$$R = A \oplus I \oplus I^2 \oplus \cdots \oplus I^n \oplus \ldots,$$

where of course I^ν is the homogenous summand of degree ν. Now we can consider the scheme

$$\pi : Z = \mathrm{Proj}(R) \longrightarrow \mathrm{Spec}(A),$$

by definition π is projective and this scheme is called the "blow up" of the subscheme Y. We have to meditate a little bit how this morphism looks like. We assume that the ring A is noetherian. We choose a system of generators $f_0, \ldots, f_r$ of the ideal I. Sending the X_i to the f_i yields a surjective homomorphism

$$F : A[X_0, X_1, \ldots, X_r] \longrightarrow A \oplus I \oplus I^2 \oplus \cdots \oplus I^n \oplus \ldots,$$

which in turn gives us a diagram

$$\begin{array}{ccc} Z & \xrightarrow{\;i\;} & X \times \mathbb{P}^r \\ {\scriptstyle\pi}\downarrow & \swarrow & \\ X & & \end{array}$$

and Z is the closed subscheme defined by the homogenous ideal $J = \bigoplus J_n$, which is the kernel of the homomorphism F.

The problem is that this ideal J might be difficult to understand. The homogenous component J_1 of degree contains the elements $f_i X_j - f_j X_i$. This shows that at any point $x \in X \setminus Y$ the stalk J_x is equal to the homogenous component of degree 1 in $A_{\mathfrak{p}_x}[X_0, X_1, \ldots, X_r]$.

This implies that the morphism

$$F' : Z \setminus \pi^{-1}(Y) \longrightarrow X \setminus Y$$

is an isomorphism.

Now we look what happens locally at a point $y \in Y$. We assume that the $f_0, \ldots, f_r$ are a minimal system of generators. It may happen, that the elements $f_i X_j - f_j X_i$ actually generate J_1 (locally at y) and even better the generate the homogenous ideal J (locally at y). (This is the case if X is an affine k-algebra over a field k and if in addition y is a smooth point on X and if Y is smooth in y (Exercise).) Under these assumptions it is clear the ideal $J \otimes A/I$ is the zero ideal and hence we see that

$$Z \times_X Y = \pi^{-1}(Y) = Y \times \mathbb{P}^r.$$

More generally we can say the following. Let $X \longrightarrow \mathrm{Spec}(k)$ be a smooth scheme and let $Y \subset X$ be a smooth sub scheme. It is defined by a sheaf of ideals $\mathcal{I}$ and we know that the sheaf $\mathcal{I}/\mathcal{I}^2$ is locally free (see 7.5.5 , it is the conormal bundle, it is dual to the normal bundle $T_{X/k}/T_{Y/k} = N_Y$. If we now blow up Y then we see that $\pi^{-1}(Y)$ is the projective bundle $\mathbb{P}(N_Y)$ attached to the conormal bundle.

Example 20. *The situation becomes much more interesting if X is not smooth and Y lies in the singular locus. We may for instance start from a field k of characteristic 0, an integer $n \geq 1$ and consider the affine scheme*

$$S = \mathrm{Spec}(k[X_1,X_2,X_3]/(X_1^2 + X_2^2 + X_3^{n+1}) = \mathrm{Spec}(k[x_1,x_2,x_3].$$

This is an integral affine scheme of dimension 2 and the singular locus consists of one point namely the origin $O = (0,0,0)$. We take for our ideal I the ideal generated by x_1,x_2,x_3 and we blow up the origin.

We give the result and strongly recommend to the reader to carry out the calculation.

If $n = 1$ then the fibre over O is a quadric in $\mathbb{P}^2$ it is given by the homogenous equation $u^2 + v^2 + w^2 = 0$ and the scheme $Z/\mathrm{Spec}(k)$ is smooth. If $n = 2$ the fibre over O is again a $\mathbb{P}^1$ but the scheme $Z = Z_1/\mathrm{Spec}(k)$ is not smooth any more, it has an isolated singularity, which lies on the fibre. But this singularity is "milder" than the original one and if we blow up Z_1 in this singular point z_1 we get a smooth scheme $Z_2 \longrightarrow Z_1$ where the fibre over z_1 is a $\mathbb{P}^1$. Then the fibre of the composition $Z_2 \longrightarrow S$ is a union of two projective lines, which intersect transversally in one point (see example17,(a1).)

We can do this for any n, but then we have to wait longer until the scheme $\pi_n : Z_n \longrightarrow S$ becomes smooth. The fibre $\pi_n^{-1}(O)$ will be a chain $Y_1,Y_2,\dots,Y_n$, where for $i = 1,\dots,i = n-1$ Y_i and Y_{i+1} intersect transversally and these are the only intersections.

We can encode this into a graph: Any of the projective lines yields a vertex and two vertices are joined by an edge if they intersect. We get the graph

$$\circ - \circ - \cdots - \circ,$$

which is the Dynkin diagram of the semi simple algebraic group (Lie algebra) of type A_n. This connection between the theory of singularities and the theory of semi simple algebraic groups is not accidental. It had been conjectured by A. Grothendieck that certain singularities are visible inside the algebraic groups and that their resolution gives rise to the Dynkin diagram. This has been proved by E. Brieskorn and P. Slodowy (See [Br] and [Slo]).

For instance in the resolution of the singularity in $\mathrm{Spec}(k[X_1,X_2,X_3]/(X_1^2 + X_2^3 + X_3^5)$ we get a configuration of projective lines, which gives rise to the Dynkin diagram of E_8:

$$\begin{array}{ccccccccccccc} & & & & \circ & & & & & & & & \\ & & & & | & & & & & & & & \\ \circ & - & \circ & - & \circ & - & \circ & - & \circ & - & \circ & - & \circ \end{array}$$

The process of blowing up points or sub schemes is the main tool to resolve singularities. By this we mean: Let X/k be an absolutely irreducible scheme of finite type, let $Y \subset X$ be the singular locus 7.5.1. We would like to construct another scheme $\pi : X' \longrightarrow X$, where π should be projective, the scheme X' should be smooth and $\pi' : X' \setminus \pi^{-1}(Y) \longrightarrow X \setminus Y$ should be an isomorphism. In case we achieved this goal, we say that we resolved the singularity.

If the ground field is algebraically closed of characteristic zero, it has been shown by Hironaka [Hi] that we always have resolution of singularities. For a more recent account we refer to [Kollar].
We may also reverse this process of blowing up a point. We consider a scheme X and a closed subscheme $Y \subset X$. Now we can ask ourselves whether it is possible to *contract* the subscheme Y to a point, this means whether we can find another scheme X' and a morphism $\phi : X \longrightarrow X'$ such that Y maps to a closed point $x \in X$,$Y = \phi^{-1}(x)$ and the morphism $\phi' : X \setminus Y \longrightarrow X' \setminus x$ is an isomorphism.
In the category of topological spaces this construction is always possible, but in algebraic geometry we need certain assumptions. We briefly discuss a special case. Let us assume that $X \longrightarrow \mathrm{Spec}(k)$ is a projective scheme, let $\mathcal{L}$ be a line bundle over X, let $V(\mathcal{L})$ be the associated vector bundle (See 6.2.3.) Let $j : X \longrightarrow V(\mathcal{L})$ be the zero section. Then we have the following result of Grauert

Theorem 8.3.5. *The zero section $j(X) \subset V(\mathcal{L})$ can be contracted to a point, if and only if the dual bundle $\mathcal{L}^\vee$ is ample.*

For the proof and a more general formulation we refer to [EGA, II. 8.9].

We consider schemes $f' : X' \longrightarrow S$,$f : X \longrightarrow S$. A S- morphism $\pi : X' \longrightarrow X$ is called a *modification* if π is projective, surjective and if we can find dense open subsets $U' \subset X'$,$U \subset X$ such that the restriction to U' induces an isomorphism $\pi : U' \longrightarrow U$.
In our previous considerations we have seen such modifications. Blowing up or contracting a closed subscheme $Y \subset X$, whose complement is dense, yields such modifications.
We can state the fundamental

Theorem 8.3.6. *(Lemma of Chow) Let S be a noetherian scheme, let $f : X \longrightarrow S$ be of finite type and separated. Then there exists a quasi projective $S-$ scheme $f' : X' \longrightarrow S$, and a modification*

$$\begin{array}{ccc} X' & \xrightarrow{\quad\pi\quad} & X \\ & {\scriptstyle f'}\searrow \quad \swarrow{\scriptstyle f} & \\ & S & \end{array}$$

If $f : X \longrightarrow S$ is proper, then then we can find a projective $f : X' \longrightarrow S$, and if X is reduced, then we can take X' also reduced.

For a proof of this theorem we refer to [EGA], II.5.6. The proof can be read rather independently from the rest of the book. The theorem gives a hint how we should think of the difference between proper morphisms and projective morphisms. For instance we may get proper schemes $X \longrightarrow \mathrm{Spec}(k)$, which are not projective if we start from a projective scheme $X' \longrightarrow \mathrm{Spec}(k)$ and contract a suitable closed sub scheme $Y \subset X$, perhaps even to a point.
After this detour we state Grothendieck's general coherence theorem

Theorem 8.3.7. *Let S be a noetherian scheme, let $f : X \longrightarrow S$ be a proper morphism of finite type. For any coherent sheaf $\mathcal{F}$ on X the higher direct images $R^q f_*(\mathcal{F})$ are again coherent $\mathcal{O}_S$-modules.*

We will not prove this theorem here, the proof is lengthy but not difficult anymore. We may make a few reduction steps and try an induction over the dimension of the support of our sheaves. The decisive idea is of course to use the Lemma of Chow and start from a coherent sheaf $\mathcal{F}'$ on X'. Then we get coherent sheaves $R^n f'_*(\mathcal{F}')$ and we have the E_2-term of the spectral sequence

$$R^p f_*(R^q\pi_*(\mathcal{F})) \Rightarrow R^n f'_*(\mathcal{F}').$$

Here the sheaves $R^q\pi(\mathcal{F})$ are coherent, but this does not immediately imply that the terms $R^p f_*(R^q\pi_*(\mathcal{F}))$ are coherent, because there may be non zero differentials. But a careful inspection of the differentials allows us to show that some of these terms must be coherent and another induction argument eventually yields the proof. For a detailed discussion we refer to Grothendieck's exposition in [Gr-EGA III], Chap. III, §3.
In this context I admit that some of the important consequences of the coherence theorem are not treated in this book. I refer to EGA loc.cit. §4 where Grothendieck discusses the implications of the coherence theorem to the comparison theorem between formal and algebraic theory (Theorem 4.1.5). This theorem treats the following situation. We consider a proper morphism $f : X \longrightarrow \mathrm{Spec}(A)$, where A is noetherian. Let $I \subset A$ an ideal, we define $\widehat{A} = \varprojlim_k (A/I^k)$ and we have the base change

$$\widehat{f} : \widehat{X} = X \times_A \widehat{A} \longrightarrow \mathrm{Spec}(\widehat{A})$$

For any sheaf coherent sheaf $\mathcal{F}$ we have its restriction $\widehat{\mathcal{F}}$ to $\widehat{X}$ and we can consider the sheaves $R^q\widehat{f}_*(\widehat{\mathcal{G}})$, these are coherent sheaves on $\mathrm{Spec}(\widehat{A}) \subset \mathrm{Spec}(A)$.
We may also consider the sheaves $\mathcal{G} \otimes A/I^k$ on X and consider their derived images $R^q f_*(\mathcal{G} \otimes A/I^k)$ on $\mathrm{Spec}(A)$. Finally we may consider $R^q f_*(\mathcal{G}) \otimes A/I^k$. Both formations yield a projective system and we get a diagram

$$\begin{array}{ccc} R^q\widehat{f}_*(\widehat{\mathcal{G}}) & \rightarrow & R^q f_*(\varprojlim_k (\mathcal{G} \otimes A/I^k)) \\ & \searrow \quad \swarrow & \\ & R^q f_*(\mathcal{G}) \otimes \varprojlim (A/I^k) & \end{array}$$

Now the theorem asserts that all three arrows are isomorphisms. For the proof the reader may also consult [Ha], Chap. III, section 11. This theorem on formal functions has very important consequences, which we want to discuss briefly, for details we refer to [Gr-EGA III], Chap. III, §4.
The connectedness theorem of Zariski (Thm. 4.3.1) says

Theorem 8.3.8. *Let $f : X \longrightarrow Y$ be a proper morphism between locally noetherian schemes. Then there exists a scheme $g : Y' \longrightarrow Y$, which is finite over Y and a morphism $f'X \longrightarrow Y'$ which is surjective and for any $y' \in Y'$ the fiber $(f')^{-1}(y')$ is connected.*

The factorization $f = g \circ f'$ is called the **Stein factorization** of f.
We can give a sketch of the proof. It it clear that the assertion is local in the base, hence we may assume that $Y = \mathrm{Spec}(A)$ and A is noetherian. Then $H^0(X,\mathcal{O}_X) = B$ is finite over A and we put $Y' = \mathrm{Spec}(B)$. This yields $f' : X \longrightarrow \mathrm{Spec}(B)$ and $g : \mathrm{Spec}(B) \longrightarrow \mathrm{Spec}(A)$. The surjectivity of f' is clear. For the rest of the statement we may assume $B = A$. We pick a prime ideal $\mathfrak{p} \in \mathrm{Spec}(A)$ and consider the fiber $f^{-1}(\mathfrak{p})$.

Assume that this fiber has several (i.e. more than one) connected components. Then it is clear that $H^0(f^{-1}(\mathfrak{p}),\mathcal{O}_{f^{-1}(\mathfrak{p})})$ is the direct sum of several fields. If we replace $\mathfrak{p}$ by a power $\mathfrak{p}^k$ and perform the base change $\mathrm{Spec}(A) \longleftarrow \mathrm{Spec}(A/\mathfrak{p}^k)$ then we get basically the same: $H^0(X \times \mathrm{Spec}(A/\mathfrak{p}^k),\mathcal{O}_{X\times \mathrm{Spec}(A/\mathfrak{p}^k)})$ is the direct sum of several local rings. This implies that the projective limit $\varprojlim_k H^0(X \times \mathrm{Spec}(A/\mathfrak{p}^k),\mathcal{O}_{X\times \mathrm{Spec}(A/\mathfrak{p}^k)})$ is a direct sum of several non zero local rings (the constant section 1 on a component does not go to zero) and we have

$$\varprojlim_k (A/\mathfrak{p}^k) \longrightarrow \varprojlim_k H^0(X \times \mathrm{Spec}(A/\mathfrak{p}^k),\mathcal{O}_{X\times \mathrm{Spec}(A/\mathfrak{p}^k)}).$$

Now we have to invoke the above mentioned theorem on formal functions, it implies that this homomorphism must be an isomorphism and this is impossible, because $\varprojlim_k (A/\mathfrak{p}^k)$ is still local.

If we are in the situation the connectedness theorem, then we may consider the set X' of points $x \in X$ which are isolated in their fiber $f^{-1}(f(x))$. Then Zariski's Main Theorem asserts that $X' \subset X$ is open and the morphism f' induces an isomorphism between X' and an open subscheme of Y'.

Again we give a brief indication, why this is true. We can easily reduce this to the case that $Y' = Y$, i.e. $f_*(\mathcal{O}_X) = \mathcal{O}_Y$. Then it follows from the connectedness theorem that for $x \in X'$ we actually have $f^{-1}(f(x)) = x$. This implies that for any open neighborhood $x \in U_x$ we can find a neighborhood $V_{f(x)}$ of $f(x)$ such that $f^{-1}(V_{f(x)}) \subset U_x$ This implies that $\mathcal{O}_{Y,f(x)} \longrightarrow \mathcal{O}_{X,x}$ is an isomorphism and the rest is clear.

In Zariski's Main Theorem we can weaken the hypothesis that $f : X \longrightarrow Y$ is proper, obviously it suffices to assume that f is quasi projective (see 8.1.6). We can easily reduce the assertion to the case of a projective morphism.

The theorem has an important application for birational isomorphisms. Let $f : X \longrightarrow Y$ be quasi projective and let us assume that X,Y are integral. Furthermore we assume that f is birational, i.e. X and Y have the same field of meromorphic functions. If in addition Y is normal then f is an isomorphism between X' and the open set $V \subset Y$, which is defined as the set $y \in Y$ for which $f^{-1}(y)$ contains an isolated point.

We do not need these results in the proofs of the theorems in the later chapters.

8.4 Base Change

We consider a projective scheme $X \longrightarrow S$, where S a noetherian base scheme, let $\mathcal{F}$ be a coherent sheaf on X. We want to study the behavior of the direct images $R^q f_*(\mathcal{F})$ under base change, i.e. we consider diagrams

$$\begin{array}{ccc} X & \xleftarrow[\mathrm{Id}\times s]{} & X \times_A S' \\ \Big\downarrow f & & \Big\downarrow f \times \mathrm{Id} \\ S & \xleftarrow[s]{} & S' \end{array}$$

and then we will see that we have a canonical homomorphism (*the base change homomorphism*) for coherent sheaves

$$\phi_{S'/S} : R^q(f \times \mathrm{Id})_*((\mathrm{Id} \times s)^*(\mathcal{F})) \longrightarrow s^*(R^q f_*(\mathcal{F})),$$

and we want to understand the properties of this homomorphism.
The question we want to study is local in the base, so there is no harm if we assume $S = \mathrm{Spec}(A)$ where A is a noetherian ring, we may also assume that $S' = \mathrm{Spec}(A')$ is affine. Finally we may assume that we have an embedding $X \hookrightarrow \mathbb{P}^n_A$, and that our sheaf on X is the restriction of a sheaf on $\mathbb{P}^n_A$. This sheaf is also called $\mathcal{F}$. Under these assumptions the sheaf $R^q f_*(\mathcal{F})$ is the quasi coherent sheaf obtained from $H^q(\mathbb{P}^n_A,\mathcal{F})$. (See proposition 6.1.18)

We make a first step to understand the behavior of the cohomology of coherent sheaves under base change. We want to simplify the notations slightly. If we have a scheme $X \longrightarrow S = \mathrm{Spec}(A)$ we allow ourselves to write X_A for $X/\mathrm{Spec}(A)$. Accordingly we write $X_{A'}$ for $X \times_A S'$.
Our sheaf is a sheaf on $\mathbb{P}^n_A$, (with support in X.) We compute the cohomology starting from the Čech complex attached to the standard covering by the U_i. We consider

$$\mathbb{P}^n_A \times_{\mathrm{Spec}(A)} \mathrm{Spec}(A') = \mathbb{P}^n_{A'} \tag{8.41}$$

If we have a sheaf $\mathcal{F}$ on X then this yields a sheaf $(\mathrm{Id} \times s)^*(\mathcal{F}) = \mathcal{F}_{A'}$ on $X_{A'}$ and by definition by $\mathcal{F}_{A'}(U_{i,A'}) = \mathcal{F}(U_i) \otimes_A A'$. This tells us that the Čech complex, which computes $H^\bullet(\mathbb{P}^n_{A'},\mathcal{F}_{A'})$, is the tensor product $C^\bullet(\mathbb{P}^n,\mathfrak{U},\mathcal{F}) \otimes_A A'$ and hence we get a map between the complexes

$$C^\bullet(\mathbb{P}^n_A,\mathfrak{U},\mathcal{F}) \longrightarrow C^\bullet(\mathbb{P}^n_{A'},\mathfrak{U},\mathcal{F}_{A'}),$$

which in turn gives a map between the cohomology groups

$$H^\bullet(\mathbb{P}^n_A,\mathcal{F}) \longrightarrow H^\bullet(\mathbb{P}^n_{A'},\mathcal{F}_{A'}).$$

Here we consider both sides as abelian groups, the left hand side carries the structure of A-modules, the right hand side cohomology groups are A'-modules. It is obvious that we get from the homomorphism above an A'-module homomorphism

$$\phi_{S'/S} : H^\bullet(\mathbb{P}^n_A,\mathcal{F}) \otimes_A A' \longrightarrow H^\bullet(\mathbb{P}^n_{A'},\mathcal{F}_{A'}). \tag{8.42}$$

This map between the cohomology groups is the **base change homomorphism.** In general we cannot say very much about this base change homomorphism.

It is very easy to see that this morphism is an isomorphism if $S \leftarrow S'$ is flat (See theorem 8.4.1 below.) The interesting case is that s is a point or even a closed point, i.e. $s : \mathrm{Spec}(k(s)) \longrightarrow S$. Then $X \times \mathrm{Spec}(k(s)) = X_s$ is a scheme over a field. Then $\dim H^q(X_s,\mathcal{F}_s)$ is a finite dimensional vector space over $k(s)$ and we may for instance ask the question how these dimensions vary with the point s. In the previous section we saw, that the fibers X_s may vary in a very irregular way. If for instance $\mathcal{F} = \mathcal{O}_X$, then we should expect some weird behavior of $\dim H^q(X_s,\mathcal{F}_s)$ unless we make some assumptions on $f : X \longrightarrow S$ or on the sheaf $\mathcal{F}$. The question is delicate, one answer is given in the theorem 8.4.5 below.

We have to make some assumptions, and here we have two options

(a) The base change $A \to A'$ is flat.

(b) The sheaf $\mathcal{F}$ is flat over A (we say it is A-flat or $S = \mathrm{Spec}(A)$-flat), i.e. for any open set $U \subset \mathbb{P}^n_A$ the $\mathcal{O}_{\mathbb{P}^n_A}(U)$-module $\mathcal{F}(U)$ is a flat A module.

We consider the case (a) first. In this case we have

Theorem 8.4.1. *If $S \leftarrow S'$ is flat, i.e. if $A \longrightarrow A'$ is flat then the base change homomorphism is an isomorphism.*

This is clear, it follows from the computation of the cohomology of the Čech-complex and lemma 8.2.3.

Now we come to the case b). For the formulation we drop the assumption that the base schemes are affine.

Theorem 8.4.2. *Let S be a locally noetherian scheme, we assume it has a finite covering by affine schemes. Let $\mathcal{F}$ be a S-flat coherent sheaf on $\mathbb{P}^n_S$. Let π be the projection $\mathbb{P}^n_S \longrightarrow S$. Then we can find an $r_0 > 0$ such that*

(1) $R^i\pi'_*(\mathbb{P}^n_{S'},\mathcal{F}_{S'}(r)) = 0$ for all $i > 0, r \geq r_0$ and all schemes $S' \longrightarrow S$.

(2) The $\mathcal{O}_S$ module $R^0\pi_*(\mathcal{F}(r))$ is locally free for $r \geq r_0$.

(3) For any $S \leftarrow S'$ the base change homomorphism

$$R^0\pi_*(\mathcal{F}(r))) \otimes_{\mathcal{O}_S} \mathcal{O}_{S'} \longrightarrow R^0\pi_*(\mathcal{F}_{S'}(r)))$$

is an isomorphism if $r \geq r_0$.

Proof: Of course we may assume that $S = \mathrm{Spec}(A)$, where A is a noetherian ring. We apply theorem 8.3.3 and choose a large enough $r > r_0$ such that $H^q(\mathbb{P}^n_A,\mathcal{F}(r)) = 0$ for all $q > 0$.
We profit from the Čech complex. We have the standard covering of $\mathbb{P}^n_A$ by the affine spaces $U_{i,A}$ and with respect to this covering we take the Čech-complex, but this time we take the alternating sub-complex (See Vol. I 4.5)

$$C^\bullet(\mathfrak{U},\mathcal{F}(r)) = 0 \longrightarrow \prod_{i=0}^{i=n} \mathcal{F}(r)(U_i) \longrightarrow \prod_{0\leq i<j\leq n} \mathcal{F}(r)(U_i \cap U_j) \longrightarrow .$$

The alternating complex becomes zero after the $n+1$-th step. We know that the inclusion of this complex into the usual Čech-complex induces an isomorphism in cohomology. Hence the cohomology groups of this complex are the cohomology groups of $\mathcal{F}(r)$. Since we assumed that the higher cohomology groups vanish we get an acyclic complex of A-modules

$$0 \longrightarrow H^0(\mathbb{P}^n_A,\mathcal{F}(r)) \longrightarrow C^0(\mathfrak{U},\mathcal{F}(r)) \longrightarrow C^1(\mathfrak{U},\mathcal{F}(r)) \longrightarrow \dots \longrightarrow C^{n+1}(\mathfrak{U},\mathcal{F}(r)) \longrightarrow 0,$$

where the $C^q(\mathfrak{U},\mathcal{F}(r))$ are A-flat and $H^0(\mathbb{P}^n_A,\mathcal{F}(r))$ is a finitely generated A-module. We show that $H^0(\mathbb{P}^n_A,\mathcal{F}(r))$ is also A-flat. We do this by induction on the length of the complex. We break the sequence and get two sequences

$$0 \longrightarrow H^0(\mathbb{P}^n_A,\mathcal{F}(r)) \longrightarrow C^0(\mathfrak{U},\mathcal{F}(r)) \longrightarrow B^0 \longrightarrow 0$$
$$0 \longrightarrow B^0 \longrightarrow C^1(\mathfrak{U},\mathcal{F}(r)) \longrightarrow \dots \longrightarrow C^{n+1}(\mathfrak{U},\mathcal{F}(r)) \longrightarrow 0$$

If our exact sequence is a short exact sequence, i.e. $n = 0$ then we see only the first sequence and have $B^0 = C^1(\mathfrak{U},\mathcal{F}(r))$ and hence B^0 is flat. We tensor this short exact sequence by an arbitrary A-module N and get the long exact sequence

$$\longrightarrow 0 \longrightarrow \mathrm{Tor}_1^A(B^0,N) \longrightarrow H^0(\mathbb{P}^n_A,\mathcal{F}(r)) \otimes N \longrightarrow C^0(\mathfrak{U},\mathcal{F}(r)) \otimes N \longrightarrow B^0 \otimes N \longrightarrow 0$$

Since B^0 is flat we have $\mathrm{Tor}_1^A(B^0,N) = 0$. Further to the left in this sequence we find some isomorphisms

$$0 = \mathrm{Tor}_{j+1}^A(B^0,N) \xrightarrow{\sim} \mathrm{Tor}_j^A(H^0(\mathbb{P}^n_A,\mathcal{F}(r)),N) \text{ for all } j \geq 1$$

and we conclude that $H^0(\mathbb{P}^n_A,\mathcal{F}(r))$ is A-flat.
Now the induction step is obvious. If $n > 0$ then the second of our sequences above becomes shorter, we get by induction assumption that B^0 is A-flat and we just proved that this implies that H^0 is A-flat.
We investigate what happens if we perform a base change $A \longrightarrow A'$. It follows from the definitions that the resulting Čech-complex is simply the tensor product of the old Čech-complex by A', we get

$$C^\bullet(\mathfrak{U},\mathcal{F}(r)) \otimes A' \xrightarrow{\sim} C^\bullet(\mathfrak{U},\mathcal{F}(r)_{A'}).$$

Since all modules are A-flat it follows that

$$H^0(\mathbb{P}^n_{A'},\mathcal{F}(r)_{A'}) \xrightarrow{\sim} H^0(\mathbb{P}^n_A,\mathcal{F}(r)) \otimes A'.$$

Finally we observe that $H^0(\mathbb{P}^n_A,\mathcal{F}(r))$ is a finitely generated A-module, since it is also flat we can conclude that it is locally free.

□

This theorem has important consequences. We go back to the proof of theorem 8.3.2 and the first remark following it. Given a coherent sheaf $\mathcal{F}$ on $\mathbb{P}^n_A$ and a point $\mathfrak{p} \in \mathrm{Spec}(A)$ we passed to a suitable neighbourhood $\mathrm{Spec}(A_g)$ and found a suitable $f \in H^0(\mathbb{P}^n_{A_g},\mathcal{O}_{\mathbb{P}^n_{A_g}}(r))$ such that we got an exact sequence

$$0 \longrightarrow \mathcal{F}_{A_g} \xrightarrow{m_f} \mathcal{F}_{A_g}(r) \longrightarrow \mathcal{F}_{A_g}(r)/f\mathcal{F}_{A_g} \longrightarrow 0.$$

Now we assume in addition that $\mathcal{F}$ is A-flat. We apply a lemma in section 8.4.1 (Lemma 8.4.8). It tells us that we can even find an f such that the quotient sheaf $\mathcal{F}_{A_g}(r)/f\mathcal{F}_{A_g}$ is again A_g-flat.
Repeating the reasoning from the proof of the coherence theorem we get:

Theorem 8.4.3. *Let $\mathcal{F}$ be an A-flat coherent sheaf on $\mathbb{P}^n_A$. Then locally in the base we find a resolution*

$$0 \longrightarrow \mathcal{F} \longrightarrow \mathcal{F}^0 \longrightarrow \mathcal{F}^1 \longrightarrow \mathcal{F}^2 \dots$$

where all the sheaves $\mathcal{F}^\nu$ are flat and "universally acyclic", i.e. $H^i(\mathbb{P}^n_{A'},\mathcal{F}_{A'}) = 0$ for all $i > 0$ and all base changes $A \longrightarrow A'$.

Remark: We come back the proof of the coherence theorem. At the end of the proof we stated a strengthening of that theorem. Here we proved that for a coherent and A-flat sheaf $\mathcal{F}$ on $\mathbb{P}^n_A$ we have -locally in the base- a quasi-isomorphism of complexes

$$\begin{array}{ccccccccccccc} & & 0 & & & & & & & & & & \\ & & \downarrow & & & & & & & & & & \\ 0 & \longrightarrow & \mathcal{F} & \longrightarrow & 0 & \longrightarrow & 0 & \longrightarrow & \cdots & & & & \\ & & \downarrow & & \downarrow & & \downarrow & & & & & & \\ 0 & \longrightarrow & \mathcal{F}^0 & \longrightarrow & \mathcal{F}^1 & \longrightarrow & \mathcal{F}^2 & \longrightarrow & \cdots, & \longrightarrow & \mathcal{F}^m & \longrightarrow & 0 \end{array}$$

where the $\mathcal{F}^i$ are A- flat and universally acyclic.
Hence we know that for any $A \longrightarrow A'$ the complex

$$0 \longrightarrow H^0(\mathbb{P}^n_{A'}, \mathcal{F}^0_{A'}) \longrightarrow H^0(\mathbb{P}_{A'}, \mathcal{F}^1_{A'}) \longrightarrow \dots$$

computes the cohomology groups $H^i(\mathbb{P}^n_{A'} \mathcal{F}_{A'})$, and we also know that

$$H^0(\mathbb{P}^n_{A'}, \mathcal{F}_{A'}) = H^0(\mathbb{P}^n_A, \mathcal{F}) \otimes_A A'.$$

Hence we see that the above flat universally acyclic resolution gives us a complex of finitely generated locally free A-modules

$$K^\bullet := 0 \longrightarrow H^0(\mathbb{P}^n_A, \mathcal{F}^0) \longrightarrow H^0(\mathbb{P}^n_A, \mathcal{F}^1) \longrightarrow \dots \longrightarrow H^0(\mathbb{P}^n_A, \mathcal{F}^n) \longrightarrow 0$$

such that for any $A \longrightarrow A'$, the complex $K^\bullet \otimes_A A'$ computes the cohomology groups $H^i(\mathbb{P}^n_{A'}, \mathcal{F}_{A'})$.
Among other things we want to understand the behaviour of the cohomology groups $H^i(\mathbb{P}^n_{A'}, \mathcal{F}_{A_\mathfrak{p}/\mathfrak{p}})$ vary if $\mathfrak{p}$ varies over the prime ideals in Spec(A). If A is integral and $\mathfrak{p}$ is not the minimal ideal (0), then $A \longrightarrow A_\mathfrak{p}/\mathfrak{p}$ is the prototype of a non flat base change.
Now we need a little bit of linear algebra.

Lemma 8.4.4. *Let k be a field, in the derived category of the category of finite dimensional k-vector spaces is any finite complex "isomorphic to its complex of cohomology groups". This means that any finite complex of finite dimensional k vector spaces*

$$C^\bullet = \longrightarrow 0 \longrightarrow C^k \longrightarrow C^{k+1} \longrightarrow \dots \longrightarrow C^m \longrightarrow 0$$

can be split into two summands $C^\bullet = H^\bullet(C^\bullet) \oplus A^\bullet(C^\bullet)$ such that the differentials split accordingly and such that the differentials on $H^\bullet(C^\bullet)$ are all zero and such that the complex $A^\bullet(C^\bullet)$ is acyclic.

Let us call a complex, in which all the differentials are zero a *cohomological complex.* We do not prove this lemma, it is simple linear algebra.
Let us assume that A is local with maximal ideal $\mathfrak{m}$. The complex $K^\bullet$ consists of free modules and we consider the homomorphism

$$K^\bullet \longrightarrow K^\bullet \otimes A/\mathfrak{m} = H^\bullet(K^\bullet \otimes A/\mathfrak{m})) \oplus A^\bullet(K^\bullet \otimes A/\mathfrak{m}))$$

where we applied the lemma above. We see easily that this yields a decomposition

$$K^\bullet = \mathcal{H}^\bullet(K^\bullet) \oplus A^\bullet(K^\bullet)$$

such that the differentials respect the direct sum decomposition, where $A^\bullet(K^\bullet)$ is still exact but where for the differential $d_{\mathcal{H}}$ on $\mathcal{H}^\bullet(K^\bullet)$ we only have

$$d_{\mathcal{H}}(\mathcal{H}^q(K^\bullet)) \subset \mathfrak{m}\mathcal{H}^{q+1}(K^\bullet)$$

In the derived category of A modules the complex $K^\bullet$ is isomorphic to $\mathcal{H}^\bullet(K^\bullet)$.
Let $\mathfrak{p}$ be another prime ideal in A let $A_\mathfrak{p}$ be the localization at $\mathfrak{p}$ and $\mathfrak{m}_\mathfrak{p}$ the maximal ideal. We have the diagram of homomorphisms

$$\begin{array}{ccc} A & \longrightarrow & A_\mathfrak{p} \\ \downarrow & & \downarrow \\ A/\mathfrak{m}. & & A_\mathfrak{p}/\mathfrak{p} \end{array}$$

and this induces a diagram in the category of complexes

$$\begin{array}{ccc} \mathcal{H}^\bullet(K^\bullet) & \longrightarrow & \mathcal{H}^\bullet(K^\bullet)\otimes A_\mathfrak{p} \\ \downarrow & & \downarrow \\ \mathcal{H}^\bullet(K^\bullet)\otimes A/\mathfrak{m} & & \mathcal{H}^\bullet(K^\bullet)\otimes A_\mathfrak{p}/\mathfrak{p} \end{array}$$

Now the complex in the lower left corner is cohomological, therefore,

$$\dim_{A/\mathfrak{m}} H^q\left(\mathbb{P}^n_{\mathrm{Spec}(A/\mathfrak{m})}, \mathcal{F}\otimes A/\mathfrak{m}\right) = \dim_{A/\mathfrak{m}}(K^q \otimes A/\mathfrak{m})$$

and this is equal to the rank of the free A module $\mathcal{H}^q(K^\bullet)$.
But $\mathcal{H}^\bullet(K^\bullet)$ is not necessarily cohomological, and hence also $\mathcal{H}^\bullet(K^\bullet)\otimes A_\mathfrak{p}/\mathfrak{p}$ is not necessarily cohomological, i.e. the differentials are not necessarily zero. Hence we get

$$\dim_{A/\mathfrak{m}} H^i\left(\mathbb{P}^n_{\mathrm{Spec}(A/\mathfrak{m})}, \mathcal{F}\otimes A/\mathfrak{m}\right) \geq \dim_{A_\mathfrak{p}/\mathfrak{m}_\mathfrak{p}} H^i\left(\mathbb{P}^n_{\mathrm{Spec}(A_\mathfrak{p}/\mathfrak{m}_\mathfrak{p})}, \mathcal{F}\otimes A_\mathfrak{p}/\mathfrak{m}_\mathfrak{p}\right)$$

We drop the assumption that A is local and summarise our findings to a theorem

Theorem 8.4.5 (Semi-continuity). **(1)** *If $\mathcal{F}$ is a coherent A-flat sheaf on $\mathbb{P}^n_{\mathrm{Spec}(A)}$ and if $\mathfrak{m}\supset\mathfrak{p}$ are two prime ideals then*

$$\dim_{A/\mathfrak{m}} H^i\left(\mathbb{P}^n_{\mathrm{Spec}(A/\mathfrak{m})}, \mathcal{F}_{A/\mathfrak{m}}\right) \geq \dim_{A_\mathfrak{p}/\mathfrak{m}_\mathfrak{p}} H^i\left(\mathbb{P}^n_{\mathrm{Spec}(A_\mathfrak{p}/\mathfrak{m}_\mathfrak{p})}, \mathcal{F}_{A_\mathfrak{p}/\mathfrak{m}_\mathfrak{p}}\right). \quad (8.43)$$

(2) *If for a given degree i and a prime ideal $\mathfrak{p}\in \mathrm{Spec}(A)$ we have $H^\nu(\mathbb{P}^n_{A_\mathfrak{p}/\mathfrak{m}_\mathfrak{p}}, \mathcal{F}_{A_\mathfrak{p}/\mathfrak{m}_\mathfrak{p}}) = 0$ for $\nu = i-1$ and $\nu = i+1$ then $H^i(\mathbb{P}^n_A, \mathcal{F})$ is locally free at $\mathfrak{p}$ and the base change homomorphism*

$$H^i(\mathbb{P}^n_A, \mathcal{F})\otimes A_{\mathfrak{p}/\mathfrak{m}_\mathfrak{p}} \longrightarrow H^i(\mathbb{P}^n_{A_\mathfrak{p}/\mathfrak{m}_\mathfrak{p}}, \mathcal{F}_{A_\mathfrak{p}/\mathfrak{m}_\mathfrak{p}})$$

is an isomorphism.

(3) *We assume that the ring A is reduced, let $\mathfrak{m} \in \mathrm{Spec}(A)$. We assume that for a given degree i and for all primes $\mathfrak{p} \subset \mathfrak{m}$ we have the equality*

$$\dim_{A/\mathfrak{m}} H^i\left(\mathbb{P}^n_{\mathrm{Spec}(A/\mathfrak{m})},\mathcal{F}_{A/\mathfrak{m}}\right) = \dim_{A_\mathfrak{p}/\mathfrak{m}_\mathfrak{p}} H^i\left(\mathbb{P}^n_{\mathrm{Spec}(A_\mathfrak{p}/\mathfrak{m}_\mathfrak{p})},\mathcal{F}_{A_\mathfrak{p}/\mathfrak{m}_\mathfrak{p}}\right). \quad (8.44)$$

Then $H^i(\mathbb{P}^n_A,\mathcal{F})$ is locally free at $\mathfrak{m}$ and the base change homomorphism

$$H^i(\mathbb{P}^n_A,\mathcal{F}) \otimes A_\mathfrak{p}/\mathfrak{m}_\mathfrak{p} \longrightarrow H^i(\mathbb{P}^n_{A_\mathfrak{p}/\mathfrak{m}_\mathfrak{p}},\mathcal{F}_{A_\mathfrak{p}/\mathfrak{m}_\mathfrak{p}})$$

is an isomorphism.

Proof: These are now rather easy consequence of our consideration above. The assertion (1) is already proved. We consider the complex $\mathcal{H}^\bullet(K^\bullet)$. The proof of (2) is easy. In this case the assumptions imply that

$$\mathcal{H}^{i-1}(K^\bullet) \otimes A_\mathfrak{p}/\mathfrak{m}_\mathfrak{p} = \mathcal{H}^{i+1}(K^\bullet) \otimes A_\mathfrak{p} = 0$$

Then the Nakayama lemma implies that

$$\mathcal{H}^{i-1}(K^\bullet) \otimes A_\mathfrak{p} = \mathcal{H}^{i+1}(K^\bullet) \otimes A_\mathfrak{p} = 0,$$

and hence

$$H^i(\mathbb{P}^n_A,\mathcal{F}) \otimes A_\mathfrak{p} = \mathcal{H}^i(K^\bullet) \otimes A_\mathfrak{p}$$

This then yields

$$H^i(\mathbb{P}^n_A,\mathcal{F}) = H^i(\mathcal{H}^\bullet(K^\bullet)).$$

To prove (3) we may assume that A is local with maximal ideal $\mathfrak{m}$. Then our assumptions imply that for all $\mathfrak{p} \subset \mathfrak{m}$ we have

$$d^{i-1} : \mathcal{H}^{i-1}(K^\bullet) \subset \mathfrak{p}\mathcal{H}^i(K^\bullet), d^i : \mathcal{H}^i(K^\bullet) \subset \mathfrak{p}\mathcal{H}^{i+1}(K^\bullet)$$

otherwise the dimension $\dim_{A_\mathfrak{p}/\mathfrak{m}_\mathfrak{p}} H^i(\mathbb{P}^n_{\mathrm{Spec}(A_\mathfrak{p}/\mathfrak{m}_\mathfrak{p})},\mathcal{F}_{A_\mathfrak{p}/\mathfrak{m}_\mathfrak{p}})$ would drop. But since we assumed that A is reduced we know that the intersection of these primes is (0) we can conclude that $d^{i-1} = 0, d^i = 0$ on $\mathcal{H}^\bullet(K^\bullet)$.

□

Usually one finds this in a slightly different form in the literature. Let S be noetherian base scheme S and an $\mathcal{F}$an S-flat coherent sheaf on a projective scheme $\pi : X \to S$. We consider the coherent sheaves $R^i\pi_*(\mathcal{F})$ on S. Then our assertion (1) in the theorem above is equivalent to

The function $s \longrightarrow \dim_{k(s)} H^i(X_s,\mathcal{F}_s)$ is upper semicontinous with respect to the Zariski topology on S.

I find it easier to remember that the function may jump up if we specialize (See 6.1.3) a point $s \to s'$. But this formulation is obviously the same as the one above because the assertion is local on S.

Theorem 8.4.6 (Invariance of the Euler characteristic)**.** *Let $\pi : X \to S$ be a projective scheme over a noetherian base scheme S and let $\mathcal{F}$ be a S flat coherent sheaf. Then the function*

$$s \mapsto \chi(X_s,\mathcal{F}) = \sum_{i=0}^{d}(-1)^i \dim_{k(s)} H^i(X_s,\mathcal{F}_s)$$

is locally constant. Here d is large enough such that the cohomology groups vanish in degrees beyond d.

Proof: This is obvious because for any $s \in S$ the Euler characteristic $\chi(X_s,\mathcal{F})$ is equal to $\sum_{i=0}^{d}(-1)^i \operatorname{Rank}_{\mathcal{O}_{k(s)}}(K^i \otimes \mathcal{O}_{k(s)})$. □

8.4.1 Flat families and intersection numbers

We now come to a very important topic, which we can treat only cursorily, a thorough treatment can be found in the book [Fu]. But we hope that it possible to illustrate some of the ideas, which play a role.

In principle we want to do the following: Let $X \longrightarrow \operatorname{Spec}(k)$ an irreducible, smooth and projective (or even only proper) scheme over a field k. Let Z_1,Z_2 be two irreducible sub schemes of codimension c_1,c_2. Let us assume that they are in complementary dimension, i.e. we have $c_1 + c_2 = d = \dim(X)$. Then we can expect that that under favourable conditions they intersect in a finite number of points and that we may define an intersection number $\#(Z_1 \cap Z_2)$. Moreover we want that this number is the sum over the points in the intersection counted with a multiplicity.

For instance it looks plausible that the intersection multiplicity of the parabola $y = x^2$ with the x-axis $x = 0$ in the point (0,0) should be equal to two. We give a more sophisticated example further down.

We do not assume that the base field k is algebraically closed, but we want that the intersection number is invariant under base change: If $k \longrightarrow \bar{k}$ is an algebraic closure of k then we want $\#(Z_1 \times_k \bar{k}) \cap (Z_2 \times_k \bar{k}) = \#(Z_1 \cap Z_2)$. If for instance the ground field is $\mathbb{Q}$ and $Z_1 = \operatorname{Spec}(\mathbb{Q}[X,Y]/(X^2 + Y^2 + 1), Z_2 = \mathbb{Q}[X,Y]/(X - Y)$ then $Z_1 \cap Z_2$ consists of one point, which will be counted with multiplicity two, because the residue field extension has degree two. If we pass to the algebraic closure, then the intersection has two points each counted with multiplicity one.

We briefly recall the analogous problem in the context of oriented manifolds (See Vol. I, 4.8.9). In this case we started with two oriented submanifolds N_1,N_2 of an oriented manifold M, which sit in complementary dimension. They provided classes $[N_1],[N_2]$ in the cohomology (with compact support). Nobody can prevent us from defining $\#(N_1 \cap N_2)$ by taking the cup product of the two classes, and then

$$[N_1] \cup [N_2] = \#(N_1 \cap N_2) \times \text{fundamental class of } M.$$

If these two manifolds intersect transversally, then we can interpret the number $\#(N_1 \cap N_2)$ is number of intersection points, where each point $p \in N_1 \cap N_2$ is counted with a sign $\epsilon(p)$, which arises from the comparison of the orientations of the two tangent spaces of N_1,N_2 at p to the orientation of the tangent space of M at p. But this intersection number is always defined, even if N_1,N_2 do not intersect transversally, for instance it may happen that $N_1 \subset N_2$ and still the intersection number is defined.

We can have an analogous situation in algebraic geometry. Let us assume that the scheme $Z_1 \cap Z_2 = Y$ is of dimension zero. If the intersection is transversal (see 7.5.21), then the intersection number will be simply the number of points in the intersection. In this case orientations do not play a role, this is clear because this notion does not make sense over an arbitrary field, for instance it does not make sense for $\mathbb{C}$ vector spaces. But if the intersection is not transversal then we have a problem.
The point is that in the theory of manifolds we can attach to the geometric object $N \subset M$ another more algebraic object $[N]$ in the cohomology ring $H^\bullet(M,\mathbb{Z})$.
And it is here where the problems in algebraic geometry start: *We need a replacement for the cohomology ring.* One option for this replacement will be the Chow ring, which will not be constructed in this book, we will only give a sketchy discussion of this ring. It should be a graded ring

$$A^\bullet(X) = \bigoplus_{\nu=0}^{d} A^\nu(X),$$

we should have a surjective homomorphism

$$A^d(X) \longrightarrow \mathbb{Z}$$

and we should have a reasonable procedure to attach to an irreducible subvariety Z_1 of codimension c_1 a class

$$[Y_1] \in A^{c_1}(X).$$

We will come back to this ring later (See 9.7.3), in this section we want to discuss some constructions, which are based on the concept of flat deformations, and which give a hint how to solve the problem of constructing such a ring.
Imagine you should compute the intersection number of the green and the blue curves in the following picture?

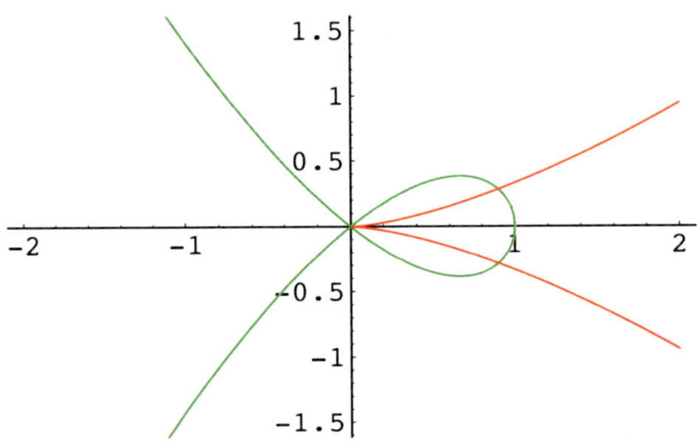

Figure 8.1 Intersecting $y^2 - x^3/3 = 0$ and $y^2 - x^2 + x^3 = 0$

We have three intersection points, the point (0,0) and two others points P_1,P_2, in which the intersection is transversal. They are all real. The intersection multiplicity in P_1 and P_2 is one, but what is the intersection multiplicity in the origin? We push the red curve to the left:
Now we see that the origin "splits" into 4 points, in which the two curves intersect transversally, this seems to indicate that the intersection number in the origin should be 4. This is in fact the right answer.

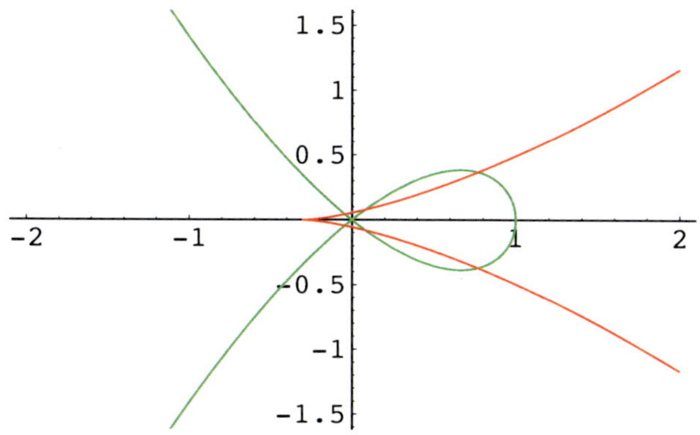

Figure 8.2 Intersecting $y^2 - (x+1/3)^3/3 = 0$ and $y^2 - x^2 + x^3 = 0$

There is another way to get this number 4. Let us localise the ring $k[X,Y]$ at the origin, as usual we denote this ring by $k[X,Y]_{\mathfrak{m}_{(0,0)}}$. Now we look at the local ring (assume $\mathrm{char}(k) \neq 3$).

$$R = k[X,Y]_{\mathfrak{m}_{(0,0)}}/(Y^2 - X^3/3, Y^2 - X^2 + X^3).$$

We check easily that the ideal $\mathfrak{m}^2_{(0,0)} \supset (Y^2 - X^3/3, Y^2 - X^2 + X^3) \supset \mathfrak{m}^3_{(0,0)}$. This implies that the ring R has dimension zero, image of the ideal $\mathfrak{m}_{(0,0)}$ is the only prime ideal in it. We get a filtration

$$R \supset \mathfrak{m}_{(0,0)} \supset \mathfrak{m}^2_{(0,0)} \supset (Y^2 - X^3/3, Y^2 - X^2 + X^3).$$

We have $\dim_k(R/\mathfrak{m}_{(0,0)}) = 1, \dim_k(\mathfrak{m}_{(0,0)}/\mathfrak{m}^2_{(0,0)}) = 2$ and $\mathfrak{m}^2_{(0,0)}/(Y^2 - X^3/3, Y^2 - X^2 + X^3)$ is of dimension one and generated by XY. Hence we conclude that $\dim_k(R/(Y^2 - X^3/3, Y^2 - X^2 + X^3)) = 4$. (Such a formula for intersection numbers is not always true, in our special case a certain flatness condition holds (See [Se3]).

We could also consider the case that $X = \mathbb{P}^2/k$ and $Z_1 = H_0, Z_2 = H_0$, i.e. we want to intersect the hyper-plane H_0 with itself. Here we replace one of the hyper-planes $X_0 = 0$ by a hyper-plane $a_0X_0 + a_1X_1 + a_2X_2 = 0$ and we and if a_1 or a_2 is not zero, then the new hyper-plane intersects the second one transversally and the intersection number is one.

Therefore, we see the principle: If we want to define intersection numbers, we deform Z_1, Z_2 (we put them into a flat family) and compute the intersection numbers of suitable members of the family.

We come to the technical construction and we prove a lemma that we used already earlier. Recall that we introduced the notation $d(\mathcal{F})$ for the maximal dimension of an irreducible component of $\mathrm{Supp}(\mathcal{F})$. As usual A is a noetherian ring. For any sheaf $\mathcal{F}$ any line bundle $\mathcal{L}$ on $\mathbb{P}^n_A$ and any section $H^0(\mathbb{P}^n_A, \mathcal{L})$ we know what it means that s is not a zero divisor in $\mathcal{F}$: We can trivialize $\mathcal{L}$ on the open sets of a suitable covering by affine sets. On any open set U of the covering we can can view s as a regular function in $\mathcal{O}_{\mathbb{P}^n_A(U)}$, it is unique up to a unit. Then $\mathcal{F}(U)$ is a module under the ring of regular functions on U and we know what it means that s is not a zero divisor in $\mathcal{F}(U)$. Then s is not a zero divisor in $\mathcal{F}$ if it is not a zero divisor for all the open sets in the covering.

Lemma 8.4.7. *Let $\mathcal{F}$ be a coherent sheaf on $\mathbb{P}^n_A$. For any point $\mathfrak{p} \in \mathrm{Spec}(A)$ can find an element $g \in A, g \notin \mathfrak{p}$, an integer $r > 0$ and a section $f \in H^0(\mathbb{P}^n_{A_g}, \mathcal{O}_{\mathbb{P}^n_{A_g}}(r))$ such that*
(i) The image $\bar{f}$ of f in $H^0(\mathbb{P}^n_{A_\mathfrak{p}/\mathfrak{p}}, \mathcal{O}_{\mathbb{P}^n_{A_\mathfrak{p}/\mathfrak{p}}}(r))$ is a non zero divisor of $\mathcal{F} \otimes k(\mathfrak{p})$, i.e. $\mathrm{Ann}_{F\otimes k(\mathfrak{p})}(\bar{f}) = 0$
(ii) The annihilator $\mathrm{Ann}_{\mathcal{F}_{A_g}}(f) = 0$
iii) We get an exact sequence

$$0 \longrightarrow \mathcal{F}_{A_g} \xrightarrow{m_f} \mathcal{F}_{A_g}(r) \longrightarrow \mathcal{F}_{A_g}(r)/f\mathcal{F}_{A_g} \longrightarrow 0.$$

and $d(\mathcal{F}) > d(\mathcal{F}_{A_g}(r)/f\mathcal{F}_{A_g})$.

We pass to the affine situation consider the restrictions of the sheaf $\mathcal{F}$ to the open set $U_i \subset \mathbb{P}^n_A$. On any of these open sets $\mathcal{F}$ becomes a $\mathcal{O}(U_i) = A[t_{i,0}, t_{i,1}, \dots t_{i,i-1}, t_{i,i+1}, \dots, t_{i,n}]$-module M_i. It suffices to prove the first assertion for any given index i. In the beginning we choose $r = 1$.The restriction of $f \in H^0(\mathbb{P}^n_A, \mathcal{O}_{\mathbb{P}^n_A}(1))$ to U_i becomes an inhomogeneous linear function $f_i = y_0 t_{i,0} + \cdots + y_i + \cdots + y_0 t_{i,n}$.
We assume that M_i is non zero, otherwise nothing has to be proved. We start from the observation that M_i contains a sub-module $M_{i,1} = \mathcal{O}(U_i)m_1$ where $m_1 \neq 0$ and where the annihilator of m_1 is a prime ideal. This is clear because if the annihilator of a given m_1 is not prime, then we find an element $m_2 \in \mathcal{O}(U_i)m_1$ whose annihilator is strictly larger, we get an ascending chain of annihilators, which must become stationary. We apply the same argument to $M_i/M_{i,1}$ and conclude that M_i has a has an ascending chain of sub-modules such that each sub-quotient is isomorphic to $\mathcal{O}(U_i)/\mathfrak{q}_{i,\nu}$ and $\mathfrak{q}_{,\nu}$ is prime. Since A and $\mathcal{O}(U_i)$ are noetherian, this chain must end with M_i after a finite number of steps.
We get a finite collection of closed sub-schemes $\mathrm{Spec}(B/\mathfrak{q}_{i,\nu}) \subset U_i, \nu = 1 \dots n_i$ in the support of M_i, such that for all $f \in H^0(\mathbb{P}^n_A, \mathcal{O}_{\mathbb{P}^n_A}(1))$ we have $\mathrm{Ann}_{M_i}(f_i) = 0$ provided $f_i \notin \mathfrak{q}_{i,\nu}$ for any ν. This can be done for any i, we can take the closures $Z_{i,\mathfrak{q}_{i,\nu}}$ of the affine schemes $\mathrm{Spec}(B/\mathfrak{q}_{i,\nu}) \subset U_i$ in $\mathbb{P}^n_A$. Then we can take the union of these sub-schemes over all i. Now we look at the fibre $\mathbb{P}^n_{A_\mathfrak{p}/\mathfrak{p}} = \mathbb{P}^n_A \times \mathrm{Spec}(k(\mathfrak{p}))$. The union of those $Z_{i,\mathfrak{q}_\nu}$, which do not meet (have empty intersection with) this fibre form a closed subscheme $Z \subset \mathbb{P}^n_A$, which does not meet the fibre. The image of the projection of Z to $\mathrm{Spec}(A)$ is a closed subscheme (See Theorem 8.1.8) $Z_1 \subset \mathrm{Spec}(A)$, which does not contain $\mathfrak{p}$. Hence can find an element $g \in A, g \notin \mathfrak{p}$, which vanishes on Z_1. We replace A by A_g and hence we may assume that all the closures $Z_{i,\mathfrak{q}_\nu}$ meet the fibre $\mathbb{P}^n_{A_\mathfrak{p}/\mathfrak{p}}$. We pick any of these $Z_{i,\mathfrak{q}_\nu}$ and take the intersection with the fibre and restrict to any of the U_j, where this intersection is non empty. This intersection is of the form $\mathrm{Spec}(\mathcal{O}(U_j)/\mathfrak{q}_{j,\nu} \otimes k(\mathfrak{p}))$. In general this will not be integral. But we apply our filtration argument again and get that there will be a finite number of prime ideals $\mathfrak{q}_{j,\alpha} \in \mathrm{Spec}(\mathcal{O}(U_j)/\mathfrak{q}_{j,\nu} \otimes k(\mathfrak{p})), \alpha = 1, \dots m_{i,j,\nu}$ such that an element $\bar{f}_j = y_0 t_{j,0} + \cdots + y_j + \cdots + y_n t_{j,n}, y_i \in k(\mathfrak{p})$, which satisfies $\bar{f}_j \notin \mathfrak{q}_{j,\alpha}$ for all α, will not be a zero divisor in $\mathcal{O}(U_j)/\mathfrak{q}_{j,\nu} \otimes k(\mathfrak{p})$. Taking the closures of these $\mathrm{Spec}(\mathcal{O}(U_j)/\mathfrak{q}_{j,\nu} \otimes k(\mathfrak{p}))$ in $\mathbb{P}^n_{k(\mathfrak{p})}$ we find a finite number of closed sub-schemes $\bar{Z}'_\beta \subset \mathbb{P}^n_{k(\mathfrak{p})}$ such that an element $\bar{f} \in H^0(\mathbb{P}^n_{k(\mathfrak{p})}, \mathcal{O}_{\mathbb{P}^n_{k(\mathfrak{p})}}(1))$, which does not vanish on any of these sub-schemes, is not a zero divisor in $\mathcal{F} \otimes k(\mathfrak{p})$ the sheaf $\mathrm{Ann}_{\mathcal{F}\otimes k(\mathfrak{p})}(\bar{f}) = 0$. If we now lift $\bar{f}$ to an element $f \in H^0(\mathbb{P}^n_A, \mathcal{O}_{\mathbb{P}^n_A}(1))$ then locally on any of the U_i this element satisfies $f \notin \mathfrak{q}_{i,\mu}$ and hence $\mathrm{Ann}_\mathcal{F}(f) = 0$.
We still have to find such an f. If we remove the zero element and consider the projection

$$p : H^0(\mathbb{P}^n_{k(\mathfrak{p})}, \mathcal{O}_{\mathbb{P}^n_{k(\mathfrak{p})}}(1)) \setminus \{0\} \longrightarrow \mathbb{P}H^0(\mathbb{P}^n_{k(\mathfrak{p})}, \mathcal{O}_{\mathbb{P}^n_{k(\mathfrak{p})}}(1)).$$

It is clear that $\bar{Z}'_\beta$ defines a proper closed sub-scheme

$$\Sigma_\beta \subset \{x \in \mathbb{P}(H^0(\mathbb{P}^n_{k(\mathfrak{p})},\mathcal{O}_{\mathbb{P}^n_{k(\mathfrak{p})}}(1)))| \; f|\bar{Z}'_\beta \equiv 0 \text{ for all } \bar{f} \in p^{-1}(x)\}.$$

If $\Sigma_{\mathcal{F}}$ is the union of these Σ_β, then any $\bar{f} \in H^0(\mathbb{P}^n_{k(\mathfrak{p})},\mathcal{O}_{\mathbb{P}^n_{k(\mathfrak{p})}}(1))$, which maps to an element in $V_{\mathcal{F}} = (\mathbb{P}(H^0(\mathbb{P}^n_{k(\mathfrak{p})},\mathcal{O}_{\mathbb{P}^n_{k(\mathfrak{p})}}(1)))(k(\mathfrak{p})) \setminus \Sigma_{\mathcal{F}})(k(\mathfrak{p}))$ and any lift to an element f yields such an element.
Now we have the minor problem that we can not show that $V_{\mathcal{F}}(k(\mathfrak{p})) \neq \emptyset$. This is rather obvious if $k(\mathfrak{p})$ is an infinite field, but if $k(\mathfrak{p})$ is finite this may not be true.
If $k(\mathfrak{p}) = \mathbb{F}_q$ is finite, then we find such an $x \in V_{\mathcal{F}}(\mathbb{F}_{q^r})$ for some r we lift it to an $\bar{f} \in p^{-1}(x)$ and now we take the product of the conjugates of this f under the Galois group $\mathrm{Gal}(\mathbb{F}_{q^r}/\mathbb{F}_q)$. We get an element in $f_1 \in H^0(\mathbb{P}^n,\mathcal{O}_{\mathbb{P}^n}(r))$. This is the reason why we need a bigger r. It is clear that that $d(\mathcal{F}' \otimes A_g) < d(\mathcal{F}_{A_g}(r)/f\mathcal{F}_{A_g}(r))$ □

We need an extension of this lemma

Lemma 8.4.8. *The notations and assumptions are the same as in the lemma above, but now we assume in addition that $\mathcal{F}$ is A flat. Then for any $x \in V(k(\mathfrak{p}))$, and a lift $\bar{f} \in p^{-1}(x)$ and any $f \in H^0(\mathbb{P}^n_A,\mathcal{O}_{\mathbb{P}^n_A}(1))$ that maps to $\bar{f}$ we can find a neighbourhood $\mathrm{Spec}(A_{g_1}) \subset \mathrm{Spec}(A_g)$ of $\mathfrak{p}$ such that $\mathcal{F}_{A_g}(r)/f\mathcal{F}_{A_g}$ is A_{g_1} flat.*

We have to show that for a suitable localization A_{g_1} and for any ideal $\mathfrak{a} \subset A_{g_1}$ the annihilator

$$\mathrm{Ann}_{\mathcal{F}\otimes A_{g_1}/\mathfrak{a}}(f) = 0.$$

We replace the sheaf by $\mathcal{F}(s)$, where s is large enough so that the global sections generate the stalks at all points and so that the assumptions in Theorem 8.4.2 are valid. We get an exact sequence for the global sections

$$0 \longrightarrow H^0(\mathbb{P}^n_A,\mathcal{F}) \xrightarrow{m_f} H^0(\mathbb{P}^n_A,\mathcal{F}(r)) \longrightarrow H^0(\mathbb{P}^n,\mathcal{F}_{A_g}(r)/f\mathcal{F}_{A_g}) \longrightarrow 0,$$

the first two modules in this sequence are locally free over A and satisfy base change. Let us assume they are indeed free and let us choose basis'es $\gamma_1,\dots,\gamma_a,\delta_1,\dots,\delta_b$ of these two modules respectively. Then the multiplication m_f is given by a (a,b) matrix $M(f)$ with coefficients in A. Since $m_{\bar{f}} \times \mathrm{Spec}(k(\mathfrak{p})) : H^0(\mathbb{P}^n_{A_\mathfrak{p}/\mathfrak{p}},\mathcal{F}\otimes k(\mathfrak{p})) \longrightarrow H^0(\mathbb{P}^n_{A_\mathfrak{p}/\mathfrak{p}},\mathcal{F}\otimes k(\mathfrak{p})(r))$ is injective we can conclude that this matrix evaluated at $\mathfrak{p}$ has rank a, hence it has a (a,a) minor whose determinant does not vanish at $\mathfrak{p}$, hence it is a unit in a suitable localization A_{g_1}. Let $\mathfrak{a}$ be an ideal in A_{g_1}, we apply base change and get a commutative diagram

$$\begin{array}{ccc} H^0(\mathbb{P}^n_{(A_{g_1}/\mathfrak{a})},\mathcal{F}\otimes(A_{g_1}/\mathfrak{a})) & \xrightarrow{m_f \times \mathrm{Spec}(A_{g_1}/\mathfrak{a})} & H^0(\mathbb{P}^n_{(A_{g_1}/\mathfrak{a})},\mathcal{F}(r)\otimes(A_{g_1}/\mathfrak{a})) \\ \downarrow & & \downarrow \\ H^0(\mathbb{P}^n_{A_{g_1}},\mathcal{F}\otimes A_{g_1})\otimes(A_{g_1}/\mathfrak{a})) & \xrightarrow{m_f \otimes A_{g_1}/\mathfrak{a}} & H^0(\mathbb{P}^n_{A_{g_1}},\mathcal{F}(r)\otimes A_{g_1})\otimes(A_{g_1}/\mathfrak{a}) \end{array}$$

The vertical arrows are base change isomorphisms, the horizontal arrow in the bottom line is injective because our (a,a) minor above has an invertible determinant. Therefore the horizontal arrow in the top line is injective. But this implies that $\mathrm{Ann}_{\mathcal{F}\otimes(A_{g_1}/\mathfrak{a})}(f) = 0$, because in the beginning we assumed that the global sections generate the stalks.

□

We translate the considerations from section 7.4.2 into the context of projective algebraic geometry. Our starting point is a noetherian ring A, the projective space $\mathbb{P}^n_A$ and a coherent sheaf $\mathcal{F}$ on the projective space. We pick a positive integer $r > 0$ and consider the free A-module $H^0(\mathbb{P}^n_A, \mathcal{O}_{\mathbb{P}^n_A}(r)) = H^0_r$ We have seen that this free A-module has as a basis the monomials $X^{\underline{\mu}} = X_0^{\mu_0} \dots X_n^{\mu_n}$ where $\sum \mu_i = \deg(\underline{\mu}) = r$. We consider the corresponding vector bundle $V(H^0_r) \longrightarrow \mathrm{Spec}(A)$, which is the affine space

$$V(H^0_r) = \mathrm{Spec}(A[\dots, C_{\underline{\mu}}, \dots]),$$

where the $C_{\underline{\mu}}$ are polynomial variables. For any A-algebra C a C-valued point in $V(H^0_r)$ is simply a section $\sum c_{\underline{\mu}} X^{\underline{\mu}} \in H^0(\mathbb{P}^n_C, \mathcal{O}_{\mathbb{P}^n_C}(r))$. Especially we may look at the base change $A \longrightarrow A[\dots, C_{\underline{\mu}}, \dots]$ and then $F = \sum C_{\underline{\mu}} X^{\underline{\mu}}$ yields the universal section

$$F \in H^0(\mathbb{P}^n \times_A V(H^0_r), \mathcal{O}_{\mathbb{P}^n \times_A V(H^0_r)}(r)).$$

Multiplication by this universal section yields an exact sequence of sheaves on $\mathbb{P}^n \times_A V(H^0_r)$

$$0 \longrightarrow \mathcal{O}_{\mathbb{P}^n \times V(H^0_r)} \xrightarrow{m_F} \mathcal{O}_{\mathbb{P}^n \times V(H^0_r)}(r) \longrightarrow \mathcal{O}_{\mathbb{P}^n \times V(H^0_r)}(r)/F\mathcal{O}_{\mathbb{P}^n \times V(H^0_r)} \longrightarrow 0.$$

We remove the zero section $\{0\}$ from $V(H^0_r)$ and we put $T^r_0 = V(H^0_r) \setminus \{0\}$, from here we have the morphism

$$\pi : T^r_0 \longrightarrow \mathbb{P}(H^0(\mathbb{P}^n_A, \mathcal{O}_{\mathbb{P}^n_A}(r))) = \mathbb{P}(H^0_r)$$

(See diagram 8.14).
We restrict the sheaves in our exact sequence to $\mathbb{P}^n \times T^r_0$. Clearly $\mathcal{O}_{\mathbb{P}^n \times V(H^0_r)}, \mathcal{O}_{\mathbb{P}^n \times V(H^0_r)}(r)$ are the quasi-coherent inverse images of $\mathcal{O}_{\mathbb{P}^n \times \mathbb{P}(H^0_r)}, \mathcal{O}_{\mathbb{P}^n \times \mathbb{P}(H^0_r)}(r)$. The multiplication by F defines a sub-sheaf $< F > \mathcal{O}_{\mathbb{P}^n \times \mathbb{P}(H^0_r)} \subset \mathcal{O}_{\mathbb{P}^n \times \mathbb{P}(H^0_r)}(r)$ and hence we get a quotient sheaf

$$\mathcal{O}_{\mathbb{P}^n \times \mathbb{P}(H^0_r)}(r) / < F > \mathcal{O}_{\mathbb{P}^n \times \mathbb{P}(H^0_r)}.$$

on $\mathbb{P}(H^r_0)$. This sheaf is the quotient of $\mathcal{O}_{\mathbb{P}^n \times \mathbb{P}(H^0_r)}(r)$ by a homogeneous ideal $< F >$, it defines a sub-scheme

$$\mathcal{H}_r \subset \mathbb{P}^n \times \mathbb{P}(H^0_r),$$

which will be called the universal hyper-surface of degree r. It is flat over $\mathbb{P}(H^0_r)$.
We consider the projective space $\mathbb{P}^n$ over a field k. Let T be an absolutely connected scheme of finite type over $\mathrm{Spec}(k)$. A *flat family of schemes over T* (we may also say parameterized by T) is a subscheme

$$\mathcal{Z} \subset \mathbb{P}^n \times_k T$$

whose structure sheaf $\mathcal{O}_{\mathcal{Z}}$ is flat over T. Let π be the projection to T. For any point $t \in T$ we can consider the fibre $\mathcal{Z}_t = \pi^{-1}(t)$, we get the members of the family. If we have a second absolutely connected scheme T' over $\mathrm{Spec}(k)$ and a morphism $\psi : T' \longrightarrow T$, then we get a flat family over T' if we simply take the inverse image $(\mathrm{Id} \times \psi)^{-1}(\mathcal{Z}) \subset \mathbb{P}^n \times T'$. Let us call this the pull-back family of $\mathcal{Z}$ via ψ.

We introduce the notion of *equivalence of two flat families* $\mathcal{Z}_1 \subset \mathbb{P}^n \times_k T_1, \mathcal{Z}_2 \subset \mathbb{P}^n \times_k T_2$. To do this we consider the two projections p_1, p_2 from $T_1 \times_k T_2$ to T_1, T_2 respectively and take the two pull-back families of $\mathcal{Z}_1, \mathcal{Z}_2$ via these two projections. We call the two families equivalent, we can find a geometric point $t = (t_1, t_2)$ such that

$$(\mathrm{Id} \times p_1)^{-1}(\mathcal{Z}_1)_t = (\mathrm{Id} \times p_2)^{-1}(\mathcal{Z}_2)_t$$

in other words if $\mathcal{Z}_{1,t_1} = \mathcal{Z}_{2,t_2}$.

The intuitive meaning of a flat family is, that the "topological type" of the members in a flat family stays constant if t moves inside the parameter space. We may for instance consider the Hilbert polynomial

$$t \mapsto (r \mapsto \chi(\mathcal{Z}_t, \mathcal{O}_{\mathcal{Z}_t}(r)))$$

and theorem 8.4.6 tells us that this function is locally constant in t, and hence constant on T, because we assumed that T should be absolutely connected. From this we conclude that the degree $t \mapsto d(\mathcal{Z}_t)$ is (locally) constant.

We have a simple process to construct flat families. Let us assume that we have a flat family

$$\mathcal{Z} \subset \mathbb{P}^n \times_k T.$$

Now we also have the universal family of degree r hyper-surfaces $\mathcal{H}_r \subset \mathbb{P}^n \times \mathbb{P}(H_r^0)$. We have the two projections

$$p_T, p_H : \mathbb{P}^n \times T \times \mathbb{P}(H_r^0) \longrightarrow \mathbb{P}^n \times T, \mathbb{P}^n \times \mathbb{P}(H_r^0)$$

we take the pullbacks of these two families $p_H^{-1}(\mathcal{Z}) = \widetilde{\mathcal{Z}}$, $p_T^{-1}(\mathcal{H}_r) = \widetilde{\mathcal{H}}_r \subset \mathbb{P}^n \times T \times \mathbb{P}(H_r^0)$ and we consider the intersection of these two sub schemes

$$\widetilde{\mathcal{Z}} \cap \widetilde{\mathcal{H}}_r \subset \mathbb{P}^n \times T \times \mathbb{P}(H_r^0)$$

This is a scheme over $T \times \mathbb{P}(H_r^0)$ and we claim

Proposition 8.4.9. *There exists a non empty open subset $U \subset T \times_k \mathbb{P}(H_r^0))$ such that the intersection $\widetilde{\mathcal{Z}} \cap \widetilde{\mathcal{H}}_r \subset \mathbb{P}^n \times T \times \mathbb{P}(H_r^0)$ is flat at any point of U, hence this intersection with $\mathbb{P}^n \times U$ is a flat family over U.*

Proof: This is our lemma 8.4.8

□

Since $U \subset T \times \mathbb{P}(H_r^0)$ is connected we see that we can define the intersection of any flat family with the universal family of hyper-planes of degree r.

This allows us to define the universal family of m-fold intersections of hyper-surfaces of degree $r_1, r_2, \ldots, r_m$. This is the intersection

$$\mathcal{H}_{r_1} \cap \mathcal{H}_{r_2} \cap \cdots \cap \mathcal{H}_{r_m} \subset \mathbb{P}^n \times \mathbb{P}(H_{r_1}^0) \times \ldots \mathbb{P}(H_{r_m}^0).$$

Our previous proposition implies that we can find a non empty open subset $\mathcal{T}_{r_1,\ldots,r_m} \subset \mathbb{P}(H_{r_1}^0) \times \ldots \mathbb{P}(H_{r_m}^0)$ such that

$$\begin{array}{c} \mathcal{H}_{r_1,r_2,\ldots,r_m} = \mathcal{H}_{r_1} \cap \mathcal{H}_{r_2} \cap \cdots \cap \mathcal{H}_{r_m} \cap \mathbb{P}^n \times \mathcal{T}_{r_1,\ldots,r_m} \\ \downarrow \\ \mathcal{T}_{r_1,\ldots,r_m} \end{array}$$

is flat.

In general it may not be so easy to check, whether a given (closed) point $\underline{f} \in \mathbb{P}(H^0_{r_1}) \times \ldots \mathbb{P}(H^0_{r_m})(k)$ is in $\mathcal{T}_{r_1,\ldots,r_m}(k)$. By definition we can interpret such a point as an array $\underline{f} = (f_1,\ldots,f_m)$ of non zero homogenous polynomials (up to homotetie) of degree $r_1,r_2,\ldots,r_m$. We know that we can check this locally on $\mathbb{P}^n$, i.e. we can restrict to $U_i \times \mathcal{T}_{r_1,\ldots,r_m} \longrightarrow \mathcal{T}_{r_1,\ldots,r_m}$. Then the point $\underline{f}$ is an array of polynomials

$$f_{a,i} = \sum_{\underline{\mu}:\deg(\underline{\mu})\leq r_i} c^{(a)}_{\underline{\mu}} t^{\mu_0}_{i,0} \ldots t^{\mu_{i-1}}_{i,i-1} t^{\mu_{i+1}}_{i,i+1} \ldots t^{\mu_n}_{i,n} \quad \text{where } a = 1,\ldots,m.$$

Now it is clear from our previous considerations that we have flatness in $\underline{f}$ if for all i the sequence $f_{1,i},f_{2,i}\ldots,f_{m,i}$ is a $\mathcal{O}(U_i)$ regular sequence and this means that for any $1 \leq b \leq m$ the element $f_{b,i}$ is not a zero divisor in $\mathcal{O}(U_i)/(f_{1,i},\ldots,f_{b-1,i})$. (For this concept and its relation to flatness see [Ei],II. 10 and II.18).

This is certainly the case if the collection of equations satisfies the Jacobi-criterion (See 7.5.4), namely that for any closed point $P \in U_i(k)$, for which $f_{1,i}(P) = f_{2,i}(P) = \cdots = f_{m,i}(P) = 0$, i.e. a point closed in $\mathrm{Spec}(\mathcal{O}(U_i)/(f_{1,i},f_{2,i}\ldots,f_{m,i})$ the Jacobi-matrix

$$\frac{\partial f_{a,i}}{\partial t_{j,i}}(P)$$

has rank m. This is so because our Theorem 7.5.4 implies that for all values $1 \leq b \leq m$ the local ring $\mathcal{O}(U_i)/(f_{1,i},f_{2,i}\ldots,f_{b,i})_{\mathfrak{m}_P}$ is integral and has dimension $n-b$ and hence $f_{b+1,i}$ has a non zero image in this ring.

To give an application we consider the case $m = n$. A point u is given by an array of n homogenous polynomials $f_i(X_0,\ldots,X_n)$ of degree r_i. We consider very special points, i.e. very special systems of equations. We make the assumption that these polynomials are products of linear forms, i.e.

$$f_i = \prod_{j=1}^{j=r_r} l_{i,j},$$

where the l_{ij} are linear. Furthermore we assume that for any choice of factors $l_{i,j_i}, i = 1,\ldots,n$ these n linear forms are linearly independent. Then it is clear that the universal family is flat over such a point.

It is also clear that for any point $P \in \mathrm{Spec}(\mathcal{O}(U_i)/(f_1,f_{2,i}\ldots,f_n))$ we find a unique choice of factors $j_1,j_2,\ldots,j_n$ such that $l_{1,j_1}(P) = l_{2,j_2}(P) = \ldots, = l_{n,j_n}(P)$ and after localization at P

$$\mathcal{O}_{\mathbb{P}^n}/(l_{1,j_1},l_{2,j_2},\ldots,l_{n,j_n})_{\mathfrak{m}_P} = \mathcal{O}_{\mathbb{P}^n}/(f_1,f_2,\ldots,f_n)_{\mathfrak{m}_P} = k.$$

We conclude that under our assumptions $\mathcal{O}_{\mathbb{P}^n}/(f_1,f_2,\ldots,f_n)$ is of dimension zero and reduced. It has exactly $r_1r_2\ldots r_n$ points and we get for the Hilbert polynomial

$$\chi(\mathcal{O}_{\mathbb{P}^n}/(f_1,f_2,\ldots,f_n)(r)) = r_1r_2\ldots r_n.$$

The Hilbert polynomial is independent of r.

But now we can apply the semi-continuity theorem, it tells us that in any point point $u \in \mathcal{T}_{r_1,\ldots,r_m}(k)$ the fibre

$$\mathcal{H}_{r_1,r_2,\ldots,r_n,u} = \mathcal{H}_{r_1,r_2,\ldots,r_n} \times k(u)$$

is of dimension zero and

$$\dim_{k(u)} H^0(\mathcal{H}_{r_1,r_2,\dots,r_n,u},\mathcal{O}_{\mathcal{H}_{r_1,r_2,\dots,r_n,u}} = \chi(\mathcal{O}_{\mathbb{P}^n}/(f_1,f_2,\dots,f_n)(r)) = r_1r_2\dots r_n.$$

The scheme for $u \in \mathcal{T}_{r_1,\dots,r_m}(k)$ the scheme $\mathcal{H}_{r_1,r_2,\dots,r_n,u}$ consists of a finite number of closed points $P_1,\dots,P_t$ and

$$\dim_k(H^0(\mathcal{H}_{r_1,r_2,\dots,r_n,u},\mathcal{O}_{\mathcal{H}_{r_1,r_2,\dots,r_n,u}}) = \sum_i \dim_k \mathcal{O}_{(\mathcal{H}_{r_1,r_2,\dots,r_n,u})\mathfrak{m}_{P_i}}$$

and the individual term $\dim_k(\mathcal{O}_{(\mathcal{H}_{r_1,r_2,\dots,r_n,u})})_{\mathfrak{m}_{P_i}}$ is called the *intersection multiplicity* of the n hyper-planes $u = (f_1,\dots,f_n)$ in the point P_i.
Hence we can say that for an array of homogenous polynomials $f_1,f_2,\dots,f_n$ of degrees $r_1,r_2,\dots,r_n$, which define a point $u \in \mathcal{T}_{r_1,\dots,r_m}(k)$, the number of points in the intersection of these hyper-planes counted with the right multiplicities is $r_1r_2\dots r_n$.

Finally we come to a classical result, which is now an easy consequence of our consideration. I want to stress that the following arguments do not depend on the lengthy considerations in the proof of the two lemmas above. We consider the projective space $\mathbb{P}^2$ over an arbitrary field k. We choose two homogenous linear forms $f_1 = \sum a_{\underline{\mu}} X^{\underline{\mu}}, f_1 = \sum b_{\underline{\mu}} X^{\underline{\mu}}$ of degree d_1,d_2. Now we exploit the fact that $k[X_0,X_1,X_2]$ has unique factorization. Then we can say that f_1,f_2 are coprime, this means that they have no common factor. This condition defines a non empty open subset $\mathcal{T}_{d_1,d_2}$ in the space of coordinates of the coefficients $(\dots a_{\underline{\mu}},\dots,b_{\underline{\mu}}\dots)$ It is clear that over this open set scheme $\mathcal{O}_{\mathbb{P}^2}/(f_1,f_2)$ is flat. Hence we get the classical

Theorem 8.4.10. *(Theorem of Bezout)*
If we have two hyper-surfaces in $\mathbb{P}^2_k$ given by homogenous polynomials

$$f_1(X_0,X_1,X_3) = \sum a_{\underline{\mu}} X^{\underline{\mu}} = \sum_{\mu_0,\mu_1,\mu_2:\sum \mu_i = d_1} a_{\mu_0,\mu_1,\mu_2} X_0^{\mu_0} X_1^{\mu_1} X_2^{\mu_2}$$

$$f_2(X_0,X_1,X_3) = \sum b_{\underline{\mu}} X^{\underline{\mu}} = \sum_{\mu_0,\mu_1,\mu_2:\sum \mu_i = d_2} b_{\mu_0,\mu_1,\mu_2} X_0^{\mu_0} X_1^{\mu_1} X_2^{\mu_2},$$

then their intersection is of dimension zero if and only if they are coprime. If this is the case the intersect in d_1d_2 points, if we count the points P in the intersection with multiplicity

$$m(P) = \dim_k(\mathcal{O}_{\mathbb{P}^2_k,\mathfrak{m}_P}/(f_1,f_2)).$$

We want to stress that the essential ingredient in the proof is the semicontinuity theorem □

The Theorem of Bertini

We want to close this section by stating a classical theorem.

Theorem 8.4.11. *Let k be field, let $X/\operatorname{Spec}(k)$ be a smooth projective scheme, which comes with an embedding i into the projective space*

$$\begin{array}{ccc} X & \xrightarrow{\quad i \quad} & \mathbb{P}^n_k \\ & {\scriptstyle f'}\searrow \quad \swarrow{\scriptstyle f} & \\ & \operatorname{Spec}(k) & \end{array}$$

Let $r > 0$. We consider the intersection of $X \times \mathbb{P}(H^0_r)$ with the hyper surface $\mathcal{H}_r \subset \mathbb{P}^n_k \times \mathbb{P}(H^0_r)$ then we can find a non empty open subset $U \subset \mathbb{P}(H^0_r)$ such that for all $u \in U$ the intersection $X \times k(u) \cap (\mathcal{H}^0_r)_u$ is smooth.

For the proof we refer to the literature (See [Ha]), but we can as well leave it as an exercise to the reader. It is also easy to prove the following extension. For is a rational point $P \in X(k)$ let $V_P \subset \mathbb{P}(H^0_r)$ of those hypersurfaces containing P. Then we can find a non empty subset $V_P^{(0)} \subset V_P$ such that the analogous assertion holds.

8.4.2 The hyperplane section and intersection numbers of line bundles

We consider a projective scheme $f : X \longrightarrow S$, where the base scheme is noetherian and connected. We assume that the scheme is flat over S. Let us assume that $\mathcal{H}$ is a very ample line bundle, it provides a projective embedding

$$\begin{array}{ccc} X & \xrightarrow{\quad i \quad} & \mathbb{P}^n_S \\ & {\scriptstyle f}\searrow \quad \swarrow{\scriptstyle \pi} & \\ & S & \end{array}$$

where $i^*(\mathcal{O}_{\mathbb{P}^n_S}(1) \xrightarrow{\sim} \mathcal{H}$.
In the previous considerations we always considered the restriction of $H^0(\mathbb{P}^n_S,\mathcal{O}_{\mathbb{P}^n_S}(r))$ to X we replace this space of sections by the more natural choice $H^0(\mathbb{P}^n_S,\mathcal{H}^{\otimes r})$.
Let d be the relative dimension of X/S and let $\mathcal{L}_1,\dots,\mathcal{L}_d$ be a collection of line bundles on X, we want to define the intersection number

$$\mathcal{L}_1 \cdot \mathcal{L}_2 \cdot \ \dots \ \cdot \mathcal{L}_d.$$

This problem was discussed in Vol. I 5.3.1 and the solution given there was satisfactory in the sense that it captured the essence of the concept, but formally it was not so satisfactory because we alluded to the cohomology theory of complex analytic varieties. Here we will demonstrate that the ideas, which we adumbrated in Volume I actually work.
Of course we want that $I(\mathcal{L}_1,\mathcal{L}_2,\dots \mathcal{L}_d) = \mathcal{L}_1 \cdot \mathcal{L}_2 \cdot \ \dots \ \cdot \mathcal{L}_d$ is commutative in the variables and that it is multilinear, i.e.

$$(\mathcal{L}_1 \otimes \mathcal{L}_1') \cdot \ \dots \ \cdot \mathcal{L}_d = \mathcal{L}_1 \cdot \mathcal{L}_2 \cdot \ \dots \ \cdot \mathcal{L}_d + \mathcal{L}_1' \cdot \mathcal{L}_2 \cdot \ \dots \ \cdot \mathcal{L}_d.$$

It is clear what we have to do if $d = 0$. In this case $X \longrightarrow S$ is finite and flat, and therefore, $\mathcal{O}_X$ is a locally free, finitely generated $\mathcal{O}_S$ module. We have the empty set of line bundles and to this empty set we attach the intersection number

$$I(\emptyset) = \mathrm{Rank}_{\mathcal{O}_S}(\mathcal{O}_X),$$

this number is well defined because we assumed the S is connected.
We proceed by induction. We replace $\mathcal{L}_d$ by $\mathcal{L}_d \otimes \mathcal{H}^{\otimes r}$ where r is sufficiently large so that the assumptions in Theorem 8.4.3 are valid for $\mathcal{H}^{\otimes r}$ and for $\mathcal{L}_d \otimes \mathcal{H}^{\otimes r}$. We apply proposition 8.4.9 to the sheaves $\mathcal{F} = \mathcal{O}_X$ and $\mathcal{F} = \mathcal{L}_d$ and consider the two schemes

$$\mathbb{P}_S(f_*(\mathcal{H}^{\otimes r})) \text{ and } \mathbb{P}_S(f_*(\mathcal{L}_d \otimes \mathcal{H}^{\otimes r}))$$

over S. To any point $s \in S$ we find open subsets $V_1 \subset \mathbb{P}(f_*(\mathcal{H}^{\otimes r})), V_2 \subset \mathbb{P}(\mathcal{L} \otimes f_*(\mathcal{H}^{\otimes r}))$, which have a non empty intersection with the fibers if we intersect them with the fibers $\mathbb{P}(f_*(\mathcal{H}^{\otimes r})) \times_S \mathrm{Spec}(k(s))$, resp. $\mathbb{P}(f_*(\mathcal{L}_d \otimes \mathcal{H}^{\otimes r})) \times_S \mathrm{Spec}(k(s))$ and such that the schemes of hyperplane sections

$$\begin{array}{ccc} \mathcal{X}_{\mathcal{L}_d} & \xrightarrow{i} & X \times_S \mathbb{P}(f_*(\mathcal{L}_d \otimes \mathcal{H}^{\otimes r})) \\ & {\scriptstyle f_{\mathcal{L}}}\searrow \quad \swarrow{\scriptstyle \pi_{\mathcal{L}}} & \\ & \mathbb{P}(f_*(\mathcal{L}_d \otimes \mathcal{H}^{\otimes r})) & \end{array}$$

and

$$\begin{array}{ccc} \mathcal{X} & \xrightarrow{i} & X \times_S \mathbb{P}(f_*(\mathcal{H}^{\otimes r})) \\ & {\scriptstyle f}\searrow \quad \swarrow{\scriptstyle \pi} & \\ & \mathbb{P}(f_*(\mathcal{H}^{\otimes r})) & \end{array}$$

are flat if we restrict them to V_2 (resp. V_1.) These schemes of hyperplane sections are now flat over V_2 (resp.) V_1 and their relative dimension is $d-1$. We take the pullbacks of the bundles to $X \times_S V_2$ (resp.) $X \times_S V_1$ and restrict them to the hyperplane sections, we get line bundles

$$\mathcal{L}'_1, \ldots, \mathcal{L}'_{d-1} \text{ on } \mathcal{X}_{\mathcal{L}_d}, \quad \mathcal{L}''_1, \ldots, \mathcal{L}''_{d-1} \text{ on } \mathcal{X}$$

then the following intersection numbers are defined and we put

$$\mathcal{L}_1 \cdot \mathcal{L}_2 \cdot \ \ldots \ \cdot \mathcal{L}_d = \mathcal{L}'_1 \cdot \cdots \cdot \mathcal{L}'_{d-1} - \mathcal{L}''_1 \cdot \ \ldots \ \cdot \mathcal{L}''_{d-1}.$$

We have to show that the definition of this intersection number neither depends on the choice of $\mathcal{H}$ nor on r. Furthermore we have to show that it is commutative and multilinear.
We consider the case $d = 1$. For a line bundle $\mathcal{H}$ on X and a section $s \in H^0(X,\mathcal{H})$, which is not a zero divisor in $\mathcal{O}_X$ we put

$$I_1(\mathcal{H}) = \text{Rank of the } \mathcal{O}_S\text{-module } \mathcal{H}/s\mathcal{O}_X,$$

we will show that this number is equal to $I(\mathcal{H})$.
We say that $\mathcal{H}$ is "ample enough" if it has non zero sections and if the set of sections, which a non zero divisors in $\mathcal{O}_X$ is a non empty open set. We have seen that for an ample bundle $\mathcal{H}$ there exists an $r > 0$ such that $\mathcal{H}^{\otimes r}$ is ample enough. (Lemma 8.4.8). If two line bundles $\mathcal{H},\mathcal{H}_1$ are ample enough, then $\mathcal{H} \otimes \mathcal{H}_1$ is ample enough, because if $s \in H^0(X,\mathcal{H}), s_1 \in H^0(X,\mathcal{H}_1)$ are non zero divisors in $\mathcal{O}_X$ then $ss_1 \in H^0(X,\mathcal{H} \otimes \mathcal{H}_1)$ is also a non zero divisor. The linearity relation $I_1(\mathcal{H} \otimes \mathcal{H}_1) = I_1(\mathcal{H}) + I_1(\mathcal{H}_1)$ means

$$\mathrm{Rank}_{\mathcal{O}_S}(\mathcal{H} \otimes \mathcal{H}_1/ss_1\mathcal{O}_X) = \mathrm{Rank}_{\mathcal{O}_S}(\mathcal{H}/s\mathcal{O}_X) + \mathrm{Rank}_{\mathcal{O}_S}(\mathcal{H}_1/s_1\mathcal{O}_X).$$

This last equality follows from the exactness of the sequence

$$0 \longrightarrow \mathcal{H} \otimes s_1\mathcal{O}_X/(s \otimes s_1)\mathcal{O}_X \longrightarrow \mathcal{H} \otimes \mathcal{H}_1/(s \otimes s_1)\mathcal{O}_X \longrightarrow \mathcal{H}_1/s_1\mathcal{H}_1 \longrightarrow 0$$

and the isomorphism $\mathcal{H} \otimes s_1\mathcal{O}_X/(s \otimes s_1)\mathcal{O}_X \xrightarrow{\sim} \mathcal{H}/s\mathcal{O}_X$, which in turn follows from the assumption that s_1 is not a zero divisor. Hence we see, that for a line bundle $\mathcal{L}$, which is ample enough, we have $I_1(\mathcal{L}) = I(\mathcal{L})$. But at the same time we see that for any line bundle $\mathcal{L}$ the above definition of the intersection number $I(\mathcal{L}) = I(\mathcal{L} \otimes \mathcal{H}^{\otimes r}) - I(\mathcal{H}^{\otimes r})$ is independent of the choice of $\mathcal{H}$ and r. Then the linearity also becomes obvious. This settles the case $d = 1$. The case $d > 1$ follows from an easy and obvious induction argument.
We will encounter these numbers $I(\mathcal{L})$ again in 9.4.1 when we discuss the degree of divisors on curves over a field.

9 Curves and the Theorem of Riemann-Roch

9.1 Some basic notions

In the following k is a field, $\bar{k}$ is an algebraic closure and $k_s \subset \bar{k}$ is the separable closure inside $\bar{k}$.
A curve over field k is a scheme C/k, which is separated, of finite type over k (See 6.2.5) and all its irreducible components are of dimension 1. In other words our scheme has a finite covering by affine schemes $U_i = \mathrm{Spec}(\mathcal{O}_C(U_i))$, where the irreducible components of $\mathcal{O}_C(U_i)$ are finitely generated k-algebras of dimension 1. We know what it means that C/k is irreducible or absolutely irreducible (See 7.2.2 and 7.2.12.)
To give simple examples we can take a non zero homogeneous polynomial of degree d say

$$f(x,y,z) = \sum a_{\nu_1\nu_2\nu_2} x^{\nu_1} y^{\nu_2} z^{\nu_3} \quad a_{\nu_1\nu_2\nu_3} \in k, \nu_1 + \nu_2 + \nu_3 = d.$$

Then the ideal (f) defines a curve

$$\begin{array}{ccc} C & \hookrightarrow & \mathbb{P}^2/k \\ & \searrow^{p_0} & \downarrow p_1 \\ & & \mathrm{Spec}(k). \end{array} \tag{9.1}$$

To see this we restrict to one of the affine planes, which cover $\mathbb{P}^2$. This means we put one of the variables equal to 1 and divide the polynomial ring in the remaining 2 variables by the principal ideal generated by the resulting polynomial. We have to show that the irreducible components of C intersected with this plane are of dimension one. This intersection is non empty if and only if this resulting polynomial is not constant. But then the irreducible components correspond to the minimal prime ideals containing the polynomial and these are of height one (Krull Hauptidealsatz) 7.1.18 and hence the quotient ring by this ideal is of dimension one.
Polynomial rings over fields are factorial (See 7.1.4). Therefore a principal ideal defined by an irreducible polynomial $F(X_1, \ldots, X_n) \in k[X_1, \ldots, X_n]$ is a prime ideal. If we apply this to our case then we see that an irreducible polynomial f defines an irreducible curve C. But irreducibility is not invariant under base change, it may happen that f is irreducible but can be factorized over a bigger field.
If for instance we take $k = \mathbb{Q}$ then $(x + y - \sqrt{2}z)(x + y + \sqrt{2}z) = (x + y)^2 - 2z^2$ is irreducible over $\mathbb{Q}$ but factors over $\mathbb{Q}[\sqrt{2}]$. If this occurs, the curve is irreducible but not absolutely irreducible. In our example the curve $C \otimes_{\mathbb{Q}} \bar{\mathbb{Q}}$ is the union of two lines, which are interchanged by the Galois group. They intersect in one point, which is not smooth. There are even worse cases. Let us assume that the ground field is not perfect and let $p > 0$ be its characteristic. Then we can find an $a \in k$, which is not a p-th power. We take $f = x^p + y^p + az^p$. Then $k(a^{1/p})$ is an inseparable extension of degree p and the equation of he curve $C \times_k k(a^{1/p})$ is simply $(x + y + a^{1/p}z)^p = 0$ In this case the curve is something like p-times a line.

It is also possible that a curve C/k is smooth, irreducible but not absolutely irreducible. This happens if we start from an absolutely irreducible curve C/k. Let k_0 be a subfield of k such that k/k_0 is a finite separable non trivial extension. Then we have $C \longrightarrow \mathrm{Spec}(k) \longrightarrow \mathrm{Spec}(k_0)$ and we can view C/k_0 as a curve over k_0. This curve will still be smooth, but it is not absolutely irreducible. To see this we simply look at $C \times_{k_0} \bar{k}$ and verify that this is a disjoint union of $[k : k_0]$ irreducible smooth curves.
There is a slightly different way of looking at this phenomenon. If we have a smooth irreducible curve C/k and an element $f \in \mathcal{O}_C(U)$, which is not in k it can happen that this element is still algebraic over k. It generates a finite extension L/k contained in $\mathcal{O}_C(U)$ and now $\mathcal{O}_C(U) \otimes_k \bar{k}$ must have zero divisors and hence our curve is not absolutely irreducible. We call such elements f, which are algebraic over k the *constant elements* or simply the *constants*(see 7.2.13). They form a finite extension of the ground field. A smooth irreducible curve is absolutely irreducible if and only if the field of constants is equal to k.

Exercise 36. We return to the case of a curve defined by a homogeneous polynomial f of degree d as above. Use the results in the previous section (See theorem 8.2.5) to compute the cohomology groups $H^\bullet(C,\mathcal{O}_C)$. Show that $H^0(C,\mathcal{O}_C)$ is a k-vector space of dimension one. Conclude that C cannot decompose into disjoint closed subschemes and conclude that C must be absolutely irreducible if it is smooth. Show that the dimension of $H^1(C,\mathcal{O}_C)$ is $\binom{d-1}{2}$.

Remark: We can construct a universal curve of degree d. If we consider forms of a fixed degree and we remove the trivial form, then we see that we can view the coefficients $(a_{\nu_1\nu_2,\dots,\nu_d})$ as the k-valued point of a projective scheme $\mathcal{S}_d = \mathrm{Proj}(\mathbb{Z}[\dots, a_{\nu_1\nu_2\dots\nu_d}, \dots])$ where we consider the $a_{\nu_1\nu_2\dots\nu_d}$ as independent variables in degree one. We can define the universal curve of degree d, which is a subscheme of

$$\begin{array}{ccc} \mathcal{C} & \hookrightarrow & \mathcal{S}_d \times \mathbb{P}^2 \\ & \searrow^{p_0} & \downarrow p_1 \\ & & \mathcal{S}_d. \end{array} \tag{9.2}$$

Exercise 37. a) Prove that there is a closed subscheme $\mathcal{S}_d^{\mathrm{sing}} \subset \mathcal{S}_d$ such that the universal curve restricted to the complement $\mathcal{S}_d^{\mathrm{smooth}} := \mathcal{S}_d \setminus \mathcal{S}_d^{\mathrm{sing}}$ is smooth and that all the fibers over $\mathcal{S}^{\mathrm{sing}}$ are singular. Show that $\mathcal{S}_d^{\mathrm{smooth}} := \mathcal{S}_d \setminus \mathcal{S}_d^{\mathrm{sing}}$ is non empty.
b) Compute $\mathcal{S}_d^{\mathrm{smooth}}(\mathbb{Z})$ for $d = 1, d = 2$. (It is a deep theorem that $\mathcal{S}_d^{\mathrm{smooth}}(\mathbb{Z}) = \emptyset$ for $d \geq 3$. Why is this not a contradiction to a)?)

A **projective curve** is a curve, which is isomorphic to a closed subscheme of some $\mathbb{P}^n/k$, our curves $V(f)$ above are projective curves. If our ground field $k = \mathbb{C}$, the field of complex numbers, then the set of $\mathbb{C}$ valued points of a smooth projective curve is the same thing as a compact Riemann surfaces. Many of the following considerations are parallel to the considerations in Chapter 5 of volume I. Some of these considerations will be easier here, because we do not have to deal with analytical difficulties. But the possibility that the ground field is not algebraically closed or even not perfect, will cause us some headaches of different kind.

9.2 The local rings at closed points

Proposition 9.2.1. *For a smooth curve C/k the local ring $\mathcal{O}_{C,\mathfrak{p}}$ at a closed point $\mathfrak{p}$ is a discrete valuation ring. For any non empty affine open set $U \subset C$ the ring $\mathcal{O}_C(U)$ is a Dedekind ring.*

Proof: Let P be a geometric point, let $\mathfrak{p}$ be the corresponding point on C. We can choose a local parameter f at P. Recall that this is an element in $f \in \mathcal{O}_{C,\mathfrak{p}}$ such that the differential df generates $\Omega^1_{C/k}$ at $\mathfrak{p}$. We have seen in the section on smooth points (See theorem 7.5.2) that we get an embedding of the polynomial ring

$$\begin{aligned} k[X] &\hookrightarrow \mathcal{O}_{C,\mathfrak{p}} \\ X &\mapsto f. \end{aligned}$$

If $\mathfrak{p}_0 = k[X] \cap \mathfrak{p}$ then the embedding of local rings

$$k[X]_{\mathfrak{p}_0} \hookrightarrow \mathcal{O}_{C,\mathfrak{p}}$$

is étale (See Definition 7.5.14)
(i) The maximal ideal $\mathfrak{p}_0 \subset k[X]_{\mathfrak{p}_0}$ generates the maximal ideal $\mathfrak{p} \subset \mathcal{O}_{C,\mathfrak{p}}$.
(ii) The extension

$$k[X]_{\mathfrak{p}_0}/\mathfrak{p}_0 \hookrightarrow \mathcal{O}_{C,\mathfrak{p}}/\mathfrak{p}$$

is a finite separable extension. (This is explained at the end of the section on smooth points, just before the section on flat morphisms).
Since the ring $k[X]$ is principal, we have $\mathfrak{p}_0 = (p(X))$ with an irreducible polynomial $p(X)$. The maximal ideal of the local ring $\mathcal{O}_{C,\mathfrak{p}}$ is also generated by $p(X)$ and the proposition follows (see definition 7.3.4.) Since we had the habit to denote a uniformizing element of the maximal ideal $\mathfrak{p}$ of a discrete valuation ring by $\pi_\mathfrak{p}$ we can choose $\pi_\mathfrak{p} = \pi_{\mathfrak{p}_0} = p(X)$. □

The above étale morphism provides in a certain sense a good approximation of $\mathcal{O}_{C,\mathfrak{p}}$ by $k[X]_{\mathfrak{p}_0}$. If we assume for instance that $k(\mathfrak{p}_0) = k(\mathfrak{p})$ then it is clear that for any n we have $k[X]/\mathfrak{p}_0^n = \mathcal{O}_{C,\mathfrak{p}}/\mathfrak{p}^n$ and hence we get an isomorphism between the completions (See theorem 7.5.2)

$$\widehat{k[X]}_{\mathfrak{p}_0} = \varprojlim k[X]/\mathfrak{p}_0^n \xrightarrow{\sim} \varprojlim \mathcal{O}_{C,\mathfrak{p}}/\mathfrak{p}^n\mathcal{O}_{C,\mathfrak{p}} = \widehat{\mathcal{O}}_{C,\mathfrak{p}}.$$

Under our assumptions the extension $k(\mathfrak{p})/k(\mathfrak{p}_0)$ is separable, but the first step $k(\mathfrak{p}_0)/k$ can be inseparable. But

Proposition 9.2.2. *If $\mathfrak{p}$ is a closed point on the smooth curve C/k and if the extension of residue fields $k(\mathfrak{p})/k$ is separable, and if $\pi_\mathfrak{p}$ is a uniformizing element at $\mathfrak{p}$, then $d\pi_\mathfrak{p}$ generates the module of differentials at $\mathfrak{p}$.*

Proof: We choose our X as above. Then our assumption implies that $p'(X)$ is a unit in $k[X]_{\mathfrak{p}_0}$ and then $d\pi_{\mathfrak{p}_0} = d\pi_\mathfrak{p} = p'(X)dX$ is a generator for the module $\Omega^1_{C,\mathfrak{p}}$ of differentials. □

9.2.1 The structure of $\widehat{\mathcal{O}}_{C,\mathfrak{p}}$

We need a little bit of information concerning the structure of $\widehat{\mathcal{O}}_{C,\mathfrak{p}}$. We have the diagram

$$\begin{array}{ccc} \widehat{\mathcal{O}}_{C,\mathfrak{p}} & \xrightarrow{\Psi} & k(\mathfrak{p}) \\ \uparrow & \nearrow & \\ k. & & \end{array}$$

If the extension of residue fields $k(\mathfrak{p})/k$ is separable then we have a unique section $s : k(\mathfrak{p}) \longrightarrow \widehat{\mathcal{O}}_{C,\mathfrak{p}}$ with $s|k = \mathrm{Id}_k$ and $\Psi \circ s = \mathrm{Id}_{k(\mathfrak{p})}$. To get this section we write $k(\mathfrak{p}) = k(\theta)$ where θ is the zero of an irreducible and separable polynomial $f(X) \in k[X]$. We lift θ to an element $\widetilde{\theta} \in \widehat{\mathcal{O}}_{C,\mathfrak{p}}$. Then $F(\theta) \equiv 0 \mod \mathfrak{p}$. Let $\pi_\mathfrak{p}$ be a uniformizing element. We modify $\widetilde{\theta}$ into $\widetilde{\theta} + \alpha\pi_\mathfrak{p}$ where $\alpha \in \widehat{\mathcal{O}}_{C,\mathfrak{p}}$. We evaluate F at this new argument and get $F(\widetilde{\theta} + \alpha\pi_\mathfrak{p}) = F'(\widetilde{\theta}) + F'(\widetilde{\theta})\alpha\pi_\mathfrak{p}$. We know that $F'(\widetilde{\theta})$ is a unit in $\widehat{\mathcal{O}}_{C,\mathfrak{p}}$ and hence we see that we can choose α in such a way that

$$F(\widetilde{\theta} + \alpha\pi_\mathfrak{p}) \equiv 0 \mod \mathfrak{p}^2.$$

We modify again by adding a $\beta\pi_\mathfrak{p}^2$ and improve our solution to a solution $\mod \mathfrak{p}^3$. This yields a sequence, which converges to an exact solution. This argument is called **Hensel's Lemma** (See [Neu]) and it is the p-adic version of Newtons method.
We identify $s(k(\mathfrak{p}))$ to $k(\mathfrak{p})$ and our diagram above becomes

$$\begin{array}{ccc} k(\mathfrak{p})[[\pi_\mathfrak{p}]] & \longrightarrow & k(\mathfrak{p}) \\ \uparrow & \nearrow & \\ k & & \end{array}$$

in other words our ring $\widehat{\mathcal{O}}_{C,\mathfrak{p}}$ is the power series ring in one variable $\pi_\mathfrak{p}$ over the residue field. The quotient field is $k(\mathfrak{p})[[\pi_\mathfrak{p}]][1/\pi_\mathfrak{p}]$, it is sometimes called the field of Laurent expansions at $\mathfrak{p}$.
If the residue field extension is not separable then the structure of $\widehat{\mathcal{O}}_{C,\mathfrak{p}}$ is not so nice and will cause us some trouble.
If the extension $k(\mathfrak{p})/k$ purely inseparable then it is clear that for any finite extension L/k the L-algebra $k(\mathfrak{p}) \otimes_k L$ is local and hence we have only one prime ideal $\mathfrak{p}'$ in the fibre over $\mathfrak{p}$. For the structure of $\widehat{\mathcal{O}}_{C,\mathfrak{p}}$ in this case we refer to [Ei], Thm 7.7.

9.2.2 Base change

We have to investigate systematically what happens if we extend the field of constants k, i.e. we choose an extension L/k and we consider the curve $C \times_{\mathrm{Spec}(k)} \mathrm{Spec}(L)$ or simply $C \times_k L$. We have a morphism (the base change morphism)

$$C \leftarrow C \times_k L.$$

We studied this question at the end of the section on affine schemes for the case $L = \bar{k}$ but it is clear that the considerations carry over to this case. The base change morphism induces a map on the closed points and the fibre over a closed point $\mathfrak{p}$ is the set of prime ideals of the finite L-algebra $k(\mathfrak{p}) \otimes_k L$. If we consider a closed point $\mathfrak{p}$ and an affine neighborhood U of $\mathfrak{p}$ then our base change morphism corresponds to the inclusion

$$A(U) \longrightarrow A(U) \otimes_k L.$$

Let us assume that L/k is finite. Then the prime ideal $\mathfrak{p} \subset A(U)$ decomposes in $A(U) \otimes_k L$, i.e.

$$\mathfrak{p}(A(U) \otimes_k L) = \mathfrak{p}_1^{e_1} \cdot \ldots \cdot \mathfrak{p}_s^{e_s}$$

or

$$(A(U) \otimes L)/\mathfrak{p}A(U) \otimes L \simeq A(U) \otimes_k L/\mathfrak{p}_1^{e_1} \oplus \ldots \oplus (A(U) \otimes_k L)/\mathfrak{p}_s^{e_s}.$$

The e_ν are called the ramification indices. There are some special cases

(1) If the residue field $k(\mathfrak{p}) = k$, then it is clear that $\mathfrak{p}$ stays prime. (This tells us also that our assumption that L/k is finite is inessential. After passing to a suitable finite extension nothing essential happens after that (if we stick to the given $\mathfrak{p}$)).

(2) If the extension $k(\mathfrak{p})/k$ is separable then the ramification exponents e_ν are one, the algebra $(A(U) \otimes L)/\mathfrak{p}A(U) \otimes L = k(\mathfrak{p}) \otimes_k L$ has no nilpotent elements.

(3) But if for instance $k(\mathfrak{p}) = k(\sqrt[p]{a})$ where $p = \operatorname{char}(k)$ and $a \notin k^p$ and $A(U) = k[X]/(X^p - a)$ and if $L = k(\sqrt[p]{a})$ then

$$k[X] \otimes_k k(\sqrt[p]{a})/_{(X^p-a)} = k(\sqrt[p]{a})[X]/(X - \sqrt[p]{a})$$

and hence we get nilpotent elements.

(4) We can always choose a finite normal extension L/k such that we can embed $k(\mathfrak{p})/k \hookrightarrow L/k$. If $k(\mathfrak{p})/k$ is purely inseparable then we may choose $L = k(\mathfrak{p})$. In this case the ramification index $e = [k(\mathfrak{p}) : k]$. In other words in the group of fractional ideals of $A(U) \otimes_k L$ we have $\mathfrak{p} = \mathfrak{p}'^{[k(\mathfrak{p}):k]}$. The completion $\mathcal{O}_{C \times_k L, \mathfrak{p}'}$ is again a power series ring.

(5) If $\mathfrak{p}$ is a separable point then we may take for L/k a normal closure of $k(\mathfrak{p})/k$ and then the closed points of $C \times_k L$, which lie over $\mathfrak{p}$ correspond to the embeddings

$$\sigma : k(\mathfrak{p})/k \hookrightarrow L/k$$

We have

$$k(\mathfrak{p}) \otimes_k L = \bigoplus_\sigma L$$

(See 7.4)

In general we can say: If we pass to the completion at $\mathfrak{p}$ and take the base extension to L/k then

$$\widehat{\mathcal{O}}_{C,\mathfrak{p}} \otimes_k L = (\varprojlim A(U)/\mathfrak{p}^n \otimes L) \xrightarrow{\sim} \mathcal{O}_{C \times_k L, \mathfrak{p}_1} \oplus \ldots \oplus \widehat{\mathcal{O}}_{C \times_k L, \mathfrak{p}_s}.$$

If the extension $k(\mathfrak{p})/k$ is separable then we get

$$\widehat{\mathcal{O}}_{C,\mathfrak{p}} \otimes_k L = k(\mathfrak{p})[[\pi_\mathfrak{p}]] \otimes_k L \simeq (k(\mathfrak{p}) \otimes_k L)[[\pi_\mathfrak{p}]] \xrightarrow{\sim} \bigoplus_\sigma L[[\pi_\mathfrak{p}]]$$

where now the points above $\mathfrak{p}$ correspond to the points of $\operatorname{Spec}(k(\mathfrak{p}) \otimes_k L)$.

9.3 Curves and their function fields

The following considerations are parallel to the reasoning in Vol. I. 5.1.7. For an irreducible curve C/k we consider the function field $k(C)$ of meromorphic functions. This is simply the stalk $\mathcal{O}_{C,\eta}$ of the sheaf $\mathcal{O}_C$ in the generic point η. For any affine non-empty open subset $U \subset C$ the ring $\mathcal{O}_C(U)$ is a one-dimensional, integral ring and $k(C)$ is its quotient field. We pick an $f \in \mathcal{O}_C(U)$, which is not constant. By definition this element cannot be algebraic over k. It yields an embedding of the polynomial ring

$$k[f] \hookrightarrow \mathcal{O}_C(U)$$

and an embedding of fields

$$k(f) \hookrightarrow k(C).$$

Proposition 9.3.1. *The field $k(C)$ is a finite extension of $k(f)$ and hence of transcendence degree one.*

This is rather clear, if we had an $g \in k(C)$, which is not algebraic over $k(f)$ then $k[f,g]$ would be a polynomial ring in two variables sitting in some Dedekind ring $\mathcal{O}_C(V)$, which is absurd. $\square$

Let us denote by $D_f \supset U$ the set of points $\mathfrak{p}$ where f is regular, i.e. we have $f \in \mathcal{O}_{C,\mathfrak{p}}$. Since $\mathcal{O}_C$ is a sheaf then we have $f \in \mathcal{O}_C(D_f)$. Now we consider the integral closure A of $k[f]$ in $k(C)$. In the section on Dedekind rings we indicated that A is again a finitely generated k-algebra and a Dedekind ring (See 7.3.8). (The extension $k(C)/k(f)$ needs not to be separable anymore). It is clear that the elements of A are integral at all points of D_f in other words we have $A \hookrightarrow \mathcal{O}_C(D_f)$. We get a diagram

$$\begin{array}{ccc} D_f & \longrightarrow & \mathrm{Spec}(A) \\ & \searrow & \\ & & C \end{array}$$

Now we assume in addition that our curve C/k is projective. Then it is clear that the morphism $D_f \hookrightarrow C$ extends uniquely to $\mathrm{Spec}(A) \longrightarrow C$. (See 8.1.10) Our diagram can be completed by a vertical arrow:

$$\begin{array}{ccc} D_f & \longrightarrow & \mathrm{Spec}(A) \\ & \searrow & \downarrow \\ & & C. \end{array}$$

If $\mathfrak{p}$ is a closed point in the image of $\mathrm{Spec}(A)$ then we have $A \subset \mathcal{O}_{C,\mathfrak{p}}$ hence we have $f \in \mathcal{O}_{C,\mathfrak{p}}$ and this implies $\mathfrak{p} \in D_f$. Hence our diagram becomes

$$\begin{array}{cccc} D_f & \longrightarrow & \mathrm{Spec}(A) & \\ & \searrow \mathrm{Id} & \downarrow & \\ & & D_f & \subset C. \end{array}$$

If we now assume that C/k is smooth then we see that the inclusion $A \hookrightarrow \mathcal{O}_C(D_f)$ must be an isomorphism.
We summarize

Proposition 9.3.2. *Let C/k be a projective smooth curve. If we pick a non constant $f \in k(C)$ and consider the set $D_f \subset C$ where f is regular. Then $\mathcal{O}_C(D_f)$ is the integral closure of $k[f]$ in $k(C)$. We can find an f such that the extension $k(C)/k(f)$ is separable.*

We may also consider $k[f^{-1}]$ and by the same procedure we get $k[f^{-1}] \subset \mathcal{O}_C(D_{f^{-1}})$. We have seen how to glue $\mathrm{Spec}(k[f])$and $\mathrm{Spec}(k[f^{-1}])$ over $\mathrm{Spec}(k[f,f^{-1}])$ and construct $\mathbb{P}^1/k$. Hence we see that our element f defines a morphism

$$\Phi_f : C \longrightarrow \mathbb{P}^1.$$

This morphism is finite, i.e. on any open affine $V \subset \mathbb{P}^1$ the morphism $\mathcal{O}_{\mathbb{P}^1}(Y) \hookrightarrow \mathcal{O}_C(\Phi_f^{-1}(V))$ is finite. If our curve is smooth and $U \subset X$ is a nonempty affine open subset then we can find an $f \in \mathcal{O}_C(U)$ such that $df \in \Omega^1_{\mathcal{O}_C(U)/k}$ is not zero (See Theorem 7.5.12). Then it is clear that under this assumption on f the morphism Φ_f is separable.

This tells us that for a smooth, projective and absolutely irreducible curve we can reconstruct the curve from its field of meromorphic functions, i.e. from $\mathcal{O}_{C,\eta} = k(C)$. The set of closed points of C can be identified to the set $\mathrm{Val}(k(C))$ of all discrete valuation rings V in $k(C)$, which contain the field of constants k and have quotient field $k(C)$. To see this we have to show that any such discrete valuation ring $V \subset k(C)$ is the stalk of $\mathcal{O}_C$ at a closed point. We pick an $f \in V$, which is not in k then $V \supset k[f]$. Then we must have $V \supset \mathcal{O}_C(D_f)$. Hence V is the discrete valuation ring at a point of D_f. We put a topology onto the set $\mathrm{Val}(k(C))$: The non empty open sets are the sets $\mathrm{Val}(k(C))_f = \{V | f \in V\}$. Then we see that the bijection $\mathrm{Val}(k(C)) \xrightarrow{\sim}$ closed points of C becomes a homeomorphism. Eventually we define a sheaf $\mathcal{O}(\mathrm{Val}(k(C))_f) = \bigcap_{V \in \mathrm{Val}(k(C))_f} V$, we get a ringed space and if we add a generic point we get an isomorphism of ringed spaces.
If we have two smooth, absolutely irreducible projective curves C_1,C_2 then we can consider morphisms $\phi : C_1 \longrightarrow C_2$. If such a morphism is not constant (i.e. it does not map C_1 to a point) then it maps the generic point to the generic point. Hence it induces a map between the function fields

$$\begin{array}{ccc} k(C_2) & \xrightarrow{{}^t\phi} & k(C_1) \\ & \searrow \quad \swarrow & \\ & k & \end{array}$$

and it is not difficult to see, that we can recover ϕ from ${}^t\phi$. Hence we see that non constant morphisms

$$\varphi : C_1 \longrightarrow C_2$$

between two smooth, absolutely irreducible projective curves are in one to one correspondence to morphisms

$${}^tf : k(C_2) \longrightarrow k(C_1)$$

between the function fields, which are the identity on the constants. This is an exceptional phenomenon in dimension one.
We may even start from a field K/k of transcendence degree one. We assume in addition that k is absolutely closed in K, i.e. any element $f \in K$, which is algebraic over k is already in k. Then we can construct a smooth, absolutely irreducible, projective curve C/k with $k(C) = K$. The set of closed points will be the set of discrete valuation rings in K, which contain k, the underlying set C is the set of closed points and the generic point. The non

empty open sets will be the sets D_f of closed points $\mathfrak{p}$ containing a given $f \in K, f \notin k$ plus the generic point. The k-algebra of holomorphic functions $\mathcal{O}_C(D_f)$ is the integral closure of $k[f]$ in K. This is a finitely generated k algebra and a Dedekind ring (See 7.3.8). It is absolutely irreducible (See Lemma 7.2.14). Since $\mathrm{Spec}(\mathcal{O}_C(D_f))/\mathrm{Spec}(k)$ is of dimension one and normal it is an affine smooth curve (See theorem 7.5.18). The set C together with the sheaf defined by the $\mathcal{O}_C(D_f)$ is a curve, a pair of elements $f, f^{-1}, f \notin k$ defines a finite morphism $\pi : C \longrightarrow \mathbb{P}^1_k$ and hence it follows from proposition 8.1.21 that C/k is projective.
Hence we can say that the category of absolutely, irreducible smooth curves over a field k is antiequivalent to the category of fields K/k of transcendence degree 1, for which k is algebraically closed in $\bar{k}$.
If the field k is not algebraically closed in K, then the field of elements, which are algebraic over k, was called the field of pseudoconstants earlier, but we could call it as well the field of constants, because we are only looking at generic points.
Remark: We just want to mention that an irreducible curve C/k, which is covered by open affine sub schemes U_i is smooth if and only if the k-algebras $\mathcal{O}(U_i)$ are normal (see 7.5.4. Hence for absolutely irreducible curves C'/k we have an easy way to desingularize them: We take their function field $K = K(C')$ and consider the smooth projective curve C/k constructed from it. We have a morphism $\pi : C \longrightarrow C'$, if we have a covering $C' = \cap V_i$ by affine schemes then the rings $\mathcal{O}_C(\pi^{-1}(V_i))$ are simple the integral closures of $\mathcal{O}_{C'}(V_i)$ in K. The curve $C \longrightarrow C'$ is called the **normalization** of C'. (See 7.1.3) .

9.3.1 Ramification and the different ideal

Let $\varphi : C_1 \to C_2$ be a separable, finite morphism between two smooth and absolutely irreducible projective curves, this means that

$$^t\varphi : k(C_2) \hookrightarrow k(C_1)$$

is a finite separable extension. For any affine open subset $U \subset C_2$ we know that

$$B = \mathcal{O}_{C_1}(\varphi^{-1}(U)) \supset \mathcal{O}_{C_2}(U) = A$$

and B is the integral closure of A in $k(C_1)$. We introduced the concept of ramification in (See definition 7.3.10), we said that B is ramified at a point $\mathfrak{p} \in \mathrm{Spec}(A)$ if and only if the $A/\mathfrak{p}$-algebra $B/\mathfrak{p}B = B \otimes_A A/\mathfrak{p}$ has nilpotent elements. This is obviously equivalent to the assertion that the trace pairing

$$\mathrm{tr}_{(B/\mathfrak{p}B)/(A/\mathfrak{p})}\ (xy)$$

is degenerate.

This leads to the definition of the fractional ideal

$$\mathfrak{D}^{-1}_{B/A} = \{x \in K_1 \mid \mathrm{tr}_{K_1/K_2}(xB) \subset A\}.$$

Since K_1/K_2 is separable we see that

$$\begin{array}{cccc} \mathrm{tr} & : & K_1 \times K_1 & \longrightarrow & K_2 \\ \mathrm{tr} & : & (x,y) & \longrightarrow & \mathrm{tr}_{K_1/K_2}(xy) \end{array}$$

is non degenerate and we conclude that we have $B \subset \mathfrak{D}_{B/A}^{-1}$ and that $\mathfrak{D}_{B/A}^{-1}$ is indeed the inverse of a non zero ideal $\mathfrak{D}_{B/A} \subset B$.

This ideal can be written as

$$\mathfrak{D}_{B/A} = \prod_{i=1}^{t} \mathfrak{p}_i^{m_i},$$

and it gives us the ramified primes and it also measures the ramification.

By definition we have a homomorphism

$$\psi : \mathfrak{D}_{B/A}^{-1} \longrightarrow \mathrm{Hom}_A(B,A)$$

$$x \mapsto \{b \mapsto \mathrm{tr}_{B/A}(xb)\}$$

and I claim that this map is an isomorphism.
To see this we pick any non zero prime ideal $\mathfrak{p}$ in A and localize at this prime ideal. If now $x \in \mathfrak{D}_{B/A}^{-1} \setminus \mathfrak{D}_{B/A}^{-1}\mathfrak{p}$ then we can find a $b \in B$ s.t. $\mathrm{tr}_{B/A}(xb) \not\equiv 0 \mod \mathfrak{p}$. Otherwise we would have $x/\pi_{\mathfrak{p}} \in \mathfrak{D}_{B/A}^{-1}$ in contradiction to our assumption on x. Hence we see that

$$\mathfrak{D}_{B/A}^{-1} \otimes A/\mathfrak{p} \hookrightarrow \mathrm{Hom}_A(B,A/\mathfrak{p})$$

and since these two vector spaces have the same dimension the claim follows.

If we vary the open set U we can put these different ideals together and get a sheaf

$$\mathfrak{D}_{C_1/C_2},$$

which measures the ramification of $\psi : C_1 \to C_2$. We consider the sheaf of differentials $\Omega^1_{C_1/k}$ and $\Omega^1_{C_2/k}$. Both of them are line bundles because we made the assumption that C_1,C_2 are smooth.
If we pull back the sheaf $\Omega^1_{C_2/k}$ via φ to the sheaf $\varphi^*(\Omega_{C_2/k})$, then we have an obvious inclusion between the two line bundles

$$\varphi^*(\Omega^1_{C_2/k}) \hookrightarrow \Omega^1_{C_1/k},$$

and we have the

Theorem 9.3.3. Riemann-Hurwitz formula. *This inclusion extends to an isomorphism*

$$\varphi^*(\Omega^1_{C_2/k}) \otimes \mathfrak{D}_{C_1/C_2}^{-1} \xrightarrow{\sim} \Omega^1_{C_1/k}.$$

Proof: This is a local formula. If we choose an affine open set $U \subset C_2$ such that $\Omega^1_{A/k}$, $\Omega^1_{B/k}$ and $\mathfrak{D}_{B/A}$ become free modules and if ω_A is generator of $\Omega^1_{A/k}$ and F a generator of $\mathfrak{D}_{B/A}$, then $F^{-1} \cdot \omega = \omega'$ is a generator of $\Omega^1_{B/k}$.

We have seen for the sheaves of differentials that they behave well under extensions of the ground field. Hence we perform the base change $\mathrm{Spec}(k) \leftarrow \mathrm{Spec}(\overline{k})$, then

$$\begin{aligned} \Omega^1_{A\otimes_k \overline{k}/\overline{k}} &= \Omega^1_{A/k} \otimes_k \overline{k} \\ \Omega^1_{B\otimes_k \overline{k}/\overline{k}} &= \Omega^1_{B/k} \otimes_k \overline{k} \end{aligned}$$

Now we look at the discriminant $\mathfrak{D}_{A/B}$. Since we assumed that our curves are absolutely irreducible the two algebras $A \otimes_k \overline{k}$, $B \otimes_k \overline{k}$ are still fields.

We showed that $\psi : \mathfrak{D}_{B/A}^{-1} \to \mathrm{Hom}_A(B,A)$ is an isomorphism, hence for any extension k'/k

$$\mathfrak{D}_{B/A}^{-1} \otimes_k k' \longrightarrow \mathrm{Hom}_{A\otimes_k k'}(B \otimes_k k', A \otimes_k k')$$

is an isomorphism. This shows

$$\mathfrak{D}_{B/A}^{-1} \otimes_k k' = \mathfrak{D}_{B\otimes_k k'/A\otimes_k k'}^{-1}.$$

Hence it suffices to prove the Riemann-Hurwitz formula under the assumption that our ground field is algebraically closed.

Since we may also pass to the completion, we are reduced to the case

$$\widehat{\mathcal{O}}_{C_2,\mathfrak{p}} = k[[x]] \longrightarrow \widehat{\mathcal{O}}_{C_1,\mathfrak{p}} = k[[y]]$$

where the extension is finite.

We can write x as a power series in y

$$x = Q(y) = y^m + a_{m+1}y^{m+1} + \cdots = y^m(1 + a_{m+1}y + \cdots),$$

which implies that $\mathrm{ord}_{(y)}(x) = m$. Then it is obvious that the elements $1,y,\cdots y^{m-1}$ form a basis of the $k[[x]]$-module $k[[y]]$, and we have an equation

$$y^m + a_1(x)y^{m-1} + \cdots a_m(x) = 0$$

where the $a_i(x) \in k[[x]]$ and $a_i(x) \equiv \mod (x)$. Looking at the order of vanishing yields $a_m(x) = \alpha_0 x + \alpha_1 x^2 \cdots$ with $\alpha_0 \neq 0$. If $P(Y) = Y^m + a_1(x)Y^{m-1} \cdots a_m(x) \in k[[x]][Y]$, we find a relation for the differentials

$$\frac{\partial P}{\partial Y}(y)dy + \left(\sum \frac{\partial a_\nu}{\partial x} y^{m-\nu}\right) dx = 0.$$

Since $\frac{\partial a_m}{\partial x} = \alpha_0 \neq 0$ we see that $\left(\sum \frac{\partial a_v}{\partial x} y^{m-v}\right) dx$ generates $\Omega^1_{A/k} \otimes B$, we see that $\Omega^1_{B/k}$ is generated by

$$\frac{dx}{\frac{\partial P}{\partial Y}(y)}.$$

Now everything boils down to show that

$$\mathfrak{D}_{B/A} = \left(\frac{\partial P}{\partial Y}(y)\right),$$

which is very easy. □

One of the important consequences of our considerations is that we can define a trace map

$$\mathrm{tr}: f_*(\Omega^1_{C_1/k}) \longrightarrow \Omega^1_{C_2/k} \tag{9.3}$$

This is clear. Locally we can write a differential $\omega = f dx$ where dx is a generator of $\Omega^1_{C_2/k}$ and $f \in \mathfrak{D}^{-1}_{C_1/C_2}$. Then

$$\mathrm{tr}_{C_1/C_2}(\omega) = \mathrm{tr}_{C_1/C_2}(f)dx$$

and $\mathrm{tr}_{C_1/C_2}(f)$ is regular by definition of the different.

9.4 Line bundles and Divisors

The following considerations are valid in a more general framework. Let X be any scheme. We consider line bundles on X, this are locally free $\mathcal{O}_X$-modules of rank one. The structure sheaf itself is a line bundle, it is called the trivial bundle. If we have two line bundles $\mathcal{L}_1$,$\mathcal{L}_2$, we can form the tensor product $\mathcal{L}_1 \otimes \mathcal{L}_2$. We can form the line bundle $\mathcal{L}^{-1} = \mathrm{Hom}_{\mathcal{O}_X}(\mathcal{L},\mathcal{O}_X)$ and we have $\mathcal{L} \otimes \mathcal{L}^{-1} \cong \mathcal{O}_X$. Hence it is rather clear that the isomorphism classes of line bundles form a group under the multiplication. It is the so-called Picard group $\mathrm{Pic}(X)$. To define this group we do not need any assumption on X.
Under certain assumptions we can identify this group to the so called divisor class group. At first we assume that X is irreducible, then we have the field of meromorphic functions on X. If we denote the generic point by η then this field is $\mathcal{O}_{X,\eta}$.
We define the **group of divisors** as the free group generated by the irreducible subschemes $\mathfrak{p}$ of codimension one. We need the concepts from dimension theory and hence we assume that our scheme is covered by a finite number of affine noetherian schemes. We are mainly interested in the case where X/k is of finite type over a field k then it is covered by a finite number of affine schemes of finite type over k. For this special case we discussed the relevant results in dimension theory in section 7.1.2

A meromorphic function f on X is simply an element in the stalk $\mathcal{O}_{X_i,\eta_i}$. We want to attach a divisor $\mathrm{Div}(f)$ to the function f. We can find a covering $X = \cup U_i$ by affine integral schemes such that on U_i we can write

$$f = \frac{g_i}{h_i}$$

with f_i,$h_i \in \mathcal{O}_X(U_i)$. We consider the prime ideals, which contain the principal ideals (f_i) resp (g_i) and among those we we consider the minimal ones. Then the Hauptidealsatz of Krull asserts that there is a finite number of such minimal prime ideals $\mathfrak{p}_1,\dots,\mathfrak{p}_s \supset (f_i)$ and $\mathfrak{q}_1,\dots,\mathfrak{q}_r \supset (g_i)$ and these have height one (See 7.1.18 ,[Ei], Thm. 10.1). But this is not enough to attach a divisor to f, we must be able to attach multiplicities to the zeroes of f_i,g_i at $\mathfrak{p}_\nu$,$\mathfrak{q}_\mu$. To define these multiplicities we make the additional assumption that the scheme X is normal, recall that this means that the affine rings $\mathcal{O}_X(U_i)$ are integrally closed in their fields of meromorphic functions (See Definition 7.1.3). This implies that for all prime ideals $\mathfrak{p}$ of height one the local rings $\mathcal{O}_{X,\mathfrak{p}}$ are discrete valuation rings. (We gave a reference and an indication of the proof of this fact in the section on low dimensional rings.)
If now $\mathfrak{p}_\nu \supset (f_i)$ (resp. $\mathfrak{q}_\mu \supset (g_i)$ then we can write the principal ideals local rings $(f_i) \subset \mathcal{O}_{X,\mathfrak{p}_\nu}$ (resp. $(g_i) \subset \mathcal{O}_{X,\mathfrak{q}_\nu}$) as

$$(f_i) = \mathfrak{p}_\nu^{\mathrm{ord}_{\mathfrak{p}_\nu}(f_i)}, (g_i) = \mathfrak{q}_\mu^{\mathrm{ord}_{\mathfrak{q}_\mu}(g_i)},$$

and we attach a divisor to the restriction of f to U_i namely

$$\sum \mathrm{ord}_{\mathfrak{p}_\nu}(f_i)\mathfrak{p}_\nu - \sum \mathrm{ord}_{\mathfrak{q}_\mu}(g_i)\mathfrak{q}_\mu.$$

Comparing these divisors on the intersections $U_i \cap U_j$ gives us a divisor on X.
Of course it can happen that we have some cancellations, i.e. that the same $\mathfrak{p}$ occurs among the $\mathfrak{p}_\nu$ and the $\mathfrak{q}_\mu$. We say that the function f has a zero (resp. pole) of order $n \geq 0$ at an irreducible subscheme $\mathfrak{p}$ of codimension 1 if it occurs among the $\mathfrak{p}_\nu, \mathfrak{q}_\mu$ and if $n = \mathrm{ord}_{\mathfrak{p}}(f_i) - \mathrm{ord}_{\mathfrak{p}}(g_i)$ resp. $n = \mathrm{ord}_{\mathfrak{p}}(g_i) - \mathrm{ord}_{\mathfrak{p}}(f_i)$ (for some i).
Hence we see that we can define the group of **principal divisors** on a noetherian, integral normal scheme as the group of divisors of meromorphic functions. We define the **divisor class group** $\mathrm{Cl}(X)$ as the group of divisors modulo principal divisors.
Under certain conditions we have an isomorphism between $\mathrm{Pic}(X)$ and the divisor class group. To get this isomorphism we have to use some results from the commutative algebra of noetherian rings, which are not in our script. In the special case of rings of dimension one they are proved in the section on "Low dimensional rings".
Assume again that X is noetherian, integral and normal. Let $\mathcal{L}$ be a line bundle on X. Then the stalk $\mathcal{L}$ at the generic point η is a one dimensional $\mathcal{O}_{X,\eta}$ - vector space. Let $s \in \mathcal{L}_\eta$ be a generator. It is a meromorphic section in the line bundle. We want to attach a divisor to this meromorphic section.
We proceed as above. We cover X by affine integral schemes U_i such that $\mathcal{L}|U_i$ becomes trivial, i.e. $\mathcal{L}(U_i) = (\mathcal{O}_X|U_i) \cdot t_i$ with $t_i \in H^0(U_i,\mathcal{L})$. Then we have for all i wee can write $s = g_i t_i$ where g_i is an element in the field of meromorphic functions $\mathcal{O}_{X,\eta}$. Again we write $g_i = \frac{f_i}{h_i}$ with $f_i,h_i \in \mathcal{O}_X(U_i)$. As before we see that there is a finite number of minimal prime ideals $\mathfrak{p}_1,\dots,\mathfrak{p}_s \supset (f_i)$. and $\mathfrak{q}_1,\dots,\mathfrak{q}_r \supset (g_i)$ and these have height one. Again we use the fact that the local rings $\mathcal{O}_{X,\mathfrak{p}_\nu},\mathcal{O}_{X,\mathfrak{q}_\nu}$ are discrete valuation rings and inside these local rings $(f_i) = \mathfrak{p}_\nu^{\mathrm{ord}_{\mathfrak{p}_\nu}(f_i)}, (g_i) = \mathfrak{q}_\nu^{\mathrm{ord}_{\mathfrak{q}_\nu}(g_i)}$. The prime ideals $\mathfrak{p}_\nu,\mathfrak{q}_\nu$ define irreducible subschemes of codimension one and we attach to $(\mathcal{L},s,t_i)$ a divisor D_i on U_i namely

$$D_i = \sum \mathrm{ord}_{\mathfrak{p}_\nu}(f_i)\mathfrak{p}_\nu - \sum \mathrm{ord}_{\mathfrak{q}_\mu}(g_i)\mathfrak{q}_\mu.$$

If we compare D_i and D_j on the intersection $U_i \cap U_j$ we see that they must coincide and hence we see that we can attach a divisor D on X to our data $(\mathcal{L},s,t_i)$. If we change then t_i then we modify g_i by a unit and D_i stays the same. If we modify the meromorphic section s, i.e. we multiply s by a meromorphic function g then the divisor is changed by the divisor of a meromorphic function.
Therefore, we can say that get a homomorphism

$$\mathrm{Pic}\ (X) \longrightarrow \text{Divisors modulo principal divisors.}$$

It can be shown that under our assumptions this homomorphism is injective: If we can choose the sections s and t_i in such a way that the divisor becomes zero, then $s = g_i t_i$ where the divisor of g_i is zero. We conclude that g_i and g_i^{-1} lie in the intersection of all discrete valuation rings $\mathcal{O}_X(U_i)_{\mathfrak{p}}$ where $\mathfrak{p}$ runs over the prime ideals of height one. Now we know that this intersection is equal to $\mathcal{O}_X(U_i)$ (See [Ei], Cor. 4.11, in dimension one also in "Low dimensional rings") Hence g_i is a unit. But then $t_i = g_i^{-1} s$ and the meromorphic section s is an element in $H^0(X,\mathcal{L})$, which generates the free $\mathcal{O}_X$-module $\mathcal{L}$.

This homomorphism is not surjective in general. If we want this we need to assume that X is **locally factorial**. This means that for any point $x \in X$ and any prime ideal $\mathfrak{p} \subset \mathcal{O}_{X,x}$ of height one we find an $f_\mathfrak{p} \in \mathcal{O}_{X,x}$ such that $\mathfrak{p} = (f_\mathfrak{p})$. This is much stronger than saying that $\mathcal{O}_{X,\mathfrak{p}}$ is a discrete valuation ring.
Under this assumption the homomorphism becomes bijective. To see this we start from a divisor $D = \sum_\mathfrak{p} n_\mathfrak{p}\mathfrak{p}$. For any point $x \in X$ we can find an open neighborhood U_x and an element $f_x \in \mathcal{O}_X(U_x)$ such that $\mathrm{Div}(f_x)|U_x = D|U_x$. Now we define the line bundle $\mathcal{O}_X(D)$: Its sections over U_x are

$$\mathcal{O}_X(U_x) = \{h \in \mathcal{O}_{X,\eta} | f_x h \in \mathcal{O}_X(U_x)\}.$$

Then it is clear that this defines a line bundle because locally at x the function $1/f_x$ trivializes the bundle. The corresponding divisor class is the class of D (Our global meromorphic section s above s is simply the constant function 1 and for the t_i we choose f_x^{-1}. The the g_i become the f_x.)
We will change the notation slightly, we denote a divisor in the form $D = \sum_i n_i Z_i$ where the Z_i irreducible closed sub schemes of codimension one, the Z_i are the closures of the $\mathfrak{p}$,$\mathfrak{q}$, which are codimension one prime ideals and define closed sub schemes on the open affine pieces.
If we have a line bundle $\mathcal{L}$ on X, which has a non zero global section $s \in H^0(X,\mathcal{L})$ then we can define **the scheme** $V(s)$ **of zeroes of** s. . We simply observe that locally $\mathcal{L}$ is trivial, hence locally s is nothing else than a non zero regular function and locally $V(s)$ is the scheme defined by the principal ideal (s). We can decompose this scheme into irreducibles whose closures are then the Z_i. Then the divisor attached to $\mathcal{L}$ is simply $\sum_i n_i Z_i$ where the n_i are the multiplicities.
Our divisor is called **effective** if all the multiplicities are ≥ 0. If D is effective then $\mathcal{O}_X(D)$ is the sheaf of germs of meromorphic functions, which may have poles of order $\leq n_i$ along the Z_i. In this case we have $1 \in H^0(X,\mathcal{O}_X(D))$ and $D = V(1)$.
It is a theorem in commutative algebra that X is locally factorial if X/k is smooth (See Prop. 7.5.19 and Thm. 7.5.20). If X is of dimension one then this is a consequence of definition 7.3.4 and the assertion contained in it.

9.4.1 Divisors on curves

Now we come back to the case where $X = C/k$ is a smooth, projective and absolutely irreducible curve. The irreducible subschemes of codimension 1 are the closed points.
The group of divisors $\mathrm{Div}(C)$ is the free abelian group generated by the closed points. We write

$$D = \sum n_\mathfrak{p}\mathfrak{p}.$$

We define the **degree of a divisor** as

$$\deg(D) = \sum n_\mathfrak{p}[k(\mathfrak{p}) : k]$$

where $[k(\mathfrak{p}) : k]$ is the degree of the extension of the residue field. This degree has also been discussed in 8.4.2, but here we want to discuss this notion independently and give more elementary treatment.

If we pick a non zero element $f \in k(C)$ then we know that for any closed point $\mathfrak{p}$ we may write $f = \varepsilon\pi_{\mathfrak{p}}^{\nu_{\mathfrak{p}}(f)}$ in the local ring $\mathcal{O}_{C,\mathfrak{p}}$. Hence we can attach a divisor to our element f namely $\mathrm{Div}(f) = \sum_{\mathfrak{p}} \mathrm{ord}_{\mathfrak{p}}(f)\mathfrak{p}$. A divisor, which has such a presentation is called **principal divisor**. The principal divisors form a subgroup of the group of all divisors.

Theorem 9.4.1. *If C/k is a smooth, absolutely irreducible projective curve, then a principal divisor has degree zero.*

This follows from the results, which we explained in the section on Dedekind rings. We assume that f is not constant. We constructed the morphism (see 9.3.2)

$$\Phi_f : C \longrightarrow \mathbb{P}^1$$

and it is clear from the construction that $f \subset k(\mathbb{P}^1)$. As an element in the function field of the projective line its divisor is $(0) - (\infty)$ where (0) is the closed point defined by $(f) \in k[f]$ and (∞) is the closed point defined by $(f^{-1}) \subset k[f^{-1}]$. This divisor has degree zero. Now we consider the divisor of f on C/k. We decompose the divisor into the divisor of zeroes and the divisor of poles:

$$\mathrm{Div}(f) = \mathrm{Div}_0(f) + \mathrm{Div}_\infty(f) = \sum_{\mathfrak{p},\,\mathrm{ord}_{\mathfrak{p}}(f)>0} \mathrm{ord}_{\mathfrak{p}}(f)\mathfrak{p} + \sum_{\mathfrak{p},\,\mathrm{ord}_{\mathfrak{p}}(f)<0} \mathrm{ord}_{\mathfrak{p}}(f)\mathfrak{p}.$$

We studied the behavior of prime ideals under extension of Dedekind rings and have the formula

$$\deg(\mathrm{Div}_0(f)) = \sum_{\mathfrak{p},\,\mathrm{ord}_{\mathfrak{p}}(f)>0} \mathrm{ord}_{\mathfrak{p}}(f)[k(\mathfrak{p}) : k] = [k(C) : k(f)]$$

(See 7.9) (Our (f) is the prime ideal $\mathfrak{p}$ there and the $\mathfrak{P}$ there correspond to the $\mathfrak{p}$ here.) The same holds for the pole divisor and the theorem is clear. □

All our assumptions are valid in the case of smooth, absolutely irreducible curves hence we define as $\mathrm{Pic}(C/k)$ the group of line bundles of our curve C/k and we identify it to the group of divisors modulo linear equivalence. Especially we can now define the degree of a line bundle, it is simply the degree of the corresponding divisor class. The degree defines a homomorphism

$$\deg : \mathrm{Pic}(C/k) \longrightarrow \mathbb{Z}.$$

The kernel of this homomorphism is denoted by $\mathrm{Pic}^0(C/k)$.
It is one of the major aims of this book to give a proof, that this group $\mathrm{Pic}^0(C/k)$ is actually the group of rational points of a so called abelian variety of dimension g over k (the Jacobian). (See Chap. X). An **abelian variety** is a connected projective variety, which in addition has a structure of a group. In the case of curves over $\mathbb{C}$ this goes back to Abel, Riemann and Jacobi. We proved this in Chapter V of volume I.

9.4.2 Properties of the degree

Let C/k be an absolutely irreducible, smooth projective curve, let us consider an effective divisor $D = \sum n_\mathfrak{p}\mathfrak{p}$. In this case the sections of $\mathcal{O}_C(D)$ over an affine open set $V \subset C$ are the meromorphic functions on U whose pole order $\mathfrak{p}$ is less or equal to $n_\mathfrak{p}$.
We notice that we have $\mathcal{O}_C \subset \mathcal{O}_C(D)$ and the quotient sheaf $\mathcal{O}_C(D)/\mathcal{O}_C$ has non zero stalk only at the points $\mathfrak{p}$ with $n_\mathfrak{p} > 0$ this set is called the support of the divisor D and sometimes denoted by $|D|$. If V is an affine open set containing the support of D then $\mathcal{O}_C(V)$ is a Dedekind ring and we can interpret D as a fractional ideal for this Dedekind ring, it is clear from the definition (see section on Dedekind rings) that $\mathcal{O}_C(D)(V) = \prod_{\mathfrak{p}\in V} \mathfrak{p}^{-n_\mathfrak{p}}$. Now it is clear that for the global sections

$$\mathcal{O}_C(D)/\mathcal{O}_C)(C) = (\mathcal{O}_C(D))/\mathcal{O}_C)(V) = \sum \mathfrak{p}^{-n_\mathfrak{p}}\mathcal{O}_{C,\mathfrak{p}}/\mathcal{O}_{C,\mathfrak{p}}.$$

This direct sum $\sum \mathfrak{p}^{-n_\mathfrak{p}}\mathcal{O}_{C,\mathfrak{p}}/\mathcal{O}_{C,\mathfrak{p}}$ is a finite dimensional vector space over k and the dimension of this vector space is $\sum n_\mathfrak{p}[k(\mathfrak{p}) : k] = \deg(D)$, we summarize

$$\dim_k(H^0(C,\mathcal{O}_C(D)/\mathcal{O}_C) = \deg(D).$$

This can be generalized to arbitrary line bundles. If we have a line bundle $\mathcal{L}$ and an effective divisor D, then we can define $\mathcal{L}(D) = \mathcal{L} \otimes \mathcal{O}_C(D)$ and we have $\mathcal{L} \subset \mathcal{L}(D)$. On the other hand if we have two line bundles one of them contained in the other $\mathcal{L} \subset \mathcal{L}_1$ then there is a unique divisor D such that $\mathcal{L}_1 = \mathcal{L}(D)$. In these cases we have the formula

$$\deg(\mathcal{L}_1) = \deg \mathcal{L}(D) = \deg(\mathcal{L}) + \deg(D) = \deg(\mathcal{L}) + \dim_k((\mathcal{L}_1/\mathcal{L})(C)) \qquad (9.4)$$

Line bundles on non smooth curves have a degree

If C/k is an absolutely irreducible projective curve, which is not necessarily smooth we still can define the homomorphism

$$\deg : \mathrm{Pic}(C) \longrightarrow \mathbb{Z}.$$

We simply consider the normalization $\pi : \widetilde{C} \longrightarrow C$ (See remark at the end of 9.3) and we get a homomorphism $\mathrm{Pic}(C) \longrightarrow \mathrm{Pic}(\widetilde{C})$ given by $\mathcal{L} \mapsto \pi^*(\mathcal{L})$ we we define $\deg(\mathcal{L}) = \deg(\pi^*(\mathcal{L}))$.
There is a different way of looking at this notion of degree. The singular locus of C/k is dimension zero hence finite (See theorem 7.5.1). We can find an affine open subset $U \subset C$ such that it contains the singular locus. A line bundle $\mathcal{L}$ on C can be restricted to U and on the expense of making smaller but still containing the singular locus we can find a section $t \in H^0(U,\mathcal{L})$, which trivializes $\mathcal{L}|U$. But the points not in U are smooth and if we trivialize $\mathcal{L}$ in a point $\mathfrak{p} \notin U$ by a section $s_\mathfrak{p}$ then $t = g_\mathfrak{p} s_\mathfrak{p}$ where g is a meromorphic function. We define the divisor $D = \sum_{\mathfrak{p}\notin U} \mathrm{ord}_\mathfrak{p}(g_\mathfrak{p})\mathfrak{p}$. But then t extends to a section in $H^0(C,\mathcal{L}(-D))$ and this extension trivializes $\mathcal{L}(-D)$. Hence it follows that $\mathcal{L} = \mathcal{O}_C(D)$. Now the map $\pi : \widetilde{C} \longrightarrow C$ is an isomorphism if we restrict it to the complement of the singular set, this means that we can say $\pi^{-1}(D) = D$ and hence $\deg(\mathcal{L}) = \deg(\pi^*(\mathcal{L}) = \deg(D)$.

Actually essentially the same reasoning shows that we only use the projectivity but not the irreducibility of the curve. If we have a line bundle $\mathcal{L}$ on an arbitrary absolutely reduced curve C/k then we base change from k to the algebraic closure $\bar{k}$ and the the degree is simply the sum of the degrees of the restriction to the irreducible components.

Base change for divisors and line bundles

If we have a base change $C \leftarrow C \otimes_k L$ then this induces a homomorphism on the group of divisors. To see this we have to check what happens for prime divisors. We discussed what happens in section 9.2.2: The divisor $\mathfrak{p}$ maps to $\sum e_i \mathfrak{p}_i$. This extends to the group of divisors and we see that this homomorphism preserves the degree. We may denote the divisor on the base extension by $D \times_k L$. If we consider any line bundle $\mathcal{L}$ on C then we have a base change of this line bundle $i^*_{L/k}(\mathcal{L})$ where $i_{L/k} : C \times \mathrm{Spec}(L) \longrightarrow C$ is the base change morphism (See 6.2.2). Clearly $i^*_{L/k}(\mathcal{O}(D)) = \mathcal{O}(D \times L)$. This implies

$$deg(i_{L/k})^*(\mathcal{L}) = \deg(\mathcal{L}) \tag{9.5}$$

9.4.3 Vector bundles over a curve

A locally free coherent $\mathcal{O}_C$−module $\mathcal{E}$ is called a vector bundle. Let n be its rank. It is an easy exercise in algebra to prove the following: If we have a subspace $V \subset \mathcal{E}_\eta$ in the $\mathcal{O}_{C,\eta}$ vector space $\mathcal{E}_\eta$ then we can find a submodule $\mathcal{F} \subset \mathcal{E}$ such that $\mathcal{F}_\eta = V$ and the quotient $\mathcal{E}/\mathcal{F}$ is again locally free. Locally on C the bundle $\mathcal{F}$ is a direct summand. This implies that our vector bundle $\mathcal{E}$ admits a complete flag $(0) = \mathcal{L}_0 \subset \mathcal{L}_1 \subset \mathcal{L}_2 \subset \cdots \subset \mathcal{L}_{n-1} \subset \mathcal{E}$ of sub bundles such that the quotient of two successive bundles is a line bundle. We can define the n-th exterior power, this is a line bundle $\det(\mathcal{E}) = \Lambda^n(\mathcal{E})$. We define the degree by $\deg(\mathcal{E}) = \deg(\det(\mathcal{E}))$ and it is clear that we can express the degree in terms of a given flag as $\deg(\mathcal{E}) = \sum \deg(\mathcal{L}_i/\mathcal{L}_{i-1})$.
For any vector bundle we can define the dual bundle as $\mathrm{Hom}_{\mathcal{O}_C}(\mathcal{E},\mathcal{O}_C) = \mathcal{E}^\vee$. It is easy to see that $\deg(\mathcal{E}) + \deg(\mathcal{E}^\vee) = 0$. Again we can derive it easily from the case of line bundles.
Our formula 9.4 generalizes to vector bundles. If we have two vector bundles $\mathcal{E} \subset \mathcal{E}_1$ of the same rank, then $\mathcal{E}_1/\mathcal{E}$ is a finitely generated torsion sheaf and

$$\deg(\mathcal{E}_1) = \deg(\mathcal{E}) + \dim_k((\mathcal{E}_1/\mathcal{E})(C)) \tag{9.6}$$

To see this we trivialize both bundles on a suitably small non empty affine open set V. Then the matrices, which transform the two bases into each other have non zero determinants in $\mathcal{O}_{C,\eta}$, which are units over a still smaller but still non empty affine open set $V_1 \subset V$. This means that $\mathcal{E}|V_1 = \mathcal{E}_1|V_1$. This shows that the stalks of $\mathcal{E}_1/\mathcal{E}$ are non zero only in the finite set $\mathfrak{p} \in C \setminus V_1$. But for these points it is clear that we can find $m_\mathfrak{p}$ such that $\mathfrak{p}^{m_\mathfrak{p}} \mathcal{E}_{1,\mathfrak{p}} \subset \mathcal{E}_\mathfrak{p}$ and hence it is clear that the quotient is a finitely generated torsion sheaf. To see the assertion concerning the degree we may proceed as follows: We choose an affine set V containing the support of $\mathcal{E}_1/\mathcal{E}$. If we take this set to be sufficiently small, then we may assume that both bundles are trivial and the theorem on elementary divisors tells us that we can find a basis $e_1,e_2,\ldots,e_n$ for $\mathcal{E}_1(V)$ and non zero elements

$a_1,a_2,\dots a_n$ in $\mathcal{O}_C(V)$ such that $a_1e_1,a_2e_2,\dots,a_ne_n$ form a basis for $\mathcal{E}(V)$. Now these bases define complete flags in both vector bundles. These are the flags, which are induced by the subspaces $V_i = \mathcal{O}_{C,\eta}e_1 \oplus \cdots \oplus \mathcal{O}_{C,\eta}e_i$ of the generic fibre. We used these flags to define the degree: The degree was the sum of the degrees of the successive quotient line bundles induced by the flag. But if we compare these quotient line bundles for both vector bundles then we see the following: If $\mathcal{M}_i$ resp. $\mathcal{M}_{1,i}$ is such a quotient line bundle obtained by the flag in $\mathcal{E}$ resp. $\mathcal{E}_1$ then the stalks $(\mathcal{M}_i)_\mathfrak{q} = (\mathcal{M}_{1,i})_\mathfrak{q}$ for all $\mathfrak{q}$ not in the support of $\mathcal{E}_1/\mathcal{E}$. For the points $\mathfrak{p}$ in the support of $\mathcal{E}_1/\mathcal{E}$ and even for the $\mathfrak{p} \in V$ we have $a_i^{-1}\mathcal{M}_{i,\mathfrak{p}} = \mathcal{M}_{1,i,\mathfrak{p}}$. Hence we see $\deg(\mathcal{M}_{1,i}) = \deg(\mathcal{M}_i) + \dim_k(a_i^{-1}\mathcal{O}_C(V)/\mathcal{O}_C(V))$. Hence we get $\deg(\mathcal{E}_1) = \deg(\mathcal{E}) + \sum_i \dim_k(a_i^{-1}\mathcal{O}_C(V)/\mathcal{O}_C(V))$. On the other hand we see that

$$\mathcal{E}_1/\mathcal{E}(C) = \mathcal{E}_1/\mathcal{E}(V) = \oplus_i a_i^{-1}\mathcal{O}_C(V)/\mathcal{O}_C(V,$$

this implies our formula.
It is also clear that for an exact sequence of vector bundles

$$0 \longrightarrow \mathcal{E}_1 \longrightarrow \mathcal{E} \longrightarrow \mathcal{E}_2 \longrightarrow 0$$

we have the relation $\deg(\mathcal{E}_1) + \deg(\mathcal{E}_2) = \deg(\mathcal{E})$.
And finally we have: If we have a vector bundle $\mathcal{E}$ over C/k and if we tensorize it by a line bundle $\mathcal{L}$ then we get the formula

$$\deg(\mathcal{E} \otimes \mathcal{L}) = \deg(\mathcal{E}) + \mathrm{Rank}(\mathcal{E})\deg(\mathcal{L}) \tag{9.7}$$

Vector bundles on $\mathbb{P}^1$

We consider vector bundles over the projective line $\mathbb{P}^1_k$. In this case we have

Theorem 9.4.2. *Any vector bundle $\mathcal{E}/\mathbb{P}^1_k$ is a direct sum of line bundles, i.e.*

$$\mathcal{E} \simeq \mathcal{O}_{\mathbb{P}^1}(d_1) \oplus \cdots \oplus \mathcal{O}_{\mathbb{P}^1}(d_n)$$

with some integers $d_1 \cdots d_n$. The integers are well defined up to order.

Proof: Clearly the assertion is insensitive to tensorization by line bundles. We consider the case rank $\mathcal{E} = 2$. Since $\dim_k H^0(\mathbb{P}^1,\mathcal{E}) < \infty$ it follows that the degree of a line sub bundle $\mathcal{L}$ is bounded (see exercises 33 , 34) we can find a line sub bundle of maximal degree. We tensorize by the inverse of this bundle and therefore, we can assume that $\mathcal{O}_{\mathbb{P}^1} \subset \mathcal{E}$ is a sub bundle of maximal degree. We have the exact sequence

$$0 \longrightarrow \mathcal{O}_{\mathbb{P}^1} \longrightarrow \mathcal{E} \longrightarrow \mathcal{E}/\mathcal{O}_{\mathbb{P}^1} \longrightarrow 0$$

the quotient is a line bundle because otherwise it had torsion and the sub bundle would not be maximal. It is isomorphic to $\mathcal{O}_{\mathbb{P}^1}(r)$ for some $r \in \mathbb{Z}$. We claim that $r \leq 0$. To see this we look at the long exact cohomology sequence for the spaces of sections. If $r \geq 0$ then $\dim_k H^0(\mathbb{P}^1,\mathcal{E}/\mathcal{O}_{\mathbb{P}^1}) = \dim_k H^0(\mathbb{P}^1,\mathcal{O}_{\mathbb{P}^1}(r)) \geq 1$. Since we have $H^1(\mathbb{P}^1,\mathcal{O}_{\mathbb{P}^1}) = 0$ we find a non zero section $s \in H^0(\mathbb{P}^1,\mathcal{E})$. This provides an embedding $\mathcal{O}_{\mathbb{P}^1} \longrightarrow \mathcal{E}$ given by $f \mapsto fs$. Now we have two possibilities. Either the two sections $1 \in H^0(\mathbb{P}^1,\mathcal{O}_{\mathbb{P}^1})$ and s generate $\mathcal{E}$ at every point or not. In the first case we get $\mathcal{E} = \mathcal{O}_{\mathbb{P}^1} \oplus \mathcal{O}_{\mathbb{P}^1}$ and we are finished. In the second case we can find a non zero section $\widetilde{s} = \alpha 1 + \beta s$, which vanishes at some point and then we found a line sub bundle $\mathcal{O}_{\mathbb{P}^1}\widetilde{s}$ of degree > 0. This is a contradiction.

Hence we get a sequence

$$0 \longrightarrow \mathcal{O}_{\mathbb{P}^1} \longrightarrow \mathcal{E} \longrightarrow \mathcal{O}_{\mathbb{P}^1}(r) \longrightarrow 0$$

where now $r < 0$. We tensorize by $\mathcal{O}_{\mathbb{P}^1}(-r)$ and get

$$0 \longrightarrow \mathcal{O}_{\mathbb{P}^1}(-r) \longrightarrow \mathcal{E}(-r) \longrightarrow \mathcal{O}_{\mathbb{P}^1} \longrightarrow 0.$$

We have the section $1 \in H^0(\mathbb{P}^1, \mathcal{O}_{\mathbb{P}^1})$, which is everywhere $\neq 0$. Again we exploit the fact that $H^1(\mathbb{P}^1, \mathcal{O}_{\mathbb{P}^1}(-r)) = 0$ and the section 1 lifts to a section $s_0 \in H^0(\mathbb{P}^1, \mathcal{E}(-r))$ which does not have any zero. This gives us a splitting of the last exact sequence.
The general case follows by induction. Let rank $(\mathcal{E}) = d$. Again we find a line sub bundle $\mathcal{L} \subset \mathcal{E}$, such that $\deg(\mathcal{L}) = d_1$ is maximal, as before we conclude that $\mathcal{E}/\mathcal{L}$ is a vector bundle. Then we have $\mathcal{L} \xrightarrow{\sim} \mathcal{O}_{\mathbb{P}^1}(r_1)$

$$0 \longrightarrow \mathcal{L} \longrightarrow \mathcal{E} \longrightarrow \mathcal{E}/\mathcal{L} \longrightarrow 0$$

and our induction hypothesis implies

$$\mathcal{E}/\mathcal{L} \quad \simeq \quad \bigoplus\nolimits_{\nu=2}^{d} \mathcal{L} \xrightarrow{\sim} \bigoplus\nolimits_{\nu=2}^{d} \mathcal{O}_{\mathbb{P}^1}(r_\nu)$$

We claim that $r_1 \geq r_\nu$ for all $\nu \geq 2$. Assume we find an index ν_0 s.t. $r_{\nu_0} > r_1$, then we consider the rank 2 bundle $\mathcal{E}' \subset \mathcal{E}$, which is the inverse image of the line sub bundle $\mathcal{O}_{\mathbb{P}^1}(r_{\nu_0})$ in $\mathcal{E}/\mathcal{L}$. This bundle decomposes

$$\mathcal{E}' = \mathcal{O}_{\mathbb{P}^1}(a) \oplus \mathcal{O}_{\mathbb{P}^1}(b)$$

where $a + b = r_1 + r_{\nu_0} = \deg(\mathcal{E}')$. One of the summands has to map non trivially to $\mathcal{E}'/\mathcal{L} = \mathcal{O}_{\mathbb{P}^1}(r_{\nu_0})$. If this is $\mathcal{O}_{\mathbb{P}^1}(a)$, then we conclude $a \leq r_{\nu_0}$. We cannot have equality because then $\deg \mathcal{O}_{\mathbb{P}^1}(a) = r_{\nu_0}$, and this contradicts the choice of $\mathcal{L}$. But then $b > r_1$ and this is again produces a sub bundle of degree $> r_1$, a contradiction.

Now we show that the sequence

$$0 \longrightarrow \mathcal{L} \longrightarrow \mathcal{E} \longrightarrow \mathcal{E}/\mathcal{L} \longrightarrow 0$$

must split. Basically we argue as in the rank 2 case but we modify our argument slightly. We can cover $\mathbb{P}^1$ by affines U_1, U_2 such that on these affines we have sections $s_i : \mathcal{E}/\mathcal{L} \to \mathcal{E}$. These local sections may differ by a homomorphism

$$\varphi_{12} : \mathcal{E}/\mathcal{L} \mid U_1 \cap U_2 \longrightarrow \mathcal{L}$$

and φ_{12} gives us a class in $H^1(\mathbb{P}^1, \mathrm{Hom}(\mathcal{E}/\mathcal{L}, \mathcal{L}))$. But this cohomology group vanishes because $\mathrm{Hom}(\mathcal{E}/\mathcal{L}, \mathcal{L}) = \bigoplus_{v=2}^{r} \mathcal{O}_{\mathbb{P}^1}(d_1 - d_\nu)$. We can bound the cocycle and get the splitting.

We add an observation. Once we have

$$\mathcal{E} = \bigoplus_{\nu=1}^{r} \mathcal{O}_{\mathbb{P}^1}(d_\nu)$$

where $d_1 = d_2 = \cdots = d_{r_0} > d_{r_0+1} \ldots$ then we can tensorize by $\mathcal{O}_{\mathbb{P}^1}(-d_1 - a)$,$a > 0$ and we get $H^0(\mathbb{P}^1, \mathrm{Hom}(\mathcal{E} \otimes \mathcal{O}_{\mathbb{P}^1}(-d_1 - a)) = 0$. But if we tensorize by $\mathcal{O}_{\mathbb{P}^1}(-d_1)$ then we get

$$\mathcal{E} \otimes \mathcal{O}_{\mathbb{P}^1}(-d_1) = \mathcal{O}_{\mathbb{P}^1}^{r_0} \oplus \bigoplus_{\nu=r_0+1}^{r} \mathcal{O}_{\mathbb{P}^1}(d_1 - d_\nu)$$

where we assumed that $d_1 = \cdots = d_{\nu_0} = d_{\nu_0}$, and the other d_v are smaller. Then we conclude that

$$H^0(\mathbb{P}^1, \mathcal{E} \otimes \mathcal{O}_{\mathbb{P}^1}(-d_{\nu_0})) = k^{r_0}$$

and that these sections generate the sub bundle $\mathcal{O}_{\mathbb{P}^1}^{r_0}$. This implies that d_{ν_0} is determined by $\mathcal{E}$ and that the sub bundle

$$\bigoplus_{\nu=1}^{r_0} \mathcal{O}_{\mathbb{P}^1}(d_{r_0})$$

is unique.

This shows: If we order these numbers

$$d_1 = \cdots = d_{r_0} > d_{r_0+1} = \cdots = d_{r_1} > d_{r_1+1} \cdots,$$

then the resulting sequence of numbers is determined by $\mathcal{E}$ and that the flag

$$\bigoplus_{\nu=1}^{r_0} \mathcal{O}_{\mathbb{P}^1}(d_\nu) \subset \bigoplus_{\nu=1}^{r_1} \mathcal{O}_{\mathbb{P}^1}(d_\nu) \subset \cdots$$

is also determined by $\mathcal{E}$. □

The theorem above and the consequences are called Grothendieck's theorem, but it occurs already in [De-We], §22.

9.5 The Theorem of Riemann-Roch

We consider line bundles on a smooth, projective, absolutely irreducible curve C/k. Such a line bundle $\mathcal{L}$ has cohomology group $H^0(C,\mathcal{L}), H^1(C,\mathcal{L}), \ldots$, which are finite dimensional k-vector spaces. We shall see that $H^i(C,\mathcal{L}) = 0$ for $i \geq 2$ and we will give a formula for

$$\chi(C,\mathcal{L}) = \dim_k H^0(C,\mathcal{L}) - \dim_k H^1(C,\mathcal{L}).$$

The first observation is

$$\dim_k H^0(C,\mathcal{O}_C) = 1.$$

If $f \in H^0(C,\mathcal{O}_C)$ then we have $f^2, f^3, \ldots \in H^0(C,\mathcal{O}_C)$ and since this is a finite dimension vector space we see that f must be algebraic over k. If $f \notin k$ then we find that

$$k(C) \otimes_k \bar{k}$$

has zero divisors and this contradicts the assumption that C/k should be absolutely irreducible.

We define the *genus* of the curve as

$$g = \dim_k H^1(C,\mathcal{O}_C).$$

Now we can state

Theorem 9.5.1. Theorem of Riemann-Roch *(first version)*
For any line bundle $\mathcal{L}$ on C we have

$$\chi(C,\mathcal{L}) = \dim_k H^0(C,\mathcal{L}) - \dim_k H^1(C,\mathcal{L}) = \deg(\mathcal{L}) + 1 - g.$$

The proof is not too difficult, it is essentially the same as the proof in the case of Riemann surfaces in Volume I 5.1 . In the case of Riemann surfaces the main difficulty was to prove that $\dim_k H^1(C,\mathcal{L})$ is finite dimensional and this required difficult analytical arguments. But here it is easier and follows from our general results in the previous chapter.

The theorem is true for $\mathcal{L} = \mathcal{O}_C$ by definition. If we have a closed point $\mathfrak{p}$ then we have an exact sequence of sheaves

$$0 \longrightarrow \mathcal{O}_C \longrightarrow \mathcal{O}_C(\mathfrak{p}) \longrightarrow \mathcal{O}_C(\mathfrak{p})/\mathcal{O}_C \longrightarrow 0.$$

The quotient is a skyscraper sheaf, its only non-zero stalk is the stalk at $\mathfrak{p}$ and there it is a one dimensional $k(\mathfrak{p})$-vector space.
Since $\mathcal{L}$ is locally free we get an exact sequence by tensoring by $\mathcal{L}$

$$0 \longrightarrow \mathcal{L} \longrightarrow \mathcal{L} \otimes \mathcal{O}_C(\mathfrak{p}) \longrightarrow \mathcal{O}_C(\mathfrak{p})/\mathcal{O}_C \longrightarrow 0.$$

We observe that

$$\dim_k(H^0(C,\mathcal{O}_C(\mathfrak{p})/\mathcal{O}_C)) = [k(\mathfrak{p}) : k]$$

and

$$H^i(C,\mathcal{O}_C(\mathfrak{p})/\mathcal{O}_C)) = 0 \quad \text{for} \quad i \geq 1.$$

We write the long exact sequence in cohomology

$$\begin{aligned} 0 \longrightarrow H^0(C,\mathcal{L}) \longrightarrow H^0(C,\mathcal{L} \otimes \mathcal{O}_C(\mathfrak{p})) \longrightarrow H^0(C,\mathcal{O}_C(\mathfrak{p})/\mathcal{O}_C) \longrightarrow \\ \longrightarrow H^1(C,\mathcal{L}) \longrightarrow H^1(C,\mathcal{L} \otimes \mathcal{O}_C(\mathfrak{p})) \longrightarrow 0 \end{aligned}$$

and and in higher degrees we get (See remark 2 after the proof of the coherence theorem)

$$H^i(C,\mathcal{L}) \xrightarrow{\sim} H^i(C,\mathcal{L} \otimes \mathcal{O}_C(\mathfrak{p})) \qquad \text{for all } i \geq 2.$$

From this we get easily that our assertions are true for $\mathcal{L} \otimes \mathcal{O}_C(\mathfrak{p})$ if and only if they are true for $\mathcal{L}$ itself. The rest is more or less clear. We have seen that $H^i(C,\mathcal{O}_C(\infty\mathfrak{p})) = 0$ for $i \geq 2$ in the section on cohomology of coherent sheaves (see exercise below). Since we know that our line bundle is isomorphic to some $\mathcal{O}_C(D)$ the theorem follows. □

Now it is an easy argument in homological algebra that we have a Riemann-Roch formula for vector bundles:

$$\chi(C,\mathcal{E}) = \dim_k H^0(C,\mathcal{E}) - \dim_k H^1(C,\mathcal{E}) = \deg(\mathcal{E}) + \text{Rank}(\mathcal{E})(1-g).$$

(Use the flag to prove it by induction)

Exercise 38. Show that the curve $C \setminus \{\mathfrak{p}\}$ is affine by applying the strategy outlined in 8.1.16.

9.5.1 Differentials and Residues

On our curve C/k we have two privileged line bundles. The first one is the structure sheaf $\mathcal{O}_C$ and the other one is the sheaf $\Omega^1_{C/k} = \Omega^1_C$ of differentials. It is called the **canonical bundle**. We know the degree of $\mathcal{O}_C$ and we computed the cohomology groups $H^0(C,\mathcal{O}_C) = k$ and $\dim_k H^1(C,\mathcal{O}_C) = g$ where the second assertion is tautological. Our next aim is to show that

$$\deg(\Omega^1_{C/k}) = 2g-2, \dim_k H^0(C,\Omega^1_C) = g \tag{9.8}$$

and that we have a canonical isomorphism: The **global residue map**

$$\mathrm{Res} : H^1(C,\Omega^1_{C/k}) \xrightarrow{\sim} k. \tag{9.9}$$

Here we mean by canonical that this map is consistent with the trace map, which we defined for separable morphisms. If we have C_1/k, C_2/k and if $f : C_1 \to C_2$ is a finite separable morphism, then we defined the $\mathrm{tr}_{C_1/C_2} : f_*(\Omega^1_{C_1/k}) \longrightarrow \Omega^1_{C_2/k}$. We require that the resulting k-linear map yields a commutative diagram

$$\begin{array}{ccc} H^1(C_1,\Omega^1_{C_1/k}) & \xrightarrow{\mathrm{tr}} & H^1(C_2,\Omega^1_{C_2/k}) \\ \mathrm{Res} \searrow & & \swarrow \mathrm{Res} \\ & k & \end{array} \tag{9.10}$$

Of course we also want compatibility with base change in the obvious sense. The existence and "uniqueness" of this form is not so easy to prove, it will take us the next 14 pages until we reach this goal. This may be considered a too long way, but I think that during our journey we will gain a lot of insights, which provide a deeper understanding of the Riemann-Roch theorem. Our approach to prove these assertions is essentially already in [De-We], for this compare the beautiful exposition by W.-D. Geyer in [Sch].

Let us accept for a moment the existence of the isomorphism Res for all curves C/k over any field whatsoever. Then we can take any line bundle $\mathcal{L}$ on C and get a pairing

$$H^0(C,\mathcal{L}) \times H^1(C,\mathcal{L}^{-1} \otimes \Omega^1_{C/k}) \longrightarrow H^1(C,\Omega^1_{C/k}) = k,$$

and it will be proved that this pairing is non degenerate (see theorem 9.5.4). If we apply the first version of the Riemann-Roch theorem to $\mathcal{L} = \mathcal{O}_C$ and $\mathcal{L} = \Omega^1_{C/k}$ and exploit the non degeneracy of the pairing then it becomes clear that

$$H^0(C,\Omega^1_{C/k}) = H^1(C,\mathcal{O}_{C/k})^\vee$$

and

$$\dim_k(H^0(C,\Omega^1_{C/k}) = g, \deg(\Omega^1_{C/k}) = 2g-2.$$

We give an outline of the construction of the above pairing. We recall how we computed $H^1(C,\mathcal{O}_C)$. We looked at effective divisors

$$D = \sum n_{\mathfrak{p}}\mathfrak{p}$$

with $n_\mathfrak{p} \geq 0$, and at the resulting exact sequence

$$0 \longrightarrow \mathcal{O}_C \longrightarrow \mathcal{O}_C(D) \longrightarrow \mathcal{L} \otimes \mathcal{O}_C(D)/\mathcal{O}_C \longrightarrow 0.$$

The quotient sheaf $\mathcal{O}_C(D)/\mathcal{O}_C$ is a collection of Laurent expansions at the points in the support of D. We introduced the notation $\mathbb{L}(D)$ for it. For any closed point $\mathfrak{p}$ in $|D|$ the stalk of $\mathbb{L}(D)$ at $\mathfrak{p}$ is

$$\{f \in k(C) | \operatorname{ord}_\mathfrak{p}(f) \geq -n_\mathfrak{p}\} = \pi^{-n_\mathfrak{p}} \mathcal{O}_{C,\mathfrak{p}}/\mathcal{O}_{C,\mathfrak{p}},$$

where $\pi_\mathfrak{p}$ is a generator of the maximal ideal $\mathfrak{p} \subset \mathcal{O}_{C,\mathfrak{p}}$. We may (or may not) pass to the direct limit

$$\varinjlim_D \mathcal{O}_C(D)/\mathcal{O}_C = \mathbb{L}$$

where $\mathbb{L}$ is the sheaf of all Laurent expansions: For an open set U its sections are given by

$$\mathbb{L}(U) = \bigoplus_{\mathfrak{p} \in U,\ \mathfrak{p} \text{ closed}} K/\mathcal{O}_{C,\mathfrak{p}}.$$

We introduce the completions of the local rings $\lim_{\leftarrow}(\mathcal{O}_{C,\mathfrak{p}}/\mathfrak{p}^n) = \widehat{\mathcal{O}}_{C,\mathfrak{p}}$ and the quotient fields of these completions $\widehat{K}_\mathfrak{p}$. We define the module of differentials $\widehat{\Omega}^1_{C_1,\mathfrak{p}} = \Omega^1_{C,\mathfrak{p}} \otimes \widehat{\mathcal{O}}_{C,\mathfrak{p}}$. Of course this is again a free module of rank one over $\widehat{\mathcal{O}}_{C,\mathfrak{p}}$ and it is not difficult to see that it is a universal separated module for continuous differentials. We may also introduce the infinitesimal meromorphic differentials $\widehat{\Omega}^1_{C,\mathfrak{p}} = \Omega^1_{C,\mathfrak{p}} \otimes \widehat{K}_\mathfrak{p}$.
Now we choose an effective divisor with $\deg(D) >> 0$ and we consider the long exact sequence in cohomology. We have seen that $H^1(C,\mathcal{O}_C(D)) = 0$ provided $\deg(D) >> 0$ and hence we get

$$0 \longrightarrow H^0(C,\mathcal{O}_C) \longrightarrow H^0(C,\mathcal{O}_C(D)) \longrightarrow H^0(C,\mathcal{O}_C(D)/\mathcal{O}_C) \longrightarrow H^1(C,\mathcal{O}_C) \longrightarrow 0$$

if $\deg(D) >> 0$, we could as well look at the same sequence where we passed to the limit over all D.
Hence we can interpret $H^1(C,\mathcal{O}_C)$ as a quotient of a space of Laurent expansions modulo the space of those Laurent expansions, which come from a meromorphic function. This means that have to understand the obstruction for a collection of Laurent expansions

$$\xi \in \oplus_\mathfrak{p} \pi_\mathfrak{p}^{-n_\mathfrak{p}} \mathcal{O}_{C,\mathfrak{p}}/\mathcal{O}_{C,\mathfrak{p}}$$

to come from a meromorphic function. The point is that the holomorphic differentials produce such obstructions. This will be made precise in the proposition below.

We identify

$$H^0(C,\mathbb{L}(D)) \simeq \mathbb{L}(D)(C).$$

For any $\mathfrak{p} \in |D|$ we have

$$\mathbb{L}(D)_\mathfrak{p} = \pi_\mathfrak{p}^{-n_\mathfrak{p}} \mathcal{O}_{C,\mathfrak{p}}/\mathcal{O}_{C,\mathfrak{p}}.$$

Our next goal is to define a **(local) residue map** : For any prime $\mathfrak{p}$ we want to define

$$\operatorname{res}_\mathfrak{p} : \lim \pi_\mathfrak{p}^{-n} \mathcal{O}_{C,\mathfrak{p}}/\mathcal{O}_{C,\mathfrak{p}} \otimes \Omega^1_{C,\mathfrak{p}} = \lim \pi^{-n} \Omega^1_{C,\mathfrak{p}}/\Omega^1_{C,\mathfrak{p}} \xrightarrow{\operatorname{res}_\mathfrak{p}} k, \tag{9.11}$$

the definition of this residue map requires some work especially if our ground field is of positive characteristic. Once we have this map, then we can define

$$\begin{aligned} \mathbb{L}(D)\otimes H^0(C,\Omega^1_C) &\longrightarrow k \\ \underline{\xi}\otimes\omega &\longrightarrow \sum_{\mathfrak{p}\in|D|} \mathrm{res}_{\mathfrak{p}}(\xi_{\mathfrak{p}}\otimes\omega) \end{aligned}$$

and we want the following property of this collection of maps:

Proposition 9.5.2. *An element $\underline{\xi}\in\mathbb{L}(D)$ is the Laurent expansion of an element $f\in H^0(C,\mathcal{O}_C(D))$ if and only if*

$$\sum_{\mathfrak{p}} \mathrm{res}_{\mathfrak{p}}(f\omega)=0$$

for all holomorphic differentials ω.

It is clear that this says that we get a non degenerate pairing

$$H^1(C,\mathcal{O}_C)\times H^0(C,\Omega^1_C)\longrightarrow k.$$

The elements

$$\omega'=f\omega\in H^0(C,\Omega^1_C(-D))$$

are called meromorphic differentials and part of the assertion above is that

$$\sum_{\mathfrak{p}} \mathrm{res}_{\mathfrak{p}}(\omega')=0$$

for any meromorphic differential. This ends the outline.

We come to the definition of the residue map 9.11 We notice that we can replace $\mathcal{O}_{C,\mathfrak{p}}$ by the completion $\widehat{\mathcal{O}}_{C,\mathfrak{p}}$ because

$$\pi_{\mathfrak{p}}^{-n}\mathcal{O}_{C,\mathfrak{p}}/\mathcal{O}_{C,\mathfrak{p}}=\pi_{\mathfrak{p}}^{-n}\widehat{\mathcal{O}}_{C,\mathfrak{p}}/\widehat{\mathcal{O}}_{C,\mathfrak{p}}.$$

At first we define this residue map only for rational points, i.e. those, for which $k(\mathfrak{p})=k$. We change the notation slightly and denote the local parameter by X, then we have (see p. 186)

$$\widehat{\mathcal{O}}_{C,\mathfrak{p}}=k[[X]],$$

the quotient field is

$$k((X))=k[[X]]\left[\frac{1}{X}\right],$$

and our differentials are of the form

$$\omega=f(X)dX=\left(\frac{a_{-n}}{X^n}+\frac{a_{n-1}}{X^{n-1}}+\cdots+\frac{a_{-1}}{X}+\cdots\right)dX.$$

Of course we want $\mathrm{res}_{\mathfrak{p}}(\frac{dX}{X})=1$ and hence we try the definition

$$\mathrm{res}_{\mathfrak{p}}(\omega)=\mathrm{res}_{\mathfrak{p}}(f\otimes dX)=a_{-1}.$$

Now we encounter a somewhat unexpected problem. We have to show that this definition does not depend on the choice of the local parameter X. Let us replace X by another local parameter

$$Y = tX + u_2X^2 + \cdots + u_nX^n + \cdots = P(X)$$

where $t \neq 0$. Then

$$\omega = \left(\frac{a'_{-n}}{Y^n} + \cdots + \frac{a'_{-1}}{Y} \cdots\right) dY,$$

and we want to show (invariance of the residue)

$$a_{-1} = a'_{-1} \tag{9.12}$$

This is indeed the case, but surprisingly difficult to prove.
Our first attempt is the naive one. If we want to prove the invariance of the residue, we have to write X as a power series in Y

$$X = \tau Y + v_2Y^2 + \cdots + v_mY^m + \cdots = Q(Y)$$

where $\tau = t^{-1}$ and the v_μ/τ are polynomials in the u_ν/t. Then we have to expand

$$\frac{dX}{X^m} = \frac{\tau + 2v_2Y + \cdots mv_mY^{m-1} + \cdots}{\tau^mY^m(1 + \frac{v_2}{\tau}Y + \frac{v_m}{\tau} \cdots Y^{m-1})^m}\,dY = \left(\cdots \frac{a'_{-1}}{Y} \cdots\right)\,dY.$$

We have to show that in this expansion the coefficient a'_{-1} of $\frac{dY}{Y}$ is

(i) equal to one if $m = 1$

(ii) it is zero if $m \geq 2$.

The first assertion is clear because the factor τ cancels. But the second assertion is not so clear. (I recommend to do the calculation for some small values of m.)
It is clear that our coefficient a_{-1} is a polynomial

$$P_m\left(\frac{v_1}{\tau}, \cdots, \frac{v_m}{\tau}\right)$$

times τ^{1-m}. The polynomial has coefficients in k and we have to show that this polynomial is identically zero if $m \geq 2$. (We can view the $\frac{v_\nu}{\tau}$ as indeterminates.) But it seems to be difficult to do this by a direct calculation. We use a trick and prove the invariance of the residue by an argument, which goes back to H. Hasse.
If the characteristic of k is equal to zero, then we can write for $m \geq 2$

$$\frac{dX}{X^m} = -\frac{1}{m-1}dX^{1-m}$$

but then is is equal to

$$-\frac{1}{m-1}dQ(Y)^{1-m} = -d(\frac{1}{\tau^{m-1}Y^{m-1}} + b_{-m+2}\frac{1}{Y^{m-2}} + \cdots + b_{-1}\frac{1}{Y} + b_0 + \dots),$$

and here the coefficient for $1/Y$ is zero. This argument fails if $\mathrm{char}(k) = p > 0$. For instance the differential $\frac{dX}{X^{p+1}}$ is not of the form $dG(X)$.

Recall that we have to show that the polynomial $P_m\left(\frac{v_1}{\tau},\cdots,\frac{v_m}{\tau}\right)$ vanishes identically. If we start from the case $k=\mathbb{Q}$, then we see easily that $P_m\left(\frac{v_1}{\tau},\cdots,\frac{v_m}{\tau}\right)$ has integer coefficients, hence it lies in

$$\mathbb{Z}\left[\frac{v_1}{\tau},\cdots,\frac{v_m}{\tau}\right].$$

Moreover it is obvious that the corresponding polynomial in arbitrary characteristic is obtained by reducing the coefficients $\mod p$, it lies in

$$\mathbb{F}_p\left[\frac{v_1}{\tau},\cdots,\frac{v_m}{\tau}\right]\subset k\left[\frac{v_1}{\tau},\cdots,\frac{v_m}{\tau}\right].$$

Our above argument shows that the polynomial in $\mathbb{Z}\left[\frac{v_1}{\tau},\cdots,\frac{v_m}{\tau}\right]$ must become the zero polynomial in $\mathbb{Q}\left[\frac{v_1}{\tau},\cdots,\frac{v_m}{\tau}\right]$, but then it must be identically zero itself. This proves (ii) in general. We proved the existence of the local residue map.

9.5.2 The special case $C=\mathbb{P}^1/k$

We observe that our problem to construct $\mathrm{res}_\mathfrak{p}$ is purely local. If we want to construct the local residue map 9.11, then we observed that we can pass to the completion of the local ring, in other words we consider the formal scheme at the point. But this formal scheme does not "know", on which curve it is lying, hence we consider first the special case that our curve is $\mathbb{P}^1/k$.
This allows to give a second construction of the local residue, which uses global arguments. I like better because it gives more insight. It will give the proposition 9.5.2 and at the same time, we will see that the proposition guides us to the definition of the residue map. We will use the fact that for differentials, which have only a first order pole the invariance of the residue is obvious. (Assertion (i) above)

We consider the special case $C/k=\mathbb{P}^1_k$ We have seen $H^1(\mathbb{P}^1,\mathcal{O}_{\mathbb{P}^1})=0$ therefore, $g=0$, we know that

$$\Omega^1_{\mathbb{P}^1}\simeq\mathcal{O}_{\mathbb{P}^1}(-2)$$

(see 8.26) and hence

$$\deg(\Omega^1_{\mathbb{P}^1})=0-2=-2.$$

We construct the canonical isomorphism

$$\mathrm{Res}:H^1(\mathbb{P}^1,\Omega^1_{\mathbb{P}^1})\longrightarrow k.$$

To do this we apply a principle, which also works in the general situation. So let us return for one moment to the case of an arbitrary smooth, projective and absolutely irreducible curve $C=C/k$. We will show later:

Proposition 9.5.3. *For any non zero effective divisor D we have $H^1(C,\Omega^1_C(D))=0$ and especially for any k rational point $a\in C(k)$ we have an isomorphism*

$$\delta_a:H^0(C,\mathbb{L}(a)\otimes\Omega^1_C)\xrightarrow{\sim}H^1(C,\Omega^1_C),$$

where the map δ_a is of course the boundary homomorphism.

This proposition is easy for $C/k = \mathbb{P}^1/k$. For any point $a \in \mathbb{P}^1(k)$ we have $\Omega^1_{\mathbb{P}^1}(a) \simeq \mathcal{O}_{\mathbb{P}^1}(-1)$ and therefore, $H^0(\mathbb{P}^1,\Omega^1_{\mathbb{P}^1}(a)) = 0$, this implies the proposition for $C = \mathbb{P}^1_k$. So we pick a point $a \in \mathbb{P}^1(k)$ and consider the sequence

$$0 \to \Omega^1_{\mathbb{P}^1} \to \Omega^1_{\mathbb{P}^1}(a) \to \mathbb{L}(a) \otimes \Omega^1_{\mathbb{P}^1} \to 0,$$

which gives us the isomorphism

$$\delta_a : \mathbb{L}(a) \otimes \Omega^1_{\mathbb{P}^1} \xrightarrow{\sim} H^1(\mathbb{P}^1,\Omega^1_{\mathbb{P}^1})$$

We noticed that for differentials with a first order pole we have defined the map

$$\mathrm{res}_a : \mathbb{L}(a) \otimes \Omega^1_{\mathbb{P}^1} \longrightarrow k,$$

and for $C/k = \mathbb{P}^1_k$ we can define the global residue map:

$$\mathrm{Res} = \mathrm{res}_a \circ \delta_a^{-1}.$$

Of course something would be wrong if we did not have the problem of well definedness again. Let us pick a point $b \in \mathbb{P}^1(k)$, which is different from a. Then we get a diagram

$$\begin{array}{ccc} \mathbb{L}(a) \otimes \Omega^1_{\mathbb{P}^1} & & \\ & \searrow^{i_a} & \\ & & \mathbb{L}(a+b) \otimes \Omega^1_{\mathbb{P}^1} \xrightarrow{\delta} H^1(\mathbb{P}^1,\Omega^1_{\mathbb{P}}) \\ & {}_{i_b}\nearrow & \\ \mathbb{L}(b) \otimes \Omega^1_{\mathbb{P}^1} & & \end{array}$$

The map δ has a nontrivial kernel and it is clear that this kernel is spanned by the form

$$\omega_{a,b} = \frac{dx}{x-a} - \frac{dx}{x-b},$$

which is a generator of $H^0(\mathbb{P}^1,\Omega^1_{\mathbb{P}^1}(a+b))$, this differential is holomorphic at ∞! Clearly $\delta_a = \delta \circ i_a$ and $\delta_b = \delta \circ i_b$ and hence

$$\mathrm{res}_a \circ \delta_a^{-1} = \mathrm{res}_b \circ \delta_b^{-1}.$$

This proves that $\mathrm{Res} : H^1(\mathbb{P}^1_k,\Omega^1_{\mathbb{P}^1_k}) \longrightarrow k$ does not depend on the choice of a.
Now we consider the special point $\infty \in \mathbb{P}^1(k)$. We have

$$\mathrm{Spec}(k[X]) \cup \mathrm{Spec}(k[Y]) = \mathbb{P}^1$$

where $XY = 1$ and ∞ is given by $0 \in \mathrm{Spec}(k[Y])$, the element Y is a local parameter at ∞. We consider the exact sequence

$$0 \to \Omega^1_{\mathbb{P}^1} \to \Omega^1_{\mathbb{P}^1}(n\infty) \to \mathbb{L}(n\infty) \otimes \Omega^1_{\mathbb{P}^1} \to 0$$

and get

$$\mathbb{L}(n\infty) \otimes \Omega^1_{\mathbb{P}^1} \xrightarrow{\delta_\infty} H^1(\mathbb{P}^1,\Omega^1_{\mathbb{P}^1}) \xrightarrow{\mathrm{Res}} 0.$$

Hence we try a new definition of the local residue map at ∞: We put

$$\mathrm{res}'_\infty = \mathrm{Res} \circ \delta_\infty$$

for the local residue map.

We want to give local formula for $\operatorname{res}'_\infty$, this is easy. We observe that the elements of $\mathbb{L}(n\infty)\otimes\Omega^1_{\mathbb{P}^1}$ can be written as

$$\omega = +\left(\frac{a_1}{Y}+\frac{a_2}{Y^2}\cdots\frac{a_n}{Y^n}\right)\otimes dY = -(a_1X+\cdots a_nX^n)\otimes\frac{dX}{X^2}.$$

By definition we have $\operatorname{res}'_\infty(\omega)=0$ if and only if the element ω is the image of

$$H^0(\mathbb{P}^1,\Omega^1_{\mathbb{P}^1}(n\infty)) \longrightarrow \mathbb{L}(na)\otimes\Omega^1_{\mathbb{P}^1},$$

and this is the case if and only if $a_1=0$. We can also say that the form

$$-(a_1X+\cdots+a_nX^n)\frac{dX}{X^2}$$

is holomorphic on $\operatorname{Spec}(k[X])\setminus\{0\}$ and has a pole of order ≤ 1 at zero with residue $-a_1$ at 0. Hence it is clear that

$$\operatorname{res}'_\infty(\omega)=a_1,$$

and this is our old definition of the residue with respect to the parameter Y at infinity.

Now we give a second proof that the residue map the residue does not depend on the choice of the local parameter.
We have seen that it suffices to show that the differentials

$$\frac{dY}{Y^m}\in\mathbb{L}(n\infty)\otimes\Omega^1_{\mathbb{P}^1},$$

which have residue zero for $m>1$ still have residue zero if we make a change of local parameters

$$Y = tU+v_2U^2\cdots+v_rU^r+\cdots$$

where $t\neq 0$ and the right hand side is a power series in U.

But we know how to characterize the elements in $\mathbb{L}(n\infty)\otimes\Omega^1_{\mathbb{P}^1}$, which have residue zero. These are the elements, which lie on the image of

$$H^0(\mathbb{P}^1,\Omega^1_{\mathbb{P}^1}(n\infty)) \longrightarrow \mathbb{L}(n\infty)\otimes\Omega^1_{\mathbb{P}^1}.$$

We approximate the power series, which defines the change of parameters by a polynomial of degree $r\geq n$, only the first n coefficients are relevant. Since the degree can be chosen to be larger, we may assume that the polynomial

$$F(U)=tU+v_2U^2+\cdots+v_rU^r$$

is separable. It provides an inclusion $k[Y]\subset k[U]$ and a morphism

$$\Phi_F:\mathbb{P}^1\longrightarrow\mathbb{P}^1$$

where the first $\mathbb{P}^1$ is $\operatorname{Spec}(k[U])\cup\operatorname{Spec}(k[U^{-1}])$ and the second one is $\operatorname{Spec}(k[X])\cup\operatorname{Spec}(k[X^{-1}])$. We can consider the pull back $\Phi_F^*(\frac{dY}{Y^m})=\omega$, this form has poles of order m in the point $U=0$ and in the other $r-1$ points $u_2\cdots u_r$ where F vanishes. (We assume that F is separable.) In the point $u=0$ its Laurent expansion is

$$\frac{F'(U)dU}{F(U)^m},$$

and we want to show that the coefficient at $\frac{1}{U}dU$ is zero.

This differential has some more poles at points $a_2 \cdots a_r$. The total order of the pole is rm. Hence ω must have a divisor of zeroes Z of degree $rm - 2$ because the degree of Ω^1 is -2. It is a global section in the line bundle

$$\omega \in H^0(\mathbb{P}^1, \Omega^1_{\mathbb{P}^1}(Z - ma_2 - \cdots - ma_r)).$$

If we multiply ω by a section $f \in H^0(\mathbb{P}^1, \mathcal{O}^1_{\mathbb{P}^1}(-Z + ma_2 \cdots + ma_r))$, then we get a meromorphic form $f\omega$, which now has only one pole at $U = 0$, the other poles are cancelled.

Of course the multiplication by f changes the expansion at $U = 0$, but we still have the choice of f at our disposal. The functions f are holomorphic at $u = 0$, we look at its expansion at $U = 0$, i.e. we map f to $\mathcal{O}_{\mathbb{P}^1,0}/(U)^{m-1}\mathcal{O}_{\mathbb{P}^1,0}$, then we get an exact sequence

$$0 \to \mathcal{O}_{\mathbb{P}^1}(-Z+ma_2+\cdots+ma_r+(m-1)(0)) \to \mathcal{O}_{\mathbb{P}^1}(-Z+ma_2+\cdots+ma_r) \to \mathcal{O}_{\mathbb{P}^1}/(U)^{m-1}\mathcal{O}_{\mathbb{P}^1} \to$$

and the line bundle on the left has degree -1, hence it has no cohomology. This yields that

$$H^0(\mathbb{P}^1, \mathcal{O}_{\mathbb{P}^1}(-Z + ma_2 + \cdots + ma_r) \xrightarrow{\sim} \mathcal{O}_{\mathbb{P}^1}/(U)^{m-1}\mathcal{O}_{\mathbb{P}^1}$$

is an isomorphism. This means that we find an $f \in H^0(\mathbb{P}^1, \mathcal{O}_{\mathbb{P}^1}(-Z + ma_2 + \cdots + ma_r))$ such that its expansion at $u = 0$ with respect to U is

$$f(U) = 1 + a_m U^m + \cdots.$$

The meromorphic differential $f\omega$ on $\mathbb{P}^1 = \mathrm{Spec}(k[U]) \cup \mathrm{Spec}(k[U^{-1}])$ has only one pole at $U = 0$ and its polar part of the Laurent expansion is the same as the polar part of the Laurent expansion of ω. We write the same expansion as above, but now in the variables $U, V = U^{-1}$

$$f\omega = +\left(\frac{a'_1}{U} + \frac{a'_2}{U^2} \cdots \frac{a'_m}{U^m}\right) \otimes dU = -(a'_1 V + \cdots a'_m V^m) \otimes \frac{dV}{V^2}$$

and since the only pole is at $U = 0$ we conclude $a'_1 = 0$.
This finishes the second proof of the invariance of the residue, it proves 9.8 and 9.9. for the case $C/k = \mathbb{P}^1_k/k$.

Historical and heuristic remark:
1) If our ground field is the field of complex numbers then the set of complex points $S = C(\mathbb{C})$ carries the structure of a compact Riemann surface. (Complex manifold in dimension one it has a natural orientation!) . In this case everything is much easier. A meromorphic differential ω, which has a pole at a point P can be written locally as

$$\omega = f(z)dz$$

where z is a uniformizing element at P and f is meromorphic. In this case we look at a small disc around P and we know that

$$\frac{1}{2\pi i}\oint \omega = \mathrm{res}_p(\omega)$$

where the left hand side is defined intrinsically as the integral over the boundary of the disc taken counterclockwise and where the right hand side is computed in terms of the Laurent expansion of $f(z)$. Hence it is clear that the right hand side does not depend on the choice of the local parameter z.
A meromorphic differential has a finite number of poles and we can compute the sum of the residues at these poles as the sum of the integrals over the boundary of these discs. Then we consider the complement U of the open discs. This is a compact manifold with boundary $\partial U=$ union of the boundaries of the discs. We can apply Stokes theorem, which says the the integral over this boundary is equal to the integral of $d\omega$ over U. But ω is holomorphic on U hence we have $d\omega = 0$, and we see that the sum over the residues is zero.

2) It may be interesting to look into the original paper by Riemann and to find out what Riemann and Roch actually proved.
For them cohomology did not exist, hence they could not define $H^1(S,\mathcal{O}_S)$ and $g = \dim H^1(S,\mathcal{O}_S)$. On the other hand they could study the cokernel of

$$H^0(S,\mathcal{O}_S(D)) \longrightarrow \mathbb{L}(D),$$

which is isomorphic to $H^1(S,\mathcal{O}_S)$ if $\deg(D) >> 0$.

Of course it was clear to them that the holomorphic 1-forms produced linear forms on this cokernel. They could define $g = \dim H^0(S,\Omega^1_S)$ and it also was clear to them that $2g$ is the the first Betti- number. It was also proved that the common kernel of linear forms produced by the differentials described the image of the map above. This is the content of Serre duality, which is much deeper. Therefore it is seems to be clear that Riemann and Roch proved more that just the first version of Riemann-Roch.

9.5.3 Back to the general case

We have to understand the sheaf $\Omega^1_{C/k}$ and to prove the two fundamental equalities 9.8. To do this we consider a finite morphism

$$\pi : C \longrightarrow \mathbb{P}^1,$$

which is induced by a choice of a function $f \in k(C)$, for which $k(C)/k(f)$ is separable. (See proposition 9.3.2 and the considerations following it.)

In this case the functor π_*, which sends coherent sheaves on C to coherent sheaves on $\mathbb{P}^1$, is acyclic because the fibers are finite. Hence we know that for any coherent sheaf $\mathcal{F}$ on C

$$H^\bullet(C,\mathcal{F}) \simeq H^\bullet(\mathbb{P}^1,\pi_*(\mathcal{F})).$$

For any line bundle $\mathcal{L}$ on C the bundle $\pi_*(\mathcal{L})$ is a vector bundle of rank $d = \deg(\pi)$ over $\mathbb{P}^1$.
We apply the Dedekind-Weber-Grothendieck theorem to this vector bundle and write

$$\pi_*(\mathcal{L}) = \mathcal{O}_{\mathbb{P}^1}(a_1) \oplus \cdots \oplus \mathcal{O}_{\mathbb{P}^1}(a_d)$$

with $a_1 \geq a_2 \cdots \geq a_d$, actually this is also the approach in [De-We]. If $\mathcal{L} = \mathcal{O}_C$ we have $H^0(C,\mathcal{O}_C) = k$ and this implies

$$a_1 = 0 \qquad \text{and} \qquad a_\nu < 0 \qquad \text{for all} \qquad \nu \geq 2.$$

Since by definition

$$\dim H^1(C,\mathcal{O}_C) = \dim H^1(\mathbb{P}^1,\pi_*(\mathcal{O}_C)) = g$$

we must have (see 8.2.5)

$$g = \sum_{\nu=2}^{d} (-a_\nu - 1) = -\sum_{\nu=2}^{d} a_\nu - (d-1).$$

We invoke the Riemann-Hurwitz formula (See 9.3.3)

$$\pi^*(\Omega^1_{\mathbb{P}^1}) \otimes \mathfrak{D}^{-1}_{C/\mathbb{P}^1} \xrightarrow{\sim} \Omega^1_C$$

By construction the vector bundle $\pi_*(\mathfrak{D}^{-1}_{C/\mathbb{P}^1})$ is dual to the bundle $\pi_*(\mathcal{O}_C)$. We wrote

$$\pi_*(\mathcal{O}_C) = \mathcal{O}_{\mathbb{P}^1} \oplus \bigoplus_{\nu=2}^{d} \mathcal{O}_{\mathbb{P}^1}(a_\nu),$$

hence we see that

$$\pi_*(\mathfrak{D}^{-1}_{C/\mathbb{P}^1}) = \mathcal{O}_{\mathbb{P}^1} \oplus \bigoplus_{\nu=2}^{d} \mathcal{O}_{\mathbb{P}^1}(-a_\nu)$$

and

$$\pi_*(\Omega^1_C) = \mathcal{O}_{\mathbb{P}^1}(-2) \oplus \bigoplus_{\nu=2}^{d} \mathcal{O}_{\mathbb{P}^1}(-a_\nu - 2).$$

Since the $a_\nu < 0$ we see $-a_\nu - 2 \geq -1$. We get

$$H^1(C,\Omega^1_C) \simeq H^1(C,\pi_*(\Omega^1_C)) = H^1(\mathbb{P}^1,\mathcal{O}_{\mathbb{P}^1}(-2)) \simeq k$$

(this is not yet what we really want) and the second of our fundamental equalities 9.8

$$\dim_k H^0(C,\Omega^1_C) = \sum_{v=2}^{d} (-a_\nu - 1) = g.$$

For any line bundle $\mathcal{L}$ on C and any effective divisor D we can consider the exact sequence

$$0 \longrightarrow \mathcal{L} \longrightarrow \mathcal{L}(D) \longrightarrow \mathbb{L}(D) \longrightarrow 0,$$

which yields the exact sequence

$$0 \longrightarrow \pi_*(\mathcal{L}) \longrightarrow \pi_*(\mathcal{L}(D)) \longrightarrow \pi_*(\mathbb{L}(D)) \longrightarrow 0$$

and $\pi_*(\mathbb{L}(D))$ is a torsion sheaf on $\mathbb{P}^1$, its space of section has

$$\dim_k \pi_*(\mathbb{L}(D)) = \deg D.$$

Hence we see that

$$\deg(\pi_*(\mathcal{L}(D)) = \deg(\pi_*(\mathcal{L})) + \deg D.$$

If we apply this to the exact sequence

$$0 \longrightarrow \mathcal{O}_C \longrightarrow \mathfrak{D}^{-1} \longrightarrow \mathfrak{D}^{-1}/\mathcal{O}_\mathcal{C} \longrightarrow 0$$

we get

$$\deg(\pi_*(\mathfrak{D}^{-1})) = \deg(\pi_*(\mathcal{O}_C)) + \dim_k(\mathfrak{D}^{-1}/\mathcal{O}).$$

On the other hand

$$\deg(\pi_*(\mathcal{O}_C)) + \deg(\pi_*(\mathfrak{D}^{-1})) = 0$$

because these bundle are dual to each other. Hence

$$\dim_k(\mathfrak{D}^{-1}/\mathcal{O}) = 2\deg(\pi_*(\mathfrak{D}^{-1})) = 2\left(-\sum_{\nu=2}^{d} a_\nu\right) = 2(g+d-1).$$

This can be read on C. We get

$$\deg(\mathfrak{D}^{-1}) = 2(g-d+1).$$

On the other hand we know that $\deg(\pi^*(\Omega^1_{\mathbb{P}^1})) = -2d$ and therefore, (the first of the fundamental equalities 9.8)

$$\deg(\Omega^1_C) = -2d + 2(g+d-1) = 2g-2.$$

The next step is to construct the canonical linear map

$$\mathrm{Res} : H^1(C,\Omega^1_C) \longrightarrow k.$$

To get this map we proceed in the same way as in the case $C = \mathbb{P}^1$. We assume that our curve has a rational point $a \in C(k)$ and consider the line bundle $\Omega^1_C(a)$. We claim that (this is again the proposition 9.5.2, which is not yet proved, but will be proved now)

$$H^1(C,\Omega^1_C(a)) = 0.$$

To see that this is so we choose a meromorphic function f such that df generates the differentials at a and we consider the map induced by f

$$\pi_f = \pi : C \longrightarrow \mathbb{P}^1,$$

it is étale at the point a. It is clear that the bundles

$$\pi_*(\mathcal{O}_C(-a)) \qquad \text{and} \qquad \pi_*(\mathfrak{D}^{-1}(a))$$

are dual to each other. We have

$$\pi_*(\mathcal{O}_C(-a)) \subset \pi_*(\mathcal{O}_C) = \mathcal{O}_{\mathbb{P}^1} \oplus \bigoplus_{v=2}^{d} \mathcal{O}_{\mathbb{P}^1}(-a_\nu),$$

and the degree of this subbundle drops by one. Since $\pi_*(\mathcal{O}_C(a))$ has no non trivial section, we conclude that

$$\pi_*(\mathcal{O}_C(a)) \simeq \mathcal{O}_{\mathbb{P}^1}(-1) \oplus \bigoplus_{v=2}^{d} \mathcal{O}_{\mathbb{P}^1}(-a_\nu)$$

and hence because of duality

$$\pi_*(\mathfrak{D}^{-1}(a)) = \mathcal{O}_{\mathbb{P}^1}(1) \oplus \bigoplus_{v=2}^{d} \mathcal{O}_{\mathbb{P}^1}(a_\nu).$$

This implies the claim

$$H^1(\mathbb{P}^1, \pi_*(\pi^*(\Omega^1_{\mathbb{P}^1}) \otimes \mathfrak{D}^{-1}(a))) = H^1(C, \Omega^1_C(a)) = 0.$$

We have the exact sequence

$$0 \longrightarrow \Omega^1_C \longrightarrow \Omega^1_C(a) \longrightarrow \mathbb{L}(a) \otimes \Omega^1_C \longrightarrow 0$$

and obtain the exact sequence in cohomology

$$H^0(C, \Omega^1_C) \to H^0(C, \Omega^1_C(a)) \to H^0(C, \mathbb{L}(a) \otimes \Omega^1_C) \xrightarrow{\delta_a} H^1(C, \Omega^1_C) \to 0.$$

The map δ_a must be an isomorphism, we have defined

$$\mathrm{res}_a : H^0(C, \mathbb{L}(a) \otimes \Omega^1_C) \xrightarrow{\sim} k$$

and define

$$\mathrm{Res} = \mathrm{res}_a \circ \delta_a^{-1}.$$

To see that this morphism does not depend on the choice of a, we choose a second point $b \in C(k)$. Now we consider the exact sequence

$$0 \to \Omega^1_C \to \Omega^1_C(a+b) \to \mathbb{L}(a+b) \otimes \Omega^1_C \to 0,$$

and find that there exists a 1-form $\omega'_{a,b}$ on $,C$ which has a simple pole at a and b and is holomorphic elsewhere. We choose an $f \in k(C)$ such that df generates Ω^1_C in the two points a,b and consider the resulting morphism

$$\pi = \pi_F : C \longrightarrow \mathbb{P}^1.$$

This morphism π induces isomorphisms between the completions $\widehat{\mathcal{O}}_{C,a} \simeq \widehat{\mathcal{O}}_{\mathbb{P}^1,\pi(a)}, \widehat{\mathcal{O}}_{C,b} \simeq \widehat{\mathcal{O}}_{\mathbb{P}^1,\pi(a)}$, and it is clear that

$$\begin{aligned} \mathrm{res}_a(\omega'_{a,b}) &= \mathrm{res}_{\pi(a)}\left(\mathrm{tr}_{C/\mathbb{P}^1}(\omega'_{a,b})\right) \\ \mathrm{res}_b(\omega'_{a,b}) &= \mathrm{res}_{\pi(b)}\left(\mathrm{tr}_{C/\mathbb{P}^1}(\omega'_{a,b})\right). \end{aligned}$$

Of course it is clear that $\mathrm{tr}_{C/\mathbb{P}^1}(\omega'_{a,b})$ is a non zero multiple of the 1-form $\omega_{\pi(a),\pi(b)}$, which we constructed on $\mathbb{P}^1$, hence we get

$$\mathrm{res}_a(\omega'_{a,b}) + \mathrm{res}_b(\omega'_{a,b}) = 0.$$

This argument shows that Res : $H^1(C,\Omega^1_C) \to k$ is well defined but it also shows that it is compatible with a map

$$\pi : C \longrightarrow \mathbb{P}^1$$

as above: By construction we habe the commutative diagram

$$\begin{array}{ccc} H^0(C,\mathbb{L}(a)\otimes\Omega^1_C) & \longrightarrow & H^1(C,\Omega^1_C) \\ \downarrow \mathrm{tr}_{C/\mathbb{P}^1} & & \downarrow \mathrm{tr}_{C/\mathbb{P}^1} \\ H^0(\mathbb{P}^1,\mathbb{L}(\pi(a))\otimes\Omega^1_{\mathbb{P}^1}) & \longrightarrow & H^1(\mathbb{P}^1,\Omega^1_{\mathbb{P}^1}), \end{array}$$

which then implies the commutativity

$$\begin{array}{ccc} H^1(C,\Omega^1_C) & & \\ & \searrow^{\mathrm{Res}} & \\ \downarrow \mathrm{tr}_{C/\mathbb{P}^1} & & k \\ & \nearrow^{\mathrm{Res}} & \\ H^1(\mathbb{P}^1,\Omega^1_{\mathbb{P}^1}) & & \end{array}$$

and then it is easy to derive the general compatibility for arbitrary separable $f : C_1 \to C_2$.

It is also clear that in the diagram above we can replace $\mathbb{L}(a)$ by $\mathbb{L}(na)$ with any $n > 0$, then we get a commutative diagram

$$\begin{array}{ccc} H^0(C,\mathbb{L}(na)\otimes\Omega^1_C) & \xrightarrow{\delta} & H^1(C,\Omega^1_C) \\ \searrow^{\mathrm{res}_a} & & \swarrow^{\mathrm{Res}} \\ & k & . \end{array}$$

If we assume that k is algebraically closed, then we always find rational points. Hence we see that under this assumption we have $H^1(C,\Omega_C(D)) = 0$ for any effective divisor $D \neq 0$. Therefore we get the diagram

$$\begin{array}{ccccc} H^0(C,\Omega^1_C(D)) & \to & H^0(C,\mathbb{L}(D)\otimes\Omega^1_C) & \to & H^1(C,\Omega^1_C) \to 0 \\ & & \sum_{a\in|D|} \mathrm{res}_a \searrow & & \swarrow \mathrm{Res} \\ & & & k & \end{array} \tag{9.13}$$

the top line is an exact sequence and the triangle in the bottom is commutative.
If k is not algebraically closed and if $D = \sum_{\mathfrak{p}} n_{\mathfrak{p}}\mathfrak{p} \neq 0$ is an effective divisor then we still have $H^1(C,\Omega^1_C(D)) = 0$ because we may extend the ground field to $\bar{k}$ and then use 8.4. This allows us to remove our assumption that k is algebraically closed.

We needed a rational point $a \in C(k)$ to construct Res : $H^1(C,\Omega^1_C) \to k$ and so far we defined the residue map

$$\mathrm{res}_{\mathfrak{p}} : \mathbb{L}(\infty\mathfrak{p})\otimes\Omega^1_{C,\mathfrak{p}} \longrightarrow k$$

only for rational points $\mathfrak{p} \in C(k)$. We shall show that we can extend the definition easily to "separable" points, i.e. points, for which $k(\mathfrak{p})/k$ is a separable extension. We know that we always have many "separable" points on C. Once we have seen this we use the same diagram 9.13 to define the global residue map.

How do we get separable points? We choose an $f \in k(C)$ such that $k(C)/k(f)$ becomes separable. Then we have the resulting $\pi_f : C \to \mathbb{P}^1$. We have a non empty affine open set $V \subset \mathbb{P}^1$ such that $\pi^{-1}(V) = U \to V$ is unramified.
Now if k is finite, then any closed point $\mathfrak{p} \in C$ is separable. Otherwise it is clear that $V(k)$ is infinite and for any $\mathfrak{p}_0 \in V(k)$ the points $\mathfrak{p} \in C$ lying over $\mathfrak{p}_0$ are separable.

Now we want to define $\mathrm{res}_\mathfrak{p}$ for a separable point $\mathfrak{p} \in C$. This is more or less clear. We have seen on page 187 that for a suitable normal separable extension L/k we have

$$\widehat{O}_{C,\mathfrak{p}} \otimes_k L = k(\mathfrak{p})[[\pi_\mathfrak{p}]] \otimes_k L \xrightarrow{\sim} (k(\mathfrak{p}) \otimes_k L)[[\pi_\mathfrak{p}]] \xrightarrow{\sim} \bigoplus_{\sigma : k(\mathfrak{p})/k \hookrightarrow L/k} L[[\pi_\mathfrak{p}]].$$

If we now have a meromorphic differential $\omega \in k(\mathfrak{p})[[\pi_\mathfrak{p}]][\frac{1}{\pi_\mathfrak{p}}] \otimes \widehat{\Omega}_{C,\mathfrak{p}}$ then we can expand it as usual

$$\omega = \left(\frac{a_{-n}}{\pi_\mathfrak{p}^n} + \frac{a_{n-1}}{\pi_\mathfrak{p}^{n-1}} + \cdots + \frac{a_{-1}}{\pi_\mathfrak{p}} + \cdots \right) d\pi_\mathfrak{p}$$

where now the $a_\nu \in k(\mathfrak{p})$. Now it it clear that we are forced to make the definition

$$\mathrm{res}_\mathfrak{p}(\omega) = \mathrm{tr}_{k(\mathfrak{p})/k}(a_1).$$

To see that this is the only reasonable definition we extend the field of scalars to L/k, where L/k is a normal closure of $k(\mathfrak{p})/k$. We have the extension of the form

$$\omega \times_k L \in (k(\mathfrak{p}) \otimes_k L)[[\pi_\mathfrak{p}]][\frac{1}{\pi_\mathfrak{p}}] \otimes \Omega_C \xrightarrow{\sim} \bigoplus_{\sigma : k(\mathfrak{p})/k \hookrightarrow L/k} L[[\pi_\mathfrak{p}]][\frac{1}{\pi_\mathfrak{p}}] \otimes \Omega_C.$$

Since we want that the residue of ω at $\mathfrak{p}$ is equal to the sum of the residues of the extended form at the points on $C \times_k L$ over $\mathfrak{p}$ we see that we must define $\mathrm{res}_\mathfrak{p}(\omega) = \sum_\sigma \sigma(a_{-1})$ and this is the definition of the trace.
We explained already that we now are able define the global residue map $\mathrm{Res} : H^1(C,\Omega_C^1) \to k$ in general: We pick a separable point $\mathfrak{p} \in C$ and consider the exact sequence in cohomology

$$H^0(C,\Omega_C^1(\mathfrak{p})) \to H^0(C,\mathbb{L}(\mathfrak{p}) \otimes \Omega_C^1) \xrightarrow{\delta} H^1(C,\Omega_C^1) \to 0.$$

Here the map δ is not necessarily an isomorphism. But if we extend our base field to the algebraic closure, then see that the kernel of our map

$$\mathrm{res} : H^0(C,\mathbb{L}(\mathfrak{p}) \otimes \Omega_C^1) \longrightarrow k$$

is equal to the kernel map given by the sum of the residues . Hence we can define Res by the diagram

$$\begin{array}{ccc} H^0(C,\mathbb{L}(\mathfrak{p}) \otimes \Omega_C^1) & \longrightarrow & H^1(C,\Omega_C^1) \\ \mathrm{res}_\mathfrak{p} \searrow & & \swarrow \mathrm{Res} \\ & k & \end{array}$$

But once we defined Res we can define $\mathrm{res}_{\mathfrak{p}}$ for arbitrary points using the same diagram. It turns out that the local residue map at non separable points is zero.

It is clear that our definition of Res has the right functoriality principles with respect to separable morphisms and extension of the ground field.

Especially we know now that for any effective divisor D the image of

$$H^0(C,\Omega^1_C(D)) \longrightarrow H^0(C,\mathbb{L}(D) \otimes \Omega^1_C)$$

consists of the elements $\underline{\xi} \otimes \omega$, for which

$$\sum_{\mathfrak{p}} \mathrm{res}_{\mathfrak{p}}(\xi_{\mathfrak{p}} \otimes \omega) = 0.$$

We can formulate the final version of the Riemann-Roch theorem. We start from a line bundle $\mathcal{L}$ on C and we pick a point $\mathfrak{p}$ (or an effective divisor). We compute the cohomology $H^1(C,\mathcal{L})$ and to do this we start from the sequence

$$0 \longrightarrow \mathcal{L} \longrightarrow \mathcal{L}(\infty\mathfrak{p}) \longrightarrow \mathcal{L}(\infty\mathfrak{p})/\mathcal{L} \longrightarrow 0.$$

We have the cohomology sequence

$$H^0(C,\mathcal{L}(\infty\mathfrak{p})) \longrightarrow H^0(C,\mathcal{L}(\infty\mathfrak{p})/\mathcal{L}) \longrightarrow H^1(C,\mathcal{L}) \longrightarrow 0.$$

Now we consider the sheaf of differentials with coefficients in $\mathcal{L}^{-1}$ namely $\mathcal{L}^{-1} \otimes \Omega^1_C$. If $\mathfrak{p} \in H^0(C,\mathcal{L}(\infty\mathfrak{p}))$ and $\omega \in H^0(\mathcal{L}^{-1} \otimes \Omega^1_{C/k})$ then $\xi_{\mathfrak{p}}\omega$ is a Laurent-expansion of a meromorphic differential at $\mathfrak{p}$ and $\mathrm{res}_{\mathfrak{p}}(\xi_{\mathfrak{p}}\omega)$ is defined. If $\xi\mathfrak{p}$ comes from a meromorphic section $s \in H^0(C,\mathcal{L}(\infty\mathfrak{p}))$ then $s\omega$ is a meromorphic differential, which is holomorphic outside $\mathfrak{p}$. Hence $\mathrm{res}_{\mathfrak{p}}(s\omega) = 0$ and we get again a pairing

$$\mathrm{Res} : H^1(C,\mathcal{L}) \times H^0(C,\mathcal{L}^{-1} \otimes \Omega^1_{C/k}) \longrightarrow k.$$

This generalizes the pairing we had for $\mathcal{L} = \mathcal{O}_C$.

Theorem 9.5.4. *(Serre Duality or final version of Riemann-Roch):*
The pairing

$$\mathrm{Res} : H^1(C,\mathcal{L}) \times H^0(C,\mathcal{L}^{-1} \otimes \Omega^1_{C/k}) \longrightarrow k$$

is non-degenerated. Especially we find

$$\dim_k(H^1(C,\mathcal{L})) = \dim_k(H^0(C,\mathcal{L}^{-1} \otimes \Omega^1_{C/k}))$$

It is quite clear that a non zero element $\alpha \in H^0(C,\mathcal{L}^{-1} \otimes \Omega^1_{C/k})$ will induce a non trivial linear form on $H^1(C,\mathcal{L})$. Hence we conclude $\dim_k H^1(C,\mathcal{L}) \geq \dim_k H^0(C,\mathcal{L}^{-1} \otimes \Omega^1_{C/k})$. This last dimension is equal to (first version of Riemann-Roch)

$$\dim_k H^1(C,\mathcal{L}^{-1} \otimes \Omega^1_{C/k}) - \deg(\mathcal{L}) + 2g - 2 + 1 - g$$

We apply our argument a second time and get $\dim_k H^1(C,\mathcal{L}^{-1} \otimes \Omega^1_{C/k}) \geq \dim_k H^0(C,\mathcal{L})$ Hence we see that

$$\dim_k H^1(C,\mathcal{L}) \geq \dim_k H^0(C,\mathcal{L}) - \deg(\mathcal{L}) + g - 1$$

.

But the first version of Riemann-Roch tells us that we must have equality in this last inequality. Hence we see that all the inequalities in between where actually equalities. But if the dimensions are equal it follows that the pairing is non degenerate. □

Finally we conclude: If we have a line bundle $\mathcal{L}$ with $\deg(\mathcal{L}) < 0$ then we have $H^0(C,\mathcal{L}) = 0$, because if we had a section $s \neq 0$ then $\deg(\mathrm{Div}(s)) = \deg(\mathcal{L})$ and $\mathrm{Div}(s)$ is effective. Now we can say: If $\mathcal{L}$ is a line bundle with $\deg(\mathcal{L}) > 2g-2$, then $H^1(C,\mathcal{L}) = H^0(C,\mathcal{L}^{-1} \otimes \Omega^1_{C/k}) = 0$ and

$$\dim H^0(C,\mathcal{L}) = \deg(\mathcal{L}) + 1 - g.$$

9.5.4 Riemann-Roch for vector bundles and for coherent sheaves.

We begin with a very general remark. If we have a projective scheme $X \longrightarrow \mathrm{Spec}(k)$ and a coherent sheaf $\mathcal{F}$ on X, then we have seen that the cohomology groups $H^i(X,\mathcal{F})$ are finite dimensional k-vector spaces and the cohomology vanishes, if $i >> 0$. This allows us to define the Euler-characteristic

$$\chi(X,\mathcal{F}) = \sum_i (-1)^i \dim_k H^i(X,\mathcal{F}).$$

A short exact sequence of sheaves

$$0 \longrightarrow \mathcal{F}' \longrightarrow \mathcal{F} \longrightarrow \mathcal{F}'' \longrightarrow 0$$

yields a long exact sequence in cohomology. It is an easy exercise in linear algebra to show that the long exact sequence provides the addivity of the Euler characteristic

$$\chi(X,\mathcal{F}) = \chi(X,\mathcal{F}') + \chi(X,\mathcal{F}'').$$

We consider the special case where $X/k = C/k$ is a smooth curve. In this case, we have the notion of $\mathrm{Rank}(\mathcal{E})$ and $\deg(\mathcal{E})$ for any locally free sheaf $\mathcal{E}$ on C. The rank is simply the dimension of the generic fibre as a vector space over $\mathcal{O}_{C,\eta}$ and the degree can be defined inductively: If $0 \longrightarrow \mathcal{E}' \longrightarrow \mathcal{E} \longrightarrow \mathcal{E}'' \longrightarrow 0$ is an exact sequence of locally free sheaves then

$$\deg(\mathcal{E}) = \deg(\mathcal{E}') + \deg(\mathcal{E}'').$$

We know (see section 9.4.3) that any locally free sheaf admits a filtration (complete flag):

$$(0) = \mathcal{F}_0 \subset \mathcal{F}_1 \subset \mathcal{F}_2 \subset \mathcal{F}_{n-1} \subset \mathcal{F}_n = \mathcal{E}$$

such that $\mathcal{F}_{i+1}/\mathcal{F}_i = \mathcal{M}_i$ is a line bundle. We reduced the definition of the degree to the case of line bundles. One has to check that if $n = \mathrm{Rank}(\mathcal{E})$

$$\Lambda^n \mathcal{E} \simeq \otimes \mathcal{F}_i/\mathcal{F}_{i-1} = \otimes \mathcal{M}_i.$$

The theorem of Riemann-Roch gives as a formula for $\chi(C,\mathcal{E})$ in terms of $\mathrm{Rank}(\mathcal{E}), \deg(\mathcal{E})$ namely

$$\chi(C,\mathcal{E}) = \deg(\mathcal{E}) + \mathrm{Rank}(\mathcal{E})(1-g).$$

To prove this we observe that both sides behave additively under exact sequences and then the existence of flags reduces the problem to the case of line bundles.

But there is a different way to look at the theorem of Riemann - Roch. We could consider all coherent sheaves on C and ask for a formula for $\chi(C,\mathcal{F})$ in terms of $\deg(\mathcal{F}), \mathrm{Rank}(\mathcal{F})$. The first problem is that we do not yet have a notion of $\deg(\mathcal{F})$ and $\mathrm{Rank}(\mathcal{F})$ for arbitrary coherent sheaves.
But let us have a closer look at the coherent sheaves on C. Locally they are finitely generated modules over Dedekind rings. If A is a Dedekind ring and M a finitely generated A-module then we have an exact sequence

$$0 \longrightarrow M_{\mathrm{tors}} \longrightarrow M \longrightarrow M/M_{\mathrm{tors}} \longrightarrow 0$$

where M_{tors} is the module of torsion elements and M/M_{tors} is locally free. Hence we see that any coherent sheaf on C sits in such a sequence

$$0 \longrightarrow \mathcal{E}_{\mathrm{tors}} \longrightarrow \mathcal{E} \longrightarrow \mathcal{E}/\mathcal{E}_{\mathrm{tors}} \longrightarrow 0$$

where $\mathcal{E}/\mathcal{E}_{\mathrm{tors}}$ is locally free. Now it is clear that a torsion M module over a discrete valuation ring $\mathcal{O}_{C,\mathfrak{p}}$ is of the form $M = \oplus\mathcal{O}_{C,\mathfrak{p}}/\mathfrak{p}^{m_i}$ and this implies that any torsion module $\mathcal{M}$ on C can be written as quotient of a vector bundle by a subbundle of the same rank. Hence it sits in an exact sequence

$$0 \longrightarrow \mathcal{E}' \longrightarrow \mathcal{E} \longrightarrow \mathcal{M} \longrightarrow 0.$$

This suggests the definition

$$\begin{aligned} \deg(\mathcal{M}) &= \deg(\mathcal{E}) - \deg(\mathcal{E}') = \dim_k H^0(C,\mathcal{M}) \\ \mathrm{Rank}(\mathcal{M}) &= \mathrm{Rank}(\mathcal{E}) - \mathrm{Rank}(\mathcal{E}') = 0. \end{aligned} \tag{9.14}$$

The first formula has been verified earlier.(See 9.6 on p. 198.) We define the degree and the rank for arbitrary coherent sheaves by

$$\begin{aligned} \deg(\mathcal{E}) &= \deg(\mathcal{E}/\mathcal{E}_{\mathrm{tors}}) + \deg(\mathcal{E}_{\mathrm{tors}}) \\ \mathrm{Rank}(\mathcal{E}) &= \mathrm{Rank}(\mathcal{E}/\mathcal{E}_{\mathrm{tors}}) \end{aligned} \tag{9.15}$$

and this gives a more general Riemann-Roch formula:

For any coherent sheaf $\mathcal{F}$ on C we have

$$\chi(C,\mathcal{F}) = \deg(\mathcal{F}) + (1-g)\,\mathrm{Rank}(\mathcal{F}).$$

The proof is almost obvious but we write it down in a slightly sophisticated form.

We introduce the group $K'(C)$. This group is generated by the isomorphism classes of coherent sheaves. For such a sheaf $\mathcal{E}$ let $[\mathcal{E}]$ be its class in $K'(C)$. The group $K'(C)$ is the free abelian group generated by the classes $[\mathcal{E}]$ divided by the following relations: For any exact sequence

$$0 \longrightarrow \mathcal{E}' \longrightarrow \mathcal{E} \longrightarrow \mathcal{E}'' \longrightarrow 0$$

we have

$$[\mathcal{E}] = [\mathcal{E}'] + [\mathcal{E}''].$$

The Euler characteristic, the degree and the rank provide homomorphisms

$$\chi : K'(C) \longrightarrow \mathbb{Z}$$
$$\deg : K'(C) \longrightarrow \mathbb{Z}$$
$$\text{Rank} : K'(C) \longrightarrow \mathbb{Z}$$

and the theorem of Riemann-Roch says

$$\chi = \deg + (1 - g) \cdot \text{Rank}\,.$$

Our proof of the theorem of Riemann-Roch can be reformulated in this new language: The previous considerations make it clear that $K'(C)$ is generated by line bundles and torsion sheaves. But since any line bundle is of the form $\mathcal{L} = \mathcal{O}_C(D)$ we see that the group $K'(C)$ is actually generated the structure sheaf and the torsion sheaves. But now it is clear that

$$\chi(\mathcal{O}_C) = \deg(\mathcal{O}_C) + (1 - g)\,\text{Rank}(\mathcal{O}_C) = 1 - g$$

and for torsion sheaves

$$\chi(\mathcal{M}) = \deg(\mathcal{M})$$

and this proves the formula.

The structure of $K'(C)$

The structure of this group is rather complicated. We have the surjective homomorphism $(\text{Rank}\,, \deg) : K'(C) \longrightarrow \mathbb{Z} \oplus \mathbb{Z}$, which means that the group has a very simple quotient. But the kernel of this group is very complicated. We leave it as an exercise to prove that the kernel is the group $\text{Pic}^0(C)$, which is the group of line bundles of degree zero.

The group $K'(X)$ can be defined for any scheme X by the same construction. These groups have been invented by A. Grothendieck. He also introduced the groups $K(X)$. They are obtained by a similar construction, but instead of looking at all coherent sheaves we consider the isomorphim classes of locally free sheaves $[\mathcal{E}]$ and we require the relation $[\mathcal{E}] = [\mathcal{E}'] + [\mathcal{E}'']$ for any exact sequence of locally free sheaves $0 \longrightarrow \mathcal{E}' \longrightarrow \mathcal{E} \longrightarrow \mathcal{E}'' \longrightarrow 0$. Of course we have a tautological homomorphism $K(X) \longrightarrow K'(X)$.
We mention the following important theorem

Theorem 9.5.5. *If X/k is a smooth quasiprojective variety over a field k then the homomorphism $K(X) \longrightarrow K'(X)$ is bijective.*

We are not giving a proof here, we refer to the article of Borel-Serre [B-S], it is the theorem 2. Our notation here differ from the notation in [B-S], our K' is what they call K our K is what they call K_1. The proof is three pages long and is essentially self contained.
Our considerations above provide a proof in the case that X/k is a smooth projective curve.

9.6 Applications of the Riemann-Roch Theorem

9.6.1 Curves of low genus

We want to begin by discussing the cases of curves with low genus. Let C/k be an absolutely irreducible, smooth and projective curve. Let us assume the genus of this curve is 0. If this curve has a rational point $P \in C(k)$, then we can consider the line bundle $\mathcal{L} = \mathcal{O}_C(P)$ and as in VII 3.1, we can consider the morphism

$$r_{\mathcal{L}} : C \longrightarrow \operatorname{Proj}\left(\bigoplus_{n=0}^{\infty} H^0(C,\mathcal{L}^{\otimes n})\right),$$

The vector space $H^0(C,\mathcal{L})$ has rank 2, it is generated by the constant function $1 \in H^0(C,\mathcal{L})$, which we call X_0 and another function X_1, which has a first order pole at P. Then it is clear that $H^0(C,\mathcal{L}^{\otimes n})$ is spanned by the homogeneous polynomials of degree n in X_0,X_1, and these form a basis. Hence we see that

$$\bigoplus_{n=0}^{\infty} H^0(C,\mathcal{L}^{\otimes n}) = k[X_0,X_1].$$

I leave it as an exercise to the reader to show that $r_{\mathcal{L}}$ provides an isomorphism

$$r_{\mathcal{L}} : C \xrightarrow{\sim} \mathbb{P}^1$$

(see theorem 8.1.20).

A curve of genus zero over an arbitrary field does not necessarily have a rational point. This can also be formulated by saying that it does not necessarily have a line bundle of degree one. But in any case we have the sheaf $\Omega^1_{C/k}$ and this is a line bundle of degree -2. Hence the dual of this line bundle has degree 2 and therefore, we can find a non zero section $t \in H^0(C,(\Omega^1_{\mathbb{P}^1})^\vee)$. This section must have zeroes. The divisor of zeroes has degree 2, hence it must be a point P of degree 2 or of the form $P_1 + P_2$ with two rational points P_1 and P_2 (which may become equal). In the second case we see that we have a rational point. In the first case the point P has a residue field $k(P)$, which is of degree 2 over k. Hence we conclude that we can always find a quadratic extension L/k such that $C \times L$ has a rational point.

To produce such an example we consider a quadratic form over a field of characteristic $\neq 2$

$$f(x,y,z) = ax^2 + by^2 + cz^2, abc \neq 0.$$

It is absolutely irreducible and it defines a curve of genus zero (see exercise 36.) Now it is clear that this curve does not have a rational point if this form does not represent zero (i.e. we can not find $(x_0,y_0,z_0) \in k^3\backslash\{0\}$ such that $ax_0^2 + by_0^2 + cz_0^2 = 0$).

In turn it is not difficult to see that the line bundle $\mathcal{L} = (\Omega^1_{\mathbb{P}^1})^\vee$ provides an embedding

$$r_{\mathcal{L}} : C \longrightarrow \mathbb{P}^2$$

where the image is described by a quadratic form.

We consider curves of genus one over a field k. Again we assume that we have a point $P \in C(k)$, and we consider the line bundle $\mathcal{L} = \mathcal{O}_C(P)$. We have the inclusions

$$\mathcal{O}_C \subset \mathcal{O}_C(P) \subset \mathcal{O}_C(2P) \subset \mathcal{O}(3P).$$

Again we consider the graded ring

$$R = \bigoplus_n H^0(C,\mathcal{L}^{\otimes n}).$$

In contrast to our precious situations this ring is not generated by elements in degree one, but we will see that this does not really matter.

The constant function 1 yields a section $x_0 \in H^0(C,\mathcal{L})$ and this section spans this space. Then we get from the Riemann-Roch theorem that we have a section $x_1 \in H^0(C,\mathcal{L}^{\otimes 2})$, which is independent of x_0^2, and we have a section $x_2 \in H^0(C,\mathcal{L}^{\otimes 3})$, which is independent of $x_0^3,x_0^2x_1$. Now it is not difficult to see that the element x_0 (in degree one), x_1 (in degree 2) and x_2 (in degree 3) generate the graded ring R. It follows from the Riemann-Roch theorem that the elements

$$\{x_0^6,x_0^4x_1,x_0^2x_1^2,x_1^3,x_0^3x_2,x_0x_1x_2,x_2^2\}$$

must be linearly dependent because the space $H^0(C,\mathcal{L}^{\otimes 6})$ has dimension 6. Therefore we find a linear relation amoung them. Before we write it down we want to derive some information about it. Let π_P be a uniformizing element at P, then

$$x_2 = \frac{a}{\pi_P^3} + \ldots \qquad a \in k^*$$

and

$$x_1 = \frac{b}{\pi_P^2} + \ldots \qquad b \in k^*.$$

It is clear that x_2^2,x_1^3 are the only terms, which have a 6-th order pole, hence our relation has to cancel that pole. We can modify x_1,x_2 by a scalar factor such that the above numbers satisfy $a = b = 1$. Then we can conclude that our relation must be of the form

$$x_2^2 + a_1x_0x_1x_2 + a_3x_0^3x_2 = x_1^3 + a_2x_0^2x_1^2 + a_4x_0^4x_1 + a_6x_0^6.$$

this is a relation among elements in $H^0(C,\mathcal{L}^{\otimes 6}) = H^0(C,\mathcal{O}_C(6P))$. Since x_0 is the constant function 1 viewed as element in $H^0(C,\mathcal{O}_C(6P))$ we see that in the monomials the exponent ν of x_0^ν does not matter. We view our x_i as elements in $H^0(C,\mathcal{L}^{\otimes 3})$ and we can modify our relation to

$$x_0x_2^2 + a_1x_0x_1x_2 + a_3x_0^2x_2 = x_1^3 + a_2x_0x_1^2 + a_4x_0^2x_1 + a_6x_0^3,$$

and this is a homogeneous relation of degree three among elements in $H^0(C,\mathcal{L}^{\otimes 3})$. (The point of this trick is that we see that we do not need all monomials of degree 3 in $x_0,x_1,x_2 \in H^0(C,\mathcal{L}^{\otimes 3})$. Hence we get the morphism

$$\begin{array}{ccccc} r_{\mathcal{L}^{\otimes 3}} & : & C & \longrightarrow & \mathbb{P}^2 \\ & & & \searrow \quad \swarrow & \\ & & & \mathrm{Spec}(k). & \end{array}$$

Again I leave it to the reader to show that this is a closed embedding.
If we do not have a rational point on C, then this does not work.

9.6.2 The moduli space

At this point we want to give an outlook to more advanced topics. Let S be a scheme of finite type over $\mathrm{Spec}(\mathbb{Z})$, we want to consider **elliptic curves** over S. An elliptic curve over S is a diagram

$$\begin{array}{c} C \\ \pi \downarrow\uparrow s \\ \\ S. \end{array},$$

where $\pi : C \longrightarrow S$ is a smooth projective scheme, all its fibers are absolutely irreducible of dimension one and of genus one and $s : S \longrightarrow C$ is a section. We will see in the last chapter that C/S has a unique structure of a groups scheme over S, for which our given section is the identity element. Given such an elliptic curve $(C,s)/S$ we will denote it by $\mathcal{E}/S$.
We want to treat the problem to construct a moduli space for elliptic curves. We discussed this kind of question in the first volume and formulated theorem 5.2.28. We pointed out that this result is not really a precise statement. This will be cured by the following considerations. We will formulate a precise statement, which asserts that a suitable functor is representable. This means that we will try to construct a moduli space for elliptic curves. In a naive sense the construction of such a moduli space means that we write down canonical equations for the curve where the coefficients of the equations are the "coordinates" of the curve. This vague formulation will become precise during the discussion below.

We return to our elliptic curve $(C,s)/S = \mathcal{E}/S$. As before, we view the image of S under the section s as a divisor P on C. It defines a line bundle $\mathcal{L} = \mathcal{O}_C(P)$ on C. If we evaluate at a point $x \in S$, then we get the curve

$$\begin{array}{c} C\times \mathrm{Spec}(k(x)) \\ \downarrow \\ \mathrm{Spec}(k(x)), \end{array} \tag{9.16}$$

the point P gives us a $k(x)$-rational point, and we are in the previous situation.

We consider the direct images of our line bundle $\pi_*(\mathcal{L}^{\otimes n})$. The Riemann-Roch theorem tells us that the $k(x)$-vector spaces

$$H^0(C \times_S k(x),\mathcal{L}^{\otimes n} \mid C \times_S k(x))$$

have dimension n if x varies. Then our semicontinouity implies that $\pi_*(\mathcal{L}^{\otimes n})$ are locally free $\mathcal{O}_S$ modules of rank n.

We localize a little bit and assume that $S = \mathrm{Spec}(A)$ and that any locally free module over A is actually free. Then $H^0(S,\pi_*(\mathcal{L}^{\otimes n})) = H^0(C,\mathcal{L}^{\otimes n})$ is a free module of rank n. We find sections

$$x_0 \in H^0(C,\mathcal{L}),x_1 \in H^0(C,\mathcal{L}^{\otimes 2}),x_3 \in H^0(C,\mathcal{L}^{\otimes 3})$$

where x_0 is the constant function with value 1 and such that these sections successively provide a basis in

$$H^0(C,\mathcal{L}) \subset H^0(C,\mathcal{L}^{\otimes 2}) \subset H^0(C,\mathcal{L}^{\otimes 3}).$$

If $J \subset \mathcal{O}_C$ is the ideal defining P the module J/J^2 is free or rank one over A. We may choose an uniformizing element π_P, which generates this A-module. We notice that the module J/J^2 is actualy isomorphic to the restriction of the line bundle $\Omega^1_{C/S}$ to our section s. Since the line bundle $\Omega^1_{C/S}$ is trivial along the fibers we see that $H^0(C,\Omega^1_{C/S})$ is a free A module of rank 1 (semicontinuity). Hence the choice of π_P is the same as the choice of a differential ω, which generates this free rank 1 module.
The elements $\pi_P^2 x_1$ and $\pi_P^3 x_2$ are elements which are regular along P, we can evaluate at P, the result is a unit in A. Now we can require

$$\pi_P^2 x_1(P) = \pi_P^3 x_2(P) = 1. \tag{Ω}$$

Then we conclude that we have a relation

$$x_2^2 + a_1 x_0 x_1 x_2 + a_3 x_0^3 x_2 = x_1^3 + a_2 x_0^2 x_1^2 + a_4 x_0^4 x_1 + a_6 x_0^6$$

with some coefficients in A. Again we perform the change of the bundle, we replace $\mathcal{L}$ by $\mathcal{L}^{\otimes 3} = \mathcal{O}_C(3P)$ and we consider x_0,x_1,x_2 as sections of this bundle, we give them the degree one and consider the homogenous relation (recall $x_0 = 1$)

$$x_0 x_2^2 + a_1 x_0 x_1 x_2 + a_3 x_0^2 x_2 = x_1^3 + a_2 x_0 x_1^2 + a_4 x_0^2 x_1 + a_6 x_0^3 \tag{$\mathcal{E}$}$$

which describes C as a closed subscheme of $\mathbb{P}^2 = \mathrm{Proj}\ A[X_0,X_1,X_2]$. The coefficients are uniquely determined by the choice of x_2,x_1,x_0. We will always choose $x_0 = 1$ as constant function 1 this is a canonical choice. If we stick to our choice of π_P (resp. ω), then we can choose $x_0' = 1, x_1', x_2'$ such that

$$\begin{array}{rcl} x_1 & = & x_1' + \alpha x_0 \\ x_2 & = & x_2' + \beta x_1' + \gamma x_0 \quad . \end{array} \tag{S}$$

Then we will get a new relation ($\mathcal{E}$) with new coefficients:

$$x_0(x_2')^2 + a_1' x_0 x_1' x_2' + a_3' x_0^2 x_2' = (x_1')^3 + a_2' x_0 (x_1')^2 + a_4' x_0^2 x_1' + a_6' x_0^3.$$

We can write formulae for the a_i' in terms of the a_i,α,β,γ This makes it clear that the coefficients of the relation are by no means determined by the curve and the choice of π_p. The following statements have to be verified by computations and a little bit of thinking. To proceed we assume that 2 and 3 are invertible in A, i.e. $\frac{1}{6} \in A$. Then we see from the formulae for the a_i' that there exists unique substitution (S) such that the new relation ($\mathcal{E}$) will be of the form

$$x_0 x_2^2 = x_1^3 + a_4' x_0^2 x_1 + a_6' x_0^3 \tag{Wei}$$

in other words $a_1' = a_3' = a_2' = 0$. This is the **Weierstrass normal form** of the equation for an elliptic curve (in the classical Weierstrass form is a factor 4 as coefficient of x_1, this is not relevant in our context).
Then a_4' and a_6' are the following expressions in the a_1,a_3,a_2,a_4,a_6:

$$\begin{aligned} a_4' &= -\frac{a_1^4}{48} - \frac{a_1^2 a_2}{6} - \frac{a_2^2}{3} + \frac{a_1 a_3}{2} + a_4 \\ a_6' &= \frac{a_1^6}{864} + \frac{a_1^4 a_2}{72} + \frac{a_1^2 a_2^2}{18} + \frac{2a_2^3}{27} - \frac{a_1^3 a_3}{24} - \frac{a_1 a_2 a_3}{6} + \frac{a_3^2}{4} - \frac{a_1^2 a_4}{12} - \frac{a_2 a_4}{3} + a_6 \end{aligned} \tag{9.17}$$

Now it is quite easy to see that the curve defined by the equation

$$x_0 x_2^2 = x_1^3 + a_4' x_1 x_0^2 + a_6' x_0^3$$

is smooth if and only if the discriminant (of the cubic polynomial $x^3 + a_4' x + a_6'$)

$$-4(a_4')^3 - 27(a_6')^2$$

is a unit in A. If we rewrite this in terms of the a_i, then we get an expression, which is a sum of monomials in the a_1,a_3,a_2,a_4,a_6. It is homogeneous of degree 12, if we give a_i the degree i. The coefficients are rational numbers, which have only powers of 2 in their denominator and where the largest denominator is 16. Hence we define

$$\Delta(a_1,a_3,a_2,a_4,a_6) = 16 \cdot (-4(a_4')^3 - 27(a_6')^2)$$

and with the help of a computer we find

$$\begin{aligned}\Delta(a_1,a_3,a_2,a_4,a_6) &= -a_1^4 a_2 a_3^2 - 8a_1^2 a_2^2 a_3^2 - 16a_2^3 a_3^2 + a_1^3 a_3^3 + 36 a_1 a_2 a_3^3 - \\ &27a_3^4 + a_1^5 a_3 a_4 + 8a_1^3 a_2 a_3 a_4 + 16 a_1 a_2^2 a_3 a_4 - 30 a_1^2 a_3^2 a_4 + 72 a_2 a_3^2 a_4 + a_1^4 a_4^2 + 8a_1^2 a_2 a_4^2 + \\ &16a_2^2 a_4^2 - 96 a_1 a_3 a_4^2 - 64 a_4^3 - a_1^6 a_6 - 12 a_1^4 a_2 a_6 - 48 a_1^2 a_2^2 a_6 - 64 a_2^3 a_6 + 36 a_1^3 a_3 a_6 + \\ &144 a_1 a_2 a_3 a_6 - 216 a_3^2 a_6 + 72 a_1^2 a_4 a_6 + 288 a_2 a_4 a_6 - 432 a_6^2.\end{aligned}$$

The coefficients of the monomials are integral, and some monomials have the coefficient ± 1.

Finally we can apply the same process to a_4' and a_6' and put

$$C_4 = 48 \cdot a_4' \quad , \quad C_6 = 864 \cdot a_6',$$

then

$$\begin{aligned} C_4 &= -a_1^4 - 8a_1^2 a_2 - 16a_2^2 + 24 a_1 a_3 + 48 a_4 \\ C_6 &= a_1^6 + 12 a_1^4 a_2 + 48 a_1^2 a_2^2 + 64 a_2^3 - 36 a_1^3 a_3 - \\ &144 a_1 a_2 a_3 + 216 a_3^2 - 72 a_1^2 a_4 - 288 a_2 a_4 + 864 a_6 \end{aligned} \tag{9.18}$$

We have $48 = 4 \cdot 12$ and $864 = 12^3/2$, therefore,

$$\Delta(a_1,a_3,a_2,a_4,a_6) = \frac{1}{12^3} \cdot (-C_4^3 - C_6^2),$$

We observe that the expressions C_4,C_6 and Δ can be written down without the assumption that $\frac{1}{6} \in A$. Furthermore it is clear that:

The expressions for C_4,C_6 and Δ are invariant under the substitutions induced on the coefficients a_1,a_3,a_2,a_4,a_6 by substitutions of the form (S).

The following theorem is almost clear from our considerations above. We drop the assumption $\frac{1}{6} \in A$ for a moment.

Theorem 9.6.1. *Let A be any commutative ring with identity. The equation*

$$x_0x_2^2 + a_1x_0x_1x_2 + a_3x_2x_0^2 = x_1^3 + a_2x_1^2x_0 + a_4x_1x_0^2 + a_6x_0^3$$

defines a projective curve C over Spec(A). *This curve is smooth if and only if*

$$\Delta(a_1,a_3,a_2,a_4,a_6) \in A^*.$$

It contains the point $P = (0,0,1)$ choosing this point as our section s makes C to an elliptic curve $\mathcal{E} = (C,s)$. The complement $C \setminus s(\mathrm{Spec}(A)) = U$ of this this section is affine, on this complement we can normalize $x_0 = 1$ and

$$U = \mathrm{Spec}(A[x_1,x_2]/(x_2^2 + a_1x_1x_2 + a_3x_2 - x_1^3 - a_2x_1^2 - a_4x_1 - a_6)$$

We have an explicit holomorphic differential ω, which on the affine part $x_0 = 1$ is given by

$$\omega = \frac{dx_2}{3x_1^2 + 2a_2x_1 + a_4 - a_1x_2} = \frac{dx_1}{2x_2 + a_1x_1 + a_3}.$$

If A is noetherian and if any locally free A-module of finite rank is free, then any elliptic curve $\mathcal{E} \longrightarrow$ Spec(A) together with a nowhere vanishing differential ω is of the form above.

The last assertion has been proved above and uses in an essential way the semi-continuity theorems. The assertion concerning smoothness is clear if $\frac{1}{6} \in A$, but it is also true without this assumption. It requires some computations, which are carried out in [Hu], Chap. 4.

We can formulate this slightly differently if we consider the a_i as indeterminates and say that the above equation defines an elliptic curve over $\mathrm{Spec}(\mathbb{Z}[a_1,a_3,a_2,a_4,a_6,\frac{1}{\Delta}])$, which comes with a nowhere vanishing differential.
At this point the reader is invited to play a little bit with this expression and to evaluate it for small values of the a_i. You will see that you never get ± 1. If we evaluate at bigger values of the a_i, then we even find rather big values for Δ. So we are tempted to believe

The diophantine equation

$$\Delta(a_1,a_3,a_2,a_4,a_6) = \pm 1,$$

has no solution in integers $a_1,a_3,a_2,a_4,a_6 \in \mathbb{Z}$

and this means that there is no elliptic curve over Spec($\mathbb{Z}$).(See exercise 37.). The assertion in exercise 37) follows from the stronger statement

There is no smooth curve $C \longrightarrow$ Spec($\mathbb{Z}$) of genus $g \geq 1$

and this has been proved by Abrashkin (for $g \leq 3$) (see [Ab]))and Fontaine for arbitrary $g \geq 1$ (see [Fo].).
But if we assume again that $\frac{1}{6} \in A$ and if we consider pairs $(\mathcal{E},\omega)$ where $\omega \in H^0(C,\Omega^1_{C/S})$ is a generator of $H^0(C,\Omega^1_{C/S})$, then the situation is different. The existence of such a form is an additional requirement, of course it exists if any locally free module of rank 1 over A is free. Let us assume this for a moment.. Then we we have seen that we have a unique choice for our sections $(x_0 = 1,x_1,x_2)$ such that (Ω) holds and such that $a_1 = a_3 = a_2 = 0$, i.e. the equation is in Weierstrass form. Then the remaining two coefficients $a_4,a_6 \in A$ are uniquely determined by the datum $(\mathcal{E},\omega)$. The pair (a_4,a_6) can be viewed as the "coordinates" of the curve.

Now we explain that this means that the following functor is representable:
We consider the base scheme $\mathrm{Spec}(\mathbb{Z}[\frac{1}{6}])$. On the category of schemes of finite type $S \longrightarrow \mathrm{Spec}(\mathbb{Z}[\frac{1}{6}])$, we define a functor

$$S \longrightarrow \mathcal{M}_{1,diff}(S) = \text{Set of isomorphism classes of pairs } (\mathcal{E}/S,\omega).$$

This is indeed a functor, if we have a $\mathrm{Spec}(\mathbb{Z}[\frac{1}{6}])$ morphism $S' \longrightarrow S$ then we get a map $\mathcal{M}_{1,diff}(S) \longrightarrow \mathcal{M}_{1,diff}(S')$ if we take the pullback of the curve, the section and the differential (see p.28, 7.5.10).

We show that this functor is representable by an affine scheme of finite type. To do this we write down a universal elliptic curve: We introduce the ring $\mathbb{Z}[\frac{1}{6}][u,v,\Delta,1/\Delta]$ where the first two variables are independent and $\Delta = -4u^3 - 27v^2$. We put $M_{1,diff} = \mathrm{Spec}(\mathbb{Z}[\frac{1}{6}][u,v,\Delta,1/\Delta])$ and our universal elliptic curve is (we perform a slight change in notations $(x_0,x_1,x_2) \longrightarrow (z,x,y)$)

$$\begin{array}{ccc} \widetilde{\mathcal{E}} : y^2z = x^3 + uxz^2 + vz^3 & \hookrightarrow & M_{1,diff} \times \mathbb{P}^2_{\mathbb{Z}}, \\ & \searrow^{p_0} & \downarrow p_1 \\ & & M_{1,diff} \end{array} \qquad (9.19)$$

and the section s is given by the point $(x,y,z) = (0,1,0)$.
This elliptic curve together with the differential $\widetilde{\omega}$ in theorem 9.6.1 is an element

$$(\widetilde{\mathcal{E}},\widetilde{\omega})_{\mathrm{unv}} \in \mathcal{M}_{1,diff}(M_{1,diff}).$$

Now the following theorem asserts that $\mathcal{M}_{1,diff}$ is representable.

Theorem 9.6.2. *For any scheme $S \longrightarrow \mathrm{Spec}(\mathbb{Z}[\frac{1}{6}])$ of finite type and any elliptic curve $(\mathcal{E}/S,\omega)$ over S we have a unique morphism $\pi : S \longrightarrow M_{1,diff}$ such that we a unique isomorphism*

$$(\pi^*(\widetilde{\mathcal{E}}),\pi^*(\widetilde{\omega})) \xrightarrow{\sim} (\mathcal{E},\omega).$$

from the pullback (see 6.2.5) of the universal curve to our given curve. The scheme $M_{1,diff} \longrightarrow \mathrm{Spec}(\mathbb{Z}[\frac{1}{6}])$ is called the moduli space of elliptic curves, which are equipped with a nowhere vanishing differential.

Most of the work has been done in our considerations above. From our elliptic curve $\pi : \mathcal{E} \longrightarrow S$ we get the locally free sheaves $\pi_*(\mathcal{O}_C) \subset \pi_*(\mathcal{O}_C(P)) \subset \pi_*(\mathcal{O}_C(2P)) \subset \pi_*(\mathcal{O}(3P))$. We cover S by open affine schemes $S_\nu = \mathrm{Spec}(A_\nu)$ such that the restriction of these sheaves to $\mathrm{Spec}(A_\nu)$ are free modules. Then we can choose unique sections $x_0^{(\nu)} = 1 \in H^0(S_\nu,\pi_*(\mathcal{L}))$, $(x_1^{(\nu)} \in H^0(S_\nu,\pi_*(\mathcal{L}^{\otimes 2}))$, $x_2^{(\nu)} \in H^0(S_\nu,\pi_*(\mathcal{L}^{\otimes 3}))$, such that the condition (Ω) holds and in the relation among these sections we have $a_1^{(\nu)} = a_3^{(\nu)} = a_2^{(\nu)} = 0$. Since these sections are unique we find that we get equality for the restrictions $x_i^{(\nu)}|S_\nu \cap S_\mu = x_i^{(\mu)}|S_\nu \cap S_\mu$. Hence they extend to sections $x_0 = 1 \in H^0(S,\pi_*(\mathcal{L}))$, ,$x_1 \in H^0(S,\pi_*(\mathcal{L}^{\otimes 2}))$, $x_2 \in H^0(S,\pi_*(\mathcal{L}^{\otimes 3})$ and we get a Weierstrass equation (Wei).

$$x_0x_2^2 = x_1^3 + a_4x_0^2x_1 + a_6x_0^3,$$

where $a_i \in H^0(S,\mathcal{O}_S)$ and $\Delta(a_4,a_6) \in H^0(S,\mathcal{O}_S)^\times$. This yields a $\mathrm{Spec}(\mathbb{Z}[\frac{1}{6}])$ morphism $\phi : S \longrightarrow M_{1,diff}$, which is defined by $u \mapsto a_4, v \mapsto a_6$ and clearly we have an isomorphism $\Phi : (\pi^*(\widetilde{\mathcal{E}}),\pi^*(\widetilde{\omega})) \xrightarrow{\sim} (\mathcal{E},\omega)$, which on the affine part (i.e. $z = 1, x_0 = 1$) is given by $x \mapsto x_1, y \mapsto x_2$.
On the other hand it is clear that the morphisms ϕ,Φ are uniquely determined by $((\mathcal{E}/S,\omega)$. To see this we can assume that $S = \mathrm{Spec}(A)$ is affine. Let us assume that we have a second pair (ϕ',Φ'). The isomorphism Φ' is determined by its value on the affine parts, and on the affine part Φ' must send $x \mapsto x_1, y \mapsto x_2$, because these elements are uniquely determined by the constraint (Ω) and the vanishing $a_1^{(\nu)} = a_3^{(\nu)} = a_2^{(\nu)} = 0$. But then ϕ has to send $u \mapsto a_4, v \mapsto a_6$ and hence we see that $(\phi',\Phi') = (\Phi,\phi)$.

□

Remark: As a byproduct of the proof we see that $(\mathcal{E}/S,\omega)$ can not have any non trivial automorphism, this follows the uniqueness of the isomorphism Φ. On the other hand the argument at the end of the proof can be used to prove the triviality of $\mathrm{Aut}(\mathcal{E}/S,\omega)$ directly. But we should also observe the the formal definition of representability does not imply the uniqueness of Φ : To prove representability in the above case it suffices to prove the uniqueness of ϕ and the existence of a Φ. The reader should keep this remark in mind during the following discussion.

At this point it seems to be natural to ask whether we can drop choice of the the form ω, this choice is somewhat arbitrary, two such choices differ by a unit $\alpha \in \mathcal{O}_S(S)^\times$. Therefore we tempted to ask whether the functor

$$\mathcal{M}_1 : \{ \text{ Schemes } S \longrightarrow \mathrm{Spec}(\mathbb{Z}[\frac{1}{6}]) \text{ of finite type} \} \longrightarrow \{ \text{ Set of Isomclasses of elliptic curves } \mathcal{E}/$$

is representable, and can be obtained from $M_{1,diff}$ by dividing by an action of the multiplicative group scheme $\mathbb{G}_m$ (see 7.5.8.)
The answer is "No" and we will explain why this is so. For the following discussion we also refer to to 10.1, where the same problem is discussed in a different context.

We drop the assumption $\frac{1}{6} \in \mathcal{O}_S(S)$ for a moment. If $(\mathcal{E},\omega)$ is an elliptic curve over S and if we replace ω by ω_1 then $\omega_1 = \alpha\omega$ with $\alpha \in \mathcal{O}_S(S)^\times$. This means that we have an action of the multiplicative groups scheme $\mathbb{G}_m/\mathrm{Spec}(\mathbb{Z})$ on the moduli scheme $M_{1,diff}$, which reflects this change of the differential. We describe this action explicitly. To do this we return to the general form $(\mathcal{E})$ of our equation. Changing the differential ω to $\alpha\omega$ is the same thing as changing the uniformizing element π_P to $\alpha\pi_P$. If sections $(x_0 = 1, x_1, x_2)$ satisfy (Ω) with respect to ω, then the sections $x_0' = x_0$, $x_1' = \alpha^{-2}x_1$ and $x_2' = \alpha^{-3}x_2$ satisfy (Ω) with respect to $\alpha\omega$. The relation $(\mathcal{E})$ among the sections (x_0,x_1,x_2) yields the relation

$$\alpha^6 x_0' x_2'^2 + \alpha^5 a_1 x_0' x_1' x_2' + \alpha^3 a_3 x_2' x_0'^2 = \alpha^6 x_1'^3 + \alpha^4 a_2 x_1'^2 x_0' + \alpha^2 a_4 x_1' x_0'^2 + a_6 x_0'^6.$$

Dividing by α^6 gives us the relation $(\mathcal{E})$ among the new sections.

$$x_0' x_2'^2 + \alpha^{-1} a_1 x_0' x_1' x_2' + \alpha^{-3} a_3 x_2' x_0'^2 = x_1'^3 + \alpha^{-2} a_2 x_1'^2 x_0' + \alpha^{-4} a_4 x_1' x_0'^2 + \alpha^{-6} a_6 x_0'^6.$$

We see that

$$\{a_1,a_3,a_2,a_4,a_6\} \longrightarrow \{\alpha^{-1}a_1,\alpha^{-3}a_3,\alpha^{-2}a_2,\alpha^{-4}a_4,\alpha^{-6}a_6\}$$

This yields a $\mathbb{G}_m$-action on $\mathrm{Spec}(\mathbb{Z}[a_1,a_3,a_2,a_4,a_6,\frac{1}{\Delta}])$. We know that C_4,C_6,Δ are homogenous of degrees $-4,-6,-12$ respectively. Hence we see that the expression

$$j(a_1,a_3,a_2,a_4,a_6) = \frac{C_4^3}{\Delta}$$

is invariant under the action of $\mathbb{G}_m$, i.e. invariant the change of the differential form. Hence it becomes clear that $j(a_1,a_3,a_2,a_4,a_6)$ only depends on the the isomorphism class of the elliptic curve $\mathcal{E}/S$. This is the famous j **-invariant** of an elliptic curve.

This seems to indicate that $\mathrm{Spec}(\mathbb{Z}[X])$ is the moduli space of elliptic curves, because for any elliptic curve $\mathcal{E} \longrightarrow S$ we found a uniquely defined morphism $j(\mathcal{E}) : S \longrightarrow \mathrm{Spec}(\mathbb{Z}[X])$, which attaches to $\mathcal{E}$ its j-invariant.

But what is next? Since we do not have-and there is no way to get it- a universal curve with j-invariant j we can not formulate the assertion in the above theorem (9.6.2).
We want to analyze this further. We revitalize our assumption $\frac{1}{6} \in \mathcal{O}_S(S)$, and we apply our consideration to $M_{1,diff} = \mathrm{Spec}(\mathbb{Z}[\frac{1}{6}][a_4,a_6,\frac{1}{\Delta}])$. For any $\mathbb{Z}[\frac{1}{6}]$-algebra B the group $B^\times = \mathbb{G}_m(B)$ acts on $\mathbb{Z}[\frac{1}{6}][a_4,a_6,\frac{1}{\Delta}](B)$ by

$$(a_4,a_6,\frac{1}{\Delta}) \mapsto (\alpha^{-4}a_4,\alpha^{-6}a_6,\alpha^{12}\frac{1}{\Delta}),$$

this means that we have an action of $\mathbb{G}_m$ on $M_{1,diff}$ (see section 7.5.8).

We come back to the apparently more natural functor $S \longrightarrow \mathcal{M}_1(S)$.
Our considerations above seem to suggest that $\mathcal{M}_1$ is representable and represented by the affine scheme $M_{1,diff}/\mathbb{G}_m$. But this is not quite right.
We see easily that $M_1 = M_{1,diff}/\mathbb{G}_m = \mathrm{Spec}(\mathbb{Z}[\frac{1}{6}][j] = \mathrm{Spec}(\mathbb{Z}[\frac{1}{6}][\frac{a_4^3}{\Delta}]$, i.e. the $\mathbb{Z}[\frac{1}{6}]$-algebra of elements of degree zero is generated by j.
We have a look at the diagram, which is provided by the general theory of $\mathbb{G}_m$-actions, here $S = \mathrm{Spec}(B)$:

$$\begin{array}{ccc}
M_{1,diff}(B) & \xrightarrow{\quad f \quad} & \\
\downarrow p_0 & & \\
\widetilde{M_{1,diff}(B)/\mathbb{G}_m(B)} & \xrightarrow{\quad g \quad} & (M_{1,diff}/\mathbb{G}_m)(B) \\
\downarrow p_1 & & \\
\mathcal{M}_1(B) & \xrightarrow{\quad h \quad} &
\end{array} \qquad , \qquad (9.20)$$

The vertical arrow p_1 is bijective, provided every locally free B-module of rank one is actually free. We apply the criteria formulated at the end of section 7.5.6 in the subsection on $\mathbb{G}_m$-actions. We see that the degrees of the action are $(4,6,-12)$. For any geometric point $(a_4,a_6,\frac{1}{\Delta})$ the last coordinate is never zero and hence at least one of the other two coordinates is non zero. Hence we can apply proposition 7.5.24 and see that $M_{1,diff}/\mathbb{G}_m$ is a geometric quotient. But we will see that g will not be a bijection, in general it will be neither injective nor surjective. We formulate two precise assertions concerning this issue.

A faithfully flat morphism $S' \longrightarrow S$ of finite type, will be called a ffft-morphism, if S',S are of finite type over $\mathrm{Spec}(\mathbb{Z})$.

(i) Let $S \longrightarrow \mathrm{Spec}(\mathbb{Z})$ be ffft. For any two two elliptic curves $\mathcal{E}_1$,$\mathcal{E}_2$ over S with $h(\mathcal{E}_1) = h(\mathcal{E}_2)$ we can find a ffft - morphism $S' \longrightarrow S$, such that $\mathcal{E}_1 \times_S S' \xrightarrow{\sim} \mathcal{E}_2 \times_S S'$.

(ii) If $[\mathcal{E}] \in \mathcal{M}_1(S)$ then we can find a ffft $S' \longrightarrow S$ such that we find a $(\mathcal{E}/S',\omega)$ with $h(\mathcal{E}/S',\omega) = [\mathcal{E}]$, then we have $p_1 \circ p_0(\mathcal{E}/S',\omega) = [\mathcal{E}]$.

For any element $\mathcal{E} \in \mathcal{M}_1(S)$ the $\mathcal{O}_S$-module $\pi_*(\Omega^1_{\mathcal{E}/S})$ is locally free, we can find a finite Zariski covering $S' = \bigsqcup_\nu \mathrm{Spec}(B_\nu)$ such that it becomes free over S' and therefore, $\mathcal{E} \times_S S'$ is in the image of $M_{1,diff}(S') \longrightarrow \mathcal{M}_1(S')$.
We consider our two curves $\mathcal{E}_1/S$,$\mathcal{E}_2/S$. To prove (i) we can apply the last argument and we may assume that $S = \mathrm{Spec}(B)$ and that we can equip both of them with a nowhere vanishing form ω_1,ω_2. Then the two pairs $(\mathcal{E}_1,\omega_1)$,$(\mathcal{E}_2,\omega_2)$ are defined by equations

$$y^2z = x^3 + a_4xz^2 + a_6,\ y^2z = x^3 + b_4xz^2 + b_6,$$

where the coefficients are in B, where $\Delta(a_4,a_6) = \Delta$, $\Delta(b_4,b_6) = \Delta_1$ are units and where the differentials are given by the expressions in Theorem 9.6.1.
If we change the differential ω_1 by a factor $\omega_1 \longrightarrow \beta\omega_1$ then we change the coefficients $a_4 \longrightarrow \beta^{-4}a_4$,$a_6 \longrightarrow \beta^{-6}a_6$. Our assumption that the two curves have the same image under h says that the two j-invariants are the same and hence

$$\frac{a_4^3}{\Delta} = \frac{b_4^3}{\Delta_1}.$$

Now Δ,Δ_1 are units, we put $u = \Delta/\Delta_1$. We consider the B-algebra $B_1 = B[\zeta] = B[X]/(\Phi_{12}(X)$, where $\Phi_{12}(X) = X^4 - X^2 + 1$ is the cyclotomic polynomial for the primitive 12-th roots of unity. Then we construct a second extension $B' = B_1[\delta] = B_1[Y]/(Y^{12} - u)$. We observe that the extension $B \hookrightarrow B'$ is finite the B-module B' is free of rank 48. It it also easily checked that the module $\Omega^1_{B'/B} = 0$, because we assumed $\frac{1}{6} \in B$. We consider the two curves $\mathcal{E}_1' = \mathcal{E}_1 \times_{\mathrm{Spec}(B)} \mathrm{Spec}(B')$,$\mathcal{E}_2' = \mathcal{E}_2 \times_{\mathrm{Spec}(B)} \mathrm{Spec}(B')$, both are equipped with a differential ω_1',ω_2'. If we now replace ω_1' by $\delta\omega_1'$, then we get a new equation for $\mathcal{E}_1'$ with coefficient $a_4' = \delta^{-4}a_4$,$a_6' = \delta^{-6}a_6$ and Δ is replaced by $u\Delta = \Delta_1$. Hence we get $(a_4')^3 = b_4^3$. Since we have the relation $-4(a_4)^3 - 27(a_6)^2 = \Delta$ we also get $(a_6')^2 = b_6^2$. Therefore we see that for $\frac{a_4'}{a_4} = \mu$ and $\frac{a_6'}{a_6} = \nu$ we have $\mu^3 = \nu^2 = 1$. From this we can conclude that we find an element $\zeta_1 \in B'$ such that $\zeta_1^4 = \mu$,$\zeta_1^6 = \nu$. (This is not entirely obvious, we leave it as an (amusing) exercise to the reader.) If we now modify our δ to $\zeta_1\delta$ then we get $a_4' = a_4$,$a_6' = a_6$.
The second assertion is easier to verify. By definition $[\mathcal{E}] \in \mathcal{M}_1(S)$ means that $[\mathcal{E}]$ is the isomorphism class of an elliptic curve $\mathcal{E}/S$. Then we even find a Zariski-covering $S' \longrightarrow S$ such that $\mathcal{E} \times_S S'$ can be equipped with a nowhere vanishing form ω and hence $[\mathcal{E}]$ is in the image of $p_1 \circ p_0$.
The following considerations form a paradigm for some much more general phenomenon in the general theory of moduli spaces.

We explain why we can not expect that g is a bijection, in general it is neither injective nor surjective. The arrow p_0 is a bijection if every locally free B-module of rank one is free. Let us assume that this is the case. We pick an elliptic curve $\mathcal{E}/S$. Now it can happen that we can find a second elliptic curve $\mathcal{E}_1/S$, which is not isomorphic to $\mathcal{E}/S$, but we can find a faithfully flat extension of finite type $S' \longrightarrow S$ such that $\mathcal{E}_1 \times_S S' \xrightarrow{\sim} \mathcal{E} \times_S S'$. If we found such a curve then clearly $h(\mathcal{E}_1) = h(\mathcal{E})$, because $M_{1,diff}/\mathbb{G}_m)(S) \longrightarrow M_{1,diff}/\mathbb{G}_m)(S')$ is injective (see theorem 6.2.17.)
Such a curve $\mathcal{E}_1/S$, which becomes isomorphic to $\mathcal{E}/S$ over a faithfully flat extension of finite type, is called an S-form of $\mathcal{E}/S$. This was already discussed in section 6.2.10 and we have explained that we have a canonical bijection

$$\{\text{Set of isomclasses of } S \text{ forms of } \mathcal{E}/S \xrightarrow{\sim} H^1(S, \mathrm{Aut}(\mathcal{E}/S)),$$

Now we compute the algebraic group $\mathrm{Aut}(\mathcal{E}/S)$, i.e. for any scheme $T \longrightarrow \mathrm{Spec}(B) = S$ we compute $\mathrm{Aut}(\mathcal{E} \times_S T/T)$. We choose a nowhere vanishing one form ω on $\mathcal{E}/S$. Then $(\mathcal{E}/S,\omega)$ is a B valued point of $(a_4,a_6,\frac{1}{\Delta(a_4,a_6)}) \in M_{1,diff}$. Any automorphism $\alpha \in \mathrm{Aut}(\mathcal{E} \times_S T/T)$ will multiply the pullback ω_T by a factor $\psi(\alpha)$. This is a homomorphism $\psi : \mathrm{Aut}(\mathcal{E} \times_S T/T) = \mathrm{Aut}(\mathcal{E}/S)(T) \longrightarrow \mathbb{G}_m(T)$. Since $\mathrm{Aut}(\mathcal{E} \times_S T/T,\omega_T)$ is trivial we get an injective homomorphism

$$\psi : \mathrm{Aut}(\mathcal{E}/S) \hookrightarrow \mathbb{G}_m/S.$$

We have to compute the image. For simplicity we assume that S is integral. The morphism $T \longrightarrow S$ yields a homomorphism $B \longrightarrow H^0(T,\mathcal{O}_T)$, let b^T be the image of $b \in B$ under this homomorphism. An element $u \in \mathbb{G}_m(T)$ induces an isomorphism between our curve $(\mathcal{E} \times_S T,\omega_T)$ with "coordinates" a_4^T,a_6^T to the curve with coordinates $u^{-4}a_4^T,u^{-6}a_6^T$. Hence it is an automorphism of our curve if and only if

$$a_4^T = u^{-4}a_4^T \text{ and } a_6^T = u^{-6}a_6^T.$$

Since Δ^T is a unit we can conclude that $u^{12} = 1$. For any integer $n > 0$ we define the sub group scheme

$$\mu_n \longrightarrow \mathrm{Spec}(\mathbb{Z}) : \ \mu_n = \mathrm{Spec}(\mathbb{Z}[U]/(U^n - 1)) \subset \mathbb{G}_m = \mathrm{Spec}(\mathbb{Z}[U,U^{-1}],$$

the comultiplication is given by $m : U \longrightarrow U \otimes U$ (see 7.5.6.) We just showed that the image of ψ is contained in μ_{12}. If we now restrict the morphisms $T \longrightarrow S$ to faithfully flat morphisms (of finite type), then the homomorphism $B \longrightarrow H^0(T,\mathcal{O}_T)$ is injective and we find

$$\mathrm{Aut}(\mathcal{E}/S)(T) = \begin{cases} \mu_2(T) & \text{if } a_4 \text{ and } a_6 \neq 0 \\ \mu_4(T) & \text{if } a_4 \neq 0 \text{ and } a_6 = 0 \\ \mu_6(T) & \text{if } a_6 \neq 0 \text{ and } a_4 = 0 \end{cases}$$

Hence we get
For any scheme S of finite type over $\mathrm{Spec}(\mathbb{Z})$ *any any elliptic curve $\mathcal{E}/S$ we have a bijection*

$$\textit{Isomclasses of } S \textit{ -forms of } \mathcal{E}/S \xrightarrow{\sim} H^1(S,\mu_n)$$

where $n = 2,4,6$ depending on $\mathcal{E}/S$.

The cohomology groups can be computed, under our two assumptions $\frac{1}{6} \in B$ and all locally free B- modules of rank 1 are free. Then it follows from a standard computation

$$H^1(\mathrm{Spec}(B),\mu_n) = B^\times/(B^\times)^n$$

We can look at this from a different point of view. For any faithfully flat morphism $S' \longrightarrow S$ our functor $\mathcal{M}_1$ yields a diagram

$$\mathcal{M}_1(S) \xrightarrow{p_0^*} \mathcal{M}_1(S') \underset{p_2^*}{\overset{p_1^*}{\rightrightarrows}} \mathcal{M}_1(S' \times_S S'),$$

and this means that $\mathcal{M}_1/S$ yields a presheaf on the ffft-topology over S. Our considerations above show that this presheaf violates the first sheaf condition (Sh1), provided B is sufficiently general. The reason for this is the non triviality of the group $\mathrm{Aut}(\mathcal{E}/S)$ and in view of theorem 6.2.17 this destroys all our hopes that $\mathcal{M}_1/S$ might be representable. But also the second condition (Sh2) will be violated. Let $p_0 : S' \longrightarrow S$ be a ffft-morphism and let $\mathcal{E}'/S'$ an elliptic curve. We ask whether the elliptic curve descends to a curve over S, this means we ask whether we can construct a curve $\mathcal{E}/S$ such that $p_0^*(\mathcal{E}/S) = \mathcal{E} \times_S S'$ is isomorphic to $\mathcal{E}'/S'$. This is a special case of a question, which has been discussed in section 6.2.8. Let us assume that we found such an $\mathcal{E}/S$, we choose an isomorphism $\phi : \mathcal{E}\times_S S' \xrightarrow{\sim} \mathcal{E}'/S'$. We have the two projections p_1,$p_2 : S'\times_S S' \longrightarrow S'$, the composition with p_0 yields a morphism $q = p_0 \circ p_1 = p_0 \circ p_2 : S' \times_S S' \longrightarrow S$. Then we get isomorphisms

$$p_1^*(p_0^*(\mathcal{E})) \xrightarrow{p_1^*(\phi)} p_1^*(\mathcal{E}'/S'),\, p_2^*(p_0^*(\mathcal{E})) \xrightarrow{p_2^*(\phi)} p_2^*(\mathcal{E}'/S'),$$

the term on the left is $q^*(\mathcal{E}/S' \times_S S') = p_1^*(p_0^*(\mathcal{E})) = p_2^*(p_0^*(\mathcal{E}))$. Therefore we get an isomorphism

$$p_2^*(\phi) \circ p_1^*(\phi)^{-1} = \phi_{12} : p_1^*(\mathcal{E}'/S') \xrightarrow{\sim} p_2^*(\mathcal{E}'/S').$$

Now we have the projections $p_{ij} : S' \times_S S' \times_S S' \longrightarrow S' \times_S S'$ these 3 projections composed with the 2 projections $p_\nu : S' \times_S S' \longrightarrow S'$ yield the three projections π_μ : $S' \times_S S' \times_S S' \longrightarrow S' \times_S S'$ (see 6.2.8). A slightly tedious computation shows that these isomorphisms satisfy a cocycle relation

$$p_{13}^*(\varphi_{12})^{-1} \circ p_{23}^*(\varphi_{12}) \circ p_{12}^*(\varphi_{12}) = \mathrm{Id}\,,$$

where factors from right to left are morphisms $\pi_1^*(\mathcal{E}'/S') \longrightarrow \pi_2^*(\mathcal{E}'/S') \longrightarrow \pi_3^*(\mathcal{E}'/S') \longrightarrow \pi_1^*(\mathcal{E}'/S')$.

Therefore we can conclude:
A first necessary condition for the curve $\mathcal{E}'$ to descend to a curve is the existence of an isomorphism

$$\phi_{12} : p_1^*(\mathcal{E}'/S') \xrightarrow{\sim} p_2^*(\mathcal{E}'/S'). \qquad (G0)$$

To such an isomorphism we can define the boundary

$$\delta(\phi_{12}) = p_{13}^*(\varphi_{12})^{-1} \circ p_{23}^*(\varphi_{12}) \circ p_{12}^*(\varphi_{12}) \in \mathrm{Aut}(\mathcal{E}' \times_{S',\pi_1} S' \times_S S' \times_S S').$$

If this boundary $\delta(\phi_{12}) = \mathrm{Id}$ then this says that ϕ_{12} is a descend datum (see Definition 6.2.20). Once we have the descend datum we need to show that it is effective. In our special case this is easy, we simply extend the argument given on p.43 to the case of projective schemes.

We assume again that our base scheme $S = \mathrm{Spec}(B)$ is integral, under this assumption we know that $\mathrm{Aut}(\mathcal{E}'/S') = \mu_d/S' = \mu_d \times_S S'$ with $d = 2{,}4{,}6$, especially we see that it is abelian and it descends in canonical way to a group scheme over S. Then it is clear that $\delta(\phi_{12})$ is a 2-cocycle (See end of section 6.2.8). We can modify ϕ_{12}) by an automorphism $h \in \mathrm{Aut}(p_1^*(\mathcal{E}'/S'))/S' \times_S S'$, this has the effect that $\delta(\phi_{12})$ gets modified by a boundary and hence we get a cohomology class

$$[\mathcal{E}'/S'] = [\delta(\phi_{12})] \in H^2(S'/S,\mu_d).$$

This class is zero if and only find a $\phi'_{12} = \phi_{12} \circ h$, which is a descend datum.

Exercise 39. Find a field K and a separable quadratic extension F/K, such that there exists an elliptic curve $\mathcal{E}/F$, which is isomorphic to its conjugate under the non trivial automorphism of F/K, and which does not descend.

We ask a weaker question. Given our curve $\mathcal{E}'/S'$, we ask whether we can find a ffft-morphism $S'' \longrightarrow S'$ such that the extension $\mathcal{E}' \times_{S'} S'' \longrightarrow S''$ descends to a curve $\mathcal{E} \longrightarrow S$. Again we have the first necessary condition:
For our curve $\mathcal{E}'/S'$ we can find a ffft-covering $\mathcal{E}'' = \mathcal{E}' \times_{S'} S'' \longrightarrow S''$, such that we can find an isomorphism

$$\phi'_{12} : (p'_1)^*(\mathcal{E}''/S'') \xrightarrow{\sim} (p'_2)^*(\mathcal{E}''/S''), \qquad (G)$$

here $p'_i : S'' \times_S S'' \longrightarrow S''$ are the two projections.
If this condition is fulfilled the elliptic curve $\mathcal{E}' \longrightarrow S'$ is called a **gerbe** of elliptic curves over S . Now we say that the gerbe descends to an elliptic curve $\mathcal{E}/S$ if we can find a ffft-covering $S'' \longrightarrow S'$ such that $\mathcal{E}' \times_{S'} S'' \longrightarrow S''$ descends to a curve $\mathcal{E} \longrightarrow S$.
If $\mathcal{E}'/S'$ is a gerbe of elliptic curves then we may proceed as before and attach a class $[\delta_{S''}(\phi'_{12})] \in H^2(S''/S,\mu_d)$ to $\mathcal{E}'/S'$. But now we may pass to still finer and finer ffft-coverings $S''' \longrightarrow S''$ and get a class

$$[\delta(\mathcal{E}'/S')] \in \lim_{S''' \longrightarrow S} H^2(S'''/S,G) = H^2(S,G),$$

We see: A gerbe $\mathcal{E}'/S'$ descends if and only if $[\delta(\mathcal{E}'/S')] = 0$.
It is quite clear that not every gerbe descends. Let $S = \mathrm{Spec}(B)$, let us assume that $\frac{1}{6} \in B$. Let $S' \longrightarrow S$ be a ffft-morphism and let $\mathcal{E}' \longrightarrow S'$ be a gerbe. Its j invariant $j(\mathcal{E}') \in B'$ and the gerbe condition (G) implies that $i_1(j(\mathcal{E}')) = i_2((j(\mathcal{E}'))$ where $i_1,i_2 : B' \longrightarrow B' \otimes_B B'$ are the two inclusions given by the first and second component. But since $B \longrightarrow B'$ is faithfully flat this implies $j(\mathcal{E}') \in B$. On the other hand our two assertion (i) and (ii) imply that given a $j \in B$ we can find a ffft-morphism $S' \longrightarrow S$ and an elliptic curve $\mathcal{E}' \longrightarrow S'$ with $j(\mathcal{E}') = j$. If we introduce some obvious notion of equivalence of gerbes then we see easily that we get a unique equivalence class of gerbes this way.
Actually we have a universal gerbe: The morphism $M_{1,diff} \longrightarrow M_1($ or in affine writing $\mathbb{Z}[\frac{1}{6}][j] \hookrightarrow \mathrm{Spec}(\mathbb{Z}[\frac{1}{6}][a_4,a_6,\frac{1}{\Delta}])$ is ffft and the universal curve over $M_{1,diff}$ has j invariant j.
We come back to our exercise 39 above: If $B = K$ is a field, then every gerbe of elliptic curves is trivial. In a less sophisticated terminology this says:

For any $x \in K$ we can find a curve with j invariant equal to x. In other words the map $M_{1,diff}(K) \longrightarrow M_1(K)$ is surjective.
Finally we see that
If we have two elliptic curves $\mathcal{E}_1,\mathcal{E}_2$ over an algebraically closed field k then they are isomorphic if and only if

$$j(\mathcal{E}_1) = j(\mathcal{E}_2)$$

This means in modern language: The affine line $\mathrm{Spec}(\mathbb{Z}[X])$ is a **coarse moduli space** for the elliptic curves. In contrast to this the scheme $M_{1,diff} \longrightarrow \mathrm{Spec}(\mathbb{Z}[\frac{1}{6}])$ is a **fine moduli space** for the elliptic curves equipped with differentials over a ffft scheme over $\mathrm{Spec}(\mathbb{Z}[\frac{1}{6}])$. We will briefly come back to this issue in the next subsection.

9.6.3 Curves of higher genus

We want to discuss some aspects of the theory of curves of genus $g \geq 2$ The following considerations are more geometric in nature, hence we assume at this point that C/k is a smooth, projective and irreducible curve over an algebraically closed field k. Let g be the genus of our curve.

Exercise 40. We pick g points $P_1,\dots,P_g$ on our curve. We form the divisor $D = \sum P_i$ and ask: Is there a non-constant meromorphic function $f \in H^0(C,\mathcal{O}_C(D))$. ?

a) What does the Riemann-Roch theorem tell us? We see that the simple version of the Riemann-Roch is not good enough.

b) What do we need to know about the behavior of the holomorphic differentials at D if we want to answer our question?

c) We want to discuss this kind of questions in a more systematic way. We consider the product of our curve by itself $C \times_k C$, then we may consider the diagonal $\Delta \subset C \times_k C$. This diagonal can locally be described by one equation and hence we can look at the line bundle $O_{C\times C}(\Delta)$.
d) We form the product

$$C \times C^g = C \times \underbrace{C \times C \times \ldots \times C}_{g-\text{times}} = X.$$

On this variety we have the line bundles $\mathcal{O}_X(\Delta_i)$, which is obtained by the diagonal in the factor C in front and the i-th factor in the product. This gives us a line bundle $\mathcal{L} = \otimes_{i=1}^g \mathcal{O}_X(\Delta_i)$ on X.
Now we look at the projection to the second factor

$$\begin{array}{ccc} C & \times & C^g \\ & & \downarrow \pi. \\ & & C^g \end{array}$$

The sheaf $\mathcal{L}$ is flat over C^g. If we pick a point $P = (P_1,\dots,P_g) \in C^g$ and restrict $\mathcal{L}$ to the fibre $\pi^{-1}(P) \simeq C$ then this restriction is exactly the line bundle $\mathcal{O}_C(P_1 \ldots + P_g)$ on C.

e) Prove that we have a non-empty Zariski open subset $U \subset C^g$ such that

$$\dim_k H^0(C,\mathcal{O}_C(P_1 + \ldots + P_g)) = 1 \quad \text{if only if} \quad P = (P_1,\ldots,P_g) \in U.$$

f) Prove that $U \neq C^g$ and show that the complement is of codimension one.

The next exercise is difficult, that is the reason why it has two stars.
g**) Assume $g \geq 2$. Show that we can find a point P, for which there exists a holomorphic differential form $\omega \neq 0$, which has a zero of order $\geq g$ in P.
Hint: We put $\Omega_1^1 = p_1^*(\Omega_C^1)$ it is a line bundle on $C \times C$, and consider the sheaf

$$\mathcal{L} = \Omega_1^1(-(g-1)\Delta),$$

this is the sheaf of sections, which have a zero of order $\geq g-1$ along the diagonal. What is the restriction of this sheaf to a fibre $C \times \{P\}$?
The Riemann-Roch theorem shows that

$$\dim_k H^0(C,\Omega_C^1(-(g-1)P)) \geq 1.$$

Show:
a) If we can find a point P_0 where $\dim_k H^0(C,\Omega_C^1(-(g-1)P_0)) \geq 2$, then we are finished. Hence we can make the additional assumption that

$$\dim_k H^0(C,\Omega_C^1(-(g-1)P)) = 1$$

for all $P \in C(k)$, i.e. the rank is constant.
b) Reformulate this conditions in terms the homomorphism

$$H^1(C,\Omega_C^1(-aP)) \longrightarrow H^1(C,\Omega_C^1)$$

where $0 < a < g$ and P is any point on C.
We apply our semicontinuity theorem 8.4.5 to the projection p_2 to the second factor. We get that the sheaf $p_{2,*}(\Omega_1^1(-(g-1)\Delta))$ is a line bundle on the curve C. We have an inclusion $p_2^*(p_{2,*}(\Omega_1^1(-(g-1)\Delta))) \hookrightarrow \Omega_1^1(-(g-1)\Delta)$.
For a point $P \in V$ a non zero section $\omega_P \in H^0(C,\Omega_C^1(-(g-1)P))$ viewed as a holomorphic 1-form has $2g-2$ zeroes, if we view it as a section in $H^0(C,\Omega_C^1(-(g-1)P))$ it has $g-1$ zeroes. We consider the subset $Z \subset C(k) \times C(k)$, which consists of pairs

$$(Q,P) \in C \times C$$

where Q is a zero of the section $\omega_P \in H^0(C,\Omega_C^1(-(g-1)P))$. This is by construction a divisor on $C \times_k C$. and it is clear that the inclusion above yields an isomorphism

$$p_2^*(p_{2,*}(\Omega_1^1(-(g-1)\Delta)) \xrightarrow{\sim} \Omega_1^1(-(g-1)\Delta)(-Z) = \Omega_1^1(-(g-1)\Delta) \otimes \mathcal{O}_{C\times C}(-Z).$$

Now we must prove that Z has a non empty intersection with the diagonal!
We have to compute the intersection number $\mathcal{L}_1 \cdot \mathcal{L}_2$ of our two line bundles on $C \times_k C$ (see 8.4.1). It is rather clear that:

If Z,Δ have no component in common then the intersection number is equal to the number of points in $Z \cap \Delta$ counted with the right multiplicities.

But how can we compute the intersection number $\mathcal{O}_{C\times C}(Z)\cdot\mathcal{O}_{C\times C}(\Delta)$? To do this we have to use the fact that the intersection product is bilinear, and reduce the problem to the computation of $p_2^*(p_{2,*}(\Omega_1^1(-(g-1)\Delta))\cdot\mathcal{O}_{C\times C}(\Delta)$ and eventually reach the trickiest part of the computation namely the computation of $p_{2,*}(\Omega_1^1(-(g-1)\Delta)$. This can be done inductively by applying the direct image functor to the exact sequences $(0<a<g)$

$$0\longrightarrow \Omega^{(1)}\otimes\mathcal{O}_{C\times C}(-a\Delta)\longrightarrow \Omega^{(1)}\otimes\mathcal{O}_{C\times C}(-(a-1)\Delta)\longrightarrow \frac{\Omega^{(1)}\otimes\mathcal{O}_{C\times C}(-(a-1)\Delta)}{\Omega^{(1)}\otimes\mathcal{O}_{C\times C}(-a\Delta)}\longrightarrow 0.$$

The term on the right is a sheaf concentrated on the diagonal, which we identify with C. It follows from the definition of the sheaf of differentials that this quotient on the diagonal is

$$\frac{\Omega^{(1)}\otimes\mathcal{O}_{C\times C}(-(a-1)\Delta)}{\Omega^{(1)}\otimes\mathcal{O}_{C\times C}(-a\Delta)}\xrightarrow{\sim}(\Omega_C^1)^{\otimes a}$$

(see also the discussion of the Riemann-Roch theorem for surfaces following later). Now we apply the functor $p_{2,*}$ to this sequence and get an exact sequence of locally free sheaves (this is still our assumption, see b) above)

$$0\longrightarrow p_{2,*}(\Omega^{(1)}\otimes\mathcal{O}_{C\times C}(-a\Delta))\longrightarrow p_{2,*}(\Omega^{(1)}\otimes\mathcal{O}_{C\times C}(-(a-1)\Delta))\longrightarrow(\Omega_C^1)^{\otimes a}\longrightarrow 0.$$

Taking suitable exterior powers we get

$$p_{2,*}(\Omega^{(1)}\otimes\mathcal{O}_{C\times C}(-(g-1)\Delta))\xrightarrow{\sim}(\Omega_C^1)^{\frac{g(g-1)}{2}}$$

and from this we get the value

$$\text{"}\#(Z\cap\Delta)\text{"}=g^3-g\text{"}$$

End of the exercise g)**

If we pick a point $P\in C(k)$ and consider the spaces $H^0(C,\mathcal{O}_C(aP))$ for $a=1,\dots,\dots$ then we get from the Riemann-Roch formula

$$\dim_k H^0(C,\mathcal{O}_C(aP))-\dim_k H^1(C,\mathcal{O}_C(aP))=a+1-g.$$

We know that $1\in H^0(C,\mathcal{O}_C(aP))$ hence $\dim_k H^0(C,\mathcal{O}_C(aP))\geq 1$.
On the other hand it is clear that

$$\dim_k H^0(C,\mathcal{O}_C((a+1)P))\leq \dim_k H^0(C,\mathcal{O}_C(aP))+1$$

the dimension may jump by one if we go from a to $a+1$ or it may stay constant. Finally we have

$$\dim H^0(C,\mathcal{O}_C((2g-1)P))=g.$$

This means that if we go up from $a=0$ to $a=2g-1$ the dimension will jump exactly $g-1$ times.

h) Show that for a point P, for which $H^0(C,\Omega_C^1(-gP))=0$ we have

$$\dim_k H^0(C,\mathcal{O}_C(aP)) = \begin{cases} 1 & \text{for } 0 \le a \le g \\ a+1-g & \text{for } g+1 \le a \end{cases}.$$

A point P, for which $\dim_k H^0(C,\Omega^1_C(-gP)) > 0$ is called a **Weierstrass-point** . In our previous exercise we have shown that Weierstrass-points always exist. Actually we have shown more: Under the assumption that $\dim_k H^0(C,\Omega^1_C(-(g-1)Q)) = 1$ for all $Q \in C(k)$ we can show that either all points are Weierstrass-points (this happens if $\Delta \subset Z$) or the number of Weierstrass-points counted with the right multiplicity is $g^3 - g$. The first case can not happen if the characteristic of k is zero (See for instance [Gr-Ha] , Chap. II, section 4).
At the same place it is shown that the number of Weierstrass-points is always $g^3 - g$ if we count them with certain weights and if the characteristic of k is zero.
I formulate some questions, for which I do not know the answer. I state them as exercise and it is likely that they have been answered in the literature.

Exercise 41. a) Of course we know that a non-zero differential cannot have a zero of order $\ge 2g-1$. But I do not know what the record for the order of vanishing for a differential is.

b) For a given g can we find a curve C/k such that $H^0(C,\Omega^1_C((-gP)) \ge 1$ for all points $P \in C(k)$? This means that all points are Weierstrass-points.

c) How early can it happen that

$$\dim_k H^0(C,\Omega^1_C(-aP)) \ge g-a+1,$$

which is the minimal a, for which this happens for a special curve or the generic curve. We consider line bundles of degree $g+1, g+2, \dots$ and ask whether they are base point free or provide an embedding of the curve into projective space.

d) Any line bundle $\mathcal{L}$ of degree $d \ge g+1$ is of the form

$$\mathcal{L} = \mathcal{O}(P_1 + \ldots + P_d).$$

What does it mean that the bundle $\mathcal{L}$ is base point free?
Prove that there is a non-empty subset $U \subset C^{g+1}$ such that $\mathcal{O}(P_1 + \ldots + P_{g+1})$ is base point free for $(P_1 \ldots, P_{g+1}) \in U$.

e) Try to prove: For a non-open set $U \subset C^{g+2}$ the line bundle $\mathcal{O}(P_1 \ldots P_{g+2})$ provides a closed immersion

$$i : C \longrightarrow \mathbb{P}^2,$$

this means that on a non-empty open set the morphism i is injective and its differential is always non-zero. (Show that not all curves can be realized as plane projective curves).

f) Prove that for a non-empty open set $U \subset C^{g+3}$ the line bundle

$$\mathcal{L} = \mathcal{O}(P_1 + \ldots + P_{g+3})$$

provides a closed embedding of C into $\mathbb{P}^3$.

The "moduli space" of curves of genus g

We may look at the questions, which we formulated in the above group of exercises, from a much more systematic point of view. In the the first part of this section we discussed the moduli space of elliptic curves, recall that this a curves of genus one together with a distinguished point. We may also ask for a moduli space $\mathcal{M}_g$ of curves of genus g, this is an algebraic variety over k, which " parameterizes" the curves of genus g. Here we are entering a dangerous terrain, such a space can not exist because curves may have non trivial automorphisms. (See also the previous discussion of the moduli space of elliptic curves .) To overcome this difficulties Deligne and Mumford introduced more general objects, namely the stacks. These stacks form a 2-category (this means that the Hom(,) do not form a set but only a category) and in this context stack the moduli stack $\mathcal{M}_g/\operatorname{Spec}(\mathbb{Z})$ of curves of genus g exists. We do not attempt to give precise definitions and statements here, but the previous discussion in the case $g = 1$ should give some impression what this means (See also 7.5.8, for an account of the theory of stacks see [L-MB].)

For a first understanding one may ignore the difference between a stack and a variety. In any case the object $\mathcal{M}_g$ is of great interest in actual research and we want to state some known results and ask some questions.

It can be shown- and was already known to Riemann- that $\mathcal{M}_g$ has dimension $3g - 3$, it is a quasi-projective variety and Deligne and Mumford constructed a compactification of it (See [De-Mu]). Hence we know that to a point $m \in \mathcal{M}(k)$ we find a curve C_m. For this curve we may ask the following question: Let d be an integer $0 < d \leq g$. Can we find a divisor $D = P_1 + \cdots + P_d$ such that $H^0(C_m,\mathcal{O}_{C_m}(P_1 + \cdots + P_d))$ contains a non constant function, i.e. is of dimension > 1.

We have seen that a non constant function $x \in H^0(C_m,\mathcal{O}_{C_x}(P_1 + P_2 \cdots + P_d)$ provides a morphism $\Phi_x : C_m \longrightarrow \mathbb{P}^1$ and during the proof of theorem 9.4.1 we saw that the degree of Φ_x is equal to d. Since $g \geq 2$ we can conclude that the case $d = 1$ does not occur.

But it may happen that $d = 2$, i.e. we can find pairs of points $P,Q \in C_m(k)$ such that we have a non constant function $x \in H^0(C_m,\mathcal{O}_{C_x}(P + Q))$. Then the morphism $\Phi_x : C \longrightarrow \mathbb{P}^1_k$ is of degree 2. Let us assume that the characteristic of k is not two and for simplicity that $P \neq Q$. Then we know that the ring of regular functions on $C_m \setminus \{P,Q\} = U$ is the integral closure of $k[x]$ in the function field of C_m. It is clear that we can find a regular function $y \in \mathcal{O}$, which satisfies an equation

$$y^2 = x^n + a_1x^{n-1} + \cdots + a_0 = f(x).$$

The polar part of the divisor of y is $-\frac{n}{2}(P + Q)$ hence we see that n must be even. The reader is invited to fill the gaps in the following reasoning:

a) There exists an unique involution $\Theta : C \longrightarrow C$, i.e. $\Theta^2 = \mathrm{Id}$, such that

$$\dim_k H^0(C_m,\Omega^1_{C_m}(-P' - Q')) = g - 1 \text{ if and only if } Q' = \Theta(P').$$

(The argument for this is not so obvious)

b) The element x is fixed by Θ and $\Theta(y) = -y$, the subfield $k(x)$ is uniquely determined by C_m, it is the fixed field of Θ.

c) The roots of $f(x)$ are pairwise different, the genus of the curve is $g = \frac{n}{2} - 1$, the holomorphic differentials are of the form

$$\omega = \frac{p(x)}{y} dx,$$

where $p(x)$ is a polynomial of degree $\leq \frac{n}{2} - 2$.

A curve C_m, for which we have an involution Θ, which satisfies a) above is called **hyperelliptic** . It is clear that the isomorphism class of a hyperelliptic curve is determined by the set of zeroes of the polynomial $f(x)$. Since variable x is not unique and can be replaced by

$$x' = \frac{ax+b}{cx+d}$$

we can assume the three of these zeroes are $0{,}1, -1$ (recall that the characteristic of k is not equal to 2!). Then the remaining $n-3$ zeroes are independent variables, which are determined by the curve up to a permutation by the curve C_m. Hence we see that the moduli space of hyperelliptic curves of genus g has dimension $2g-1$ and the theorems on semi-continuity imply that the hyperelliptic curves of genus g form a closed subvariety of this dimension in $\mathcal{M}_g$. Only if $g = 2$ all curves are hyperelliptic.
If $g > 2$ then we can consider the open subvariety $\mathcal{M}^{\text{nonhyp}}$ of non hyperelliptic curves, i.e. we need $d \geq 3$. Again we can ask for those curves, for which we find three points P_1, P_2, P_3 such that we can find a non constant $x \in H^0(C_m, \mathcal{O}_{C_x}(P_1 + P_2 + P_3))$. Again these curves will form a subvariety $\mathcal{M}_g^{(3)}$ in the moduli space.
So we found a procedure to construct sub varieties in the moduli space and it is an interesting problem to understand these subvarieties.
We may also play the same game with Weierstrass points, in the exercise g**) we made the assumption that the first jump happens if we go from $a = g-1$ to $a = g$. Again it is rather clear that the curves C_m, for which this is true form an open subset U (non empty?) and we can define subvarieties by describing certain certain pattern of jumps. (For further developments see [Arb] [ACGH].)

9.7 The Grothendieck-Riemann-Roch Theorem

The Riemann-Roch Theorem has been generalized. We want to discuss and explain this generalization but in a rather informal way. We skip the precise discussion of some of the concepts needed, and we do not give the proofs. But we will give a detailed exposition in a non trivial special case and hopefully in this exposition some of the general ideas will be visible. For a very condensed treatment we also refer to [Fa2].
The first step is of course formulate a question for higher dimensional, smooth projective schemes:
Let $X \longrightarrow \text{Spec}(k)$ be a projective algebraic variety, let $\mathcal{F}$ be a line bundle (or a divisor or even a vector bundle) on X, how can we compute the Euler characteristic $\chi(X, \mathcal{F}) = \sum_\nu (-1)^\nu H^\nu(X, \mathcal{F})$?

The answer has to be expressed in terms of certain data, which we attach to the coherent sheaf. In the case of curves these data where the degree and the rank, which are numbers or at least look like numbers. Hence we encounter a first problem, namely we have to attach such data to our sheaf in the case of higher dimensional varieties.
For surfaces, i.e $\dim(X) = 2$ the classical algebraic geometers proved a Riemann-Roch theorem for divisors. If $X = \mathbb{P}^n_k$ and $\mathcal{F} = \mathcal{O}_{\mathbb{P}^n_k}(r)$ we have a formula for the Euler-characteristic if $\mathcal{F} = \mathcal{O}_{P^n_k}(r)$ in terms of n,r (see theorem 8.2.5).
If the ground field is $\mathbb{C}$, then the set of complex valued points $X(\mathbb{C})$ is a complex manifold and therefore, a topological space. If $\mathcal{F}$ a holomorphic vector bundle, then we have the theory of Chern classes, these are cohomology classes $c_i(\mathcal{F}) \in H^{2i}(X(\mathbb{C}),\mathbb{Z})$, they are topological invariants attached to the bundle. Then the Riemann-Roch formula of Hirzebruch (See [Hi] and exercise 43) expresses the Euler characteristic in terms of the Chern classes of $\mathcal{F}$ and the Chern classes of the tangent bundle of X.
A. Grothendieck formulated and proved a still more general and certainly much more systematic version of the Riemann-Roch Theorem. He considers a morphism $f : X \longrightarrow Y$ between smooth projective varieties over k. Let $\mathcal{F}$ be a coherent sheaf on X. Then the higher direct images $R^\nu f_*(\mathcal{F})$ are again coherent. Recall that we have defined the K-groups $K(\ )$,$K'(\ )$ in (9.5.4 on p. 220). They provide elements $[R^\nu f_*(\mathcal{F})] \in K'(Y)$ and we can define $R^\bullet f_*(\mathcal{F}) = \sum_\nu (-1)^\nu [R^\nu f_*(\mathcal{F})] \in K'(Y)$. The Grothendieck-Riemann-Roch Theorem provides information concerning this element in $K'(Y)$.
We want to illustrate and prove the Grothendieck-Riemann-Roch Theorem in a special case.

9.7.1 A special case of the Grothendieck -Riemann-Roch theorem

Let C/k be a smooth absolutely irreducible curve. We form the product of our curve C/k by itself, we put $X = C \times_k C$, we have the two projections p_1,$p_2 : X \rightrightarrows C$. Let $\mathcal{L}$ be a line bundle on X. Of course we may ask again for a formula for the Euler characteristic

$$\chi(X,\mathcal{L}) = \sum \dim_k(-1)^\nu H^\nu(X,\mathcal{L})$$

in terms of certain data, which are attached to the line bundle $\mathcal{L}$. The answer is given by the classical Riemann-Roch theorem for surfaces. It is treated in Hartshorn's book in Chapter IV for arbitrary smooth projective surfaces X. Here we discuss the special case where X is the product of a curve by itself, but we want to discuss a stronger theorem, which is a special case of a Riemann- Roch theorem in the sense of Grothendieck.
In the supplementary section on the properties of the degree function (See p. 218) we introduced the group $K'(C)$ and I showed that the formula of Riemann-Roch gave an expression for the Euler characteristic of an element in $K'(C)$ in terms of its degree and its rank. Here we recall that we interpreted the Euler characteristic, the degree and the rank as homomorphisms

$$\chi_C : K'(C) \longrightarrow \mathbb{Z}$$
$$\mathrm{Rank}_C : K'(C) \longrightarrow \mathbb{Z}$$
$$\deg_C : K'(C) \longrightarrow \mathbb{Z}.$$

The Grothendieck-Riemann-Roch theorem will give an answer to the the following question: Let $\mathcal{L}$ be a line bundle on X, we consider the coherent sheaves $R^\nu p_{2*}(\mathcal{L})$ on the curve. How can we compute the Euler characteristic

$$R^\bullet p_{2*}(\mathcal{L}) = \sum (-1)^\nu [R^\nu p_{2*}(\mathcal{L})] \tag{9.21}$$

as an element in $K'(C)$?
If we are able to "compute" this element, then we can also compute $\chi(X,\mathcal{L})$: We use the spectral sequence with E_2-term $(H^p(C,R^q p_{2*}\mathcal{L})),d_2) \Rightarrow H^n(X,\mathcal{L})$ and as explained in Vol.I. 4.6 we get

$$\chi_C(R^\bullet p_{2*}\mathcal{L})) = \chi(X,\mathcal{L}) = \sum \dim_k (-1)^\nu H^\nu(X,\mathcal{L})$$

Hence it is clear that we have a better theorem if we have a formula for $R^\bullet p_{2*}(\mathcal{L})$, such a formula will imply the classical Riemann-Roch formula for $C \times_k C$.

We explain how we can get such a formula in a special situation. Let us assume that our line bundle is of the form $\mathcal{L} = \mathcal{O}_X(Z)$ where Z is a smooth irreducible curve in X, which is not one of the fibers. Therefore the restrictions of the two projections $p_1 : Z \longrightarrow C, p_2 : Z \longrightarrow C$ are finite, we denote their degrees by $d_1(Z), d_2(Z)$. We proceed in exactly the same way as in the case of curves. We write the exact sequence

$$0 \longrightarrow \mathcal{O}_X \longrightarrow \mathcal{O}_X(Z) \longrightarrow \mathcal{O}_X(Z)/\mathcal{O}_X \longrightarrow 0.$$

This gives us the formula

$$\chi(X,R^\bullet p_{2*}(\mathcal{L})) = \chi(X,R^\bullet p_{2*}(\mathcal{O}_X)) + \chi(R^\bullet p_{2*}(\mathcal{O}_X(Z)/\mathcal{O}_X))$$

The situation is similar (but much more complicated) as in the case of curves. The sheaf $\mathcal{O}_X(Z)/\mathcal{O}_X)$ is supported on Z and since the morphism $p_{2*} : Z \longrightarrow C$ is finite the higher direct images vanish. Therefore we have $\chi(R^\bullet p_{2*}(\mathcal{O}_X(Z)/\mathcal{O}_X) = [p_{2*}(\mathcal{O}_X(Z)/\mathcal{O}_X)]$ and since it is clear that $\chi(X,R^\bullet p_{2*}(\mathcal{O}_X)) = -[\mathcal{O}_C^{g-1}]$ our formula simplifies

$$\chi(X,R^\bullet p_{2*}(\mathcal{L})) = -[\mathcal{O}_C^{g-1}] + [p_{2*}(\mathcal{O}_X(Z)/\mathcal{O}_X)].$$

Hence we are left with the "computation" of $[p_{2*}(\mathcal{O}_X(Z)/\mathcal{O}_X)]$.
Here we have to stop for a second. What does it mean to compute this element in $K'(C)$? This group is much to complicated to identify this individual object. But remember what we actually want. Eventually we want to compute the Euler characteristic $\chi_C(\chi(X,R^\bullet p_{2*}\mathcal{L}))$ and to do this we only need to know the degree and the rank of $\chi(X,R^\bullet p_{2*}\mathcal{L})$. Therefore we will be content if we can compute the two numbers

$$\deg p_{2*}(\mathcal{O}_X(Z)/\mathcal{O}_X) \text{ and } \operatorname{Rank} p_{2*}(\mathcal{O}_X(Z)/\mathcal{O}_X).$$

In principle we learned how to do that. We did this in the special case when we discussed the degree of the module of differentials in section 9.5.3. It is quite clear that the rank $\operatorname{Rank} p_{2*}(\mathcal{O}_X(Z)/\mathcal{O}_X)$ is degree $d_2(Z)$ of the morphism $p_{2*} : Z \longrightarrow C$. The computation of the degree is more subtle and will be done in the following section.

9.7.2 Some geometric considerations

The following considerations are valid for arbitrary smooth surfaces X and a smooth curve $Z \subset X$. The $\mathcal{O}_X$ - module sheaf $\mathcal{N} = \mathcal{O}_X(Z)/\mathcal{O}_X$ is a line bundle when we restrict it to Z. To see this we recall that we can write Z locally by one equation and hence the ideal $\mathcal{I}(Z)$, which defines Z is a line bundle, it is the line bundle $\mathcal{O}_X(-Z)$. Then $\mathcal{I}(Z) \otimes \mathcal{O}_Z = \mathcal{I}(Z)/\mathcal{I}(Z)^2$ is a line bundle on Z, it is the conormal bundle (see 7.5.5). The bundle $\mathcal{N} = \mathcal{O}_X(Z)/\mathcal{O}_X$ is obviously dual to $\mathcal{I}(Z)/\mathcal{I}(Z)^2$.

If we restrict the tangent bundle T_X to Z then we get an exact sequence

$$0 \longrightarrow T_Z \longrightarrow T_X|Z \longrightarrow T_X|_Z/T_Z \longrightarrow 0$$

But now one sees that $I(Z)/\mathcal{I}(Z)^2$ is the dual bundle to the quotient $T_X|Z/T_Z$. This is so because the differentials df for $f \in \mathcal{I}(Z)$ can be evaluated on the tangent vectors $t_z \in T_{X,z}$ they vanish on the subspace $T_{Z,z}$ and this gives the pairing. Hence it is clear that $\mathcal{N} = T_X|_Z/T_Z$ is the normal bundle. The degree of the line bundle $\mathcal{N}$ on Z is equal to the intersection number $Z \cdot Z = \mathcal{O}_X(Z) \cdot \mathcal{O}_X(Z)$. (see 8.4.2)
We get a formula

$$\Lambda^2(T_X|_Z) = T_Z \otimes \mathcal{N},$$

which is called the **adjunction formula**. Very often is is written in dual form. The second exterior power $\Lambda(T_X)$ is the dual bundle to the so called **canonical bundle** $K_X = \Lambda^2(\Omega^1_X)$ and we get

$$K_X|_Z = \Omega^1_Z \otimes \mathcal{I}(Z)/\mathcal{I}(Z)^2.$$

This yields in terms of degrees

$$\deg(K_X|_Z) = \deg(\Omega^1_Z) + \deg(\mathcal{I}(Z)/\mathcal{I}(Z)^2)$$

and this can be interpreted in terms of intersection numbers as (see the considerations following below)

$$K_X \cdot Z = 2g_Z - 2 - Z \cdot Z$$

.

We return to our discussion of the Grothendieck-Riemann-Roch theorem, we assume again that $X = C \times_k C$. Hence we have in $K'(X)$ that $[\mathcal{N}] = [\mathcal{O}_Z] + [\mathcal{M}]$, where $\mathcal{M}$ is a virtual torsion sheaf of degree $Z \cdot Z$. Hence we get

$$p_{2*}(\mathcal{O}_X(Z)/\mathcal{O}_X) = p_{2*}(\mathcal{N}) = [p_{2*}(\mathcal{O}_Z)] + [p_{2*}\mathcal{M}]$$

The second term is torsion and consequently its rank is zero and its degree is $Z \cdot Z$. We are not yet at the end, we need to compute $p_{2*}(\mathcal{O}_Z)$. This can be done by the same method, which we applied when we computed the degree of the sheaf of differentials. Again we introduce the "different"-module $\mathcal{D}_{Z/C}$ using the same definition as before, we replace $\mathbb{P}^1$ by C and C by Z. We have the perfect duality of $\mathcal{O}_C$-modules

$$p_{2*}(\mathcal{O}_Z) \times p_{2*}(\mathcal{D}^{-1}_{Z/C}) \longrightarrow \mathcal{O}_C,$$

which implies that the degrees of these two modules add up to zero. Again we have the Hurwitz formula

$$\Omega^1_Z = p_2^*(\Omega^1_C) \otimes \mathcal{D}^{-1}_{Z/C}$$

This gives a formula for the degree of $\mathcal{D}^{-1}_{Z/C}$ and this formula yields

$$\deg(p_{2*}(\mathcal{O}_Z)) = -\frac{1}{2}\deg(\mathcal{D}^{-1}_{Z/C}) = d_2(Z)(g-1) - (g_Z - 1)$$

This formula is still not completely satisfactory, we have to compute the genus g_Z of the curve Z. To get this we recall that we have seen that $T_Z \otimes \mathcal{N} = \Lambda^2(T_X|_Z)$, which implies $2g_Z - 2 = Z \cdot Z - \deg(\Lambda^2(T_X|_Z)$

Collecting all the terms gives us

$$(\mathrm{Rank}\,,\deg)(\mathcal{O}_X(Z)/\mathcal{O}_X) = (d_2(Z),d_2(Z)(g-1) + \frac{1}{2}(Z\cdot Z + \deg(\Lambda^2(T_X|_Z)))$$

The tangent bundle T_X is the direct sum of the two pullbacks of the tangent bundle on C hence it is clear that $\deg(\Lambda^2(T_X|_Z)) = (2-2g)(d_1(Z)+d_2(Z))$ and our formula simplifies to

$$(\mathrm{Rank}\,,\deg)(\mathcal{O}_X(Z)/\mathcal{O}_X) = (d_2(Z), -d_1(Z)(g-1) + \frac{1}{2}Z\cdot Z)$$

To get the final formula we remember that $\mathcal{O}_X(Z)/\mathcal{O}_X$ was only one term in our exact sequence. For our original line bundle $\mathcal{L} = \mathcal{O}_X(Z)$ we get

$$(\mathrm{Rank}\,,\deg)(R^\bullet p_{2*}(\mathcal{O}_X(Z))) = (-(g-1) + d_2(Z), -d_1(Z)(g-1) + \frac{1}{2}Z\cdot Z).$$

Recall that we now have a formula for $R^\bullet p_{2*}(\mathcal{L})$ (see 9.21) for a line bundle $\mathcal{L} = \mathcal{O}_X(Z)$, where Z is a smooth curve. But now it is clear how to get a formula for arbitrary line bundles on X. On the product X we have the two special divisors $H_1 = x_0 \times C, H_2 = C \times x_0$ where x_0 is just an arbitrary point. Then it is clear that the degrees $d_1(Z) = Z\cdot H_1, d_2(Z) = Z\cdot H_2$. Hence the general formula should be

$$(\mathrm{Rank}\,,\deg)(R^\bullet p_{2*}(\mathcal{L})) = (-(g-1) + \mathcal{L}\cdot H_2, -(g-1)\mathcal{L}\cdot H_1 + \frac{1}{2}\mathcal{L}\cdot\mathcal{L}) \qquad (9.22)$$

To prove it we consider exact sequences $(a>0)$

$$0 \longrightarrow \mathcal{L} \longrightarrow \mathcal{L}(aH_1) \longrightarrow \mathcal{L}(aH_1)/\mathcal{L} \longrightarrow 0$$

$$0 \longrightarrow \mathcal{L} \longrightarrow \mathcal{L}(aH_2) \longrightarrow \mathcal{L}(aH_2)/\mathcal{L} \longrightarrow 0$$

and we apply $R^\bullet p_{2*}$ to both sequences. It is not difficult to show that in both cases the resulting long exact sequence shows that the formula above is true for $\mathcal{L}$ if and only if it is true for $\mathcal{L}(aH_1)$ resp. $\mathcal{L}(aH_2)$.

Then it is clear that it suffices to prove the Riemann-Roch formula for $\mathcal{L}(a_1H_1 + a_2H_2)$ where $a_1,a_2 >> 0$. But then the bundle $\mathcal{L}(a_1H_1 + a_2H_2)$ will be very ample and provide a projective embedding of $C\times_k C$. Then we get $\mathcal{L}(a_1H_1 + a_2H_2) = \mathcal{O}_X(Z)$ where Z is a section with a hyperplane. Now we invest the theorem of Bertini, which says that we can choose the hyperplane so that Z is smooth and now we are in the case, which we treated above.

Of course we can now easily derive the formula for the Euler characteristic of $H^\bullet(X,\mathcal{L})$:

$$\sum(-1)^\nu \dim_k H^\nu(X,\mathcal{L}) = (g-1)^2 - (g-1)\mathcal{L}\cdot(H_1+H_2) + \frac{1}{2}\mathcal{L}\cdot\mathcal{L}.$$

It is quite clear that our approach still has a defect. In our argument we used that we have a Riemann -Roch formula for arbitrary coherent sheaves on curves. More precisely we used the following. If we have two smooth curves C_1,C_2 and a finite morphism $f: C_1 \longrightarrow C_2$ and if we have a coherent sheaf $\mathcal{F}$ on C_1 then we a formula for $[f_*(\mathcal{F})]$ as an element in the group $K'(C_2)$ in terms of $[\mathcal{F}] \in K'(C_1)$. We used the fact that it suffices to compute $f_*(\mathcal{O}_{C_1})$ and this is done by the Riemann-Hurwitz formula (See 9.3.3).

Hence we should be much more consequent and ask for a formula for $R^\bullet p_{2*}(\mathcal{F})$ for any coherent sheaf on X and not only for line bundles. We can define the group $K'(X)$ by the same construction as in the case of curves p. 218 and then $\chi_f : \mathcal{F} \longrightarrow R^\bullet p_{2*}(\mathcal{F}))$ is a homomorphism $\chi_f : K'(X) \longrightarrow K'(C)$. This situation is analogous to the situation discussed on p. 218. We have to attach certain coarser data to the element $[\mathcal{F}]$, from which we can compute the corresponding coarser data of $R^\bullet p_{2*}(\mathcal{F})$, namely the degree and the rank. To define these coarser data attached to $[\mathcal{F}]$ we have to introduce the Chow ring, this will be done in the next section.

9.7.3 The Chow ring

For a more detailed exposition of the following we refer to the article of Borel and Serre [Bo-Se] and to the book of Fulton [Fu].
We discuss the **Chow ring**

$$A^\bullet(X) = A^0(X) \oplus A^1(X) \oplus A^2(X) \oplus \dots$$

for a smooth projective, absolutely irreducible variety X/k of dimension d. .
The Chow ring is an associative, commutative graded ring. The graded pieces $A^\nu(X)$ are abelian groups and any irreducible reduced sub scheme $Z \subset X$ of codimension ν gives us a class $[Z] \in A^\nu(X)$. The graded piece $A^\nu(X)$ is generated by these classes, i.e. any element in $A^\nu(X)$ can be written as a finite sum $\sum n_Z[Z]$. If $\nu = 1$ then this says that $A^1(X)$ is a quotient of $\mathrm{Div}(X)$.
We want to define a ring structure on $A^\bullet(X)$ by defining an intersection product

$$A^\nu(X) \times A^\mu(X) \longrightarrow A^{\nu+\mu}(X),$$

which should basically be given by intersecting cycles $[Z_1]\cdot[Z_2] = [Z_1 \cap Z_2]$. We explained already that this is too naive and requires some extra reasoning. We have to introduce an equivalence relation on the cycles, which allows us to choose representing cycles $Z_i' \in [Z_i]$ such that Z_1',Z_2' lie in a "nice position" relative to each other. For instance we can require that for each irreducible component Y of $Z_1 \cap Z_2$ the intersection is transversal in Y (see 7.5.21). In this last case we define

$$[Z_1] \cdot [Z_2] = \sum_{\text{irred comp } Y \text{ in} Z_1' \cap Z_2'} Y.$$

Of course we have to show that this product is well defined. But we also want it to be non trivial. To get this non triviality we consider $A^d(X)$. Then an element in $A^d(X)$ is of the form $c = \sum_{\mathfrak{p}} n_{\mathfrak{p}} \mathfrak{p}$ where $\mathfrak{p}$ are closed points. We require that the homomorphism

$$c \mapsto \sum_{\mathfrak{p}} n_{\mathfrak{p}}[k(\mathfrak{p}) : k] = \deg(c)$$

induces a homomorphism $\deg : A^d(X) \longrightarrow \mathbb{Z}$.
Such equivalence relations exist, actually we have several options for choosing them. Therefore we get different versions of a Chow ring, the resulting Chow ring will of depend on the choice of the equivalence relation. For instance we may take as equivalence relation the **linear equivalence** of cycles (See [Bo-Se], [Fu].) Two cycles $A = \sum n_Z Z$,$B = \sum n_{Z'} Z'$ of codimension ν are linear equivalent if we can find a smooth sub scheme

$W \subset X$ of codimension $\nu - 1$, which contains the support of these cycles- hence the cycles can be viewed as divisors on W- and a meromorphic function f on W whose divisor is $A - B$. This relation of linear equivalence on cycles generates the relation of linear equivalence on the free group of cycles.
This is a very fine equivalence relation and this has the effect that in some sense the $A^\nu(X)$ become very "big" and cannot be controlled. In any case the "moving lemma" (see ([Fu])) guarantees that we can define the intersection product and hence a ring structure on $A^\bullet_{\mathrm{lin}}(X)$. Let us call this Chow ring $A^\bullet_{\mathrm{lin}}(X)$. Our considerations in 9.4 allow us to describe the group $A^1_{\mathrm{lin}}(X)$, in this case the codimension 1 cycles are simply the divisors and the relation of linear equivalence is exactly the linear equivalence relation for divisors. Hence we see that

$$A^1_{\mathrm{lin}}(X) \xrightarrow{\sim} \mathrm{Pic}(X) \tag{9.23}$$

If we work with this equivalence relation, then we have to pay for it. If X/k is irreducible then we clearly have $A^0(X) = \mathbb{Z}$, the group is generated by X itself. If we hope for some kind of duality then we may wish that also $A^d_{\mathrm{lin}}(X) = \mathbb{Z}$ or more precisely we hope that the homomorphism $\deg : A^d_{\mathrm{lin}}(X) \longrightarrow \mathbb{Z}$ might be an isomorphism. In general it will not be surjective, but this is not so bad. For instance if our ground field is algebraically closed, then it will be surjective. But in general it will have a big kernel. Already in the case of projective smooth curves C/k the kernel is the group $\mathrm{Pic}^0(C/k)$ (see 9.4.1), which may be very big and difficult to understand. So our hope fails completely.
We may also start from the algebraic equivalence of cycles: Two irreducible sub-varieties Z_1,Z_2 are called **algebraically equivalent** if we can find a connected scheme T/k of finite type and a flat family schemes $\mathcal{Z} \subset X \times_k T$ such that we can find two points t_1,t_2 such that $\mathcal{Z}_{t_1} = Z_1, \mathcal{Z}_{t_2} = Z_2$. We can extend this equivalence relation to the free group of cycles and define $A^\nu_{\mathrm{alg}}(X)$ as the group of cycles of codimension ν modulo algebraic equivalence. This may open the option to define the Chow ring $A^\bullet_{\mathrm{alg}}(X)$, but I do not know a reference for this.
But for the relation of algebraic equivalence we have a better understanding of $A^d_{\mathrm{alg}}(X)$. Let us assume that k is algebraically closed. Then $A^d_{\mathrm{alg}}(X)$ generated by zero dimensional cycles $\sum_P n_P P$. If $P,Q \in X(k)$ then the theorem of Bertini asserts, that we can find to smooth hypersurface sections $X \cap H_1, X \cap H_2$ with $P \in H_1(k), Q \in H_2(k)$. We can find a point $Q_1 \in X \cap H_1 \cap H_2$. Iterating this shows that we can find a finite collection of smooth curves $C_1, C_2, \dots, C_r$ and points $Q_1, \dots, Q_{r-1}$ such that $P,Q_1 \in C_1(k), Q_1,Q_2 \in C_2(k), \dots, Q_{r-1}, Q \in C_r(k)$ and this clearly implies that $[P] = [Q] \in A^d_{\mathrm{alg}}$, and from this we get that $deg : A^d_{\mathrm{alg}}(X) \xrightarrow{\sim} \mathbb{Z}$. If k is not algebraically closed a cycle represents the trivial class if and only iff $\sum_{\mathfrak{p}} n_{\mathfrak{p}}[k(\mathfrak{p}) : k] = 0$. If in addition X/k has a rational point $P \in X(k)$ and if $d = \dim(X)$ then $A^d_{\mathrm{alg}}(X) = \mathbb{Z}$ and is generated by the class of this point. Hence we see that we get an injection $A^d_{\mathrm{alg}}(X) \longrightarrow \mathbb{Z}$. Certain diophantine problems amount to the computation of the index of the image.
If we accept the ring structure on $A^\bullet_{\mathrm{lin}}(X)$, then we can introduce the very coarse equivalence relation of numerical equivalence: Two $\nu-$ codimensional cycles Z_1,Z_2 are numerically equivalent, if for all cycles Z_3 in codimension $d-\nu$ we have $\deg(Z_1 \cdot Z_3) = \deg(Z_2 \cdot Z_3)$. If we divide the by this relation of numerical equivalence then we get quotients $A^\nu_{\mathrm{num}}(X)$ and clearly the ring structure on $A^\bullet_{\mathrm{lin}}(X)$ defines a product structure on $A^\bullet_{\mathrm{num}}(X)$. For this version of the Chow ring the pairing

$$A^{\nu}_{\mathrm{num}}(X) \otimes \mathbb{Q} \times A^{d-\nu}_{\mathrm{num}}(X) \otimes \mathbb{Q} \longrightarrow \mathbb{Q}$$

is non degenerate by definition.
Finally we mention that for a smooth surface X/k we have defined the intersection product

$$A^1_{\mathrm{lin}}(X) \times A^1_{\mathrm{lin}}(X) \xrightarrow{\mathrm{int}} A^2_{\mathrm{lin}}(X) \xrightarrow{deg} \mathbb{Z}.$$

Here we only need to recall that $A^1_{\mathrm{lin}}(X) \xrightarrow{\sim} \mathrm{Pic}(X)$ and to apply our results in the end of 8.4.1. To be a little bit more precise, we only constructed the composition deg $\circ$ int but a closer look at the reasoning shows that we actually also constructed int.
For projective, smooth surfaces X/k we have defined the non degenerate pairing

$$A^1_{\mathrm{num}}(X) \times A^1_{\mathrm{num}}(X) \xrightarrow{I} \mathbb{Z}.$$

If the ground field is $k = \mathbb{C}$ then we can attach to any irreducible smooth subscheme $Z \subset X$ of codimension ν a class $[Z] \in H^{2\nu}(X(\mathbb{C}),\mathbb{Z})$. This is the so called **cycle class**. This has been explained in Vol. I 4.8.8: We consider $X(\mathbb{C})$ as an oriented $2d$ dimensional $\mathcal{C}_\infty$ manifold and the set of complex valued points $Z(\mathbb{C})$ is a $2d-2r$ dimensional submanifold and for this situation we constructed the fundamental class $[Z(\mathbb{C})] \in H^{2\nu}(X(\mathbb{C}),\mathbb{Z})$ and this will be our $[Z]$.
If we drop the assumption that $Z/\mathbb{C}$ is smooth then it is still possible to attach to it a cycle class $[Z] \in H^{2\nu}(X(\mathbb{C}),\mathbb{Q})$. We will not construct this class in detail but we give the basic idea behind this construction and explain the geometric meaning of this class.
We recall Poincaré-duality, it gives us a non degenerate pairing

$$H^{2\nu}(X(\mathbb{C}),\mathbb{Q}) \times H^{2d-2\nu}(X(\mathbb{C}),\mathbb{Q}) \xrightarrow{\cup} \mathbb{Q},$$

hence we know that we know the class $[Z]$ if we know the values $[Z] \cup \xi$ for all $\xi \in H^{2d-2\nu}(X(\mathbb{C}),\mathbb{Q})$. In Vol. I 4.8.6 we gave a somewhat sketchy argument that we have a canonical isomorphism $PD : H_{2\nu}(X(\mathbb{C}),\mathbb{Q}) \xrightarrow{\sim} H^{2d-2\nu}(X(\mathbb{C}),\mathbb{Q})$ and then we also basically explained in 4.8.9 that for a class $c \in H_{2\nu}(X(\mathbb{C}),\mathbb{Q})$ and for $[Z]$ smooth we have $PD(c) \cup [Z] = c \cdot Z(\mathbb{C})$, where we think of $c = \sum m_\sigma \sigma$ as being represented by a $\mathcal{C}_\infty$ singular cycle also called c. We can choose c in such a way that it intersects $Z(\mathbb{C})$ in transversally in a finite number of points then $c \cdot Z(\mathbb{C})$ is the intersection number as in Vol. I 4.8.9.
This means we identify $[Z]$ by the intersection numbers of $Z(\mathbb{C})$ with singular cycles in the complementary dimension. But if Z is singular then the singular locus Z_{sing} has a larger codimension. A simple reasoning using some Mayer-Vietoris sequences shows that we may choose the singular cycle in such a way that it avoids the singular locus. And hence way may define $[Z]$ by the intersection relation

$$PD(c) \cup [Z] = c \cdot Z(\mathbb{C}).$$

This shows us that we may define a version $A^{\bullet}_{\mathrm{coh}}(X)$ of the Chow ring, it is simply the image generated by the cycle classes in $H^{2\bullet}(X(\mathbb{C}),\mathbb{Q})$. Of course the intersection product in the Chow ring goes to the cup product in cohomology.

At the very end of this book we will discuss the étale cohomology groups $H^\bullet(X \times \bar{k}, \mathbb{Q}_\ell)$ of algebraic varieties over an arbitrary field. They also allow the construction of a cycle class (See [De1], [Cycle]) we get cohomological versions-depending on the cohomology theory, which we choose, of the Chow ring. Cycles, which are algebraically to zero map to zero under the cycle map. Hence we get a sequence of surjective maps

$$A^\bullet_{\text{lin}}(X) \longrightarrow A^\bullet_{\text{alg}}(X) \longrightarrow A^\bullet_{\text{coh}}(X) \subset H^{2\bullet}(X).$$

We can conclude that a class of a cycle $[Z] \in A^\nu_{\text{coh}}(X)$ is zero if its cup product with all classes in $H^{2d-2\nu}(X)$ is zero. (Poincaré duality is valid for our cohomology theories). This implies that we have a natural homomorphism

$$r : A^\bullet_{\text{coh}}(X) \longrightarrow A_{\text{num}}(X).$$

But interestingly enough it is not clear that this homomorphism r is an isomorphism. Because if we want to test the vanishing of the class $[Z]$ we are only allowed to take the cup product with classes $[Z']$ of cycles and in general $A^\bullet_{\text{coh}}(X)$ is a proper subspace of $H^{2\bullet}(X)$.

It is one of the major problems in algebraic geometry (or arithmetic algebraic geometry) to achieve some understanding of the subspace generated by the cycles classes. If there is some space left, we come back to this problem at the end of this book. We have two major conjectures- namely the Hodge and the Tate conjecture-, which once they have been proved yield a description of the image.

The Chow ring has a very intuitive geometric meaning, so I ask the reader to accept these geometrical ideas and to believe that they can by rigorously justified.

Finally we want to mention that for the projective space $\mathbb{P}^n/k$ the Chow ring becomes very simple. We have the class H of a hyperplane in $A^1(\mathbb{P}^n)$ and then

$$A^\bullet(\mathbb{P}^n) = \mathbb{Z}[H]/(H^{n+1}).$$

Similar results hold for Grassmann manifolds.

Base extension of the Chow ring

Sometimes it useful to pass to the geometric situation: If X/k is absolutely irreducible then we can choose an algebraic closure $\bar{k}$ and consider the ring $A^\bullet(X \times_k \bar{k})$. Of course we have to understand the relationship between these two rings. We construct a homomorphism from $i : A(X) \longrightarrow A(X \times_k \bar{k})$.

Let us consider a reduced and irreducible cycle $Z \subset X$. We want to attach an element in $A(X \times_k \bar{k})$ to it. In section 7.2.1 we investigated what can happen if we extend the field of scalars. We learned that an extension $Y \times_k L$ may not be irreducible anymore and it it may also become non reduced. We have seen that we can find an affine non empty open subset $V \subset Z$ such that the ring $A = \mathcal{O}_Z(U)$ contains the field L/k of pseudoconstants. This is a finite extension of k. Then we saw that

$$L \otimes_k \bar{k} = \sum_\sigma L_\sigma$$

where $\sigma \in \text{Hom}_k(L,\bar{k})$ and where L_σ are finite local algebras over $\bar{k}$. We have $L_\sigma = L \otimes_k \bar{k} e_\sigma$ where the e_σ are the orthogonal idempotents. Then $A \otimes_k \bar{k} = \sum_\sigma (A \otimes_k \bar{k})e_\sigma$, where now $(A \otimes_k \bar{k})e_\sigma$ is an absolutely irreducible $\bar{k}$ algebra. If we divide $(A \otimes_k \bar{k})e_\sigma$ by its radical, then defines an irreducible reduced cycle $Y_\sigma \subset X \times_k \bar{k}$.

But the L_σ may not be reduced. We know from elementary algebra that L/k has a unique maximal sub extension L_i/k, which is purely inseparable over k. It is the field consisting of those elements $x \in L$, for which $\sigma(x) = \tau(x)$ for all pairs $\sigma,\tau \in \mathrm{Hom}_k(L,\bar{k})$. Then the extension L/L_i becomes separable, we get $[L : L_i] = \#\mathrm{Hom}_k(L,\bar{k})$ and $\dim_{\bar{k}} L_\sigma = [L_i : k]$. Then it becomes clear that we should define

$$i(Z) = \sum_\sigma (\dim_{\bar{k}} L_\sigma) Y_\sigma = [L_i : k] \sum_\sigma Y_\sigma. \tag{9.24}$$

The following example should convince the reader that this definition is natural. Again we consider the irreducible curves C_{f_1},C_{f_2} defined by the equations (see 9.1) $f_1 = (x+y)^2 - 2z^2$ and $f_2 = x^p + y^p + az^p$.
The first curve is irreducible over $\mathbb{Q}$. For our set U we can take the open set where $z \neq 0$. Then $u = (x+y)/z$ is a pseudoconstant and the field of pseudoconstants is $L = \mathbb{Q}[u]$, which is a separable extension of degree 2. If we extend the field of scalars to L, then our curve is the union of two different absolutely irreducible, absolutely reduced smooth curves. The sum of these to curves is a cycle on $\mathbb{P}^2 \times L$, it is the image of C_{f_1}.
For the second curve the field of pseudoconstants L/k is generated by $v = (x+y)/z$ and u satisfies the equation $u^p = a$. This is a purely inseparable extension of degree p. Now $C_{f_2} \times_k L$ is absolutely irreducible but not reduced. In this case we have only one σ and $Y_\sigma = (A \otimes_k \bar{k})e_\sigma)_{\mathrm{red}}$(See 7.2.1).The image of C_{f_2} under the base change to L will then be $p((A \otimes_k \bar{k})e_\sigma)_{\mathrm{red}}$.

We also see that the Chow ring changes if we extend the ground field. Let us consider the curve C/k defined by the quadratic form $f = ax^2 + by^2 + cz^2$ considered on p.221. Then it is clear that for this curve C/k the index of $A(C)$ in $\mathbb{Z}$ is two, if the quadratic form does not represent zero. It is one otherwise. If we are in the first case then we see that $A^1(C)$ has index two in $A^1(C \times k(\sqrt{a}))$.
The intermediate groups $A^\nu(X)$ may become very mysterious. But we observe that we have a good definition for $A^1(X)$. If we have two irreducible cycles $Y \subset X$ of codimension one and Z of dimension one then we can define the line bundle $\mathcal{O}(-Y))$ it is the ideal sheaf defining Y. We know that this is a line bundle 9.4. We can restrict this line bundle to the curve Z. The curve may be singular.

In the case where $\dim(X) = 2$ it is clear how to define the Chow ring. In this case we have to say what $A^1(X)$ is and we have to define the intersection product $A^1(X) \times A^1(X) \longrightarrow A^2(X) = \mathbb{Z}$. As equivalence relation we take the linear equivalence of divisors to define $A^1(X)$ (See 9.23) and then we define the intersection numbers in 8.4.2. In our special case that $X = C \times_k C$ the group is $A^1(X)$ is essentially the group of correspondences in 10.3.

We can relate the Chow ring to topology if our smooth projective variety over $\mathbb{C}$. Then we can define a homomorphism

$$A^\bullet(X) \longrightarrow H^{2\bullet}(X(\mathbb{C}),\mathbb{Z}).$$

To get this homomorphism we pick an irreducible cycle $Z \subset X$ of codimension ν. The set of complex points $Z(\mathbb{C})$ is not necessarily a complex variety, but it is possible to show that the singularities do not matter, and we can attach a fundamental class $[Z] \in H^{2p}(X,\mathbb{Z})$ to a cycle of codimension p. Then it is possible to show-with some technical effort- that this defines a homomorphism of rings.

9.7.4 The formulation of the Grothendieck-Riemann-Roch Theorem

The next thing we have to accept is the theory of **Chern classes** for vector bundles on smooth projective varieties X/k. If we have a line bundle $\mathcal{L}$, then we know that $\mathcal{L} \xrightarrow{\sim} \mathcal{O}(D)$, where D is a divisor, we write it as a linear combination $D = \sum_i n_i Y_i$ where the Y_i are irreducible subvarieties of codimension one. Then the first Chern class of $\mathcal{L}$ is simply this divisor considered as an element in $A^1(X)$, i.e. $c_1(\mathcal{L}) = D$. We also define the total Chern class

$$c(\mathcal{L}) = (1,c_1(\mathcal{L}),0,0,\dots) \in A^0(X) \oplus A^1(X) \oplus \dots.$$

To any vector bundle $\mathcal{E}$ on X we can attach a Chern class

$$(c_1(\mathcal{E}),c_2(\mathcal{E}),\dots) \in A^1(X) \oplus A^2(X) + \dots.$$

We encode these Chern classes by writing the Chern polynomial

$$P(\mathcal{E},t) = 1 + c_1(\mathcal{E})t + c_2(\mathcal{E})t^2 + \dots$$

The fundamental properties of these Chern classes are:

(1) If we have an exact sequence of vector bundles

$$0 \longrightarrow \mathcal{E}' \longrightarrow \mathcal{E} \longrightarrow \mathcal{E}'' \longrightarrow 0$$

then

$$P(\mathcal{E},t) = P(\mathcal{E}',t)P(\mathcal{E}'',t),$$

here we use the ring structure of $A^\bullet(X)$.

(2) The Chern class of a line bundle is given by the above rule.

The Chern classes are defined uniquely by these two conditions.
We briefly mention Grothendieck's formula for the Chern classes: If $\mathcal{E}$ is a vector bundle of rank d over X then we can define the bundle $\mathbb{P}(\mathcal{E})$ (see end of section 8.1.6). Since this bundle is locally trivial we can define the hyperplane class $H \in A^d(\mathbb{P}(\mathcal{E}))$ which restricted to each fiber yields the hyperplane class. If restrict to an open set $U \subset X$ such that $BP(\mathcal{E})|U = U \times \mathbb{P}^d$ then $H_U^d = 0$. But globally on X we get the relation

$$H^d - c_1(\mathcal{E})H^{d-1} + c_2(\mathcal{E})H^{d-1} + \dots + (-1)dc_d(\mathcal{E}) = 0.$$

(See [Gr-Ch], 3.) (The sign comes from a different convention to define $\mathbb{P}(\mathcal{E})$, our $\mathbb{P}(\mathcal{E})$ is Grothendieck's $\mathbb{P}(\mathcal{E}^\vee)$.)
We know that for a smooth variety X/k the group $K'(X)$ is isomorphic to $K(X)$. (See Theorem 9.5.5). Each of these two groups has some advantages over the other. For instance we can define a ring structure on $K(X)$ using the tensor product of vector bundles, this works because an exact sequence of vector bundles stays exact of we tensorize it by a vector bundle. This construction does not work on $K'(X)$), but we can define the product structure via the isomorphism. On the other hand we can not define the homomorphism $R_*^\bullet$ on the group $K(X)$ directly.
Now we define a homomorphism of rings (the **Chern character**)

$$\mathrm{ch} : K(X) = K'(X) \longrightarrow A^\bullet(X)$$

by the following rule:
For a vector bundle $\mathcal{E}$ of rank $(\mathcal{E}) = d$, we factorize the Chern polynomial "symbolically"

$$1 + c_1(\mathcal{E})t + \dots c_d(\mathcal{E})t^d = (1 + \lambda_1 t)\dots(1 + \lambda_d t),$$

i.e. we think of the Chern classes as being the elementary symmetric functions in the λ_i. Then we put

$$\mathrm{ch}(\mathcal{E}) = (\sum_{i=1}^d \lambda_i^0, \sum_{i=1}^d \lambda_i, \frac{1}{2!}\sum_{i=1}^d \lambda_i^2, \dots, \frac{1}{n!}\sum_{i=1}^d \lambda_i^n) \in A^\bullet(X)$$

in other words

$$\mathrm{ch}(\mathcal{E}) = (\mathrm{Rank}(\mathcal{E}), + c_1(\mathcal{E}), \frac{1}{2}(c_1(\mathcal{E})^2 - 2c_2(\mathcal{E})), \frac{c_1^3(\mathcal{E}) - 3c_1(\mathcal{E})c_2(\mathcal{E}) + 3c_3(\mathcal{E})}{6}, \dots).$$

This can also be written in the form

$$\mathrm{ch}(\mathcal{E}) = \mathrm{Rank}(\mathcal{E})t^0 + c_1(\mathcal{E})t + \frac{1}{2}(c_1(\mathcal{E})^2 - 2c_2(\mathcal{E}))t^2 + \frac{c_1^3(\mathcal{E}) - 3c_1(\mathcal{E})c_2(\mathcal{E}) + 3c_3(\mathcal{E})}{6}t^3 \dots$$

For an exact sequence of vector bundles

$$0 \longrightarrow \mathcal{E}' \longrightarrow \mathcal{E} \longrightarrow \mathcal{E}'' \longrightarrow 0$$

the property 1) translates into

$$\mathrm{ch}(\mathcal{E}) = \mathrm{ch}(\mathcal{E}') + \mathrm{ch}(\mathcal{E}'').$$

This shows that ch is a homomorphism for the additive structure.
For a line bundle $\mathcal{L} = \mathcal{O}_X(D)$ we have by definition

$$\mathrm{ch}(\mathcal{L}) = (1, D, \frac{1}{2}D \cdot D, \dots, \frac{1}{n!}D^n, \dots) = \sum_k \frac{D^k}{k!}t^k$$

If we now have two line bundles $\mathcal{L}_1, \mathcal{L}_2$ then it is clear from the construction that $\mathrm{ch}(\mathcal{L}_1 \otimes \mathcal{L}_2) = \mathrm{ch}(\mathcal{L}_1)\,\mathrm{ch}(\mathcal{L}_2)$. Then it follows from general principles that for any two vector bundles $\mathcal{E}_1, \mathcal{E}_2$ we have $\mathrm{ch}(\mathcal{E}_1 \otimes \mathcal{E}_2) = \mathrm{ch}(\mathcal{E}_1)\,\mathrm{ch}(\mathcal{E}_2)$ and then it is clear that ch is a ring homomorphism.
If D is effective then we had the sequence

$$0 \longrightarrow \mathcal{O}_X \longrightarrow \mathcal{O}_X(D) \longrightarrow \mathcal{O}_X(D)/\mathcal{O}_X \longrightarrow 0$$

and we find

$$\mathrm{ch}(\mathcal{O}_X(D)/\mathcal{O}) = (0, D, \frac{1}{2}D \cdot D, \dots).$$

We observe that for a coherent sheaf $\mathcal{F}$ on X (Now X maybe arbitrary again), whose support has codimension p, the value $\mathrm{ch}(\mathcal{F})$ in $A^\bullet(X)$ has entry zero in the first p components.
We still assume that $X/k, C/k$ are smooth projective and f is an arbitrary morphism.

We can define a homomorphism of groups

$$f^\bullet : A^\bullet(X) \longrightarrow A^\bullet(C).$$

If $Z \subset X$ is irreducible and of codimension ν, then the closure $f(\bar{Z})$ of $p_2(Z)$ is also an irreducible subvariety. We put $p_2^p(Z) = 0$ if $\dim(Z) > \dim(f(\bar{Z}))$. If we have equality of dimensions then the restriction $f : Z \longrightarrow p_2(Z)$ is finite if we restrict it to a suitably non empty open set (or to the generic point.) Then we put

$$f^\nu(Z) = [k(Y) : k(p_2(Z)]f(\bar{Z}).$$

The homomorphism sends $A^\nu(X)$ to $A^{\nu+\dim(C)-\dim(X)}(X)$.
We get a diagram

$$\begin{array}{ccc} K'(X) & \xrightarrow{\text{ch}} & A^\bullet(X) \\ R^\bullet f_* \downarrow & & f^\bullet \downarrow . \\ K'(C) & \xrightarrow{\text{ch}} & A^\bullet(C) \end{array}$$

This diagram does not commute! We have a deviation from commutativity. To understand this deviation we introduce the **Todd genera** $\mathcal{T}(T_X)$ resp $\mathcal{T}(T_C)$ of the tangent bundles $\mathcal{T}_X$ resp $\mathcal{T}_C$. Its general definition is as follows: We write as above the Chern class of the tangent bundle as

$$1 + c_1(T_X)t + \dots c_d(T_X)t^d = (1 + \lambda_1 t)\dots(1 + \lambda_d t)$$

and then

$$\mathcal{T}(T_X) = \prod \frac{\lambda_i t}{(1 - e^{-t\lambda_i})} = \prod_i (1 + \frac{1}{2}\lambda_i t + \frac{1}{12}\lambda_i^2 t^2 - \frac{1}{720}\lambda_i^4 t^4 + \dots) =$$

$$1 + \frac{c_1(T_X)}{2}t + \frac{c_1(T_X)^2 + c_2(T_X)}{12}t^2 + \frac{c_1(T_X)c_2(T_X)}{24}t^3$$

$$+ \frac{(-c_1(T_X)^4 + c_1(T_X)c_3(T_X) + 4c_1(T_X)^2 c_2(T_X) + 3c_2(T_X)^2 - c_4(T_X))}{720}t^4$$

Now can write down Grothendieck's Riemann-Roch formula for an arbitrary morphism $f : X \longrightarrow C$ between two smooth projective varieties over a field:

$$f^\bullet(\text{ch}(\mathcal{F}) \cdot (\mathcal{T}(T_X)) = \text{ch}(R^\bullet f_*(\mathcal{F})) \cdot \mathcal{T}(T_C) \tag{GRR}$$

The Todd genera $\mathcal{T}(T_X)$,$\mathcal{T}(T_C)$ are units in the rings $A^\bullet(X)$,$A^\bullet(C)$.
We make the point that the general GRR is of striking simplicity and the formula has its own aesthetic beauty. Only if you start to produce explicit formulae it becomes complicated. A large part of the ingenuity required to prove this theorem is to find the right correcting terms.

9.7.5 Some special cases of the Grothendieck-Riemann-Roch-Theorem

Let us consider the case that X/k is a smooth, absolutely irreducible projective curve. Then we have $A^\bullet(X) = A^0(X) \oplus A^1(X) \subset \mathbb{Z} \oplus \mathbb{Z}$ and the Chern character is given by

$$\mathrm{ch} : K'(X) \longrightarrow \mathbb{Z} \oplus \mathbb{Z}, \mathcal{F} \mapsto (\mathrm{Rank}(\mathcal{F}), \deg(\mathcal{F})),$$

we introduced this homomorphism on p. 220.
Now we a separable finite morphism $f : C_1 \longrightarrow C_2$ between two smooth, absolutely irreducible projective curves. Then we have

$$\begin{array}{ccc} K'(C_1) & \xrightarrow{\mathrm{ch}} & A^\bullet(C_1) \\ f_* \downarrow & & f^\bullet \downarrow \\ K'(C_2) & \xrightarrow{\mathrm{ch}} & A^\bullet(C_2) \end{array}$$

and we apply GRR to the trivial sheaf $\mathcal{O}_{C_1}$. We have $\mathrm{ch}(\mathcal{O}_{C_1}) = 1$ and $\mathcal{T}(T_{C_1}) = 1 + (1 - g_{C_1})t$. Then $f^\bullet(\mathrm{ch}(\mathcal{O}_{C_1}) \cdot \mathcal{T}(T_{C_1})) = \deg(f) + (1 - g_{C_1})t$. On the other hand we have $\mathrm{ch}(f_*(\mathcal{O}_{C_1})) = \deg(f) + \deg(f_*(\mathcal{O}_{C_1}))t$ and the GRR yields the equality

$$\big(\deg(f) + \deg(f_*((\mathcal{O}_{C_1})t)\big(1 + (1 - g_{C_2})t\big) =$$

$$\deg(f) + (\deg(f_*(\mathcal{O}_{C_1})) + \deg(f)(1 - g_{C_2}))t = \deg(f) + (1 - g_{C_1})t.$$

The non trivial information in this formula is the equality

$$\deg(f_*(\mathcal{O}_{C_1})) + \deg(f)(1 - g_{C_2}) = 1 - g_{C_1}$$

and this a slight generalization of the Riemann-Hurwitz formula. (See Thm. 9.3.3)

Another special case is that $C = \mathrm{Spec}(k)$. In this case $\mathrm{ch} : K'(C) \xrightarrow{\sim} A^0(C) = \mathbb{Z}$ and $R^\nu f_*(\mathcal{F}) = H^\nu(X,\mathcal{F})$. Therefore we get for the right hand side of the formula the Euler characteristic $\chi(X,\mathcal{F}) = \sum_\nu (-1)^\nu \dim_k(H^\nu(X,\mathcal{F}))$. On the left hand side we compute the product $\mathrm{ch}(\mathcal{F}) \cdot \mathcal{T}(X) \in A^\bullet(X)$. and then we have to apply $f^\bullet$ to the result. But here it is clear that $f^\nu = 0$ unless we are in the top dimension, i.e. $\nu = d = \dim_k(X)$. In this degree the component $\mathrm{ch}(\mathcal{F})\mathcal{T}(X)^{(d)}$ is a zero dimensional cycle $\sum n_\mathfrak{p}\mathfrak{p}$, which is a linear combination of intersections of Chern classes of $\mathcal{E}$ and cycle classes in the coefficients of $\mathcal{T}(T_X)$. Now it is the definition that $f^d(\sum n_\mathfrak{p}\mathfrak{p}) = \sum_\mathfrak{p} n_\mathfrak{p}[k(\mathfrak{p}) : k]$ and the Riemann-Roch theorem says that

$$\chi(X,\mathcal{F}) = f^d(\mathrm{ch}(\mathcal{F}) \cdot \mathcal{T}(T_X)) = f^d(\sum n_\mathfrak{p}\mathfrak{p})$$

We leave it as an exercise to the reader to verify that for a smooth, projective and absolutely irreducible curve X/k this is our old Riemann-Roch theorem. But we see a small subtlety: In our first version of the Riemann-Roch theorem the genus g entered as the dimension of $H^1(X,\mathcal{O}_X)$ where in the version above it is defined by the equality $\deg(T_X) = 2 - 2g$.

9.7.6 Back to the case $p_2 : X = C \times C \longrightarrow C$

We want to show that we actually almost proved the Grothendieck-Riemann-Roch-Theorem for this special case. Of course it is not entirely clear what it means that we proved it, since already the statement depends on several concepts and results (the theory of Chern classes and the equality $K(X) = K'(X)$), which we did not prove here).

But in the special case of a smooth, projective surface X/k we have defined $A^1(X)$ and the intersection product $A^1(X) \times A^1(X) \longrightarrow A^2(X) \subset \mathbb{Z}$. (See 8.4.2). It this point I recommend to work with the relation of algebraic equivalence on cycles. We have the Chow ring in this case and the degree map gives us an inclusion $A^2(X) \hookrightarrow \mathbb{Z}$.
For a coherent sheaf $\mathcal{F}$ on X we want to define (compute) the value of the Chern character $\mathrm{ch}(\mathcal{F})$. We have to do some maneuverings. We have the additional problem, that we have not proved that $K(X) = K'(X)$ and we propose a strategy to solve both problems at once.
The support of the sheaf is a closed subscheme, we define $\mathrm{ch}(\mathcal{F})$ by induction on the dimension and the number of components of maximal dimension in its support. Basically this is the same strategy as in the discussion of the Riemann-Roch theorem for curves.
Let us assume that $\mathcal{F}$ let $Y \subset \mathrm{supp}(\mathcal{F})$ be an irreducible component of maximal dimension of the support. We can find a non empty open subset $V \subset Y$ such that $\mathcal{F}$ restricted to V becomes trivial, i.e. we have sections $t_1, \ldots, t_d \in H^0(V, \mathcal{F})$, which trivialize $\mathcal{F}|V$. We take an ample sheaf $\mathcal{L}$ on X and a global section $s \in H^0(X,)$, which is not vanishing on Y. Let D be the divisor of zeroes of s then we get an embedding $\mathcal{F} \hookrightarrow \mathcal{F} \otimes \mathcal{O}_X(rD)$ for any $r > 0$. We have seen in the proof of Thm. 8.3.3 that for $r >> 0$ our sections t_i extend to sections of $\mathcal{F} \otimes \mathcal{O}_X(rD)$ and hence we get an embedding $\mathcal{O}_Y^d \hookrightarrow \mathcal{F} \otimes \mathcal{O}_X(rD)$ and therefore, an exact sequence

$$0 \longrightarrow \mathcal{O}_Y^d \longrightarrow \mathcal{F} \otimes \mathcal{O}_X(rD) \longrightarrow \mathcal{F}' \longrightarrow 0.$$

The support of $\mathcal{F}'$ is strictly smaller than the support of $\mathcal{F}$, and we have

$$\mathrm{ch}(\mathcal{F})\,\mathrm{ch}(\mathcal{O}_X(rD)) = d\,\mathrm{ch}(\mathcal{O}_Y) + \mathrm{ch}(\mathcal{F}').$$

This tells us that $K'(X)$ as an abelian group is generated by line bundles on X, the restriction of line bundle $\mathcal{L}$ on X to (one dimensional) irreducible sub schemes $Y \subset X$ and closed points. But for an irreducible sub scheme of dimension 1 we have the sequence

$$0 \longrightarrow \mathcal{O}_X(-Z) \longrightarrow \mathcal{O}_X \longrightarrow \mathcal{O}_Y \longrightarrow 0$$

and hence

$$0 \longrightarrow \mathcal{O}_X(-Z) \otimes \mathcal{L} \longrightarrow \mathcal{L} \longrightarrow \mathcal{L} \otimes \mathcal{O}_Y \longrightarrow 0,$$

and hence the restriction of a line bundle to Y is also in the group generated by line bundles on X. Therefore, in our special case it suffices to prove GRR for line bundles on X and sheaves with support of dimension zero. Now we show that for line bundles our formula 9.22 is equivalent to (GRR) and then we for points..
Our formula 9.22 yields

$$R^\bullet p_2(\mathcal{L}) = [R^0 p_2(\mathcal{L})] - [R^1 p_2(\mathcal{L})] = ((1-g) + \mathcal{L} \cdot H_2) + \big((1-g)\mathcal{L} \cdot H_1 + \frac{1}{2}\mathcal{L} \cdot \mathcal{L}\big)t.$$

As before we have $\mathcal{T}(T_C) = 1 + (1-g)t$ and hence we get on the right hand side

$$\mathrm{ch}(R^\bullet p_2(\mathcal{L}))\cdot \mathcal{T}(T_C) = ((1-g)+\mathcal{L}\cdot H_2) + \big((1-g)\mathcal{L}\cdot H_1 + \frac{1}{2}\mathcal{L}\cdot\mathcal{L} + (1-g)^2\big)t$$

To compute the left hand side we know that $\mathrm{ch}(\mathcal{L}) = 1 + c_1(\mathcal{L})t + \frac{1}{2}c_1(\mathcal{L})\cdot c_1(\mathcal{L})t^2$. For the tangent bundle have have the identity $T_X = p_1^*(T_C)\oplus p_2^*(T_C)$. We know that $c_1(p_1^*(T_C)) = (2-2g)[H_1]$,$c_1(p_2^*(T_C)) = (2-2g)[H_2]$, where $[H_i]$ is the image of H_i in $A^1(X)$. We get for the Chern polynomials

$$P(p_1^*(T_C),t) = 1+(2-2g)[H_1]t,\ P(p_2^*(T_C),t) = 1+(2-2g)[H_2]t$$

and in accordance with our above rule we get

$$P(T_X,t) = (1+(2-2g)[H_1]t)(1+(2-2g)[H_2]t) = 1+((2-2g)([H_1]+[H_2]))t+4(g-1)^2t^2.$$

This means that $c_1(T_X) = p_1^*(c_1(T_C))+p_2^*(c_1(T_C))$ and $c_2(T_X) = p_1^*(c_1(T_C))\cdot p_2^*(c_1(T_C)) = 4(g-1)^2$. Hence the Todd genus is

$$\mathcal{T}(T_X) = 1+\frac{p_1^*(c_1(T_C))+p_2^*(c_1(T_C))}{2}t+\frac{(p_1^*(c_1(T_C))+p_2^*(c_1(T_C)))^2+p_1^*(c_1(T_C))\cdot p_2^*(c_1(T_C))}{12}t^2$$

$$1+((1-g)([H_1]+[H_2]))t+(g-1)^2t^2.$$

We multiply this by $\mathrm{ch}(\mathcal{L})$ and get

$$\big(1+c_1(\mathcal{L})t+\frac{1}{2}c_1(\mathcal{L})\cdot c_1(\mathcal{L})t^2\big)\cdot\big(1+((1-g)([H_1]+[H_2]))t+(g-1)^2t^2\big) =$$

$$1+\big(c_1(\mathcal{L})+(1-g)([H_1]+[H_2])\big)t+\big(g-1)^2+c_1(\mathcal{L})\cdot\big((1-g)([H_1]+[H_2])\big)+\frac{1}{2}\cdot c_1(\mathcal{L})\big)t^2$$

Now we have to apply $p_2^\bullet$ to this expression. The constant term vanishes because the fibre has dimension one. The linear term yields the constant term, $c_1(\mathcal{L})$ maps to $\mathcal{L}\cdot H_2$ the class $[H_2]$ maps to zero and $[H_1]$ maps to one. The coefficient of the quadratic term is a number, which then becomes the coefficient of the linear term. Hence we see that somewhat miraculously (or not?) we get

$$p_2^\bullet(\mathrm{ch}(\mathcal{L})\cdot\mathcal{T}(T_X)) =$$

$$((1-g)+\mathcal{L}\cdot H_2)+\big((1-g)\mathcal{L}\cdot H_1+\frac{1}{2}\mathcal{L}\cdot\mathcal{L}+(1-g)^2\big)t =$$

$$\mathrm{ch}(R^\bullet p_2(\mathcal{L}))\cdot\mathcal{T}(T_C)$$

Hence our previous calculations yield a proof of GRR for the case of $p_2 : C\times C\longrightarrow C$ and line bundles $\mathcal{L}$ on it. But this also yields GRR for divisors on $C\times C$, hence it remains to prove GRR for sheaves with zero dimensional support. Let us look at this last case. For the moment X can be any smooth irreducible projective variety of dimension d. If the dimension of the support of $\mathcal{F}$ is zero, then our sheaf is a skyscraper sheaf, which is direct sum over a finite set of closed points $\mathcal{F} = \oplus_\mathfrak{p}\mathcal{S}_\mathfrak{p}$, where $\mathcal{S}_\mathfrak{p}$ is an $\mathcal{O}_{X,\mathfrak{p}}-$ module of finite length. (This means that it is annihilated by a suitable power $\mathfrak{m}_\mathfrak{p}^N$ of the maximal ideal and the successive quotients $\mathfrak{m}_\mathfrak{p}^m\mathcal{S}_\mathfrak{p}/\mathfrak{m}_\mathfrak{p}^{m+1}\mathcal{S}_\mathfrak{p}$ are finite dimensional vector spaces of dimension $d_{\mathfrak{p},m}$ over $k(\mathfrak{p})$). Then we claim that the value of the Chern character on $\mathcal{F}$ is given by

$$\operatorname{ch}(\mathcal{F}) = (d-1)! \sum_{\mathfrak{p},m} [k(\mathfrak{p}) : k] d_{\mathfrak{p},m} t^d.$$

This is not entirely obvious. If our base field is algebraically closed, it says that for all closed points $x \in X(k)$ we have $\operatorname{ch}(\mathcal{O}_X/\mathfrak{m}_x) = (d-1)!t^d$. If we accept this fact, then GRR becomes true for sheaves with zero dimensional support (see [Gr-Ch], 16).
We give a hint how this can be proved for surfaces X/k. Let us choose two irreducible subvarieties $Z_1, Z_2 \subset X$, which intersect transversally in smooth points. If I_1, I_2 are the ideals defining these subvarieties, we get an exact sequence of sheaves

$$0 \longrightarrow \mathcal{O}_X/\mathcal{O}_X(-Z_1 - Z_2) \longrightarrow \mathcal{O}_X/\mathcal{O}_X(-Z_1) \oplus \mathcal{O}/\mathcal{O}_X(-Z_2) \longrightarrow J \longrightarrow 0$$

where J is a torsion sheaf whose stalk is non zero only at the intersection points of Z_1, Z_2. At these points we have

$$J = \mathcal{O}_{X,x}/(I_1, I_2) = \oplus_{x \in Z_1 \cap Z_2} \mathcal{O}_{X,x}/\mathfrak{m}_x.$$

Hence our exact sequence yields

$$\operatorname{ch}(J) = \operatorname{ch}(\mathcal{O}_X/I_1) + \operatorname{ch}(\mathcal{O}_X/I_2) - \operatorname{ch}(\mathcal{O}_X/\mathcal{O}_X(Z_1 + Z_2)) =$$

$$tZ_1 - \frac{1}{2} Z_1 \cdot Z_1 t^2 + tZ_2 - \frac{1}{2} Z_2 \cdot Z_2^2 - (t(Z_1 + Z_2) - \frac{1}{2}(Z_1 + Z_2) \cdot (Z_1 - Z_2)t^2) = t^2 Z_1 \cdot Z_2$$

and hence we get $\sum_{x \in Z_1 \cap Z_2} \operatorname{ch}(\mathcal{O}_{X,x}/\mathfrak{m}_x) = \#(Z_1 \cap Z_2)$ and our assertion above follows if we accept that $\operatorname{ch}(\mathcal{O}_{X,x})$ does not depend on $x \in X$.
In any case in our special case $X = C \times_k C$ we know that any point $(x,y) \in C \times C$ is the transversal intersection of two cycles $\{x_0\} \times H_2 \cap H_1 \times \{y_0\}$ and the above claim follows. Hence we proved GRR for the projection $p_2 : C \times C \longrightarrow C$.

Exercise 42. Consider the case $X = \mathbb{P}^d/k$, write a point x as intersection of d hyperplanes and prove the above formula for $\operatorname{ch}(\mathcal{O}_{\mathbb{P}^d,x}/\mathfrak{m}_x)$ using the same strategy. Once you have done this you proved another special case of GRR.

Exercise 43. Let us consider a smooth, projective and irreducible scheme $X \longrightarrow \operatorname{Spec}(k)$, i.e. Y is a point, let $\mathcal{F}$ be a coherent sheaf. Write the GRR in this case. This is the Hirzebruch-Riemann-Roch-formula.

Exercise 44. Prove the Hirzebruch-Riemann-Roch formula for $\mathbb{P}^n/k$ and $\mathcal{F} = \mathcal{O}_{\mathbb{P}^n}(r)$ using 8.2.3.

The GRR is a marvelous example for the paradigm that theorems become easier to prove, concepts become clearer if we formulate them in greater generality. In a nutshell this is already visible in [De-We], in principle they discuss vector bundles over curves and this allows them to make use of the morphisms $C \longrightarrow \mathbb{P}^1$.
We want to discuss a very important application of the GRR theorem. We still consider the case where $X = C \times_k C$ is the product of a smooth, projective, absolutely irreducible curve by itself. We have the non degenerate pairing

$$A^1_{\mathrm{num}}(X) \otimes \mathbb{Q} \times A^1_{\mathrm{num}}(X) \otimes \mathbb{Q} \longrightarrow \mathbb{Q},$$

which is given by the intersection pairing. In $A^1_{\mathrm{num}}(X)$ we have the two classes given by H_1, H_2, it is clear that $H_1 \cdot H_1 = H_2 \cdot H_2 = 0$ and $H_1 \cdot H_2 = 1$. Hence $\mathbb{Q}H_1 \oplus \mathbb{Q}H_2$ is a hyperbolic plane in $A_{\mathrm{num}}(X) \otimes \mathbb{Q}$. Hence it has an orthogonal complement $A^1_0(X)$ and

$$A^1_{\text{num}}(X) \otimes \mathbb{Q} = \mathbb{Q}H_1 \oplus \mathbb{Q}H_2 \oplus A^1_0(X).$$

We get the famous **Hodge index theorem**:

Theorem 9.7.1. *The intersection form is negative definite on $A^1_0(X)$.*

Let $\mathcal{L}$ be a line bundle, which corresponds to a class $\xi \in A^1_0(X)$. Then we get from the Riemann-Roch theorem that

$$\mathrm{ch}(R^\bullet p_{2*}(\mathcal{L})) = 1 - g + \frac{1}{2}\mathcal{L} \cdot \mathcal{L}t$$

We consider the degree and get

$$\deg(R^0 p_{2,*}(\mathcal{L})) - \deg(R^1 p_{2,*}(\mathcal{L})) = \frac{1}{2}\mathcal{L} \cdot \mathcal{L}..$$

Since $R^0 p_{2,*}(\mathcal{L})$ has no torsion it is a locally free sheaf. We assume for simplicity that C/k has rational points $x_0, y_0, \ldots$. The fibre $\mathcal{L}|C \times \{x_0\}$ has degree zero, hence we get $\dim_k H^0(C \times \{x_0\}, \mathcal{L}|C \times \{x_0\}) \leq 1$ and is equal to one if and only if $\mathcal{L}|C \times \{x_0\})$ is trivial. We can conclude that either $R^0 p_{2,*}(\mathcal{L})$ is a line bundle $\mathcal{M}$ or it is zero (See Theorem 8.4.5). The adjointness of $p_{2,*}$ and p_2^* yields a homomorphism $i : p_2^*(p_{2,*})(\mathcal{L}) \longrightarrow \mathcal{L}$ (See I,3.4.1), which in the second case is an isomorphism. In other words we have have $p_{2,*}(\mathcal{L}) = 0$ or $\mathcal{L} = p_2^*(\mathcal{M})$. We restrict $p_2^*(\mathcal{M})$ to a $\{y_0\} \times C$, which represents the class $[H_2]$., Then we see, that this restriction has degree zero, hence in any case $\deg(R^0 p_{2,*}(\mathcal{L})) = 0$. Therefore,

$$-\deg(R^1 p_{2,*}(\mathcal{L})) = \frac{1}{2}\mathcal{L} \cdot \mathcal{L}$$

Now we show that we can find a constant $m \geq 0$ such that $\deg(R^1 p_{2,*}(\mathcal{L})) \geq -m$ independently of $\mathcal{L}$. To see this we consider the tensor product $\mathcal{L} \otimes \mathcal{O}_X(r(y_0 \times C) = \mathcal{L}(r(y_0 \times C)$ and the resulting exact sequence

$$0 \longrightarrow \mathcal{L} \longrightarrow \mathcal{L}(r(y_0 \times C)) \longrightarrow \mathcal{L}(r(y_0 \times C))/\mathcal{L} \longrightarrow 0.$$

We restrict $\mathcal{L}(r(y_0 \times C))$ to the fibers $(C \times \{t\})$. The degree of these restriction is r. Hence we know: If $r > 2g-2$ then $H^1(C \times \{t\}), \mathcal{L}(r(y_0 \times C))) = 0$. Therefore the semicontinuity theorem yields $R^1 p_{2,*}(\mathcal{L}(r(y_0 \times C))) = 0$, and we get the exact sequence

$$0 \longrightarrow R^0 p_{2,*}(\mathcal{L}) \longrightarrow R^0 p_{2,*}(\mathcal{L}(r(y_0 \times C))) \longrightarrow$$

$$R^0 p_{2,*}(\mathcal{L}(r(y_0 \times C))/\mathcal{L}) \longrightarrow R^1 p_{2,*}(\mathcal{L}) \longrightarrow 0.$$

If we denote by $\mathcal{L}_{y_0}$ the restriction of $\mathcal{L}$ to $\{y_0\} \times C$ and observe that the projection p_2 induces an isomorphism $\{y_0\} \times C \xrightarrow{\sim} C$. Then

$$R^0 p_{2,*}(\mathcal{L}(r(y_0 \times C))/\mathcal{O}_X) \xrightarrow{\sim} \mathcal{L}^r_{y_0}.$$

We want to have a bound from below for the degree of $R^1 p_{2,*}(\mathcal{L})$. Let $\mathcal{N}$ be be the image of $R^0 p_{2,*}(\mathcal{L} \otimes \mathcal{O}_X(r(y_0 \times C)) \longrightarrow \mathcal{L}^r_{y_0}$, this means that we have an exact sequence

$$0 \longrightarrow \mathcal{N} \longrightarrow \mathcal{L}^r_{y_0} \longrightarrow R^1 p_{2,*}(\mathcal{L}) \longrightarrow 0.$$

The degree of $\mathcal{L}_{y_0}$ is zero, hence we get

$$\deg(R^1 p_{2,*}(\mathcal{L})) = -\deg(\mathcal{N}).$$

We need an estimate for the degree of $\mathcal{N}$ from above. Since $\dim H^0(C,\mathcal{L}^r_{y_0}) \leq r$ and $H^0(C,\mathcal{N}) \subset H^0(C,\mathcal{L}^r_{y_0})$, we get an estimate for the dimension of the space of sections of $\mathcal{N}$ and then the Riemann-Roch formula in section 9.5.4 yields an estimate for the degree of $\mathcal{N}$ from above.
We find that the degree of $\mathcal{L} \cdot \mathcal{L}$ is bounded from above, hence $\mathcal{L}^{\otimes n} \cdot \mathcal{L}^{\otimes n} = n^2 \mathcal{L} \cdot \mathcal{L}$ is bounded from above. This implies $\mathcal{L} \cdot \mathcal{L} \leq 0$. Since the form is non degenerate we must have $\mathcal{L} \cdot \mathcal{L} < 0$ if the class ξ of the bundle in $A^1_{\mathrm{num}}(X)$ is not zero.

□

9.7.7 Curves over finite fields.

We consider the special situation of a projective, smooth and absolutely irreducible curve $C/\mathbb{F}_q$, where $\mathbb{F}_q$ is the field with q elements. This special case is historically the origin for the theory of curves over arbitrary fields. The classical geometers always studied the case of curves over $\mathbb{C}$ and the theory of Riemann surfaces.

Let $K = \mathbb{F}_q(C)$ be the function field , this is a finite separable extension of a rational function field $\mathbb{F}_q[f]$ (see Prop. 9.3.1). Since $C/\mathbb{F}_q$ is absolutely irreducible we know that the field of constants is equal to $\mathbb{F}_q$. We can recover the curve $C/\mathbb{F}_q$ from the field K as we explained in section 9.3.

These function fields attracted the attention of number theorists since they are analogous to number fields. The closed points $\mathfrak{p}$ of the curve C are in one to one correspondence to the discrete valuation ring $\mathcal{O}_\mathfrak{p} \subset K$. If we remove one point ∞ from C, then $C \setminus \{\infty\}$ is the spectrum of the Dedekind ring $\mathcal{O}_C(C \setminus \{\infty\})$, and this has to be seen in analogy to the number field case where we have $\mathrm{Spec}(\mathcal{O})$ where $\mathcal{O}$ is the ring of integers.

For instance we can attach a ζ-function to our curve, which we define as

$$\zeta_K(s) = \zeta_C(s) = \prod_{\mathfrak{p}:\text{closed point}} \frac{1}{1 - \frac{1}{N\mathfrak{p}^s}}$$

where $N\mathfrak{p} = \#(\mathcal{O}_\mathfrak{p}/\mathfrak{p})$ is the number of elements in the residue field. This ζ-function is analogous to the Dedekind ζ-function of a number field, and it is easy to see that the product converges for $\mathrm{Re}\,(s) > 1$.

We will show that the Riemann-Roch theorem implies that this ζ-function is a rational function in the variable $t = q^{-s}$ and more precisely we have

$$\zeta_C(s) = \frac{P_1(q^{-s})}{(1 - q \cdot q^{-s})(1 - q^{-s})}$$

where $P_1(q^{-s}) = 1 + a_1 q^{-s} \cdots + q^g \cdot q^{-2gs}$ is a polynomial of degree $2g$ in q^{-s} with integer coefficients. We will see that the theorem of Riemann-Roch implies that we have a functional equation

$$\zeta(s) = q^{(2-2g)s} q^{1-g} \zeta(1-s).$$

This is in a perfect analogy to the situation of a number field.

It has been observed by Artin (see [Art2] that we can formulate the analogue of the Riemann hypothesis, which would say that the zeroes of $\zeta(s)$ have imaginary part $\mathrm{Re}\,(s) = \frac{1}{2}$. This assertion can be formulated in terms of the polynomial $P_1(q^{-s}) = P_1(t)$. We may factor it over $\overline{\mathbb{Q}} \subset \mathbb{C}$ and get

$$P_1(t) = \prod_{i=1}^{2g} (1 - \omega_i t)$$

where the ω_i are algebraic integers. It is easy to see that the functional equation allows a grouping of these numbers ω_i so that we have

$$\omega_1 \cdots \omega_g \ , \ \omega_{g+1} \cdots \omega_{2g}$$

and $\omega_\nu \cdot \omega_{\nu+g} = q$.
Then the Riemann hypothesis is equivalent to the assertion that all the ω_i have absolute value

$$|\omega_i| = q^{\frac{1}{2}}.$$

This has been conjectured by Artin in his thesis, and he verified it in several cases. But actually Gauss knew it as special cases in a somewhat disguised form and also Artin's thesis advisor Herglotz had proved in it in a special case.

The Riemann hypothesis was then proved by Hasse in 1934 ([Has]) for curves of genus one and A. Weil announced the proof in the general case in 1941([We3]). The final proof appeared in 1948, and it is based on the theory of the Jacobian of curves.

Elementary properties of the ζ-function.

In this section we call elementary properties of the ζ-function those, which follow from the Riemann-Roch theorem for the curve. (See [Scm]). As in the case of the Riemann

ζ-function we can expand the product

$$\prod_{\mathfrak{p}} \frac{1}{1 - \frac{1}{N\mathfrak{p}^s}} = \sum_{\mathfrak{a}} \frac{1}{N(\mathfrak{a})^s}$$

where $\mathfrak{a}$ runs over the effective divisors

$$\mathfrak{a} = \sum n_i \mathfrak{p}_i \qquad n_i \leq 0,$$

where $N(\mathfrak{a}) = \prod (N\mathfrak{p}_i)^{n_i} = q^{\deg(\mathfrak{a})}$. For the following discussion we have to assume that our curve has a divisor of degree one. This assumption is equivalent to the assumption that the degrees $f_{\mathfrak{p}}$ of the prime divisors are coprime. It is a theorem of F. K. Schmidt that this is always the case (see [Scm]).

If we make this assumption (or if we believe F. K. Schmidt's theorem), then it is clear that the divisor classes of degree n for any n form a principal homogeneous space under $\mathrm{Pic}^0(C)(\mathbb{F}_q)$, this is the group of divisor classes of degree 0. The number h of these divisor classes is finite and it is called the class number. We get

$$\zeta_C(s) = \sum_{n=0}^{\infty} \frac{c(n)}{q^{ns}}$$

where $c(n) = \#$ of effective divisors of degree n. (This number is clearly finite.) For $n > 2g-2$ we have a formula for $c(n)$: If $\mathfrak{a}$ is an effective divisor of degree n, then we can consider the line bundle $\mathcal{L} = \mathcal{O}_C(\mathfrak{a})$, and we have

$$\dim H^0(C,\mathcal{O}_C(\mathfrak{a})) = n+1-g.$$

A non zero section $s \in H^0(C,\mathcal{L})$ defines the divisor $\mathfrak{b} = \mathrm{Div}(s)$. and we see that the divisors $\mathfrak{b}$, which we get if s varies are exactly those divisors, which are linearly equivalent to $\mathfrak{a}$. Then

$$c(n) = h\ \frac{q^{n+1-g}-1}{q-1}.$$

(As a byproduct we proved the finiteness of the class number.) We make the substitution $q^{-s} = t$, and write $\zeta_C(s) = Z_C(t)$. Then we define a new function $Z^*_C(t)$ by

$$Z_C(t) = Z^*_C(t) + \frac{h}{q-1} \cdot \left(q^g \frac{t^{2g-1}}{1-qt} - \frac{1}{1-t} \right),$$

and

$$Z^*_C(t) = \sum_{n=0}^{2g-1} c^*(n)t^n.$$

The coefficients $c^*(n)$ are equal to zero for $n \geq 2g-1$, and we have

$$c(n) = c^*(n) - \frac{h}{q-1} \quad \text{for} \quad 0 \leq n \leq 2g-2.$$

The correcting term

$$R(t) = \frac{h}{q-1} \left(q^g \frac{t^{2g-1}}{1-qt} - \frac{1}{1-t} \right)$$

satisfies the functional equation

$$R(t) = t^{2g-2}q^{1-g}R(\frac{1}{qt}),$$

and hence we have to show that $Z^*_C(t)$ satisfies this functional equation.

Now we observe that we have a precise formula for $c(n)$. We sum over the classes of bundles of degree n and count the effective divisors in a class. If $\mathcal{L}$ is a line bundle of degree n, then the number of effective divisors in this class is

$$\frac{1}{q-1} \left(q^{\dim H^0(C,\mathcal{L})} - 1 \right),$$

and hence

$$c(n) = \frac{1}{q-1} \sum_{\mathcal{L}:\deg(\mathcal{L})=n} \left(q^{\dim H^0(C,\mathcal{L})} - 1\right).$$

Now we assume that for all n this sum is not empty. This follows from the theorem of F. K. Schmidt that we can always find a divisor of degree one. Then we can conclude that the number of terms in the sum is equal to h and hence

$$c^*(n) = \sum_{\mathcal{L}:\deg(\mathcal{L})=n} q^{\dim H^0(C,\mathcal{L})}.$$

Our problem is that we do not know the dimension of the space of sections. But we have in involution n, the set of divisor classes of degree between 0 and $2g-2$.

$$\mathcal{L} \longrightarrow \mathcal{L}^{-1} \otimes \Omega = \mathcal{L}',$$

and then $\deg(\mathcal{L}) \to 2g-2-\deg(\mathcal{L})$. We see that

$$c^*(2g-2-n) = \sum_{\mathcal{L}':\deg(\mathcal{L}')=2g-2-n} q^{\dim H^0(C,\mathcal{L}')}.$$

If now $\mathcal{L}' = \mathcal{L}^{-1} \otimes \Omega$, then

$$\dim H^0(C,\mathcal{L}) - \dim H^0(C,\mathcal{L}') = n+1-g,$$

hence this difference depends only on the degree and not on the class. This implies

$$c^*(n) = q^{n+1-g} c^*(2g-2-n),$$

and this implies the functional equation for $Z^*_C(t)$. We can conclude that

$$Z_C(t) = \frac{1 + a_1 t \cdots + q^g t^{2g}}{(1-t)(1-qt)}.$$

The coefficients a_i must be integers, since the $c(n)$ are integers. We can write

$$P_1(t) = \prod_{i=1}^{2g} (1 - \omega_i t),$$

the $\omega_i \in \mathbb{C}$ are algebraic and the functional equation implies that the collection

$$\{\omega_1 \cdots \omega_{2g}\}$$

is invariant under the substitution

$$\omega_i \longrightarrow \omega_i^{-1} q.$$

At this moment it is neither clear that they are algebraic integers nor that the Riemann hypothesis

$$|\omega_i| = q^{\frac{1}{2}} \qquad \text{for} \qquad i = 1 \cdots 2g$$

holds.

These so called elementary results are all due to F. K. Schmidt (See [Scm]). By the way F. K. Schmidt was clever enough to remove the assumption that we have a divisor of degree one. Can you also prove it?

The Riemann hypothesis.

We recall the multiplicative definition of the ζ-function, and we remember that

$$N\mathfrak{p}^{-s} = q^{-f_{\mathfrak{p}}s} = t^{f_{\mathfrak{p}}}$$

where $[\mathbb{F}_q(\mathfrak{p}) : \mathbb{F}_q] = q^{f_{\mathfrak{p}}}$, i.e. $f_{\mathfrak{p}}$ is the degree of the residue field at $\mathfrak{p}$. Now we consider the expression

$$t \cdot \frac{Z_C'(t)}{Z_C(t)} = \sum_{\mathfrak{p}} \frac{f_{\mathfrak{p}} t^{f_{\mathfrak{p}}}}{1 - t^{f_{\mathfrak{p}}}}.$$

If we expand the expression on the right, we get an infinite sum

$$\sum \ a_n t^n$$

and clearly

$$a_n = \sum_{f_{\mathfrak{p}} | n} \ f_{\mathfrak{p}}.$$

But $f_{\mathfrak{p}}$ is exactly the number of geometric points over $\mathfrak{p}$, and these geometric points lie in

$$C(\mathbb{F}_{q^{f_{\mathfrak{p}}}}) \subset C(\mathbb{F}_{q^n}),$$

and hence we see that the right hand side must be

$$t \cdot \frac{Z_C'(t)}{Z_C(t)} = \sum \#C(\mathbb{F}_{q^n}) t^n.$$

Going back to our expression for $Z_c(t)$ as a rational function we find

$$t \cdot \frac{Z_C'(t)}{Z_C(t)} = \frac{qt}{1 - qt} + \frac{t}{1 - t} - \sum_{i=1}^{2g} \frac{\omega_i t}{1 - \omega_i t},$$

and from here we get

$$\#C(\mathbb{F}_{q^n}) = q^n + 1 - \sum_{i=1}^{2g} \ \omega_i^n$$

for all n.

(This implies that $\sum_{i=1}^{2g} \omega_i^n$ is an integer for all numbers n, and this implies that the ω_i must be algebraic integers).

The following proof of the Riemann hypothesis is due to A. Grothendieck (see [Gr-RH]) and is based on intersection theory and the Hodge index theorem.

Our curve $C/\mathbb{F}_q$ is defined over $\mathbb{F}_q$, and this allows us to define the Frobenius morphism

$$\Phi_q : C \longrightarrow C.$$

To define this morphism we consider an affine open subset $U \subset C$, then the ring of regular function $\mathcal{O}_c(U)$ is an $\mathbb{F}_q$-algebra and the map

$$\Psi_q : \mathcal{O}_c(U) \longrightarrow \mathcal{O}_c(U),$$

which sends $f \in \mathcal{O}_c(U)$ to f^q is an $\mathbb{F}_q$-algebra homomorphism. This defines the restriction of Φ_q to U, and since we can cover C by such open affine sets, we have defined it every where. If we have an embedding of our curve

$$\begin{array}{ccc} i : C & \longrightarrow & \mathbb{P}^n_{\mathrm{Spec}(\mathbb{F}_q)} \\ & \searrow \ \swarrow & \\ & \mathrm{Spec}(\mathbb{F}_q). & \end{array}$$

Then the action of the Frobenius Φ_q on the geometric points is given by

$$\underline{x} = (x_0, \cdots, x_n) \longrightarrow x^q = x^q = (x_0^q, \cdots, x_n^q)$$

for any point $\underline{x} \in \mathbb{P}^n(\mathbb{F}_p)$.

Now we pass to a geometric situation, we replace C by $\overline{C} = C \times_{\mathbb{F}_q} \overline{\mathbb{F}}_q$ and our information that $\overline{C}$ comes from the curve $C/\mathbb{F}_q$ is encoded in the datum of the Frobenius morphism

$$\Phi_q : \overline{C} \longrightarrow \overline{C},$$

which also can be viewed as given by its graph

$$\Gamma_\Phi \subset \overline{C} \times \overline{C} = \overline{X}.$$

Then it is clear by definition that

$$C(\mathbb{F}_q) = \Gamma_\Phi \cap \Delta \subset C(\overline{\mathbb{F}}_q) \times C(\overline{\mathbb{F}}_q).$$

The tangent space of $\overline{X}$ in any geometric point $p = (x,y)$ is given by

$$T_{\overline{X},p} = T_{\overline{C},x} \oplus T_{\overline{C},y},$$

and the tangent space of Γ_Φ at $p = (x,\Phi(x))$ consist of vectors in

$$T_{\Gamma_\Phi,p} = T_{\overline{C},x}.$$

The tangent space of Δ is an point (x,x) is the diagonal, hence we see that Γ_Φ and Δ intersect transversally, and this gives us

$$\#C(\mathbb{F}_q) = \Gamma_\Phi \cdot \Delta.$$

If we pick any two points $x_0 \in C(\overline{\mathbb{F}}_q)$, $y_0 \in C(\overline{\mathbb{F}}_q)$, then we may consider the divisors

$$H_1 = \{x_0\} \times \overline{C} \ , \ \overline{C} \times \{y_0\} = H_2$$

on $\overline{C} \times \overline{C}$. We get an orthogonal decomposition of the Chow group

$$A^1(\overline{X}) = \widetilde{A}^1(\overline{X}) \oplus \mathbb{Z}[H_1] \oplus \mathbb{Z}[H_2].$$

Since $H_1 + H_2$ is ample on $\overline{X}$, we can conclude that the intersection form restricted to $\widetilde{A}^1(\overline{X})$ is negative definite. We write in $A^1(\overline{X})$

$$[\Delta] = [\Delta'] + [H_1] + [H_2],$$

then $[\Delta'] \in \widetilde{A}^1(\overline{X})$. More generally we can say: If $\Gamma \subset \overline{C} \times \overline{C}$ is the graph of a morphism $f : \overline{C} \to \overline{C}$, then

$$\begin{aligned} \Gamma \cdot (\{x_0\} \times \overline{C}) &= \Gamma \cdot H_2 = 1 \\ \Gamma \cdot (\overline{C} \times \{y_0\}) &= \Gamma \cdot H_1 = \deg(f) \end{aligned}$$

and

$$[\Gamma'] = [\Gamma] - (\deg f)[H_2] - [H_1] \in \widetilde{A}(\overline{X}).$$

We need the self intersection $\Gamma \cdot \Gamma$. To get this number we recall the adjunction formula. The tangent bundle $T_{\overline{X}} \simeq p_1^*(T_{\overline{C}}) + p_2^*(T_{\overline{C}})$ and hence

$$\Lambda^2 T_{\overline{X}} = p_1^*(T_{\overline{C}}) \otimes p_2^*(T_{\overline{C}}).$$

Now we have

$$\Gamma \cdot \Lambda^2 T_{\overline{X}} = \Gamma \cdot p_1^*(T_{\overline{C}}) + \Gamma \cdot p_2^*(T_{\overline{C}}) = (2g-2) \cdot (\deg(f) + 1).$$

Then the adjunction formula yields

$$\Gamma \cdot \Lambda^2 T_{\overline{X}} = 2g_\Gamma - 2 - \Gamma \cdot \Gamma,$$

and $g_\Gamma = g$ because the projection p_1 provides an isomorphism $p_1 : \Gamma \to \overline{C}$. Hence we get

$$\Gamma \cdot \Gamma = (2-2g) \cdot \deg(f)$$

and this yields

$$\Gamma' \cdot \Gamma' = -2g \deg(f).$$

We look at the case $\Gamma = \Gamma_\Phi$. We have

$$\Gamma_\Phi \cdot \Delta = \Gamma_\Phi \cdot ([\Delta'] + [H_1] + [H_2]) = \Gamma_\Phi \Delta' + q + 1,$$

and hence we get

$$-\sum_{i=1}^{2g} \omega_i = \Gamma_\Phi \cdot \Delta' = \Gamma'_\Phi \cdot \Delta'.$$

We know that

$$(n\Gamma'_\Phi + m\Delta') \cdot (n\Gamma'_\Phi + m\Delta') \leq 0$$

for all integers $n.m$, and this says that

$$-n^2 \cdot 2gq + 2nm\Gamma'_\Phi \cdot \Delta' - m^2 \cdot 2g \leq 0$$

for all n, m. This is equivalent to the inequality

$$(\Gamma'_\Phi \cdot \Delta')^2 = \left(\sum_{i=1}^{2g} \omega_i\right)^2 \leq 4g^2 q,$$

and hence we get the estimate

$$|\sum_{i=1}^{2g} \omega_i| \leq 2g\sqrt{q}.$$

This is in principle the Riemann hypothesis. The only thing we have to say is that we can replace $\mathbb{F}_q$ by any field $\mathbb{F}_{q^r}$, i.e. we get the inequality

$$|\sum_{i=1}^{2g} \omega_i^r| \leq 2g\sqrt{q^r}$$

for all $r = 1,2,\cdots$. Since we have

$$\prod_{i=1}^{2g} \omega_i = q^g,$$

this can be the case only if

$$|\omega_i| = \sqrt{q}$$

for all $i = 1 \cdots 2g$.

10 The Picard functor for curves and their Jacobians

Introduction:

In the last chapter of volume I we constructed the Jacobian of a compact Riemann surface S. The Jacobian was defined as the group of isomorphism classes of holomorphic line bundles on S. Our main result asserted that the Jacobian had the structure of a complex torus, and assuming theorems of Lefschetz and Chow we proved that this torus is a projective algebraic variety. We heavily relied on transcendental methods.
We formulate the goal of this chapter. For any smooth, projective and absolutely irreducible curve C/k over an arbitrary field k we want to construct a Jacobian $J_C/k = J/k$. This Jacobian should be the " variety" of line bundles of degree zero on C. It is not really clear what this means. A certain minimal requirement might be that J/k is a group scheme and for any field extension L/k the set $J(L)$ is canonically isomorphic to to the group isomorphism classes of line bundles over $C \times_k L$. But it is still not really clear what this means.
We get a closer understanding if we recall the situation in Chapter V of volume I. At the end of this chapter we constructed a line bundle $S \times J$ and formulated a universality property for the line bundle $\mathcal{N}$, which we did not prove in all detail. This universality property will become the basic principle for the construction of J/k. We will use the concept of representable functors.
Once we constructed the Jacobian J/k we will show that is a projective algebraic algebraic, it carries the structure of an algebraic group. Again want to investigate the group of line bundles on J/k, this means that we want to construct the dual $J^\vee$ of J/k and we want to show that we have a canonical identification $J \longrightarrow J^\vee$.

10.1 The construction of the Jacobian

10.1.1 Generalities and heuristics :

Let S be a scheme. For any scheme X (of finite type) over S, we can define the Picard functor

$$\begin{array}{rcl} \mathcal{PIC}_{X/S} : \text{Schemes of finite type}/S & \longrightarrow & \text{Abelian groups} \\ T & \longrightarrow & \text{Pic}(X \times_S T) \end{array},$$

where $\mathcal{PIC}_{X/S}(T) = \text{Pic}(X \times_S T)$ is the group of isomorphism classes of line bundles over $X \times_S T$. We may ask whether this functor is representable (in the category of schemes of finite type over S) (See proposition 6.2.18).
We recall what this means: We can find a group scheme $\text{Pic}_{X/S}/S$, (which is of finite type over S) and a "universal" line bundle

$$\mathcal{P} \quad \text{on} \quad X \times_S \text{Pic}_{X/S}/S$$

such that for any scheme $T \to S$ of finite type and any line bundle $\mathcal{L}$ over $X \times_S T$ we have a unique S-morphism

$$\psi(\mathcal{L}) : T \to \mathrm{Pic}_{X/S}$$

such that we can find an isomorphism

$$\eta : \mathcal{L} \xrightarrow{\sim} (\mathrm{Id} \times \psi(\mathcal{L}))^*(\mathcal{P}).$$

The answer to this question is simply "No". It is important to understand why this is so.

Actually we know already why the functor can not be representable. Let us assume in addition that X/k is projective, reduced and connected. Let us assume we constructed a $\mathrm{Pic}_{X/S}$ and a line bundle $\mathcal{P}$ on $X \times \mathrm{Pic}_{X/S}$. The point is that a line bundle $\mathcal{L}$ on $X \times_S T$ has non trivial automorphisms, the group of automorphisms is $\mathcal{O}(X \times_S T)^*$ and under our assumptions this is $\mathcal{O}(T)^\times$. But if now $T = S' \longrightarrow S$ is faithfully flat (and of finite type) then the homomorphism

$$\mathcal{PIC}_{X/S}(S) \longrightarrow \mathcal{PIC}_{X/S}(S')$$

is not necessarily injective. We have seen in 6.2.11, 6.48 and in the section on the moduli space of elliptic curves 9.6.2 that the kernel is given by the cohomology group $H^1(S'/S, \mathbb{G}_m)$, which is not trivial in general. This means that our functor does not satisfy the first sheaf condition (Sh1) and our functor can not be representable.

Instead of asking for representability of the functor we can pose a weaker question: Let us now assume that $S = \mathrm{Spec}(k)$ is the spectrum of a field. Let X/k a smooth projective scheme. The question is: Can we find a scheme $\mathrm{Pic}_{X/S}$ over k and a line bundle $\mathcal{P}$ on $X \times_k \mathrm{Pic}_{X/k}$ such that for any scheme T/k of finite type and any line bundle $\mathcal{L}$ on $X \times_k T$ we find a unique morphism

$$\psi(\mathcal{L}) : T \to \mathrm{Pic}_{X/k}$$

such that we can find a covering $T = \cup_\alpha T_\alpha, i_\alpha : T_\alpha \hookrightarrow T$ by Zariski open subsets such that

$$(\mathrm{Id} \times i_\alpha)^*(\mathcal{L}) \xrightarrow{\sim} \psi_\alpha^*(\mathcal{P}).$$

This formulation should be compared to the analogous statement in Volume I (Prop. 5.2.10).

In general we will call a line bundle $\mathcal{L}$ on $X \times T$ **locally trivial in** T if we can cover T by open sets T_α such that $\mathcal{L}|X \times T_\alpha$ becomes trivial. We call two line bundles $\mathcal{L}_1, \mathcal{L}_2$ on $X \times T$ **locally isomorphic** in T if we can cover T by open sets T_α such that $\mathcal{L}_1|X \times T_\alpha \xrightarrow{\sim} \mathcal{L}_2|X \times T_\alpha$

We formulate a more modest goal. We show that our functor is **locally representable**: This means that we want to construct a scheme $\mathrm{Pic}_{X/k}$ over k and a line bundle $\mathcal{P}$ on $X \times \mathrm{Pic}_{X/k}$ such that for any bundle $\mathcal{L}$ on $X \times T$ with T of finite type over k we find a unique $\psi : T \to \mathrm{Pic}_{X/k}$ such that

$$(\mathrm{Id} \times \psi)^*(\mathcal{P}) \sim_T \mathcal{L},$$

and this means that the bundles are locally isomorphic in T.

Once we have achieved this goal we will say that $\mathcal{PIC}_{X/k}$ **is locally represented** by the pair $(\mathrm{Pic}_{X/k}, \mathcal{P})$.

The condition $(\mathrm{Id} \times \psi)^*(\mathcal{P}) \sim_T \mathcal{L}$ can be reformulated: Let $q_2 : X \times_S T \longrightarrow T$ be the projection to the second factor: Then there exists a line bundle $\mathcal{N}$ on T such that

$$\mathcal{L} \otimes q_2^*(\mathcal{N}) \xrightarrow{\sim} (\mathrm{Id} \times \psi)^*(\mathcal{P}).$$

Remark: It is clear by essentially the same reasoning that the line bundle $\mathcal{P}$ is only unique up to a twist by a line bundle $p_2^*(\mathcal{M})$, where $\mathcal{M}$ is a line bundle on $\mathrm{Pic}_{X/k}$. Then our above equation becomes

$$(\mathrm{Id} \times \psi)^*(\mathcal{P} \otimes p_2^*(\mathcal{M})) \xrightarrow{\sim} \mathcal{L} \otimes q_2^*(\mathcal{N}) \otimes (\mathrm{Id} \times \psi)^*(\mathcal{M}).$$

Of course we will need some finiteness conditions on T . The condition for T/k to be of finite type is natural, but we should also allow localizations of such schemes.

Rigidification of $\mathcal{PIC}$

Now we apply the same method, which was applied in the discussion of moduli spaces of elliptic curves. We considered elliptic curves $\mathcal{E} \longrightarrow S \longrightarrow \mathrm{Spec}(\mathbb{Z})[1/6]$ and since these objects still have automorphisms we equipped them with a nowhere vanishing 1-form ω. After that these objects have no non trivial automorphisms anymore and we proved the representability of the functor $\mathcal{M}_{1,diff}$ (See 9.6.2).
In our situation here we assume that X is projective, absolutely reduced and connected over a field k and it comes with a distinguished point $P \in X(k)$. Let $S = \mathrm{Spec}(k)$. We consider line bundles $\mathcal{L}$ on $X \times_S T$ whose restriction $\mathcal{L}_P$ to $\{P\} \times_S T$ is trivial. We define the functor

$$\mathcal{PIC}_{X/k,P}(T) = \{(\mathcal{L},s) | \mathcal{L} \text{ line bundle on } X \times T, s \in H^0(\{P\} \times_S T) \text{ trivializes } \mathcal{L}_P\}$$

This means we look at the group of isomorphism classes of line bundles on $X \times T$, which are trivial on $\{P\} \times T$, and which are equipped with an isomorphism

$$\eta_P : \mathcal{L}|\{P\} \times T \xrightarrow{\sim} \mathcal{O}_{\{P\} \times T}.$$

where we require that $\eta_P(s) = 1$.

If we have two such pairs $(\mathcal{L}_1,s_1)$ and $(\mathcal{L}_2,s_2)$ with isomorphic line bundles then we have exactly one isomorphism, which sends s_1 into s_2. Hence we can say that the two pairs $(\mathcal{L}_1,s_1)$, $(\mathcal{L}_2,s_2)$ are not only isomorphic, **they are even equal.** Here we used of course that $X/\mathrm{Spec}(k)$ is projective, reduced and connected. Especially we see that the pairs $(\mathcal{L},s)$ are rigid, i.e. their group of automorphisms is trivial. Hence our modified functor $\mathcal{PIC}_{X/k,P}$ is in fact a sheaf for the faithfully flat topology, and therefore, we can hope for representability the functor $\mathcal{PIC}_{X/k,P}$. Passing from $\mathcal{PIC}_{X/k}$ $\mathcal{PIC}_{X/k,P}$ means that we put some additional structure on our objects such that these do not have automorphism. This process is called **rigidification**.

Let us assume that we have proved the representability of $\mathcal{PIC}_{X/k,P}$. We recall that representability of $\mathrm{Pic}_{X/k,P}$ gives us a universal triplet

$$(X \times \mathrm{Pic}_{X/k,P}, \mathcal{P}, s)$$

where is $\mathrm{Pic}_{X/k,P}$ a scheme, $\mathcal{P}$ a line bundle on $X \times \mathrm{Pic}_{X/k,P}$ and $s_0 \in H^0(\{P\} \times \mathrm{Pic}_{X/k}, \mathcal{P})$ a section, which is everywhere non zero, and such that this triplet satisfies the universality property.

We drop the section s and claim that

$$(X \times \mathrm{Pic}_{X/k,P}, \mathcal{P})$$

provides a local presentation of $\mathcal{PIC}_{X/k}$.

To see this we start from a line bundle on $X \times_S T$. Let $\mathcal{L}_P$ its restriction to $\{P\} \times T$. We have the structural morphism $t : X \longrightarrow \mathrm{Spec}(k)$, from this we get the line bundle $(t \times \mathrm{Id})^*(\mathcal{L}_P)$ on $X \times_k T$ and we consider the bundle $\mathcal{L} \otimes ((t \times \mathrm{Id})^*(\mathcal{L}_P)))^{-1}$ on $X \times_k T$. This bundle has a canonical trivializing section if we restrict it to $\{P\} \times_k T$, because $\mathcal{L}_P \otimes ((t \times \mathrm{Id})^*(\mathcal{L}_P)^{-1})|\{P\} \times T = \mathcal{O}_T$, namely the element $1 \in H^0(\{P\} \times_k T, \mathcal{O}_T)$. Hence we see that $\mathcal{L} \otimes (t \times \mathrm{Id})^*(\mathcal{L}_P)^{-1}$ is an object in $\mathcal{PIC}_{X/k,P}(T)$, and provides a unique morphism

$$\psi : T \longrightarrow \mathrm{Pic}_{X/k,P}$$

such that we have a unique isomorphism

$$\eta : (\mathcal{L} \otimes (t \times \mathrm{Id})^*(\mathcal{L}_P)^{-1}) \xrightarrow{\sim} (\mathrm{Id} \times_k \psi)^*(\mathcal{P}),$$

which maps the given sections into each other. But now it is clear that we have

$$\mathcal{L} \sim_T (\mathrm{Id} \times_k \psi)^*(\mathcal{P}).$$

Therefore our claim is proved if we can show that ψ is uniquely determined provided it only satisfies this last relation. Assume that we have a second morphism ψ_1, for which we know

$$\mathcal{L} \sim_T (\mathrm{Id} \times_k \psi_1)^*(\mathcal{P}).$$

Hence we can find a line bundle $\mathcal{N}$ and an isomorphism η_1

$$\eta_1 : \mathcal{L} \otimes q_2^*(\mathcal{N}) \xrightarrow{\sim} (\mathrm{Id} \times_k \psi_1)^*(\mathcal{P}).$$

But we must have $q_2^*(\mathcal{N}) \xrightarrow{\sim} ((t \times \mathrm{Id})^*(\mathcal{L}_P)))^{-1}$. and choosing the right section in $H^0(\{p\} \times_k, \mathcal{L} \otimes q_2^*(\mathcal{N}))$ we can conclude that $(\psi, \eta) = (\psi_1, \eta_1)$.

We may also go in the opposite direction, let us assume we have a local representation $(\mathrm{Pic}_{X/k}, \mathcal{P})$. Using the same argument as above we may assume that $\mathcal{P}|\{P\} \times_k \mathrm{Pic}_{X/k}$ is trivial. Then we choose a trivializing section $s \in H^0(\{P\} \times_k \mathrm{Pic}_{X/k}$ and then the triplet $(\mathrm{Pic}_{X/k}, \mathcal{P}, s)$ represents $\mathcal{PIC}_{X/k,P}$.
All this tells us that constructing a pair $(\mathrm{Pic}_{X/k}, \mathcal{P})$, which yields a local representation of $\mathcal{PIC}_{X/k,P}$ or the construction a representation of $\mathcal{PIC}_{X/k,P}$ are actually the same problem.

In a sense $\mathcal{PIC}_{X/k,P}$ is a better functor since it is representable. But it depends on the somewhat arbitrary choice of a point $P \in X(k)$, which may even not exist. The more natural functor $\mathcal{PIC}_{X/k}$ is not representable but locally representable. The underlying schemes $\mathrm{Pic}_{X/k}$ and $\mathrm{Pic}_{X/k,P}$ are canonically isomorphic, we can even say that they are equal.

If we analyze the concept of local representability then we see that we do not need the line bundle $\mathcal{P}$ on $X \times \mathrm{Pic}_{X/k}$, we get along with something less: It suffices to give this bundle locally in $\mathrm{Pic}_{X/k}$. This means that we need a Zariski covering $\mathrm{Pic}_{X/k} = \cup U_\alpha$ and line bundles $\mathcal{P}_\alpha$ on $X \times U_\alpha$ such that

$$\mathcal{P}_\alpha | X \times (U_\alpha \cap U_\beta) \sim_{U_\alpha \cap U_\beta} \mathcal{P}_\beta | X \times (U_\alpha \cap U_\beta).$$

i.e. the two restrictions differ by the pullback of a line bundle on $U_\alpha \cap U_\beta$.

Now we say that $\mathrm{Pic}_{X/k}$ together with the family of line bundles $\{\mathcal{P}_\alpha \text{ on } X \times U_\alpha\}_\alpha$ is a weak local representation of $\mathcal{PIC}_{X/k}$ if for any line bundle $\mathcal{L}_1$ on $X \times T$ we find a unique $\psi : T \longrightarrow \mathrm{Pic}_{X/k}$ such that for any α and $T_\alpha = \psi^{-1}(U_\alpha)$ and $\psi_\alpha : T_\alpha \longrightarrow U_\alpha$ we have $(\mathrm{Id} \times \psi_\alpha)^*(\mathcal{P}_\alpha) \sim_{U_\alpha} \mathcal{L}_1 | X \times T_\alpha$.

It is clear that we also may require the existence and uniqueness of the $\psi_\alpha : T_\alpha \longrightarrow U_\alpha$ such that we have the above relation. Then it is clear that these ψ_α coincide on the intersections $U_\alpha \cap U_\beta$ and fit together to a morphism ψ.

We denote the datum $\mathrm{Pic}_{X/k}$ together with the covering and the $\mathcal{P}_\alpha$ simply by $(\mathrm{Pic}_{X/k}, \underline{\mathcal{P}})$ and call $\underline{\mathcal{P}}$ a $\mathrm{Pic}_{X/k}$ gerbe (See 9.6.2)

If our scheme X/k has a rational point $P \in X(k)$ then we may argue as before: We can modify our $\mathcal{P}_\alpha$ such that $\mathcal{P}_\alpha = \mathcal{O}_{\{P\} \times \mathrm{Pic}_{X/k}}$, i.e. they are trivial and equipped with a nowhere vanishing section. Then we can glue them together to a line bundle $\mathcal{P}$, whose restriction to $\{P\} \times \mathrm{Pic}_{X/k}$ has a nowhere vanishing global section. This shows us that weak local representability of $\mathcal{PIC}_{X/k}$ implies representability of $\mathcal{PIC}_{X/k,P}$.

It is quite clear that the concept of weak local representability is more natural than local representability

The next 28 pages are devoted to the proof of weak local representability of $\mathrm{Pic}_{C/k}$ where C/k is a smooth, projective, absolutely irreducible curve over k.

10.1.2 General properties of the functor $\mathcal{PIC}$

The locus of triviality

We consider a reduced, projective and connected scheme X/k and line bundles on $X \times T$ where T should be of finite type. In the following considerations it is always possible to replace these schemes T by the affine schemes $\mathrm{Spec}(A)$ since our questions will be local in T. We will also allow ourselves to pass to local rings at points in T.

We need some finiteness condition for T because we have to apply the finiteness theorem 8.3.2. Since the Picard functor will be of finite type our condition above seems to be natural. But the passage to local rings does not hurt.

Of course the following lemma would be clear if we had local representability.

Lemma 10.1.1. *Let X/k be an irreducible, reduced projective scheme, let $T \to \mathrm{Spec}(k)$ a scheme of finite type . Let $\mathcal{L}$ be a line bundle on $X \times_{\mathrm{Spec}(k)} T$. Then there exists a "largest" closed subscheme $T_1 \subset T$ such that $\mathcal{L}_1 = \mathcal{L}|X \times_{\mathrm{Spec}(k)} T_1$ is locally trivial in T_1 and "largest" means for any closed subscheme $T' \subset T$, for which $\mathcal{L}|X \times_k T'$ is locally trivial in T', the inclusion $T' \hookrightarrow T$ factors over T_1.*

The subscheme T_1 will be called the **locus of triviality of** $\mathcal{L}$.
Let $T' \subset T$ be a subscheme such that $\mathcal{L}_{T'} = \mathcal{L}|X \times_k T'$ is locally trivial. Our assumptions on X imply that the projection $p : X \times_k T' \to T'$ yields two line bundles on T' namely $p_*(\mathcal{L}_{T'})$, and $p_*(\mathcal{L}_{T'}^{-1})$. These bundles are locally constant in the following sense: For any closed subscheme $T'' \subset T'$ the natural restriction provides isomorphisms

$$p_*(\mathcal{L}_{T'}) \otimes \mathcal{O}_{T''} \xrightarrow{\sim} p_*(\mathcal{L}_{T''}), p_*(\mathcal{L}_{T'}^{-1}) \otimes \mathcal{O}_{T''} \xrightarrow{\sim} p_*(\mathcal{L}_{T''}^{-1}).$$

But if in turn for a subscheme $T' \subset T$ these two sheaves are locally constant of rank one, then $\mathcal{L}_{T'}, \mathcal{L}_{T'}^{-1}$ are locally trivial in T'. To see this, we can consider two local generators

$$s \in H^0(V_t, p_*(\mathcal{L}_{T'}))\ ,\ s_- \in H^0(V_t, p_*(\mathcal{L}_{T'}^{-1})).$$

(Here V_t is an open neighborhood of a point $t \in T'$) . Their product gives an element

$$ss_- \in H^0(X \times V_t, \mathcal{O}_{X \times V_t}),$$

By definition of local constancy we know that the restrictions of these sections to the special fibre over t generate the spaces of sections $H^0(X \times \{t\}, \mathcal{L}_t)$ and $H^0(X \times \{t\}, \mathcal{L}_t^{-1})$, which are both of dimension one. Then the product ss_- restricted to the fibre is not zero, but since $X \times \{t\}$ is projective and connected, it follows that they never vanish. The closed subset in $X \times V_t$ where ss_- vanishes does not meet the fibre $X \times \{t\}$. Its projection to V_t is closed in V_t (see theorem 8.1.8) and does not contain t. Hence we find an open neighborhood W_t of t such that s, s_- are nowhere zero on $X \times W_t$ and hence $\mathcal{L}_{T'}$ and $\mathcal{L}_{T'}^{-1}$ are trivial on $X \times W_t$.

We assume that we have a point $t \in T$ such that $\mathcal{L}|X \times \{t\}$ is trivial. Let A be the local ring $\mathcal{O}_{T,t}$. We restrict our sheaf to $X \times \mathrm{Spec}(A)$, let $\mathfrak{m}$ be the maximal ideal. We want to show that there is a smallest ideal I_0 such that our sheaf restricted to $X \times \mathrm{Spec}(A/I_0)$ becomes trivial. By our previous considerations this means that $p_*(\mathcal{L}|X \times_A (A/I_0))$ is free of rank one.

For any ideal $I \subset A$ we compute $p_*(\mathcal{L}|X \times_A (A/I))$. In the section (semicontinuity) we have seen that we can construct a resolution of $\mathcal{L}$ on $X \times_k T$

$$0 \to \mathcal{L} \to \mathcal{E}_0 \to \mathcal{E}_1 \to \cdots,$$

such that the $\mathcal{E}_i$ are coherent, flat and acyclic for p_* and such that the $p_*(\mathcal{E}_i) = \mathcal{M}_i$ are locally free on T (see Thm. 8.4.3. We restrict to $\mathrm{Spec}(A) \subset T$, so that the $\mathcal{M}_i$ are actually free, and we get

$$0 \to p_*(\mathcal{L}) \to \mathcal{M}_0 \to \mathcal{M}_1 \to \cdots.$$

We have $\mathcal{M}_0 = A^{N_0}$, $\mathcal{M}_1 = A^{N_1}$ and the map $\alpha_0 : \mathcal{M}_0 \to \mathcal{M}_1$ is given by a $N_0 \times N_1$-matrix.

The complex

$$0 \longrightarrow \mathcal{M}_0 \otimes (A/I) \longrightarrow \mathcal{M}_1 \otimes (A/I) \longrightarrow \dots$$

computes the cohomology $H^\bullet(X \times \mathrm{Spec}(A/I), \mathcal{L}|X \times \mathrm{Spec}(A/I))$ especially we know that the kernel $(A/\mathfrak{m})^{N_0} \longrightarrow (A/\mathfrak{m})^{N_1}$. is of rank 1. Now we choose an element $m_1 \in A^{N_0}$, which reduces to a basis element of this kernel. We can assume that m_1 is the first element of a basis of A^{N_0}. We consider m_1 as a column vector

$$m_1 = \begin{pmatrix} 1 \\ 0 \\ \vdots \\ 0 \end{pmatrix}.$$

The linear map $A^{N_0} \to A^{N_1}$ is given by a matrix $M(\alpha_0)$. Our matrix operates by multiplication from the left, then it will be of the form

$$M(\alpha_0) = \begin{pmatrix} m_{11} & * & \cdots & * \\ m_{12} & & & \\ \vdots & & B & \\ m_{1,N} & & & \end{pmatrix}$$

where the $m_{1,\nu}$ are in the maximal ideal $\mathfrak{m}$. We reduce $\mod \mathfrak{m}$. The matrix $B \mod \mathfrak{m}$ maps the space of vectors

$$\begin{pmatrix} 0 \\ x_2 \\ \vdots \\ x_{N_0} \end{pmatrix} \qquad \text{with} \qquad x_i \in A/\mathfrak{m}$$

injectively into $(A/\mathfrak{m})^{N_1}$. This implies that B contains an invertible $(N_0 - 1) \times (N_0 - 1)$ submatrix. This allows us to modify the basis in the target such that the vectors

$$\begin{pmatrix} 0 \\ 1 \\ \vdots \\ \vdots \\ 0 \end{pmatrix} \cdots \begin{pmatrix} 0 \\ \vdots \\ 1 \\ 0 \\ 0 \end{pmatrix}$$

are the images of the $(N_0 - 1)$-vectors

$$\begin{pmatrix} 0 \\ 1 \\ \vdots \\ 0 \end{pmatrix}, \dots, \begin{pmatrix} 0 \\ 0 \\ \vdots \\ 1 \end{pmatrix}$$

and then our matrix B is of the form

$$\begin{pmatrix} m_{11} & 0 & \cdots & 0 \\ m_{12} & 1 & 0 & \\ \vdots & 0 & \ddots & 1 \\ m_{1,N_1} & & & \end{pmatrix}.$$

Now we take an ideal $I \subset A$ and ask: When is the kernel

$$(A/I)^{N_0} \longrightarrow (A/I)^{N_1}$$

free of rank one and surjects to the kernel of

$$(A/\mathfrak{m})^{N_0} \longrightarrow (A/\mathfrak{m})^{N_1},$$

i.e. $\mathcal{L}$ is locally constant of rank one on $\mathrm{Spec}(A/I)$.
Clearly this is the case if and only if this kernel is spanned by the element

$$\begin{pmatrix} 1 \\ -m_{12} \\ \vdots \\ -m_{1,N_0} \end{pmatrix}$$

The first condition to be satisfied is that $m_{11} \in I$. Then the second, third up to N_0-th component of the image is zero. Now the vanishing of the $N_0 + 1$-th up to the N_1-th entry means that the $m_{1,\mu}$ with $\mu = N_0 + 1 \ldots$ to N_1 satisfy

$$m_{1,\mu} - \sum_{j=2}^{j=N_0} b_{\mu j} m_{1,j} = 0.$$

This gives us a collection of elements, which must be in I and in turn if I contains these elements, then the kernel of $(A/I)^{N_0} \to (A/I)^{N_1}$ is free of rank one.
Hence we see that we can take for I_0 the ideal generated by these elements

$$\left\{ m_{11}, m_{1,\mu} - \sum b_{\mu j} m_{1,j} \right\} = I_0$$

and $\mathrm{Spec}(A/I_0)$ is the largest subscheme of $\mathrm{Spec}(A)$, on which $\mathcal{L}|X \times \mathrm{Spec}(A/I_0)$ is locally constant of rank one. If we do this reasoning also for $\mathcal{L}^{-1}$ then we find a second ideal I_0' and then our argument above shows that the ideal generated by these two ideals define the largest subscheme, on which $\mathcal{L}$ becomes locally trivial.

□

10.1.3 Infinitesimal properties

In our chapter on Riemann surfaces we exploited the exact sequence of sheaves

$$0 \to \mathbb{Z} \to \mathcal{O}_S \to \mathcal{O}_S^* \to 1,$$

which give rise to the sequence in cohomology

$$0 \longrightarrow H^1(S,\mathbb{Z}) \to H^1(S,\mathcal{O}_S) \to H^1(S,\mathcal{O}_S^*) \to H^2(S,\mathbb{Z}) \longrightarrow$$

(actually we can do this over any compact complex algebraic variety). From this sequence we get that $H^1(S,\mathcal{O}_S^*)$ is a complex analytic group with tangent space $H^1(S,\mathcal{O}_S)$. We showed that $H^1(S,\mathbb{Z})$ is a lattice in $H^1(S,\mathbb{Z})$ and $H^1(S,\mathcal{O}_S)/H^1(S,\mathbb{Z}) = \mathrm{Pic}^0(S)$ is a compact complex analytic group. The homomorphism

$$H^1(S,\mathcal{O}_S) \to H^1(S,\mathcal{O}_S)/H^1(S,\mathbb{Z}) = \mathrm{Pic}^0(S)$$

should be viewed as the exponential map from the tangent space to this complex analytic group.

We want to save some parts of this argument to our abstract situation. For instance we want to make it clear that the "additive" coherent cohomology group $H^1(X,\mathcal{O}_X)$ is the tangent space of our functor $\mathrm{Pic}_{X/k}$.

To understand the infinitesimal properties we assume that $T = \mathrm{Spec}(A)$ where A is a local artinian k-algebra. Its maximal ideal $\mathfrak{m}$ is nilpotent hence $A = k \oplus \mathfrak{m}$. We want to consider the line bundles on $X \times_k \mathrm{Spec}(A)$, which are trivial on the special fibre $X = X \times_k A/\mathfrak{m}$. This means that we want to understand the kernel

$$\mathrm{Pic}(X \times_k \mathrm{Spec}(A))_e = H^1(X \times_k A,\mathcal{O}_{X\times A}^*)_e = \ker[H^1(X \times_k A,\mathcal{O}_{X\times A}^*) \longrightarrow H^1(X,\mathcal{O}_X^*)].$$

For any integer n we put $A_n = A/\mathfrak{m}^n$ and hence $k = A_1$. We put $X_n = X \times_k \mathrm{Spec}(A_n)$, then our assumption implies that $H^0(X_n,\mathcal{O}_{X_m}^*) = (A/\mathfrak{m}^n)^*$.

For any pair of integers $n > n_1$ we can consider the embedding

$$X_{n_1} \hookrightarrow X_n$$

as a closed subscheme, it is an isomorphism on the topological spaces and we have the restriction

$$(\mathcal{O}_{X_n})^* \longrightarrow (\mathcal{O}_{X_{n_1}})^*,$$

which is surjective. For $n = n_1 + 1$ we can see easily that the kernel is isomorphic to $\mathcal{O}_X \otimes \mathfrak{m}^n/\mathfrak{m}^{n+1}$. Since $H^0(X_n,\mathcal{O}_{X_n}^*) = (A/\mathfrak{m}^n)^* \longrightarrow H^0(X_{n-1},\mathcal{O}_{X_{n-1}}^*) = (A/\mathfrak{m}^{n-1})^*$ is surjective, we get an exact sequence

$$0 \to H^1(X,\mathcal{O}_X \otimes \mathfrak{m}^{n-1}/\mathfrak{m}^n) \to H^1(X_n,\mathcal{O}_{X_n}^*) \to H^1(X_{n-1},\mathcal{O}_{X_{n-1}}^*) \longrightarrow H^2(X,\mathcal{O}_X \otimes \mathfrak{m}^{n-1}/\mathfrak{m}^n)$$

We have

$$H^1(X,\mathcal{O}_X \otimes \mathfrak{m}^{n-1}/\mathfrak{m}^n) \xrightarrow{\sim} H^1(X,\mathcal{O}_X) \otimes \mathfrak{m}^{n-1}/\mathfrak{m}^n$$

and since we know that $H^1(X,\mathcal{O}_X)$ is finite dimensional, we can conclude that

The abelian group $\mathrm{Pic}(X \times_k \mathrm{Spec}(A))_e$ *is an extension of finite dimensional k- vector spaces.*

The sum of the dimensions, which we want to call the **size** *of* $\mathrm{Pic}(X \times_k \mathrm{Spec}(A))_e$*, can be estimated by* $H^1(X,\mathcal{O}_X)$ *and the structure of A. Especially if $A = k[\epsilon]$ is the k algebra of dual numbers then*

$$\mathrm{Pic}(X(k[\epsilon])_e \xrightarrow{\sim} H^1(X,\mathcal{O}_X).$$

We can reformulate this and say that $H^1(X,\mathcal{O}_X)$ is the tangent space of the Picard functor at e.

Of course in general we do not know whether

$$H^1(X_n,\mathcal{O}^*_{X_n}) \longrightarrow H^1(X_{n-1},\mathcal{O}^*_{X_{n-1}})$$

is surjective. We could say that the functor is **smooth** if this is the case.

Differentiating a line bundle along a vector field

Let us assume that X/k is smooth, then we have seen 7.5.5 that a vector field $V \in H^0(X,T_X)$ can be viewed as a $X[\epsilon] = X \times_k k[\epsilon]$ valued point

$$\begin{array}{ccc} X & \overset{V}{\nwarrow} & \\ \downarrow & & \\ X & \longleftarrow & X[\epsilon]. \end{array}$$

The morphism $p : X[\epsilon] \longrightarrow X$ is given by the trivial vector field. Let $\mathcal{L}$ be a line bundle on X, then we can consider line bundle

$$V^*(\mathcal{L}) \otimes p^*(\mathcal{L})^{-1},$$

this is a line bundle on $X[\epsilon]$, which is trivial on $X \subset X[\epsilon]$, hence it is an element in $\mathrm{Pic}(X(k[\epsilon])_e = H^1(X,\mathcal{O}_X)$. Therefore, any line bundle $\mathcal{L}$ defines a map

$$\delta_{\mathcal{L}} : H^0(X,T_X) \longrightarrow H^1(X,\mathcal{O}_X).$$

It is easy to describe this map in terms of cocycles. If $\mathcal{L}$ is given by a cocycle $\{g_{\alpha,\beta}\}_{(\alpha,\beta)\in I\times I}$ with respect to an open covering $\bigcup_{\alpha\in I} U_\alpha$.
We can differentiate the sections $g_{\alpha,\beta}$ with respect to the restriction of the vector field V to $U_\alpha \cap U_\beta$ and we put

$$h_{\alpha,\beta} = g_{\alpha,\beta}^{-1} V g_{\alpha,\beta} \in \mathcal{O}(U_\alpha \cap U_\beta).$$

Clearly this is a 1-cocycle with values in $\mathcal{O}_X$ and it follows from the definition that the class of this cocycle is $\delta_{\mathcal{L}}(V)$.

The theorem of the cube.

We combine the above results and Lemma 10.1.1. Again we consider a line bundle $\mathcal{L}$ on $X \times \mathrm{Spec}(A)$, where A is still local. Let $I_1 \subset A$ be the ideal, which defines the maximal subscheme $\mathrm{Spec}(A/I_1)$ such that $\mathcal{L}|X \times \mathrm{Spec}(A/I_1)$ is trivial. Let $\mathfrak{m} \subset A$ be the maximal ideal, then $I_1/\mathfrak{m}I_1$ is a finite dimensional $A/\mathfrak{m}$-vector space. Any non-zero linear form $\Psi : I_1/\mathfrak{m}I_1 \to A/\mathfrak{m}$ defines an ideal $I_\Psi \subsetneq I_1$ and $\mathcal{L}|X \times \mathrm{Spec}(A/I_\Psi)$ is not trivial anymore.

We want to compute the obstruction to trivialize $\mathcal{L}|X \times \mathrm{Spec}(A/I_\Psi)$, i.e. we try to extend a trivialization to this larger scheme, and we will get a non zero element $\xi_\Psi \in H^1(X \times \mathrm{Spec}(A/\mathfrak{m}),\mathcal{O}_X)$, which tells us that this is impossible.

Let I be any ideal such that $\mathcal{L}|X \times \mathrm{Spec}(A/I)$ is trivial. Let $I' = I\mathfrak{m}$. We have the inclusion

$$X \times \mathrm{Spec}(A/I) \hookrightarrow X \times \mathrm{Spec}(A/I'),$$

and our line bundle $\mathcal{L}|X \times \mathrm{Spec}(A/I)$ has a section $s \in H^0(X \times \mathrm{Spec}(A/I),\mathcal{L})$, which is a generator at all points. We can cover $X \times \mathrm{Spec}(A/I')$ by affine open sets U_i such that the section s extends to a section $s_i \in H^0(U_i,\mathcal{L})$, which then will be a generator at all points. We have

$$s_i = g_{ij}s_j$$

with $g_{ij} \in H^0(U_i \cap U_j,\mathcal{O}^*_{U_i\cap U_j})$. These g_{ij} are of the form

$$g_{ij} = 1 + h_{ij}$$

where $h_{ij} \in I$. Since we compute modulo I', we have $I/I' \simeq k(t_0)$ and $\{h_{ij}\}$ defines a class in $H^1(X \times k(t_0),\mathcal{O}_{X\times k(t_0)})$. This makes it clear that for any point $t_0 \in \mathrm{Spec}(A/I)$ we get a linear map

$$\delta_{\mathcal{L}} : I/\mathfrak{m}_{t_0}I \longrightarrow H^1(X \times k(t_0),\mathcal{O}_{X\times k(t_0)}).$$

We may for instance take as our ideal I simply the maximal ideal attached to a point $t_0 \in \mathrm{Spec}(A)$ and get a linear map

$$\delta_{\mathcal{L}} : \mathfrak{m}_{t_0}/\mathfrak{m}_{t_0}^2 \longrightarrow H^1(X \times k(t_0),\mathcal{O}_{X\times k(t_0)}),$$

which can be interpreted as the map from the tangent space of $\mathrm{Spec}(A)$ in the point t_0 to the "tangent space" of $\mathrm{Pic}(X)(k(t_0))$ in the "point" $\mathcal{L}$.

It is clear: If $I \subset A$ is the maximal ideal such that $\mathcal{L}|X \times \mathrm{Spec}(A/I)$ is trivial, then the map

$$\delta_{\mathcal{L}} : I/\mathfrak{m}_{t_0}I \longrightarrow H^1(X \times k(t_0),\mathcal{O}_{X\times k(t_0)})$$

is injective for all $t_0 \in \mathrm{Spec}(A/I)$. This has the following consequence

Theorem 10.1.2. *(***Theorem of the cube***) Let X,Y be projective schemes over k. Let T be a connected scheme of finite type over k, and let $\mathcal{L}$ be a line bundle on*

$$X \times Y \times T.$$

Let us assume that we have points $x_0 \in X(k)$, $y_0 \in Y(k)$, $t_0 \in T(k)$ such that $\mathcal{L}$ restricted to

$$x_0 \times Y \times T \;\; X \times y_0 \times T \;, \; X \times Y \times t_0$$

becomes trivial. Then the bundle itself is trivial.

Let A be the local ring at t_0,We have $\mathrm{Spec}(A) \subset T$. Let us consider the locus of triviality of $\mathcal{L}$ in $\mathrm{Spec}(A)$, it is of the form

$$\mathcal{L}|X \times Y \times \mathrm{Spec}(A/I).$$

We want to show $I = 0$. If not, then $I/\mathfrak{m}_{t_0}I \neq 0$, and we get an injective map

$$\delta_{\mathcal{L}} : I/\mathfrak{m}_{t_0}I \longrightarrow H^1(X \times Y \times k(t_0),\mathcal{O}_{X\times Y}).$$

But now we have the Künneth formula (See 8.2.2), which tells us

$$\begin{aligned} H^1(X \times Y \times k(t_0),\mathcal{O}_{X\times Y\times t_0}) &= H^1(X \times k(t_0),\mathcal{O}_{X\times t_0}) \oplus H^1(Y \times k(t_0),\mathcal{O}_{Y\times t_0}) \\ &= H^1(X \times y_0 \times k(t_0),\mathcal{O}_{X\times t_0}) \oplus H^1(X \times y_0 \times k(t_0),\mathcal{O}_{Y\times t_0}). \end{aligned} \quad (10.1)$$

But $\mathcal{L}|X \times y_0 \times T$ and $\mathcal{L}|x_0 \times Y \times T$ are assumed to be trivial. This means $\delta_{\mathcal{L}} = 0$. This is a contradiction.
Now we have seen that we can find an open neighborhood V of t_0 such that $\mathcal{L} \mid X\times Y\times V$ is trivial. This proves that the locus of triviality is open, on the other hand it is closed. Since it is non empty it follows that it is equal to T. Since X,Y are projective it follows that $\mathcal{L} = p_3^*(\mathcal{M})$ where $\mathcal{M}$ is a line bundle on T. This line bundle restricted to $\{x_0\}\times\{y_0\}\times T$ is trivial, hence $\mathcal{M}$ is trivial.

□

The Picard functor has two natural sub functors. We say that a line bundle $\mathcal{L}$ on our projective scheme X/k is **numerically equivalent to zero**, if its restriction to any curve $C \hookrightarrow X/k$ has degree zero.

We say that $\mathcal{L}$ is **algebraically equivalent to zero** , if we can find a connected scheme of finite type T/k with two points t_0,$t_1 \in T(k)$ on it and a line bundle $\widetilde{\mathcal{L}}$ over $X \times_k T$ such that

$$\begin{aligned} &\widetilde{\mathcal{L}}|X \times t_0 \quad \text{is trivial} \\ &\widetilde{\mathcal{L}}|X \times t_1 \quad \text{is isomorphic to } \mathcal{L}. \end{aligned} \quad (10.2)$$

The intuitive meaning of this concept is: A line bundle is algebraically equivalent to zero, if it can be "deformed" into the trivial bundle.

Proposition 10.1.3. *If a line bundle $\mathcal{L}$ on X is algebraically equivalent to zero, then it is numerically equivalent to zero.*

It suffices to consider the case of a non singular projective curve C/k. If we have a connected family $\widetilde{\mathcal{L}}$ on $C \times T$ with $\widetilde{\mathcal{L}}_{t_0} \simeq \mathcal{O}_c$, $\widetilde{\mathcal{L}}_{t_1} \simeq \mathcal{L}$ then the function

$$t \longrightarrow \dim_{k(t)} H^0(C \times k(t),\widetilde{\mathcal{L}}) - \dim_{k(t)} H^1(C \times k(t),\widetilde{\mathcal{L}})$$

is constant on T (see Thm. 8.4.6). But the right hand side is equal to $\deg(\widetilde{\mathcal{L}}_t) + 1 - g$, therefore, the degree is constant and equal to zero because $\deg(\widetilde{\mathcal{L}}_{t_0}) = 0$. □

We define sub functors. We put

$$\mathcal{PIC}^0_{X/k}(T) = \text{Isomorphism classes of line bundles on } X \times_k T$$
$$\text{such that } \mathcal{L}_t \text{ is numerically equivalent to zero for all } t \in T. \quad (10.3)$$

The second functor is defined on the category of connected schemes T/k of finite type, equipped with a base point $t_0 \in T(k)$ and the morphisms respecting the base point. Then

$$\mathcal{PIC}^{00}_{X/k}(T) = \text{Isomorphism classes of line bundles } \mathcal{L} \text{ on } X \times_k T,$$
$$\text{for which } \mathcal{L}|X \times \{t_0\} \simeq \mathcal{O}_X \quad (10.4)$$

It is a central point is that this last functor is "linear":

Proposition 10.1.4. *For a product $X \times Y$ of two projective schemes, which are also equipped with base points $x_0 \in X(k)$ and $y_0 \in Y(k)$, we have*

$$\mathcal{PIC}^{00}_{X\times Y/k}(T) \simeq \mathcal{PIC}^{00}_{X/k}(T) \oplus \mathcal{PIC}^{00}_{Y/k}(T)$$

This is a direct consequence of the theorem of the cube. If we have a line bundle $\mathcal{L}$ over $X \times Y \times T$, then the restrictions

$$\begin{array}{rcl} \mathcal{L}_X & = & \mathcal{L}|X \times y_0 \times T \\ \mathcal{L}_Y & = & \mathcal{L}|x_0 \times Y \times T \end{array}$$

provide line bundles on $X \times T$ and $Y \times T$. Now we consider the two projections

$$\begin{array}{rcccl} p_X & : & X \times Y \times T & \to & X \times T \\ p_Y & : & X \times Y \times T & \to & Y \times T, \end{array}$$

and we consider

$$\mathcal{L} \otimes p_X^*(\mathcal{L}_X)^{-1} \otimes p_Y^*(\mathcal{L}_Y)^{-1}.$$

This line bundle on $X \times Y \times T$ is trivial by the theorem of the cube. □

Assume we have proved (local) representability for these functors. Then we denote the resulting schemes by $\mathrm{Pic}_{X/k}$, $\mathrm{Pic}^0_{X/k}$, $\mathrm{Pic}^{00}_{X/k}$, it follows from their universal properties that they are group schemes.

We could think of $\mathrm{Pic}^0_{X/k}$ as being the "connected component " of the identity of our functor, the quotient $\mathrm{Pic}_{X/k} / \mathrm{Pic}^0_{X/k}$ gives us discrete invariants. These invariant can viewed as elements in some second cohomology group $H^2(X)$. (See Vol. I, 5.2.1.) But this group is not yet defined at this stage. In any case the class $c_1(\mathcal{L})$ of a line bundle $\mathcal{L}$ on X/k in $\mathrm{Pic}_{X/k} / \mathrm{Pic}^0_{X/k}$ will be called the Chern class of $\mathcal{L}$. Again the quotient $NS(X) = \mathrm{Pic}_{X/k}(k)/\mathrm{Pic}^0_{X/k}(k)$ is called **Neron-Severi group** of X/k.

Let $\mathcal{L}$ be a line bundle on a product $X \times Y$. Since the second cohomology of a product $X \times Y$ is bigger than the sum of the second cohomology groups of X and Y (Künneth-formula), we can not test the vanishing of $c_1(\mathcal{L})$ by restricting it to $\{x_0\} \times Y$ and $X \times \{y_0\}$ and this is the reason why we have the theorem of the cube and not the theorem of the square. The Picard functor is quadratic and not linear.

10.1.4 The basic principles of the construction of the Picard scheme of a curve.

The principal aim of this chapter is the construction of $\mathrm{Pic}_{C/k}$ for a smooth, projective and absolutely irreducible curve C/k, i.e we want to construct the scheme, which together with appropriate line bundle $\mathcal{P}$ on $C \times \mathrm{Pic}_{C/k}$ weakly represents the functor $\mathcal{PIC}_{C/k}$. At the same time we will construct $\mathcal{PIC}_{C/k,P}$ if $P \in C(k)$ is a k-rational point.

It is of course clear that this functor is a "disjoint union"

$$\mathcal{PIC}_{C/k} = \bigsqcup \mathcal{PIC}^r_{C/k},$$

where $\mathcal{PIC}^r_{C/k}$ is the functor of line bundles of degree r on C.
Let $\mathrm{Pic}^r_{C/k}$ be the corresponding components. It will turn out that $\mathrm{Pic}^0_{C/k}$ is an abelian variety, it is called $J = J_C$, the Jacobian of the curve. We will show that J and the $\mathrm{Pic}^r(C)$ are smooth, projective and irreducible.

In the second section of this chapter we will also discuss the Picard functor of J.

The basic idea of the construction is simple and goes back to Jacobi. The first object, which we will construct is $\mathrm{Pic}^g_{C/k}$. Let us assume that $k = \overline{k}$ is algebraically closed. Let us consider line bundles $\mathcal{L}$ of degree g on C. The Riemann-Roch theorem says

$$\dim H^0(C,\mathcal{L}) = g + 1 - g + \dim H^1(C,\mathcal{L}) = 1 + \dim H^1(C,\mathcal{L}).$$

We introduce the notion of a "generic" line bundle of degree g. A bundle of degree g is generic if $\dim H^1(C,\mathcal{L}) = 0$, which is equivalent to $\dim H^0(C,\mathcal{L}) = 1$.
The leading principle will be that a "generic" line bundle $\mathcal{L}$ has a non zero global section $s \in \dim H^0(C,\mathcal{L})$, which is unique up to a scalar. The set of zeroes of s is a divisor

$$\mathrm{Div}(s) = P_1 + \cdots + P_g,$$

where $P_i \in C(k)$. This collection of g points is unique up to permutations. Therefore we can view it as a point in $C^g/\Sigma_g(k)$ where C^g/Σ_g is the quotient of C^g divided by the symmetric group Σ_g. (We will discuss the construction of the quotient C^g/Σ_g and its properties in the following section, we anticipate the obvious properties of this construction). On the other hand a point $D \in C^g/\Sigma_g(k)$ can be lifted to a point $(P_1, \dots, P_g) \in C^g$ hence we can say $D = P_1 \cdots + P_g$ and this point yields a line bundle $\mathcal{O}(P_1 + \cdots + P_g)$, which comes with a non vanishing section namely the constant 1. We say that D is generic if $\mathcal{O}(D)$ is generic. We have seen in exercise 40 that the set of generic divisors $D \in C^g/\Sigma_g(k)$ is the set of geometric points of an open, non empty subscheme $U_{\mathrm{gen}} \subset C^g/\Sigma_g$.
Therefore the following becomes clear: If we have a line bundle $\mathcal{L}$ on $C \times_k T$ such that $\mathcal{L}_t = \mathcal{L}|C \times \{t\}$ is of degree g and generic for $t \in T(k)$, then we have a unique map

$$\psi : T(k) \longrightarrow U_{\mathrm{gen}}(k)$$

such that $\mathcal{L}_t \xrightarrow{\sim} \mathcal{O}(\psi(t))$.
But we know that this is not yet what we want. We have to show that ψ is in fact induced by a morphism and we have to construct a line bundle on $C \times_k U_{\mathrm{gen}}$ whose pullback is locally in T isomorphic to $\mathcal{L}$. Only if we have established these two points we have proved the local representability of $\mathcal{PIC}^{g,\mathrm{gen}}_{C/k}$. The detailed proof is a little bit technical and lengthy.

To remove the restriction to generic bundles we will show that there is a finite number of line bundles $\mathcal{F}_0 = \mathcal{O}_C, \mathcal{F}_1, \ldots, \mathcal{F}_r$ such that for any line bundle $\mathcal{L}$ on C there exists a index $0 \leq i \leq r$, for which the line bundle $\mathcal{L} \otimes \mathcal{F}_i$ is generic. Hence we define $U^{(i)}_{\text{gen}}$ to be the open subset, for which

$$U^{(i)}_{gen}(k) = \{t \in C^g/\Sigma_g(k) \mid \mathcal{O}(t) \otimes \mathcal{F}_i \text{ is generic }\}$$

Clearly these $U^{(i)}_{gen}$ yield local representations of appropriate "sub functors" $\mathcal{PIC}^{g,i}_{C/k}$ of $\mathcal{PIC}^{g}_{C/k}$, we have to glue them together to get our $\mathrm{Pic}_{C/k}$. Eventually we apply the methods of Galois descend (see 6.2.9) to remove the restriction that k is algebraically closed.

10.1.5 Symmetric powers

Let k be a field and let X/k be a quasi-projective scheme. Let $\bar{k}$ be an algebraic closure of k. For any integer r we can form the r-fold product $X^r = X \times_k X \cdots \times X$ (r factors) and on this product we have an action of the symmetric group Σ_r. We will construct a quotient X^r/Σ_r together with a projection

$$\pi : X^r \longrightarrow X^r/\Sigma_r$$

such that we have the obvious universal property: For any scheme T and any morphism $h : X^r \to T$, which commutes with the action of Σ_r (i.e. $h \circ \sigma = h$ for all $\sigma \in \Sigma_r$) we have a unique morphism $\bar{h} : X^r/\Sigma_r \to T$ such that $h = \bar{h} \circ \pi$.
The construction of this quotient is easy if $X = \mathrm{Spec}(A)$, where A is an affine k-algebra. We have $X^r = \mathrm{Spec}(A^{\otimes r})$ and

$$X^r/\Sigma_r = \mathrm{Spec}\left((A^{\otimes r})^{\Sigma_r}\right).$$

If X/k is quasi-projective, then we write

$$\begin{array}{ccc} X & \hookrightarrow & \mathbb{P}^n \\ \searrow & & \swarrow \\ & \mathrm{Spec}(k) & \end{array}$$

and we observe that a geometric point $P \in X^r(\bar{k})$ gives us an r-tuple of points $(P_1, \cdots, P_r)$ with $P_i \in X(\bar{k})$. Since all these point lie already in $X(L)$ for a suitable normal finite extension L/k we get a finite set of conjugates P_i^σ where σ runs through the elements of $\mathrm{Gal}(L/k)$. We can find a hyperplane $H \subset \mathbb{P}^n$, which is defined over a separable, normal extension F/k such that $P_i^\sigma \notin H(\bar{k})$ for all i,σ. The complement U of the union of the conjugates of the hyperplanes $\cup_\tau H^\tau, \tau \in \mathrm{Gal}(\bar{k}/k)$ is a non empty affine subscheme $U/k \subset X/k$ and $P \in U^r(\bar{k})$. Hence we can cover X^r by a finite number of open affine subsets $V_\nu = U_\nu \cdots \times \ldots U_\nu$, where U_ν is affine and defined over k. We can form the quotients U_ν^r/Σ_r and now it is obvious how to glue (see 6.2.1) these pieces together to get X^r/Σ_r.

We claim that this construction has the following properties:

(i) It commutes with extensions of the field of scalars, i.e. for any $k \to L$ we have

$$(X \times_k L)^r/\Sigma_r = X^r/\Sigma_r \times_k L.$$

(ii) The scheme X^r/Σ_r is again quasi-projective, it is affine if X/k is affine and it is projective if X/k is projective.

To see the first assertion we assume that $X = \mathrm{Spec}(A)$ where A is an affine k-algebra. This algebra is free as a k-vector space, let $e_0, e_1, \cdots, e_m, \cdots$ be a basis, we choose $e_0 = 1$. The elements

$$e_{\underline{i}} = e_{i_1} \otimes \cdots e_{i_r} \qquad \text{for} \qquad (i_1, \cdots, i_r) \in \mathbb{N}^r$$

form a basis of the k-vector space $A^{\otimes r}$. The group Σ_r acts upon $\mathbb{N}^r$, let $\Sigma_{r,i}$ be the stabilizer of $\underline{i}$ in Σ_r. We put

$$E_{\underline{i}} = \sum_{\sigma \in \Sigma_r/\Sigma_{r,\underline{i}}} e_{\sigma(\underline{i})},$$

and these elements form a basis of the k-vector space $(A^{\otimes r})^{\Sigma_r}$. Since the $e_0, \cdots, e_m, \cdots$ also form a basis of $A \otimes_k L$ over L, our assertion (i) is clear for affine k-algebras. But then it is obviously also true for any quasi-projective X/k. The assertion (ii) is obvious for affine schemes. It suffices to prove it for projective schemes. But now we may write for instance

$$X = \mathrm{Proj}(k[x_0, \cdots, x_n])$$

where $k[x_0, \cdots, x_n]$ is a graded k-algebra and where the x_i have degree one. We have seen that we can write

$$X^r = \mathrm{Proj}(k[\cdots u_{\underline{\alpha}} \cdots])$$

where $\underline{\alpha} = (\alpha_1, \cdots, \alpha_r) \in [0,n]^r$ and

$$u_{\underline{\alpha}} = x_{\alpha_1} \otimes x_{\alpha_2} \cdots \otimes x_{\alpha_r}.$$

Then we get

$$X^r/\Sigma_r = \mathrm{Proj}\left((k[\cdots x_\alpha \cdots])^{\Sigma_r}\right).$$

Proposition 10.1.5. *If A/k is an affine k-algebra, then the algebra $(A^{\otimes r})^{\Sigma_r}/k$ is generated by the elementary symmetric functions*

$$\sigma_\mu(f) = \sum 1 \otimes \cdots f \otimes 1 \cdots \otimes f \cdots \otimes 1,$$

where we sum over all the possible placements of $\mu (\leq r)$ factors f in the above tensor product.

To see this we start from our basis $e_0, e_1, \cdots e_m \cdots$ recall that $e_0 = 1$. We consider a basis element $E_{\underline{i}}$ as above, and we want to show that it is in the algebra generated by the elementary symmetric functions. We proceed by downward induction on the number of times 0 (i.e. $e_0 = 1$) occurs in $\underline{i} = (i_1, \cdots, i_r)$. If 0 occurs r-times the assertion is clear. Now we write

$$\underline{i} = (\underbrace{0, \cdots, 0}_{\nu_0-\text{indices}}, \underbrace{i, \cdots, i}_{\nu_1-\text{indices}}, \cdots, \underbrace{j, \cdots, j}_{\nu_m-\text{indices}})$$

where $0,i,\cdots,j$ are pairwise different. Then

$$e_{\underline{i}} = (1\otimes\cdots e_i\otimes\cdots\otimes e_i\otimes 1\cdots 1)\cdot(1\otimes\cdot\otimes e_i\otimes e_i\cdots e_i\otimes 1\cdots\otimes 1)(1\otimes 1\cdots 1\otimes e_j\otimes\cdots e_j).$$

Consider

$$E_{\underline{i}} - \sigma_{\nu_1}(e_i)\cdots\sigma_{\nu_m}(e_j),$$

which is in $(A^{\otimes r})^{\Sigma r}$. If we look carefully at

$$(+\cdots 1\otimes\cdots\otimes e_i\otimes\cdots 1\otimes e_i\otimes 1\cdots)\cdots(\cdots 1\otimes e_j\otimes\cdots 1\otimes e_j\cdots\otimes 1\cdots),$$

and expand, then we see that $E_{\underline{i}}$ is exactly that part of the expansion where we multiply tensors, which have the $e_i\cdots e_j$ at different places. So this part cancels. But for the remaining terms we see that we get some

$$\cdots\otimes e_\nu e_\mu\otimes\cdots$$

in the product. This $e_\nu e_\mu$ can be expressed by linear combinations of the $e_i\cdots e_j$ and hence the number of 1's in the tensor product goes up in these terms. □

It is clear that the morphism

$$\pi : X^r \longrightarrow X^r/\Sigma_r$$

is finite. This implies that we have a surjective map on the set of geometric points

$$X^r(\overline{k}) \longrightarrow X^r/\Sigma_r(\overline{k}).$$

It is obvious that points, which are equivalent under the action of the symmetric group map to the same point in $X^r/\Sigma_r(\bar{k})$ hence the map factorizes

$$X^r(\overline{k})/\Sigma_r \longrightarrow X^r/\Sigma_r(\overline{k})$$

Exercise 45. Show that the above map is a bijection.

But over an arbitrary field the map $X^r(k) \to X^r/\Sigma_r(k)$ may not be surjective. A point $Q \in X^r/\Sigma_r(k)$ can be lifted to a point $P = (P_1,\cdots,P_r) \in X^r(\overline{k})$ but then we have to answer the following question

When does a geometric point $P = (P_1,\cdots,P_r) \in X^r(\overline{k})$ give a point $\pi(P) \in X^r/\Sigma_r(k)$?

To answer this question we may assume that $X = \mathrm{Spec}(A)$. Then our point $P = (P_1\cdots P_r) = (\psi_1,\cdots,\psi_r)$ where the ψ_i are geometric points of X and this is the same as $\otimes\psi_i : B^{\otimes r} \longrightarrow \overline{k}$. The image of $(\psi_1,\cdots,\psi_r)$ lies in $X^r/\Sigma_r(k)$ if and only if the restriction $\otimes\psi_i : (B^{\otimes r})^{\Sigma_r} \longrightarrow k$. Now it follows from proposition 10.1.5

Proposition 10.1.6. *We have* $\otimes\psi_i : (B^{\otimes r})^{\Sigma_r} \longrightarrow k$ *if and only if for all* $f \in A$ *the polynomial*

$$(Y - \psi_1(f))\cdots(Y - \psi_r(f)) \in k[Y].$$

Now we want to assume that our scheme is a smooth projective curve C/k. A point $P = (P_1, \cdots, P_r) \in C^r(\overline{k})$ gives us a divisor

$$\overline{D} = \sum_{i=1}^{r} P_i = \sum_{P \in C(\overline{k})} n_P P$$

on the curve $C \times_k \overline{k}$. We want to discuss the question whether this divisor is the base extension of a divisor on C/k

$$D = \sum_{\mathfrak{p}} n_{\mathfrak{p}} \mathfrak{p}.$$

In this case we say that our divisor is rational over k. Of course we must have

$$\deg(\overline{D}) = \sum n_P = \deg(D) = \sum n_{\mathfrak{p}}[k(\mathfrak{p}) : k].$$

We have the action of the Galois group $\mathrm{Gal}(\bar{k}/k)$ on the points in the support of our divisor. An orbit of this action consist of the points P, which lie above one point $\mathfrak{p}$, i.e. of the points $P \longrightarrow \mathfrak{p}$. It is clear that we can assume that we have just one orbit, which means that all the points P_i lie over one closed point $\mathfrak{p}$ on C/k.

Hence we get: The base extension of the divisor $\mathfrak{p}$ on C/k to the geometric curve is

$$n_{\mathfrak{p}}^{(i)} \Big(\sum_{P \to \mathfrak{p}} P \Big),$$

where the Galois acts transitively on the $P \to \mathfrak{p}$ and where $n_{\mathfrak{p}}^{(i)}$ is the degree of inseparability of the extension $k(\mathfrak{p})/k$ (See 247, 9.24)

We conclude: If we look at our divisor $\overline{D} = \sum_{P \in C(\overline{k})} n_P P$, and if we divide the points in its support under the action of the Galois group, then the orbits correspond to points $\mathfrak{p}$ on C/k. Our divisor

$$\overline{D} = \sum_{\mathfrak{p}} n_{\mathfrak{p}} \Big(\sum_{P \to \mathfrak{p}} P \Big),$$

is rational over k if and only if for all $\mathfrak{p}$ we have $n_{\mathfrak{p}}^{(i)} | n_{\mathfrak{p}}$.

We want to show that the condition $n_{\mathfrak{p}}^{(i)} | n_{\mathfrak{p}}$ for all (p) is equivalent to the condition that

$$(\cdots \underbrace{P, \cdots, P}_{n_{\mathfrak{p}} \text{ times}} \cdots) \in C^r/\Sigma_r(k).$$

To see that this is the case we can assume that we have only one $\mathfrak{p}$ and our points lie in an affine open subset $U \subset C$ and $U = \mathrm{Spec}(A)$. Let $k \hookrightarrow \bar{k}$ be an algebraic closure, let k_s,k_i be the maximal separable (inseparable) sub extension respectively. Our points are k-homomorphism $\psi_\nu : A \longrightarrow \bar{k}$, they form an orbit under the action of $\mathrm{Gal}(\bar{k}/k)$. For any $f \in A$ the polynomial

$$\prod_{\nu} (Y - \psi_\nu(f)) \in k_i[Y]$$

because it is invariant under the action of the Galois group.

Lemma 10.1.7. *The degree of inseparability $n_{\mathfrak{p}}^{(i)} = p^e$ is the smallest power of p, for which all the $\psi_\nu(f)^{p^e} \in k_s$.*

Let us assume this Lemma. Then we know that p^e is the smallest power of p such that for all $f \in A$ the polynomial

$$\prod_\nu (Y - \psi_\nu(f))^{p^e} \in k[Y],$$

and now it follows from proposition 10.1.6 that p^e is the smallest power of p such that

$$(\cdots \underbrace{P, \cdots, P}_{p^e \text{ times}} \cdots)_{P \longrightarrow \mathfrak{p}} \in C^r/\Sigma_r(k).$$

It remains to prove the Lemma. We choose an $F \in A$ such that the differential dF generates the module of differentials at $\mathfrak{p}$. Then we know that the extension $k[F] \hookrightarrow A$ is etale at $\mathfrak{p}$. This implies that $\psi_\nu(f)^{p^e} \in k_s$ for all $f \in A$ if we have $\psi_\nu(F)^{p^e} \in k_s$ (see prop. 7.5.15). But now $\mathfrak{p} \cap k[F] = (p(F))$ where $p(F) \in k[F]$ is an irreducible polynomial. The field $k[F]/(p(F))$ has degree of inseparability equal to p^e. Then it follows from elementary algebra that

$$p(F) = F^{p^e n} + a_1 F^{p^e(n-1)} + \ldots a_{n-1} F^{p^e} + a_n$$

with some coefficients in k. This polynomial is equal to

$$\prod_\nu (F - \psi_\nu(F))^{p^e} \in k[F].$$

If we know remember that this polynomial is irreducible we get the assertion of the Lemma. Hence we come to the conclusion

Proposition 10.1.8. *We have a canonical bijection between the effective divisors on C/k of degree r and the points on $C^r/\Sigma_r(k)$.*

In our special case we have

Theorem 10.1.9. *For a smooth curve C/k the quotient C^r/Σ_r is again smooth.*

We may assume that k is algebraically closed. We can pick a geometric point $P \in C^r/\Sigma_r$, we find a finite number of points $Q_1, \ldots, Q_s$ in $C^r(k)$ lying above this point. We can write $Q_1 = (P_1, \ldots, P_r)$ and the other Q_i are obtained by permuting the coordinates of Q_1. Now we have two possibilities to proceed. The first one is to pick a meromorphic function f on C, which is regular at these points $P_1, \ldots, P_r$, and which has the property that df is a generator of the differentials in these points. We can find an affine scheme $\mathrm{Spec}(A) \subset C$ such that our points lie in $\mathrm{Spec}(A)$. Now we know that $k[f] \hookrightarrow A$ is etale and from this we conclude that $(k[f]^{\otimes r})^{\Sigma_r} \hookrightarrow (A^{\otimes r})^{\Sigma_r}$ (boring argument omitted). Hence we see that we have to prove our theorem for the special case $C = \mathrm{Spec}(k[f])$. But now the theorem of the elementary symmetric functions says

$$(k[f]^{\otimes r})^{\Sigma_r} \xrightarrow{\sim} k[X_1, \ldots, X_r]^{\Sigma_r} = k[\sigma_1, \ldots, \sigma_r]$$

where the σ_i are the elementary symmetric functions. This is a polynomial ring and hence smooth.

The second possibility is to investigate the completion of the local ring $\mathcal{O}_{C^r/\Sigma_r,P}$ at the point P. We have two extreme possibilities. If all the coordinates of $P = (P_1,\dots,P_r)$ are pairwise different then it clear that for the completions

$$\widehat{\mathcal{O}_{C^r/\Sigma_r,P}} = \widehat{\mathcal{O}_{C^r,(P_1,\dots,P_r)}}$$

and smoothness becomes clear. The opposite case is that all the coordinates are equal say to P'. Then this point $(P',\dots,P')$ is the only point in $C^r(k)$, which lies over P and it is clear

$$(\widehat{\mathcal{O}_{C^r,(P',\dots,P')}})^{\Sigma_r} = \widehat{\mathcal{O}_{C^r/\Sigma_r,P}}.$$

But now the ring on the left is the power series ring in r variables and then the ring of invariants is the power series ring in the elementary symmetric functions. To treat the general case one has to mix the arguments. □

10.1.6 The actual construction of the Picard scheme of a curve.

We start from a smooth, projective and connected curve C/k. Our next aim is to prove

Theorem 10.1.10. *Let C/k be smooth, projective and connected and let g be the genus of C/k. The functor $T \longrightarrow \mathcal{PIC}^g_{C/k}(T)$ is locally representable.*

Of course this implies that we construct a scheme $\mathrm{Pic}^g_{C/k}$ and a line bundle $\mathcal{P}$ (or a $\mathrm{Pic}^g_{C/k}$ gerbe $\{\mathcal{P}_\alpha,\psi_{\alpha,\beta}\}$) on $C \times \mathrm{Pic}^g_{C/k}$, such that the universal property is fulfilled. Our strategy has been outlined in 10.1.4. We begin with the construction of a line bundle on $C \times C^g/\Sigma_g$.

In exercise 40 in section 9.6.3 we constructed a line bundle on

$$C \times C^g,$$

which was the product

$$\mathcal{L}' = \otimes \mathcal{O}_{C\times C^g}(\Delta_i),$$

where Δ_i was the inverse image of the diagonal under the projection

$$p_{0i} : C \times C^g \longrightarrow C \times C$$

to the zeroe'th and the i'th factor. The symmetric group Σ_g acts on the second factor and we have the quotient map

$$\mathrm{Id} \times \Pi : C \times C^g \longrightarrow C \times C^g/\Sigma_g.$$

We construct a line bundle $\mathcal{Q}$ on $C \times C^g/\Sigma_g$ whose inverse image $\Pi^*(\mathcal{Q}) \xrightarrow{\sim} \mathcal{L}'$. If $P \in C \times C^g/\Sigma_g$ is a closed point and if we lift this point to a point $\widetilde{P} \in C \times C^g$, then we can project it to the components and we get a $g+1$ tuple $(P_0,P_1,\dots,P_g)$. We can find a function in $k(C)$, which is holomorphic at all the points $P_0,\dots,P_g$, and which has the additional property that df generates the differentials at all these points. In a suitable neighborhood $V_{\widetilde{P}}$ of $\widetilde{P}$ the inverse image of the diagonal $\Delta \subset C \times C$ under the projection p_{0i} is defined by the ideal generated by

$$F_i = f \otimes 1 \cdots \otimes 1 - 1 \otimes 1 \cdots \otimes f \cdots \otimes 1,$$

where we put the f into the i-th place. The inverse of this function trivializes the bundle $\mathcal{O}_{C\times C^g}(\Delta_i)$ in our given neighborhood. If we start from another point $\widetilde{Q}$ and G_i in a neighborhood $V_{\widetilde{Q}}$ then the quotient F_i/G_i will be a unit on $V_{\widetilde{Q}} \cap V_{\widetilde{P}}$. This means that $\mathcal{L}'$ is trivialized locally at $\widetilde{P}$ by the product $(\prod_{i=1}^{i=g} F_i)^{-1}$. This is a meromorphic function on $C \times C^g/\Sigma_g$. This system of trivializing sections defines $\mathcal{Q}$.

We left it as an exercise (See exercise 40)to the reader to show that the set

$$U' = \{u \in C^g | \dim_{k(u)} H^0(C \times_k k(u), \mathcal{Q}_u) = 1\} \subset C^g.$$

is a non empty open subset in C^g. It is clearly invariant under the action of the symmetric group on $C \times U'$, we can form the quotient by this action and get an open subset $U = U'/\Sigma_g \subset C^g/\Sigma_g$. We introduced this open set already earlier and called it U_{gen} Our line bundle $\mathcal{Q}$ over $C \times C^g/\Sigma_g$, has a specific properties: If we project $\pi_0 : C \times C^g/\Sigma_g \longrightarrow C^g/\Sigma_g$, and if we restrict this projection to $C \times U_{\text{gen}} \longrightarrow U_{\text{gen}}$ then the sheaf $(\pi_0)_*(\mathcal{Q})$ will be locally free of rank one over U_{gen}. (See Theorem 8.4.5 (2). We want to denote the restriction of $\mathcal{Q}$ to $C \times U_{\text{gen}}$ by $\mathcal{P}'$. Since our sheaf $\mathcal{P}'$ contains the structure sheaf $\mathcal{O}_{C\times C^g/\Sigma_g}$ we can even say that the $\mathcal{O}_U$ module $(\pi_0)_*((\mathcal{P}')$ is generated by the element $1 \in H^0(C \times U_{\text{gen}}, \mathcal{O}_{C\times U_{\text{gen}}})$.

The following proposition says that our line bundle $\mathcal{P}'$ has a universal property.

Proposition 10.1.11. *Let T be a scheme of finite type over k, let us assume that we have a line bundle $\mathcal{L}_1$ over $C \times_k T$, such that for any point $t \in T$ we have* $\deg \mathcal{L}_{1,t} = \deg \mathcal{L}_1 | C \times_k k(t) = g$ *and*

$$\dim_{k(t)} H^0(C \times_k k(t), \mathcal{L}_{1,t}) = 1.$$

Then there is a unique morphism

$$\psi : T \longrightarrow U_{gen},$$

such that

$$(\mathrm{Id} \times \psi)^*(\mathcal{P}') | C \times T \sim_T \mathcal{L}_1 | C \times T.$$

We want to comment on this proposition. It gives us already a large part of our theorem 10.1.10, it says that the sub functor of $\mathcal{PIC}^g_{C/k}$ defined by the generic bundles is locally represented by $(C \times U_{\text{gen}}, \mathcal{P}')$. We will consider U_{gen} to be a open subset of C^g/Σ_g and of $\mathrm{Pic}_{C/k}$ as well.
The second comment is that we proved the proposition in a special case. This will be explained in the following exercise.

Exercise 46. Assume that our scheme T is integral then it has a field $L = k(T)$ of meromorphic function, which is the residue field of the generic point. We can restrict our line bundle to $C \times \mathrm{Spec}(L)$.
a) Show that our considerations in 1.4. imply that this restriction of ψ gives us an L valued point in $C^g/\Sigma_g(L)$.

b) Show that this L valued point gives us a morphism $\psi' : T' \longrightarrow C^g/\Sigma_g$ from some non empty open $T' \subset T$ such that $(\mathrm{Id} \times \psi')^*(\mathcal{P}')|C \times T' \xrightarrow{\sim} \mathcal{L}_1|C \times_k T'$.
c) Let us assume we have a line bundle $\mathcal{L}_0$ on C/k, which is generic, it yields a point $P \in C^g/\Sigma_g(k)$. Assume that this point lifts to a point $P = (P_1,P_2,\cdots,P_g) \in C^g(k)$ were all the components are pairwise different. We consider line bundles $\mathcal{L}$ on $T = \mathrm{Spec}(k[\epsilon])$ whose restriction to $C \times \mathrm{Spec}(k)$ give $\mathcal{L}_0$. Clearly these $\mathcal{L}$ form a torsor under the group of line bundles on $C \times T$, which are trivial on $C \times \mathrm{Spec}(k)$ and this group is isomorphic to to $H^1(C,\mathcal{O}_C)$. The tangent space to C^g/Σ_g at P is (see 10.1.3) the direct sum of tangent spaces T_{P_i} at the points $P_i \in C(k)$. Hence our proposition tells us that we should have an isomorphism

$$H^1(C,\mathcal{O}_C) \xrightarrow{\sim} \sum T_{P_i}.$$

Write down this isomorphism!

We come to the proof of the proposition. We consider the projection

$$p_2 : C \times_k \mathrm{Spec}(A) \to \mathrm{Spec}(A).$$

Our assumption implies that for all $t \in \mathrm{Spec}(A)$ the $k(t)$ vector space $\pi_*(\mathcal{L}_{1,t}) = H^0(C \times \{t\},\mathcal{L}_{1,t})$ is one-dimensional and $H^1(C \times \{t\},\mathcal{L}_{1,t}) = 0$. The the semi-continuity theorem (See Theorem 8.4.5 (2)) implies that $\pi_*(\mathcal{L}_1)$ is a locally free module of rank 1 over A. After passing to a neighborhood of t_0 we may assume that it is free.

Let $V(\mathcal{L}_1)$ be the bundle of one-dimensional vector space over $C \times \mathrm{Spec}(A)$ obtained from $\mathcal{L}_1$ (see p. 20) let p be the projection morphism. Then we have the zero section in this bundle

$$\begin{array}{ccc} C \times_k \mathrm{Spec}(A) \times \{0\} & \overset{i_0}{\hookrightarrow} & V(\mathcal{L}_1) \\ p \circ i_0 \searrow & & p \downarrow \\ & & C \times_k \mathrm{Spec}(A) \end{array}$$

and we use the isomorphism $p_0 \circ i_0$ to identify the zero section and $C \times_k \mathrm{Spec}(A)$. Our section s defines a subscheme in $V(\mathcal{L}_1)$, which is defined locally by one equation. This subscheme intersected with the zero section defines via our identification a subscheme

$$[s = 0] \subset C \times_k \mathrm{Spec}(A).$$

We know that this subscheme is of degree g in any fibre, this means that for any $t \in \mathrm{Spec}(A)$ the subscheme

$$[s_t = 0] := [s = 0] \times_k \mathrm{Spec}(k(t)) \subset C \times_k \mathrm{Spec}(k(t))$$

is finite and of degree g (see Theorem 8.1.8). But we have something much more precise

Lemma 10.1.12. *After passing to a neighborhood of t_0 the scheme $[s = 0]$ is flat over* $\mathrm{Spec}(A)$ *and therefore, the A-algebra $B_0 = B \otimes A/I$ is free of rank g (as an A-module).*

For this proof we may and do replace A by the local ring at t_0. In a first step we show that $[s = 0]$ is a finite affine scheme over $\mathrm{Spec}(A)$. We consider a finite subscheme $F \subset C$, $F \neq \emptyset$ such that $F \cap [s_{t_0} = 0] = \emptyset$. We put $V = C \setminus F$, this is affine and we consider

$$(F \times_k \mathrm{Spec}(A)) \cap [s = 0].$$

This must be empty, because if we project it to Spec(A) we get a closed subscheme, which does not contain t_0 . Since A is local, it follows that this closed subscheme is empty.

We see that the scheme $[s = 0]$ is an affine scheme. But of course the restriction $\pi_0 : [s = 0] \longrightarrow \mathrm{Spec}(A)$ is also projective we know that $\pi_{0,*}(\mathcal{O}_{[s=0]})$ is coherent and therefore, $[s = 0]$ is finite over Spec(A). We put $B = \mathcal{O}_C(V)$, then $V = \mathrm{Spec}(B) \subset C$, and

$$[s = 0] = \mathrm{Spec}(B \otimes A/I),$$

where I is an ideal, which is locally principal.

In the second step we prove the flatness. Assume we know that $\mathrm{Tor}_1^A(B\otimes A/I, k(t_0)) = 0$. Then we choose a basis for the $k(t_0)$ vector space $(B \otimes A/I) \otimes k(t_0)$ and lift these basis elements to elements $h_1 \cdots, h_g$ in $B\otimes A/I$, which then will be generators by Nakayama's lemma. Hence we get a surjective homomorphism and an exact sequence

$$0 \to R \to A^g \to A \otimes B/I \to 0,$$

where R is the A-module of relations. If we tensorize by $A/\mathfrak{m}_0 = k(t_0)$ the sequence is still exact, because of the vanishing of the Tor^1. The arrow $(A/\mathfrak{m}_0)^g \to (B\otimes A/I)\otimes k(t_0)$ becomes an isomorphism and hence we conclude that $R \otimes A/\mathfrak{m}_0 = 0$. Again we apply Nakayama and get the result.

Now we have to show the vanishing of the Tor_1. We consider the sequence

$$0 \to I \to B \otimes A \to B \otimes A/I \to 0,$$

and since $B \otimes A$ is free over A we get an exact sequence

$$0 \to \mathrm{Tor}_1^A(B \otimes A/I, A/\mathfrak{m}_0) \to I \otimes A/\mathfrak{m}_0 \to B \otimes A/\mathfrak{m}_0 \to .$$

We are through, if we show that

$$I \otimes A/\mathfrak{m}_0 \to B \otimes A/\mathfrak{m}_0$$

is injective. We may assume that I is generated by an element f. (We may take our V above so small that $\mathcal{L}_1$ becomes trivial on an open neighborhood of $V \times \mathrm{Spec}(A/\mathfrak{m})$). Then we have to consider elements

$$yf \otimes 1 \in I \otimes A/\mathfrak{m}_0,$$

whose image yf in $B\otimes A/\mathfrak{m}_0$ is zero. But $B\otimes A/\mathfrak{m}_0$ is integral and since f is non-zero in $B \otimes A/\mathfrak{m}_0$, we can conclude y goes to zero in $B \otimes A/\mathfrak{m}_0$. But then it follows that $yf \otimes 1$ is zero and this is the injectivity. □

It is clear that we can recover the line $\mathcal{L}_1$ from $[s = 0]$, we have $\mathcal{L}_1 = \mathcal{O}_{C\times_k \mathrm{Spec}(A)}([s = 0])$ or equivalently $\mathcal{L}_1 = I^{-1}$. The same consideration applies tp $[\Delta = 0] \subset C \times_k C^g/\Sigma_g$ and the line bundle $\mathcal{P}'$. Hence we see that our requirement on ψ in proposition 10.1.11 can be reformulated

$$(\mathrm{Id} \times \psi)^*(\mathcal{P}')|C \times T \sim_T \mathcal{L}_1|C \times T \iff (\mathrm{Id} \times \psi)^{-1}([\Delta = 0]) = [s = 0]$$

At this point we introduce the notion of a **relative divisor** on $C \times \mathrm{Spec}(A)$. By this we mean a subscheme $Y \subset C \times \mathrm{Spec}(A)$, which is finite and flat over $\mathrm{Spec}(A)$ and where the sheaf of ideals defining Y is locally principal. The argument in the lemma above shows that such a subscheme is always contained in an open subscheme $\mathrm{Spec}(B) \times \mathrm{Spec}(A)$ where B is the algebra of regular functions on a suitable affine subset $V \subset C$, and hence

$$Y = \mathrm{Spec}(B \otimes A/I)$$

where the ideal I is locally principal. The $A/\mathfrak{m}_0$ algebra $(B \otimes A/I) \otimes (A/\mathfrak{m}_0)$ has a rank r, which is called the *degree* of the relative divisor. Our A-algebra $(B \otimes A)/I$ is free of rank r. (Lemma 1.5.5.)

Our aim is to show that a relative divisor of degree r is nothing else than a $\mathrm{Spec}(A)$ valued point on C^r/Σ_r. We observe that in our construction of the line bundle $\mathcal{L}$ on $C \times_k C^g/\Sigma_g$ we can replace g by any integer $r \geq 0$. We get a line bundle $\mathcal{Q}_r$ on $C \times_k C^r/\Sigma_r$, which has the constant function 1 as a global section, let us call this section $s_r \in H^0(C \times_k C^r/\Sigma_r, \mathcal{Q}_r)$. This section has zeroes and the locus of this zeroes

$$[s_r = 0] = [\Delta_r = 0] \subset C \times_k C^r/\Sigma_r$$

is a relative divisor of degree r.
If we now have a scheme $T \longrightarrow \mathrm{Spec}(k)$ of finite type and a morphism $\psi : T \longrightarrow C^r/\Sigma_r$ then we consider $\mathrm{Id} \times_k \psi : C \times_k T \longrightarrow C \times_k C^r/\Sigma_r$ and clearly

$$(\mathrm{Id} \times_k \psi)^{-1}([\Delta_r = 0]) = Y \subset C \times_k T$$

is a relative divisor of degree r. It defines a line bundle $\mathcal{L}' = \mathcal{O}_{C \times_k T}(Y)$ and by definition we have

$$\mathcal{L}' = (\mathrm{Id} \times_k \psi)^*(\mathcal{Q}_r).$$

Now it is clear what we have to prove: We have to show that $[\Delta_r = 0]$ is the universal divisor of degree r. The following assertion (B) makes this precise and it is clear that this assertion applied to the case $r = g$ implies proposition 10.1.11. We formulate it under the assumption that $T = \mathrm{Spec}(A)$ is affine, the assertion (B) is local in T anyway.

(B) *Let $Y \subset C \times_k \mathrm{Spec}(A)$ be a relative divisor of degree r. There exists a unique morphism*

$$\psi : \mathrm{Spec}(A) \to C^r/\Sigma_r,$$

such that in the diagram

$$\begin{array}{ccc} C \times \mathrm{Spec}(A) & \xrightarrow{\mathrm{Id} \times \psi} & C \times C^r/\Sigma_r \\ \cup & & \cup \\ Y & & [\Delta_r = 0] \\ \downarrow & & \downarrow \\ \mathrm{Spec}(A) & \xrightarrow{\psi} & C^r/\Sigma_r \end{array}$$

the scheme Y is the pull back of $[\Delta_r = 0]$ under $\mathrm{Id} \times \psi$*, i.e.*

$$(\mathrm{Id} \times \psi)^{-1}[\Delta_r = 0] = Y. \qquad (*)$$

The proof of (B) is a little bit technical. We observed already that we can find an affine sub scheme $\mathrm{Spec}(B) \subset C$ sucht that $Y \subset \mathrm{Spec}(B) \times_k \mathrm{Spec}(A)$. We may have to localize at a given point t_0. We write down the affine version of our assertion, we get a diagram with reversed arrows (ψ is also the homomorphism between the affine rings)

$$\begin{array}{ccc} B \otimes (B^{\otimes r})^{\Sigma_r} & \xrightarrow{\mathrm{Id} \times \psi} & B \otimes_k A \\ \downarrow & & \downarrow j_1 \\ B \otimes (B^{\otimes r})^{\Sigma_r}/I_{\Delta_r} & & B \otimes_k A/I_Y \\ \uparrow & & \uparrow i_1 \\ (B^{\otimes r})^{\Sigma_r} & \xrightarrow{\psi} & A \end{array}$$

and the condition is that I_Y is the image of I_{Δ_r} under $\mathrm{Id} \times \psi$. This means that the image of a local generator F_r of I_Y under $\mathrm{Id} \times \psi$ is a local generator of I_Y. We will say that ψ satisfies the condition (*).

Given to us is the right column in our diagram, we have to find ψ and prove uniqueness. We consider the case $r = 1$. This means that the arrow i_1 is an isomorphism, it can be inverted and we get a homomorphism

$$B \longrightarrow B \otimes A \xrightarrow{j_1} B \otimes_k A/I_Y \xrightarrow{i_1^{-1}} A$$

and this is our homomorphism ψ. We can insert it into the top line and we have to show that I_Y is the image of I_{Δ_1} under $\mathrm{Id} \times \psi$, and that it is uniquely determined by this requirement. But this is clear: The ideal I_{Δ_1} is locally generated by an element of the form $f \otimes 1 - 1 \otimes f$, this is mapped to $f \otimes 1 - 1 \otimes \psi(f)$ and this is by construction a local generator of I_Y (see prop. 7.5.16). This shows that ψ has the right property but it is also clear that ψ is uniquely determined.

We come to the general case, it is not so easy. We think at this point a moment of meditation is in order. We want to show that our relative divisor $Y \subset C \times_k T$ is the same as a T-valued point, i.e. an element of $\psi \in C^r/\Sigma_r(T)$. But how do we get such elements ψ. The map $C^r(T) \longrightarrow C^r/\Sigma_r(T)$ is not surjective in general (we have seen this already in exercise 46 in the case that $T = \mathrm{Spec}(k)$ where k is a non algebraically closed field). But we can find a faithfully flat extension $T' \longrightarrow T$ such that the image $\psi' \in C^r/\Sigma_r(T')$ is in the image of $C^r(T') \longrightarrow C^r/\Sigma_r(T')$. We apply theorem 6.2.17 and we see:

We get a point $\psi \in C^r/\Sigma_r(T)$ if we find a point

$$\begin{array}{ccc} \widetilde{\psi}' = (\phi_1, \dots, \phi_r) & \in & C^r(T') \\ \downarrow & & \downarrow \\ \psi' & \in & C^r(T')/\Sigma_r \end{array},$$

which in addition satisfies $p_1(\psi') = p_2(\psi')$ where p_1, p_2 are the two maps obtained from the two projections $p_1, p_2 : T' \times_T T' \rightrightarrows T'$.

To make use of this principle of construction points in $C^r/\Sigma_r(T)$ we can ask ourselves, under which conditions a relative divisor Y is given by a morphism $\psi : T \longrightarrow C\times_k C^r/\Sigma_r$, which lifts to morphism $\widetilde{\psi} = (\phi_1,\dots,\phi_r) : T \longrightarrow C^r$. We recall that the bundle $\mathcal{L}$ on $C \times C^r/\Sigma_r$ has a pullback $\mathcal{L}' = (\mathrm{Id}\times\Pi)^*(\mathcal{L}) = \bigotimes_{i=1}^r \mathcal{O}_{C\times C^g}(\Delta_i)$. Therefore,

$$\mathcal{L}_1 = (\mathrm{Id}\times\psi)^*(\mathcal{L}) = (\mathrm{Id}\times\widetilde{\psi})^*(\mathrm{Id}\times\Pi)^*(\mathcal{L}) = \bigotimes_i (\mathrm{Id}\times\phi_i)^*(\mathcal{O}_{C\times C})(\Delta).$$

Hence we see that

$$\mathcal{L}_1 = \bigotimes_i \mathcal{O}_{C\times T}(\mathrm{Id}\times\phi_i)^{-1}(\Delta)) = \bigotimes_i \mathcal{O}_{C\times T}(Y_i).$$

Clearly the Y_i are relative divisors of degree one.

This leads us to introduce the notion of a *decomposable* relative divisor. A relative divisor Y of degree r decomposable if the ideal I_Y can be written as a product $I_Y = \prod_{i=1}^r I_i$ where I_i is locally given by one equation and where $Y_i \subset C\times T$ is a relative divisor of degree 1.

In a first step we prove (B) for decomposable divisors. The existence of ψ is obvious, we have proved (B) for the case $r = 1$, we apply this to the Y_i this gives us a $\widetilde{\psi} = (\phi_1,\dots,\phi_r)$ and ψ is the image of $\widetilde{\psi}$ under $C^r(T) \longrightarrow C^r/\Sigma_r(T)$. But we also have to show uniqueness. We choose an element $f \in B$ whose differential df has no zeroes on Y, this is possible locally in $\mathrm{Spec}(A)$. Then we know (from the case $r = 1$) that $\prod(f\otimes 1 - 1\otimes\phi_i(f))$ is a local generator of I_Y at all points (t,x) where $df_x \neq 0$. Now let $\psi_1 : (B^{\otimes r})^{\Sigma_r} \longrightarrow A$ be a homomorphism such that $(\mathrm{Id}\times\psi_1)^{-1}(I_{\Delta_r})(B\otimes A) = I_Y$. Let $f^i = 1\otimes\cdots\otimes f\cdots\otimes 1 \in B^{\otimes r}$ be the element where the factor f is at spot i. The element

$$\prod(f\otimes 1 - 1\otimes f^{\{i\}}) = f^r\otimes 1 + f^{r-1}\otimes\sigma_1(f) + \dots 1\otimes\sigma_r(f)$$

lies in I_{Δ_r} and it is mapped to

$$f^r\otimes 1 + f^{r-1}\otimes\psi_1(\sigma_1(f)) + \dots 1\otimes\psi_1(\sigma_r(f)) \in I_Y.$$

and hence it is a multiple of $\prod(f\otimes 1 - 1\otimes\phi_i(f))$ in the ring $k[f,\frac{1}{P}]\otimes A$ where P describes the locus where df vanishes. But then it is clear that we must have

$$\prod(f\otimes 1 - 1\otimes\phi_i(f)) = f^r\otimes 1 + f^{r-1}\otimes\sigma_1(\phi_1(f),\dots,\phi_r(f)) + \dots 1\otimes\sigma_r(\phi_1(f),\dots,\phi_r(f)),$$

i.e.

$$\psi_1(\sigma_\mu(f)) = \sigma_\mu(\phi_1(f),\dots,\phi_r(f))$$

This proves the uniqueness in assertion (B) in the case of decomposable divisors.

In a second step we show that for any relative divisor $Y \subset C\times_k T$ we can find a faithfully flat $T' \longrightarrow T$ of finite type such that the base change $Y\times_T T' \subset C\times_k T'$ becomes decomposable. This can be done by an easy induction argument. We know that $Y \longrightarrow T$ is flat. Therefore we can consider the base change diagram

$$\begin{array}{ccc} Y\times_T Y & \subset & C\times_T Y \\ & & \downarrow \\ & & Y \end{array},$$

which gives us a relative divisor over Y. This relative divisor is the sum of two relative divisors namely the diagonal $\Delta_Y \subset Y\times_T Y$ and its complement $Y' \subset C\times_T Y$, this complement is a relative divisor of degree $r-1$. If we apply the same process to Y' and after $r-1$ faithfully flat base changes we have decomposed out relative divisor.
The rest is clear. Let $Y\subset C\times_k T$ be a relative divisor, we choose a faithfully flat $T'\longrightarrow T$ such that $Y\times_T T' \subset C\times_k T'$ decomposes. Then we find a unique $\psi' : T' \longrightarrow (C^{\otimes r})^{\Sigma_r}$, which satisfies (*) . We have the two projections $p_1,p_2 : T'\times_T T' \rightrightarrows T'$. and this gives us the two $T'\times_T T'$ valued points $\psi\circ p_1,\psi\circ p_2$. But they they also satisfy (*) and hence they must be equal. But this says that there is a unique $\psi : T\longrightarrow (C^{\otimes r})^{\Sigma_r}$ whose base extension is ψ', and which satisfies (*). This proves (B) and hence proposition 10.1.11.

The gluing

Our goal is to construct a scheme $\mathrm{Pic}^g_{C/k}$ and a line bundle $\mathcal{P}$ over $C\times \mathrm{Pic}^g_{C/k}$ such that this pair of data provides a local representation of $\mathcal{PIC}_{C/k}$.

Our proposition 10.1.11 tells us that we reached this goal for a certain sub functor: We considered only those families of bundles of degree g, for which $\dim H^0(C\times k(t),\mathcal{L}_t)=1$.

Our field k is still an arbitrary field. We choose an algebraic closure $\bar{k}$ and inside it we have the separable closure k_s.

Proposition 10.1.13. *For any smooth, projective, absolutely irreducible curve C/k of genus g we can find a finite family of degree zero line bundles*

$$\mathcal{F}_1,\cdots,\mathcal{F}_r$$

on $C\times_k k_s$ such that for any line bundle $\mathcal{L}'$ on $C\times_k \bar{k}$ of degree g we can find an index i such that

$$\dim H^0(C\times_k L,\mathcal{L}'\otimes\mathcal{F}_i)=1$$

i.e. $\mathcal{L}'\otimes\mathcal{F}_i$ is generic.

Our proof is based on a general principle namely that the line bundles of degree g on C/k form a *bounded family.* In our case this means that we have the bundle $\mathcal{Q}$ on $C\times_k C^g/\Sigma_g$ and for any line bundle $\mathcal{L}_1$ of degree g on on $C\times\bar{k}$ we can find a point $v\in C^g/\Sigma_g(\bar{k})$ such that $\mathcal{Q}_v \xrightarrow{\sim} \mathcal{L}_1$. (Every line bundle $\mathcal{L}_1$ "occurs" (perhaps several times) in the family $\mathcal{Q}$ on $C\times C^g/\Sigma_g$).To see this we simply choose a non zero section in $H^0(C\times\bar{k},\mathcal{L}_1)$ and look at the divisor of zeroes of this section, this gives the point v. If now $\mathcal{L}_1$ is given then we can find a line bundle $\mathcal{F}$ on $C\times k_s$ such that $\mathcal{L}_1\otimes\mathcal{F}$ is generic. We can find an open set $V\subset C^g/\Sigma_g$ containing the point v such that for all $v'\in V$ the line bundle $\mathcal{Q}_v\otimes\mathcal{F}$ is generic. This yields a covering of C^g by open sets V_i and bundles $\mathcal{F}_i$ on $C\times k_s$ such that $\mathcal{Q}\otimes\mathcal{F}_i$ is generic on $C\times V_i$). This covering has a finite sub covering and gives us a finite list of $\mathcal{F}_i$. For $i=1$ we choose $V_1=U$ and $\mathcal{F}_1=\mathcal{O}_C$.

Now we come back to our line bundle $\mathcal{L}'$ on $C \times L$. Of course we can assume that L is the quotient field of a finitely generated k-algebra B, and that $\mathcal{L}'$ is obtained by base change from a line bundle $\mathcal{L}^*$ on $C\times\mathrm{Spec}(B)$. We have the projection $p_1 : C\times\mathrm{Spec}(B) \longrightarrow C$. The pullbacks of $\mathcal{F}_i$ by p_1 are still called $\mathcal{F}_i$. Now we consider a closed point $t_0 \in \mathrm{Hom}_k(B,k)$ and consider the line bundle $\mathcal{L}^*_{t_0}$ on C/k. We can find an index i such that $\mathcal{L}^*_{t_0} \otimes \mathcal{F}_i$ is generic. Then it follows from semicontinuity that $\mathcal{L} \otimes \mathcal{F}_i$ is generic. □

This makes it clear how to proceed with the construction of $\mathrm{Pic}^g(C)$. If we have a line bundle $\mathcal{L}_1$ on $C \times_k T$, where T is of finite type, then we have a finite covering $T = \cup\, T_i$ by open sets such that $\mathcal{L}_1 \otimes \mathcal{F}_i$ restricted to $C \times T_i$ is generic in all points $t \in T_i$. Then we find a

$$\psi_i : T_i \longrightarrow U$$

such that $(\mathrm{Id} \times \psi_i)^*(\mathcal{P}')$ is locally isomorphic to $\mathcal{L}_1 \otimes \mathcal{F}_i$ or in other words

$$(\mathrm{Id} \times \psi_i)^*(\mathcal{P}' \otimes \mathcal{F}_i^{-1}) \sim_{T_i} \mathcal{L}_1 \mid C \times T_i .$$

Hence we do the following: We consider the open set $U \subset C^g/\Sigma_g$ and the line bundle $\mathcal{P}'$ on $C \times U$. We consider r copies of $C \times U$ and on these copies we put the line bundles $\mathcal{P}' \otimes \mathcal{F}_i^{-1}$. Our previous consideration shows that $C \times U$ together with $\mathcal{P}' \otimes \mathcal{F}_i^{-1}$ is a universal bundle for families of line bundles of degree g, for which $\mathcal{L}_1 \otimes \mathcal{F}_i$ is generic.

We form the disjoint union

$$\bigcup_{i=1}^{r} C \times U = C \times (U \times [1,\ldots,r]),$$

on which we consider the line bundle $\mathcal{L}_i = \mathcal{P}' \otimes \mathcal{F}_i^{-1}$ on the i'-th component. Remember that for $i = 1$ we put $\mathcal{F}_1 = \mathcal{O}_C$. Let us put $U_i = U \times \{i\}$.

For any of the components $(C \times U_i, \mathcal{P}' \otimes \mathcal{F}_i^{-1})$ and for any other index j we may consider the open subset $C \times U_{ij} \subset C \times U_i$ where $u \in U_{ij}$ if and only if $\mathcal{P}'_u \otimes \mathcal{F}_i^{-1} \otimes \mathcal{F}_j$ is generic. Then it follows from our previous arguments that we have unique isomorphisms

$$\psi_{ij} : U_{ij} \xrightarrow{\sim} U_{ji} \subset U_j$$

such that we have

$$(\mathrm{Id} \times \psi_{ij})^*(\mathcal{P}' \otimes \mathcal{F}_j^{-1}) \sim_{U_{ij}} \mathcal{P}' \otimes \mathcal{F}_i^{-1}.$$

It is clear that this family of morphisms satisfies $\psi_{ik} \circ \psi_{ij} = \psi_{ik}$, and

$$\psi_{ji} \circ \psi_{ij} = \mathrm{Id}\,.$$

We get an equivalence relation on our disjoint union and this allows us to glue these copies of U via the identifications ψ_{ij} to a scheme $\mathrm{Pic}^g_{C/k}$. By construction we have a covering of $\mathrm{Pic}^g_{C/k}$ by open sets U_i and we have the line bundles $\mathcal{L}_i$ on the products $C \times U_i$. The restrictions of $\mathcal{L}_i$,$\mathcal{L}_j$ are locally isomorphic on $C \times (U_i \cap U_j)$. (See proposition 10.1.11)

This means that we have proved the weak local representability of $\mathcal{PIC}^g_{C_L/L}$ if L/k is a finite separable extension, over which all the $\mathcal{F}_i$ are defined.

Theorem 10.1.14. *If C/k has a rational point then the functor $\mathcal{PIC}_{C/k,P}$ is representable, it is the disjoint union of all $\mathcal{PIC}^r_{C/k,P}, r \in \mathbb{Z}$. We denote the representing objects by $(C \times \mathrm{Pic}^r_{C/k,P}, \mathcal{P}_r, s)$.*

Start with $r = g$. We pass to a finite, normal and separable extension L/k over which the $\mathcal{F}_i$ are defined. Then we have weak local representability and we have seen that this implies representability for $\mathcal{PIC}_{C_L/L,P}$. We have a universal object $(\mathrm{Pic}_{C_L/L}, \mathcal{P}, s)$. Now we can apply the general principles in 6.2.8. We put $S = \mathrm{Spec}(k), S' = \mathrm{Spec}(L)$, let $\mathrm{Pic}^g_{C/k,P}$ the representing scheme. For any $\sigma \in \mathrm{Gal}(L/k)$ we get a unique morphism

$$\phi_\sigma : ((\mathrm{Pic}^g_{C_L,P})^\sigma, \mathcal{P}^\sigma, s^\sigma) \longrightarrow (\mathrm{Pic}^g_{C_L,P}, \mathcal{P}, s)$$

and because of uniqueness it has to satisfy the cocycle relation, and hence we get a descend datum. We have to prove that it is effective. Here we have to anticipate 10.2.2, where we prove that $\mathrm{Pic}^g_{C/L,P}$ is projective. This implies that we can apply the criterion on page 48.
Now we take r arbitrary. Again we choose a finite separable extension L/k such that we can find a second point $Q \in C(L), Q \neq P$. Now we pass to C_L and use this point to identify the functors $\mathcal{PIC}^r_{C/k,P}$ for the different values of r. For any integer ν we have the line bundle $\mathcal{O}(\nu Q)$ on C_L. This bundle restricted to a neighborhood V of P is canonically trivial, we have the section $1 \in H^0(V, \mathcal{O}(\nu Q))$ provided $Q \notin V$. For any $T \longrightarrow \mathrm{Spec}(k)$ of finite type the bundle $p_2^*(\mathcal{O}(\nu Q))$ is a bundle on $C \times T$ whose restriction to $\{P\} \times T$ is equipped with a trivialization. If we have a line bundle $\mathcal{L} \in \mathcal{PIC}^r_{C/k,P}(T)$ then $\mathcal{L} \otimes p_2^*(\mathcal{O}(\nu Q))$ is a line bundle on $C \times T$ with a trivialization at $\{P\} \times T$ and hence

$$\mathcal{L} \otimes p_2^*(\mathcal{O}(\nu Q)) \in \mathcal{PIC}^{r+\nu}_{C_L/L,P}(T).$$

Therefore, we get a bijection $\mathcal{PIC}^r_{C_L/L,P}(T) \xrightarrow{\sim} \mathcal{PIC}^{r+\nu}_{C_L/L,P}(T)$ and this makes it clear that all $\mathcal{PIC}^r_{C_L/L,P}$ are representable once we have representability for $r = g$. But now we apply the argument from 6.2.8 and see that already $\mathcal{PIC}^r_{C/k,P}$ is representable for all values of r. □

The scheme $\bigsqcup_{r\in\mathbb{Z}} \mathrm{Pic}^r_{C/k,P} = \mathrm{Pic}_{C/k,P}$ is not of finite type, but this does not really matter, it is a disjoint union (indexed by the integers) of schemes of finite type.

We pick two integers r,s and we consider the scheme

$$C \times \mathrm{Pic}^r{}_{C/k,P} \times \mathrm{Pic}^s{}_{C/k,P}.$$

We have the two projections p_{12} to the first and second factor and p_{23} to the first and third factor. We get a new bundle

$$p_{12}^*(\mathcal{P}_r) \otimes p_{13}^*(\mathcal{P}_s)$$

on $C \times \mathrm{Pic}^r{}_{C/k,P} \times \mathrm{Pic}^s{}_{C/k,P}$ and this line bundle has degree $r + s$ in any point $y \in \mathrm{Pic}^r(C) \times \mathrm{Pic}^s(C)$. Therefore, we get a unique morphism

$$m : \mathrm{Pic}^r{}_{C/k,P} \times \mathrm{Pic}^s{}_{C/k,P} \longrightarrow \mathrm{Pic}^{r+s}{}_{C/k,P}$$

such that we get a unique isomorphism

$$\eta : p_{12}^*(\mathcal{P}_r) \otimes p_{13}^*(\mathcal{P}_s) \xrightarrow{\sim} (\mathrm{Id} \times m)^*(\mathcal{P}_{r+s}),$$

here the trivializations along $\{P\} \times$ *second factor* are of course important.
It is clear that this defines a structure of a group scheme on

$$\mathrm{Pic}_{C/k,P} = \bigsqcup \ \mathrm{Pic}^r{}_{C/k,P},$$

which is a scheme over k, which is not of finite type.
As a special case this gives us a group scheme structure on $\mathrm{Pic}^0_{C/k,P}$, and all the $\mathrm{Pic}^r_{C/k,P}$ are principal homogeneous spaces under the action of $\mathrm{Pic}^0_{C/k,P}$.

Proposition 10.1.15. *The schemes* $\mathrm{Pic}^r_{C/k,P}/k$ *are smooth, separated and absolutely irreducible.*

It suffices to consider the case $r = g$. The open dense subset U_{gen} is absolutely irreducible, since our curve is absolutely irreducible this implies the same assertion for $\mathrm{Pic}^g_{C/k}$. We write $\mathrm{Pic}^g_{C/k} = X$. Of course the non empty open subscheme $U_{\mathrm{gen}} \subset X$ is smooth and separated. The non smooth points form a closed subset in $Z \subset X$. If $Z \neq \emptyset$ then $Z(\bar{k}) \neq \emptyset$. Since $\mathrm{Pic}^0_{C/k}(\bar{k})$ acts transitively on $X(\bar{k})$ and since $Z(\bar{k})$ is invariant under this action it follows that $Z(\bar{k}) = \emptyset$ because $U_{\mathrm{gen}}(\bar{k})) \neq \emptyset$. A similar argument works for separatedness. We have to show that the diagonal Δ_X is closed in $X \times X$. Assume it is not, then we can find a geometric point $(P,Q) \in \bar{\Delta}_X(\bar{k}) \subset X(\bar{k}) \times X(\bar{k})$ with $P \neq Q$. We have the diagonal action of $\mathrm{Pic}^g_{C/k}(\bar{k})$ on $X(\bar{k}) \times X(\bar{k})$ and clearly $\Delta_X(\bar{k})$ and $\bar{\Delta}_X(\bar{k})$ are invariant under this action. We can find an element $u \in \mathrm{Pic}^0_{C/k}(\bar{k})$ such that $T_u(P,Q) \in U_{\mathrm{gen}} \times U_{\mathrm{gen}}$. But since U_{gen}/k is separated we see that the diagonal is closed in $U_{\mathrm{gen}} \times U_{\mathrm{gen}}$. This is a contradiction. □

10.1.7 The local representability of $\mathcal{PIC}^g_{C/k}$

We want to drop the assumption that we have a rational point $P \in C(k)$. Now we only have the functor $\mathcal{PIC}^r_{C/k}$ and we want to investigate what happens to this functor.
Of course we proceed as before and choose a finite, normal and separable extension L/k such that C_L has a rational point. Hence we know that $\mathcal{PIC}^r_{C_L/L}$ is locally representable, this provides a scheme $\mathrm{Pic}_{C_L/L}$ and a line bundle $\mathcal{P}_r$ on $C \times \mathrm{Pic}^r_{C_L/L}$. We want to discuss to what extend this object over L descends to an object over k. We will find a canonical effective descend datum for $\mathrm{Pic}^r_{C_L/L}$, and hence we construct a scheme $\mathrm{Pic}^r_{C_k/k}$. But in general the bundle may not descend.

The descend datum for $\mathrm{Pic}_{C_L/L}$ is almost obvious. We pick an element $\sigma \in \mathrm{Gal}(L/k)$ an conjugate by σ, we consider $\mathcal{P}_r^\sigma$ on $C \times (\mathrm{Pic}^r_{C_L/L})^\sigma$ (See 6.2.9). Then we find a unique morphism $f_\sigma : (\mathrm{Pic}^r_{C_L/L})^\sigma \longrightarrow \mathrm{Pic}^r_{C_L/L}$ such that $(\mathrm{Id} \times f_\sigma)^*(\mathcal{P}_r) \sim \mathcal{P}_r^\sigma$.
Since these morphisms between the schemes are unique we get the cocycle relation

$$f_{\tau\sigma} = f_\tau \circ f_\sigma^\tau.$$

Hence we see

The interpretation as locally representing scheme yields a canonical effective descend datum on $\mathrm{Pic}^r_{C_L/L}$ *and hence we constructed a scheme* $\mathrm{Pic}^r_{C/k}/k$ *for any* r.

But in general we will not be able to construct a bundle (or even only a gerbe) $\mathcal{P}_r$ on $C \times \mathrm{Pic}^r_{C/k}$, which provides local representability. We want to explain briefly why this is so. We consider the function field $k(\mathrm{Pic}^r_{C/k}) = K_r$ and we have the generic point $\eta : \mathrm{Spec}(K_r) \longrightarrow \mathrm{Pic}^r_{C/k}$. If we have a bundle (or only a gerbe) $\mathcal{P}_r$ on $C \times \mathrm{Pic}^r_{C/k}$ it will introduce a line bundle $\mathcal{P}^\eta_r$ on $C \times \eta$. We know that we can find such a bundle $\mathcal{P}^\eta_{r,L}$ on $C_L \times \eta_L$ after extending the scalars by a finite, normal separable extension. For any $\sigma \in \mathrm{Gal}(L/k)$ we find an isomorphism $\alpha_\sigma : (\mathcal{P}^\eta_{r,L})^\sigma \longrightarrow \mathcal{P}^\eta_{r,L}$ of line bundles over $C_L \times \eta_L$. These α_σ are not unique they can be modified by an element $c_\sigma \in (K_r \otimes L)^\times$. Now there is no reason why these α_σ satisfy the cocycle relation, which would make them into a descend datum. In general we will get a map

$$(\sigma,\tau) \mapsto t_{\sigma,\tau} = \alpha_{\tau\sigma}\alpha_\tau^{-1}(\alpha_\sigma^\tau)^{-1},$$

which will be a 2-cocycle, which defines a class $[c_r] \in H^2(\mathrm{Gal}(K_r \otimes L/K_r),(K_r \otimes L)^\times)$ and there is no reason why this class should vanish. Hence there is no reason that we may be able to change the α_σ into a descend datum.
I want to stress the analogy between the situation here and the discussion in 9.6.2, in both cases the violation of the second sheaf condition for a functor leads to obstructions in certain second cohomology groups. We became modest and wanted to construct a gerbe $\underline{\mathcal{P}}_r$, which was a gerbe for the Zariski-topology on $\mathrm{Pic}^r_{C/k}$ but now we see that we only can construct a gerbe for the étale topology on $\mathrm{Pic}^r_{C/k}$.
To give a simple example let us consider a curve C/k of genus one, which does not have a rational point, i.e. $C(k) = \emptyset$. If we take $r = 1$ then we have $\mathrm{Pic}^1_{C/k} = C$ and $\mathcal{P}_1 = \mathcal{O}_{(C\times C)(\Delta)}$ (See 10.1.1).
Now ask ourselves whether we have a $\mathcal{P}_0$ on $C \times \mathrm{Pic}^0_{C/k}$. We have an action of $\mathrm{Pic}^0_{C/k}$ on $\mathrm{Pic}^1_{C/k} = C$ and it is clear that $\mathrm{Pic}^1_{C/k}$ is indeed a $\mathrm{Pic}^0_{C/k}$-torsor. 6.2.11. We explained that the isomorphism classes of these torsors correspond to the elements in $H^1(\mathrm{Gal}(k_s/k), \mathrm{Pic}^0_{C/k})$ and that the class $[C]$ is zero if and only if C/k has a k-rational point. It follows from a relatively simple computation with exact sequences that this class $[C]$ maps under a boundary map to a class in $\delta([C]) \in H^2(\mathrm{Gal}(K_r \otimes L/K_0),(K_0 \otimes L)^\times)$. This class is exactly the obstruction class $[c_0]$ and it vanishes if and only if $[C]$ vanishes. Hence we can conclude

For a curve C/k *of genus one we have a bundle* $\mathcal{P}_0$ *on* $C \times \mathrm{Pic}^0_{C/k}$ *if and only if* C/k *has a* k*-rational point.*

Finally we come to the proof of Theorem 10.1.10. We remains to be proved is that we can construct a line bundle $\mathcal{P}_g$ on $C \times \mathrm{Pic}^g_{C/k}$ such that $(\mathrm{Pic}^g_{C/k},\mathcal{P}_g)$ is a local representation of $\mathcal{PIC}^g_{C/k}$. The case $r = g$ is special because we have the open subset $U_{\mathrm{gen}} \subset \mathrm{Pic}^g_{C/k}$ and we have the line bundle $\mathcal{P}'$ on $C \times U_{\mathrm{gen}}$. On this open subset we may choose our searched for $\mathcal{P}_g$ to be $\mathcal{P}'$.
Now we need a simple proposition

Proposition 10.1.16. *Our open set* $U_{gen} \subset \mathrm{Pic}^g_{C/k}$ *has a complement of codimension* ≥ 2 *and the line bundle* $\mathcal{P}'$ *on* $C \times U_{gen}$ *extends uniquely to a line bundle* $\mathcal{P}$ *on* $C \times \mathrm{Pic}^g_{C/k}$. *This line bundle* $\mathcal{P}$ *has the property that* $\mathcal{P}|C \times U_i \sim_{U_i} \mathcal{L}_i$

Consider the morphism $\psi : C^g/\Sigma_g \to \mathrm{Pic}^g_{C/k}$. The fibre of a point $u \in \mathrm{Pic}^g_{C/k}$ consists of those points $(P_1 \cdots P_g) \in C^g/\Sigma_g$, for which

$$\mathcal{L}_u \simeq \mathcal{O}(P_1 + \cdots + P_g).$$

Hence we see that the fibre is simply the space of lines in $H^0(C \times k(u),\mathcal{L}_u)$, this is the space $\mathbb{P}(H^0(C \times k(u),\mathcal{L}_u)^\vee)$. Hence its dimension is

$$\dim H^0(C \times k(u),\mathcal{L}_u) - 1.$$

The map ψ identifies the open set $U_{\mathrm{gen}} \subset C^g/\Sigma_g$ with its image in $\mathrm{Pic}^g_{C/k}$ by construction, hence we see that for points $u \in \mathrm{Pic}^g_{C/k} \setminus U_{\mathrm{gen}}$ the dimension of the fibre is ≥ 1. We apply the reduction process in the beginning of 7.4.3 to the morphism $C^g/\Sigma_g \setminus U_{\mathrm{gen}} \longrightarrow \mathrm{Pic}^g_{C/k} \setminus U_{\mathrm{gen}}$ and then the first assertion follows from proposition 7.4.6.
To prove the extendability we use a general principle, which says that on smooth, irreducible schemes isomorphism between line bundles or sections in line bundles extend over closed subset of codimension ≥ 2.
To be more precise: Let V/k be a smooth, irreducible scheme of finite type, let $\mathcal{L}$ be a line bundle over V and let $U \subset V$ be an open subset such that the complement has codimension ≥ 2. Let $s \in H^0(U,\mathcal{L})$ be a section. Then this section extends to a unique section on V. To see that this is so we pick a point $p \in V \setminus U$ and we choose a local section $s_p \in H^0(V_p,\mathcal{L})$, which is a generator. Then we have $s = hs_p$ over $V_p \cap U$. We write h as a ratio of two elements in the local ring at p. Since this local ring factorial (See [Ei], Thm. 19.19) the denominator must be unit and h extends uniquely to a regular function in $\mathcal{O}_X(V_p)$.
If we have two line bundles $\mathcal{L}_1,\mathcal{L}_2$ over V, which are isomorphic over U then this isomorphism is given by non zero sections in $s_1 \in H^0(U,\mathcal{L}_1 \otimes \mathcal{L}_2^{-1}),s_2 \in H^0(U,\mathcal{L}_2 \otimes \mathcal{L}_1^{-1})$, whose product is one. Our argument above shows that these sections extend to sections on V and their product is still one. This implies that an extension-if it exists- from $\mathcal{P}'$ on $C \times U$ to a line bundle $\mathcal{P}$ on $C \times \mathrm{Pic}^g(C)$ is unique.
Finally we show that $\mathcal{P}'$ extends from $C \times U$ to a line bundle $\mathcal{P}$ on $C \times \mathrm{Pic}^g(C)$. Let us assume that we extended $\mathcal{P}'$ to a line bundle-still called $\mathcal{P}'$- on $C \times U'$ where $U' \supset U$ is open. If this open set is not yet $\mathrm{Pic}^g_{C/k}$ then we pick a point $a \in \mathrm{Pic}^g(C) \setminus U'$. We find an index i such that $a \in U_i$. We consider the open subset $U_i' \subset U' \cap U_i$ where $\mathcal{P}' \otimes \mathcal{F}_i$ is generic. By construction we have $U_i' \subset U_i$ and $\mathcal{P}'|C \times U_i' \sim_{U_i'} \mathcal{L}_i|C \times U_i'$. (The morphism ψ, which realizes $\mathcal{P}'|C \times U_i'$ is now the inclusion). But now we can say that we can find a line bundle $\mathcal{M}$ on U_i' such that

$$\mathcal{P}'|C \times U_i' \xrightarrow{\sim} \mathcal{L}_i|C \times U_i' \otimes p_2^*(\mathcal{M})$$

(see 10.1.1) and hence we can glue the bundles $\mathcal{P}'$ on $C \times U'$ and $\mathcal{L}_i \otimes p_2^*(\mathcal{M})$ on $C \times U_i$ over $C \times U_i'$ to a larger extension of $\mathcal{P}'$. This process stops after a finite number of steps. □

This finishes the proof of theorem 10.1.10, we simply may take the extension of $\mathcal{P}'$ to $\mathrm{Pic}^g_{C/k}$ as our $\mathcal{P}_g$.

10.2 The Picard functor on X and on J

Some heuristic remarks

We want to get a better understanding of the scheme $\mathrm{Pic}_{C/k}$. Since all its connected components are isomorphic, we can concentrate on the components. For us it is sometimes convenient to consider $\mathrm{Pic}^g_{C/k}$ and $\mathrm{Pic}^0_{C/k}$. Therefore we introduce the notations

$$\begin{aligned} X &= \mathrm{Pic}^g_{C/k} \\ J &= \mathrm{Pic}^0_{C/k} \end{aligned}$$

We will see that X/k is a smooth projective scheme. Then J is also smooth projective, and it is a group scheme, hence an abelian variety. Notice that is always defined, even if we do not have a rational point. It is already clear that $\mathrm{Pic}^0_{C/k}$ is a proper scheme because we have the surjective morphism

$$\pi : C^g/\Sigma_g \longrightarrow \mathrm{Pic}^g_{C/k},$$

and that we have identifications $\mathrm{Pic}^0_{C/k} \simeq \mathrm{Pic}^g_{C/k}$. The morphism π is birational (see 7.4.3), because it induces an isomorphism on the open subset U_{gen}. The scheme J/k is called the **Jacobian** of the curve.

As in the transcendental case our main objects objects of interest will now be the Picard functors $\mathrm{Pic}_{X/k}$ and $\mathrm{Pic}_{J/k}$. We will prove local representability as in volume I Chapter V. The understanding of the structure of these resulting Picard schemes is the key to many beautiful results.

10.2.1 Construction of line bundles on X and on J

We remind the reader that the fundamental tool for this study was the existence of the polarization and the resulting line bundles. (See Volume I, V. 5.2) The following results are more geometric in nature, therefore, we assume for a while that our base field k is algebraically closed. We assume that we picked a point $P_0 \in C(k)$.

We can restrict the bundle $\mathcal{P}$ to $\{P_0\} \times X \simeq X$ and this gives us a line bundle $\mathcal{P}_{P_0}$ on X. This bundle will play the role of - or is - the principal polarization. We will denote it by Θ and it is called the **Theta bundle** or **Theta divisor**. It depends on the choice of P_0 but its class in the Neron-Severi group $NS(X)$ is independent of this choice. Actually it is this class, which is the relevant object.
We have the action of J on X we denote it by $m : X \times J \longrightarrow X$. This allows us to translate line bundle and divisors. For any $x \in J(k)$ we have the translation

$$T_x : X \longrightarrow X, T_x : y \mapsto m(x,y) \text{ or occasionally } y \mapsto x + y$$

and we can consider the translated bundle $T_x^*(\Theta)$ and compare it to Θ by forming the quotient

$$T_x^*(\Theta) \otimes \Theta^{-1}.$$

We can view x as a variable or better we can construct the bundle

$$\widetilde{\Theta} = m^*(\Theta) \otimes p_2^*(\Theta^{-1}) \text{ on } X \times J \tag{10.5}$$

This bundle evaluated at x gives the above bundle. The scheme J comes with a distinguished point namely the identity (or zero) element $e \in J(k)$ and $\widetilde{\Theta}_e \xrightarrow{\sim} \mathcal{O}_X$. Hence we can view $\widetilde{\Theta}$ as an object in $\mathcal{PIC}^{00}_{X/k}(J)$. (See 10.4) The bundle $T_x^*(\Theta) \otimes \Theta^{-1}$ itself is in $\mathcal{PIC}^0_{X/k}(k)$.
One of our aims is to prove that this pair $(X \times J, \widetilde{\Theta})$ gives us a local representation of the functor $\mathcal{PIC}^{00}_{X/k}$.

The homomorphisms $\phi_{\mathcal{M}}$

Let us consider any line bundle $\mathcal{M}$ on X. We apply our construction above to this bundle and form the bundle $\mathcal{N}_{\mathcal{M}} = m^*(\mathcal{M}) \otimes p_2^*(\mathcal{M}^{-1})$ on $X \times_k J$. Hence for any scheme $T \longrightarrow k$ (of finite type) and any k-morphism $\psi : T \longrightarrow J$, i.e. $\psi \in J(T)$ we get the restricted bundle $(\mathrm{Id} \times_k \psi)^*(\mathcal{N}_{\mathcal{M}})$ on $X \times_k T$, i.e. a point in $\mathcal{PIC}^0_{X/k}(T)$. This gives us a functorial map

$$\phi_{\mathcal{M}} : J(T) \longrightarrow \mathcal{PIC}^0_{X/k}(T),$$

i.e. a morphism from the functor J to the functor $\mathcal{PIC}^0_{X/k}$ We want to show that $\phi_{\mathcal{M}}$ is a homomorphism.
Here we are in an amusing trap. We will see that our assertion is a consequence of the theorem of the cube applied to a suitable line bundle on $X \times J \times J$. But recall that that the basic ingredients of the proof of the theorem of the cube are the finiteness results for higher direct images of coherent sheaves under projective morphisms and the resulting semi continuity results. But we don't know yet that X, J are projective, they are only proper. We will prove projectivity later, but in the proof we will need that $\phi_{\mathcal{M}}$ is a homomorphism.
Now we have three options for the reader. The first option is to apply theorem 8.3.7, which is not proved in this book. The second option is to look up the proofs of the theorems 8.3.6 and 8.3.7, in [Gr-EGA II] and [Gr-EGA III], Chap. III, §3.). This option is highly recommendable anyway.

The third option is to follow me and read the direct proof that $\phi_{\mathcal{M}}$ is a homomorphism if we restrict $\phi_{\mathcal{M}}$ to the geometric points. We prove

Proposition 10.2.1. *The map*

$$\phi_{\mathcal{M}} : J(k) = \mathrm{Pic}^0_{C/k}(k) \to \mathcal{PIC}^0_{X/k}(k)$$

defined by

$$\phi_{\mathcal{M}}(x) = T_x^*(\mathcal{M}) \otimes \mathcal{M}^{-1},$$

is a homomorphism and if $\mathcal{M}$ is algebraically equivalent to zero then this homomorphism is trivial.

We reduce the proof to the proof of the second assertion. Let us pick a point x, then $\mathcal{L} = T_x^*(\mathcal{M}) \otimes \mathcal{M}^{-1}$ is algebraically equivalent to zero. The second assertion says that $T_y^*(\mathcal{L}) \simeq \mathcal{L}$ therefore,

$$T_y^*(T_x^*(\mathcal{M}) \otimes \mathcal{M}^{-1}) \simeq T_x^*\mathcal{M} \otimes \mathcal{M}^{-1}$$

rearranging yields

$$T_{x+y}^*(\mathcal{M}) \otimes \mathcal{M}^{-1} \simeq T_x^*(\mathcal{M}) \otimes \mathcal{M}^{-1} \otimes T_y^*(\mathcal{M}) \otimes \mathcal{M}^{-1},$$

and this is the reduction to the second assertion. Hence we consider the case of an $\mathcal{M}$, which is algebraically equivalent to zero. This means that we have a line bundle

$$\widetilde{\mathcal{M}}|X \times Z$$

where Z is of finite type and connected over k and such that

$$\begin{array}{rcccl} \widetilde{\mathcal{M}}|X & \times & z_1 & \simeq & \mathcal{M} \\ \widetilde{\mathcal{M}}|X & \times & z_0 & \simeq & \mathcal{O}_X. \end{array}$$

for some points z_1,$z_0 \in Z(k)$. We can pick a point $x_0 \in X(k)$ and tensorize $\widetilde{\mathcal{M}}$ by $p_2^*(\widetilde{\mathcal{M}}|x_0 \times Z)^{-1})$; then $\widetilde{\mathcal{M}}|x_0 \times Z \simeq \mathcal{O}_Z$. We have two morphisms

$$\begin{array}{rcccc} p_{13} & : & X \times J \times Z & \longrightarrow & X \times Z \\ m_{12} \times \mathrm{Id}_Z & : & X \times J \times Z & \longrightarrow & X \times Z, \end{array}$$

the first one is the projection, the second one multiplication in the first to factors (times identity). We consider the bundle

$$(m_{12} \times \mathrm{Id}_Z)^*(\widetilde{\mathcal{M}}) \otimes p_{13}^*(\widetilde{\mathcal{M}})^{-1} = \mathcal{N},$$

which evaluated at a point $x \in J(k)$ and $z \in Z(k)$ gives us

$$T_x^*(\widetilde{\mathcal{M}}_z) \otimes \widetilde{\mathcal{M}}_z^{-1}$$

where $\widetilde{\mathcal{M}}_z$ is of course $\widetilde{\mathcal{M}}|X \times \{z\}$. Now we know that

$$\mathcal{N}|x_0 \times J \times Z \qquad \text{and} \qquad \mathcal{N}|X \times e \times Z \qquad \text{and} \qquad \mathcal{N}|X \times J \times z_0$$

are trivial. Now we are in the situation to apply the theorem of the cube. But for this we need to know that X,J are projective. Since we don't know this yet, we have to make a slight detour at this point.

We remember that we have a morphism

$$\pi \times \mathrm{Id}_Z : C^g/\Sigma_g \times C^g/\Sigma_g \times Z \longrightarrow X \times J \times Z,$$

which is birational and where the fibers are product of projective spaces. We pull our line bundle back and apply the theorem of the cube upstairs. To do this we need a little

Lemma 10.2.2. *For any line bundle $\mathcal{L}$ on $X \times J \times Z$ we have $(\pi \times \mathrm{Id}_Z)_*(\pi \times \mathrm{Id}_Z)^*(\mathcal{L}) \xrightarrow{\sim} \mathcal{L}$*

We consider the pullback of the bundle

$$\widetilde{\mathcal{L}} = (\pi \times \mathrm{Id}_Z)^*(\mathcal{L}) \mid C^g/\Sigma_g \times C^g/\Sigma_g \times Z.$$

The identity in $\mathrm{Hom}((\pi\times\mathrm{Id}_Z)^*(\mathcal{L}),(\pi\times\mathrm{Id}_Z)^*(\mathcal{L}))$ yields via the adjointness formula (See Vol. I, 3.4.1) a morphism

$$j:\mathcal{L}\longrightarrow(\pi\times\mathrm{Id}_Z)_*((\pi\times\mathrm{Id}_Z)^*(\mathcal{L})).$$

We want to show that this is an isomorphism. This is a local question on $X\times J\times Z$ and therefore, we can assume that $\mathcal{L}$ is the trivial bundle $\mathcal{O}_{X\times J\times Z}$. Then the pullback is by definition the structure sheaf $\mathcal{O}_{C^g/\Sigma_g\times C^g/\Sigma_g\times Z}$. We have to show that

$$(\pi\times\mathrm{Id}_Z)_*(\mathcal{O}_{C^g/\Sigma_g\times C^g/\Sigma_g\times Z})=\mathcal{O}_{X\times J\times Z}$$

Of course this follows if we can show the corresponding assertion for the morphism

$$\pi:C^g/\Sigma_g\times C^g/\Sigma_g\longrightarrow X\times J$$

because the sheaves in question are simply the pullbacks via the projection p_{12} to the first and second factor. Now we have seen that the morphism π is projective and therefore, we can conclude that $\pi_*(\mathcal{O}_{C^g/\Sigma_g\times C^g/\Sigma_g})$ is a coherent sheaf on $X\times J$ (See Thm. 8.3.2). Hence we see that for any open subset $V\subset X\times J$ the algebra $\pi_*(\mathcal{O}_{C^g/\Sigma_g\times C^g/\Sigma_g})(V)$ is finite over $\mathcal{O}_{X\times J}(V)$. If we pass to the stalks at a point x then we have

$$\varinjlim_{V:x\in V}\mathcal{O}_{X\times J}(V)=\mathcal{O}_{X\times J,x}\subset\varinjlim_{V:x\in V}(\mathcal{O}_{C^g/\Sigma_g\times C^g/\Sigma_g}(\pi^{-1}(V))$$

and this is a finite extension. The fibers of π are products of projective spaces and therefore, connected.
We claim that the limit on the right is a local ring: An element

$$f\in\varinjlim_{V:x\in V}(\mathcal{O}_{C^g/\Sigma_g\times C^g/\Sigma_g}(\pi^{-1}(V))$$

restricted to the fiber is constant, because the fiber is connected. If the value of this constant is not zero, then the set $V(f)$ of zeroes of f is closed and does not meet the fiber. Hence its image under π is closed in $\mathrm{Spec}(\mathcal{O}_{X\times J,x})$ and does not contain x and hence empty. So $V(f)$ is empty. This shows that f is invertible and this implies the claim. The two local rings have the same field of fractions, namely the field of meromorphic functions on $X\times J$. But since $X\times J$ is smooth we know that the local ring $\mathcal{O}_{X\times J,x}$ is regular, hence factorial and hence integrally closed in its field of fraction (see prop.7.5.19, Thm. 7.5.20 and exercise 19, 1.), we conclude that the two local rings are equal and this proves the Lemma. □

Now we know that the first two factors are projective. If we take inverse images of our points $\widetilde{x}_0$ and $\widetilde{e}$, then we still have the triviality conditions. Now we can apply the theorem of the cube and conclude $\widetilde{\mathcal{N}}$ is locally trivial in Z, i.e. we can cover Z by open schemes Z_α s.t.

$$\widetilde{\mathcal{N}}\mid C^g/\Sigma_g\times C^g/\Sigma_g\times Z_\alpha$$

is trivial. But then the lemma above yields that already $\mathcal{N}$ is locally trivial in Z. □

We consider the special case of $\mathcal{M}=\mathcal{P}_{P_0}$. The tangent bundle T_J is trivial, a global section is determined by its value at e. Hence $H^0(J,T_J)=H^1(C,\mathcal{O}_C)$. If we differentiate we get a homomorphism $\delta_{\mathcal{P}_{P_0}}:H^1(C,\mathcal{O}_C)\longrightarrow H^1(X,\mathcal{O}_X)$. Now the Künneth formula (See 8.2.2) tells us that $H^1(X,\mathcal{O}_X)=H^1(C,\mathcal{O}_C)$ hence

$$\delta_{\mathcal{P}_{P_0}} : H^1(C,\mathcal{O}_C) \longrightarrow H^1(C,\mathcal{O}_C)$$

is an endomorphism.

Proposition 10.2.3. *The endomorphism $\delta_{\mathcal{P}_{P_0}}$ is the identity*

We recall the definition of $\mathcal{P}$ by a cocycle and compute its derivative. □

10.2.2 The projectivity of X and J

The above proposition 10.2.1 helps us to prove that the scheme X/k is projective. To do this we construct a line bundle on X with many sections.
Our bundle $\mathcal{P}$ on $C \times_k \mathrm{Pic}^g_{C/k}$ has a non zero global section: The constant function 1 is a global section of the restriction of $\mathcal{P}$ to $C \times_k U_{\mathrm{gen}}$. But since the complement of U_{gen} in $\mathrm{Pic}^g_{C/k}$ has codimension ≥ 2 it is clear that this section extends.
Now we consider the bundle Θ. We can restrict the above global section 1 to $\{P_0\} \times_k \mathrm{Pic}^g_{C/k}$, let us call this section s_Θ. I has an effective divisor D of zeroes and

$$\Theta = \mathcal{O}_X(D).$$

(The divisor D has as its support $|D|$ those points $(P_1,\dots,P_g)$ where one of the entries is equal to P_0.) We have

$$\begin{aligned} T_x^*\Theta &= \mathcal{O}_x(T_x^*D) = \Theta \otimes \mathcal{L}_x \\ T_{-x}^*\Theta &= \mathcal{O}_X(T_{-x}^*D) = \Theta \otimes \mathcal{L}_{-x} \end{aligned}$$

and if we apply proposition 10.2.1we get

$$T_x^*\Theta \otimes T_{-x}^*\Theta \simeq \Theta \otimes \mathcal{L}_x \otimes \Theta \otimes \mathcal{L}_{-x} \simeq \Theta^{\otimes 2}.$$

These translated bundles have sections $T_x(s_\Theta)$,$T_{-x}(s_\Theta)$ and via the above isomorphism we get sections $T_x(s_\Theta) \cdot T_{-x}(s_\Theta) \in H^0(X,\Theta^{\otimes 2})$. We can pick any point $u \in X$ and we find an $x \in \mathrm{Pic}^0_{C/k}$ such that

$$u \notin T_x^*|D| \cup T_{-x}^*|D|.$$

Hence we find a global section of $\Theta^{\otimes 2}$, that does not vanish at u. This means that $\Theta^{\otimes 2}$ has no base point (see definition 8.1.18).

Then we have seen (see Thm. 8.1.8) that we get a morphism

$$r_{\Theta^{\otimes 2}} : X \longrightarrow \mathbb{P}^N = \mathbb{P}(H^0(X,\Theta^{\otimes 2})).$$

We claim that this morphism has finite fibers. We assume that we have a fibre $r^{-1}_{\Theta^{\otimes 2}}(y)$, which has positive dimension. This fibre contains an irreducible curve Z. We can find a section $s \in H^0(X,\Theta^{\otimes 2})$, which does not vanish at a particular point $x_1 \in Z(k)$ since we do not have base points, but then this section will not vanish at any point of Z, which implies that $\Theta^{\otimes 2} \mid Z$ is the trivial bundle.

Therefore we know that Θ and all its translates $T_x^*(\Theta)$ have degree zero on Z. Since $\Theta = \mathcal{O}_X(D)$ we see that for any translate of D we have

$$T_x^*(D) \cap Z = \begin{cases} \emptyset \\ Z \end{cases}$$

or in other words we have $T_x^*(Z) \subset |D|$ or $T_x^*(Z) \cap |D| = \emptyset$.
Now we show that for any two points $z_1,z_2 \in Z(k)$ we have

$$T_{z_2-z_1}^*(|D|) = |D|.$$

Let $d_0 \in |D|(k)$ then $d_0 = T_{d_0-z_1}^*(z_1)$ and it follows

$$T_{d_0-z_1}^*(Z) \subset |D|.$$

Now $T_{d_o-z_1}^*(z_2) = T_{d_0-z_1}^*(T_{z_2-z_1}^*(z_1)) = T_{z_2-z_1}^*(T_{d_0-z_1}^*(z_1)) = T_{z_2-z_1}^*(d_0) \in |D|$.

But this is impossible since it would imply that $T_{z_2-z_1}^*(\Theta) \simeq \Theta$ for all $z_2,z_1 \in Z$. Now we pick a smooth point z_1 and take $z_2 = z_1 + \epsilon V$, where V is a non zero tangent vector at Z in z_1. Then we get $\delta_{\mathcal{P}_{P_0}}(V) = 0$ and this contradicts our proposition above.
This implies that the fibers are zero dimensional and we can apply theorem 8.1.20. To do this we have to verify that the assumption a1) is true. We know that the fibers are finite. But since X is a homogenous space under the action of J it is also clear that any of the fibers is contained in an affine open subscheme. To see this let us consider any finite set $\{a_1,a_2,\dots,a_N\}$ of points in $X(k)$. Let $U \subset X$ be a non empty open subset. For any index i we can define the non empty open set $V_i = \{y \in J(k) \mid y + a_i \in U\}$. Since J is irreducible the intersection of these open subsets is non empty. For any point $b \in J(k)$, which lies in this intersection we have $\{a_1,a_2,\dots,a_N\} \subset T_b(U)$.
We have proved the projectivity of X and J.

The morphisms $\phi_{\mathcal{M}}$ are homomorphisms of functors

We defined the morphism between functors

$$\begin{array}{ccc} \phi_{\mathcal{M}} : J(T) & \longrightarrow & \mathcal{PIC}^0_{X/k}(T) \\ \phi_{\mathcal{M}} : y & \mapsto & T_y^*(\mathcal{M}) \otimes \mathcal{M}^{-1} \end{array}$$

and now we have

Proposition 10.2.4. *These morphisms $\phi_{\mathcal{M}}$ between functors with values in abelian groups are homomorphisms.*

To see this we apply the theorem of the cube. We consider

$$X \times J \times J,$$

we have three morphisms to $X \times J$, namely, the projections p_{12},p_{13} and $\mathrm{Id}_X \times m_{23}$ where m_{23} is the multiplication in the second and third factor. Then we have the two morphisms $m,p_1 : X \times J \longrightarrow X$, where m is the action of J on X. For any line bundle $\mathcal{M}$ on X we can consider the line bundle

$$\begin{aligned} \mathcal{N} = (m \circ m_{23})^*(\mathcal{M}) \otimes (m \circ p_{12})^*(\mathcal{M})^{-1} \otimes (m \circ p_{13})^*(\mathcal{M})^{-1} \otimes \\ \otimes (p_1 \otimes m_{23})^*(\mathcal{M})^{-1} \otimes (p_1 \circ p_{12})^*(\mathcal{M}) \otimes (p_1 \circ p_{13})^*(\mathcal{M}) \end{aligned}$$

on $X \times J \times J$.
We pick a point $x_0 \in X(k)$ and evaluate this bundle on

$$x_0 \times J \times J \qquad X \times e \times J \qquad X \times J \times e,$$

and find that its restrictions to the second and the third subscheme are trivial. Let $\mathcal{N}_{x_0}$ be its restriction to the first subscheme, which we identify to $J \times J$. Let

$$\widetilde{\mathcal{N}}_{x_0} = p_{23}^*(\mathcal{N}_{x_0})$$

then we see that

$$\mathcal{N} \otimes p_{23}^*(\mathcal{N}_{x_0})^{-1} = \mathcal{N} \otimes \widetilde{\mathcal{N}}_{x_0}^{-1}$$

is trivial on all three subschemes. Hence it follows from the theorem of the cube that

$$\mathcal{N} \simeq p_{23}^*(\mathcal{N}_{x_0}).$$

This can be formulated differently by saying $\mathcal{N} \sim_{J \times J} \mathcal{O}_{X \times J \times J}$.
If now x,$y \in J(T)$ then $x \times y : T \longrightarrow J \times J$ is the product of two T valued points. Hence we have the morphism

$$\mathrm{Id} \times x \times y : X \times T \longrightarrow X \times J \times J$$

and we compute $(\mathrm{Id} \times x \times y)^*(\mathcal{N})$. The observation we just made implies $(\mathrm{Id} \times x \times y) * (\mathcal{N}) \sim_T \mathcal{O}_{X \times T}$. Exploiting the definition we get

$$(\mathrm{Id} \times x \times y) * (\mathcal{N}) = T_{x+y}^*(\mathcal{M}) \otimes T_x^*(\mathcal{M})^{-1} \otimes T_y^*(\mathcal{M})^{-1} \otimes \mathcal{M}^{-1} \otimes \mathcal{M} \otimes \mathcal{M}.$$

Since the left hand side is locally trivial in T it follows that

$$\phi_{\mathcal{M}}(x + y) = \phi_{\mathcal{M}}(x) + \phi_{\mathcal{M}}(y)$$

where of course + means taking the tensor product. □

We will apply this to the special bundle Θ. We can now interpret this as a homomorphism of functors

$$\phi_\Theta : J \longrightarrow \mathcal{PIC}^0(X)$$

and our goal is -in a certain sense- to show that this is an isomorphism. Recall that in the definition of Θ we had to choose a point $P_0 \in C(k)$. But obviously two different choices of this point yield divisors, which are algebraically equivalent. Hence proposition 10.2.1 implies that these two divisors define the same ϕ_Θ.

10.2.3 Maps from the curve C to X, local representability of $\mathcal{PIC}_{X/k}$, $\mathcal{PIC}_{J/k}$ and the self duality of the Jacobian

To reach this goal we construct a homomorphism in the opposite direction. To get such homomorphisms we construct certain morphisms from the curve C into X and study the restriction (via these maps) of Θ and of other line bundles on X to C. We have seen in Vol. I. 5.2.3 that such morphisms explain the self duality of the Jacobian.

We keep our point $P_0 \in C(k)$. The naive construction of such morphisms is easy to explain. We choose an array of points

$$\underline{Q} = (Q_1,\cdots,Q_g) \in C^g(k)$$

From these data we construct a morphism

$$j_{P_0,\underline{Q}} : C \longrightarrow X,$$

which sends a point $P \in C(k)$ to the line bundle $\mathcal{O}_C(-P + P_0 + Q_1 + \cdots + Q_g)$, (just a reminder, we allow first order poles at P_0 and the Q_i and require a zero at P.) The point P_0 is fixed, we consider $\underline{Q}$ as variable.
We say that $(Q_1,\cdots,Q_g) \in \overline{C^g}(k)$ is **generic** if the images of the points

$$(Q_1,\cdots,Q_g) \qquad \text{and} \qquad (P_0,Q_1,\cdots,\widehat{Q}_i,\cdots,Q_g)$$

are all in U_{gen}.
To give the correct definition of this morphism we consider the curve $C \times C$ with its two projections p_1,p_2 to C. On this surface we have the line bundle

$$p_1^*(O_C(P_0 + Q_1 + \cdots + Q_g)) \otimes O_{C\times C}(-\Delta) = \mathcal{L}_{\underline{Q}},$$

where $\Delta \subset C \times C$ is the diagonal. This gives us a family of line bundles on the first factor, which is parameterized by the second factor. For a point $P \in C(k)$ in the second factor the restriction of $\mathcal{L}_{\underline{Q}}$ to $C \times \{P\}$ is the bundle $\mathcal{O}_C(-P + P_0 + Q_1 - \cdots + Q_g)$. Hence we have a unique morphism - and this is our $j_{P_0,\underline{Q}}$ - such that locally in the second component

$$(\mathrm{Id} \times j_{P_0,\underline{Q}}(\mathcal{P})) \sim \mathcal{L}_{\underline{Q}}.$$

We want to compute the line bundle

$$j^*_{P_0,\underline{Q}}(\Theta),$$

this is a line bundle on C. We assume that $\underline{Q}$ is generic. We claim that under our above assumptions the morphism $j_{P_0,\underline{Q}} : C \to \overline{\mathrm{Pic}^g}(C)$ factors through the open subset U_{gen}. To see this we have to show that

$$\dim H^0(C,\mathcal{O}_C(-P + P_0 + Q_1 + \cdots + Q_g)) = 1$$

for all points $P \in C(k)$. This is of course clear if P is one of our points P_0 or $Q_1 \cdots Q_g$ because this is our assumption. If P is not equal to any of these points, and if we have

$$\dim H^0(C,\mathcal{O}_C(-P + P_0 + Q_1 \cdots + Q_g)) \geq 2,$$

we can conclude that

$$\dim H^0(C,\mathcal{O}_C(P_0 + Q_1 \cdots + Q_g)) \geq 3,$$

because this space contains 1 and 1 is not zero at P. But then we can conclude that

$$\dim H^0(C,\mathcal{O}_C(+Q_1 + \cdots + Q_g)) \geq 2$$

and this is again a contradiction to our assumption.
The morphism $j_{P_0,\underline{Q}}$ sends the point P to a point $j_{P_0,\underline{Q}}(P) = (Q_1(P),\cdots,Q_g(P))$ where $(Q_1(P)\cdots Q_g(P)) \in U_{\text{gen}}$, which is well defined up to an element in the symmetric group. It is determined by the relation

$$\mathcal{O}_C(-P + P_0 + Q_1 \cdots + Q_g) \simeq \mathcal{O}_C(Q_1(P) + \cdots + Q_g(P)),$$

or to say it differently $Q_1(P)\cdots + \ldots Q_g(P)$ is the divisor of zeroes a a non zero section $s \in H^0(C,\mathcal{O}_C(-P + P_0 + Q_1 + \cdots + Q_g))$.

Now we recall the definition of $\mathcal{P}$ and Θ. The restriction of $\mathcal{P}$ to $\{P_0\} \times U$ is the line bundle, which is induced by the divisor

$$D' = \Sigma \;\; C \times \cdots C \times P_0 \times C \cdots C \tag{10.6}$$

on C^g. This divisor descends to a divisor D on C^g/Σ_g and the line bundle $\Theta|U_{\text{gen}} = \mathcal{O}_{C^g/\Sigma_g}(D)$.

Let us assume that all the points $Q_1,Q_2,\ldots,Q_g$ are pairwise different. By definition a point $(Q_1(P),\ldots,Q_g(P))$ lies on D if and only if for some index i_0 we have $Q_{i_0}(P) = P_0$. Then we get an isomorphism between the two line bundles

$$\mathcal{O}_C(-P + Q_1 \cdots + Q_g) \xrightarrow{\sim} \mathcal{O}_C(+Q_1(P) + \cdots + Q_{i_0-1}(P) + Q_{i_0+1}(P) + \cdots + Q_g(P)).$$

Moving the $-P$ to the right hand side we get

$$\mathcal{O}_C(Q_1 \cdots + Q_g) \xrightarrow{\sim} \mathcal{O}_C(+Q_1(P) + \cdots + Q_{i_0-1}(P) + P + Q_{i_0+1}(P) + \cdots + Q_g(P)).$$

Since we assumed the $\underline{Q} \in U_{\text{gen}}$ we can conclude that $\underline{Q} = (Q_1(P),\ldots,P,\ldots,Q_g(P))$ (up to an element in the symmetric group.)
This yields the innocent looking but fundamental relation

$$j^*_{P_0,\underline{Q}}(\Theta) \simeq \mathcal{O}_C(Q_1 + \cdots + Q_g) \tag{10.7}$$

We consider the point $\underline{Q} \in U_{\text{gen}}$ as a variable, more precisely we consider it as a point in X. We look at the diagram

$$\begin{array}{cccccc} & \xrightarrow{p_{12}} & C \times X & & \\ C \times X \times C & & & & \\ & \xrightarrow{p_{13}} & C \times C & \xrightarrow{p_1} & C \end{array}$$

and the line bundle

$$p^*_{12}(\mathcal{P}) \otimes p^*_{13}(p^*_1(\mathcal{O}_C(P_0)) \otimes \mathcal{O}_{C\times C}(-\Delta)) = \widetilde{\mathcal{L}}$$

on $C \times X \times C$. If we evaluate at the point $\underline{Q} \in U(k) \subset X(k)$, then we get our line bundle $\mathcal{L}$ above. We view this as a family of degree g bundles on the first factor, which is parameterized be the product of the second and third factor. The universal property gives us a unique morphism

$$j_{P_0} : X \times C \longrightarrow X$$

such that $(\mathrm{Id}_C \times j_{P_0})^*(\mathcal{P})$ is locally in $X \times C$ isomorphic to the line bundle $\widetilde{\mathcal{L}}$.
Of course it is quite clear to describe this map on geometric points. If (u,P) is a geometric point on $X \times C$ then u is the isomorphism class of a line bundle $\mathcal{L}_u$ of degree g on $C \times \bar{k}$. Then $\mathcal{L}_u \otimes \mathcal{O}_C(P - P_0)$ is a line bundle of degree g on C, hence a geometric point on X and this point is the image. From this description we get

Proposition 10.2.5. *The group scheme J acts on both sides, by the action $m \times \mathrm{Id}$ on the left hand side and by m on the right hand side. The morphism j_{P_0} is J invariant for these actions.*

This map j_{P_0} gives us a line bundle

$$j_{P_0}^*(\Theta)$$

on $X \times C$. We consider this as a family of line bundles on C, which now is parameterized by the first factor and find a unique morphism

$$\psi_{P_0} : X \longrightarrow X$$

such that $(\psi_{P_0} \times \mathrm{Id}_C)^*(\mathcal{P}) \sim_X j_{P_0}(\Theta)$ where in this case $\sim$ means that the bundles are locally isomorphic in the first variable X.

But for a dense set of geometric points $\underline{Q} \in U(k)$ we have shown that

$$\psi_{P_0}(\underline{Q}) = \underline{Q}$$

and since X is reduced we can conclude that $\psi_{P_0} = \mathrm{Id}_X$. Of course $(\mathrm{Id}_X \times \mathrm{Id}_C)^*(\mathcal{P}) \simeq \mathcal{P}$, and we conclude

$$\mathcal{P} \sim j_{P_0}^*(\Theta)$$

locally in X on $X \times C$.

We have constructed the homomorphism

$$\phi_\Theta : \mathrm{Pic}^0_{C/k} \longrightarrow \mathcal{PIC}^0_{X/k}$$

and we have the restriction

$$j_{P_0,\underline{Q}}^* : \mathcal{PIC}^0_{X/k} \longrightarrow \mathrm{Pic}^0_{C/k} .$$

The composition of these to homomorphisms is the identity. It suffices to check this on the set of geometric points. Actually we only check this on the non empty Zariski open subset of pairs $(\underline{Q},x)$ where $\underline{Q} \in U_{\mathrm{gen}}(k)$,$T_x(\underline{Q}) \in U_{\mathrm{gen}}(k)$. We to show

$$j_{P_0,\underline{Q}}^*\left(T_x^*(\Theta) \otimes \Theta^{-1}\right) \simeq \mathcal{L}_x$$

where $\mathcal{L}_x$ is a line bundle corresponding to x. For the computation of left hand side we use the invariance under translations (proposition 10.2.5) and our formula 10.7

$$j_{P_0,\underline{Q}}^*(T_x^*(\Theta)) \otimes j_{P_0,\underline{Q}}^*(\Theta)^{-1} = j_{P_0,T_x(\underline{Q})}^*(\Theta) \otimes j_{P_0,\underline{Q}}^*(\Theta)^{-1} = \mathcal{O}_C(T_x(\underline{Q}) - \underline{Q}) \xrightarrow{\sim} \mathcal{L}_x$$

and this is the claim.

In section 10.2.1 we constructed the line bundle $\widetilde{\Theta}$ on $X \times J$. (See 10.5.) The restriction $j_{P_0}^*(\widetilde{\Theta})$ is a bundle on $C \times J$ and hence we find a unique morphism $\psi_{P_0} : J \longrightarrow J$ such that $(\mathrm{Id} \times \psi_{P_0})^*(\mathcal{P}_0) \sim_J j_{P_0}^*(\widetilde{\Theta})$. Our computation above shows that on geometric points $x \in J(k)$ we have $\psi_{P_0}(x) = x$ hence we can conclude that

$$\mathcal{P}_0 \sim_J j_{P_0}^*(\widetilde{\Theta}). \tag{10.8}$$

Now we can state and prove

Theorem 10.2.6. *The scheme $X \times J$ together with the line bundle $\widetilde{\Theta}$ provides a local representation of the functor $\mathcal{PIC}^0_{X/k}$: For any line bundle on*

$$\mathcal{L} \mid X \times T,$$

where T is connected and of finite type such that $\mathcal{L} \mid X \times t_0 \simeq \mathcal{O}_X$ for some point $t_0 \in T$ we have a unique morphism

$$\psi : T \longrightarrow J$$

such that $\mathcal{L} \sim (\mathrm{Id} \times \psi)^*(\widetilde{\Theta})$.

We start from a line bundle $\mathcal{L}$ on $X \times T$. We just saw that for any $j_{\underline{Q},P_0} : C \to X$ the pullback of the line bundle $\widetilde{\Theta}$ to $C \times J$ is the universal bundle on $C \times J$ (See 10.8). Hence we can take the pullback of $\mathcal{L}$ on $X \times T$ via $j_{P_0,\underline{Q}}$ to $C \times T$, and we see that we have a unique morphism

$$\psi : T \longrightarrow J$$

such that

$$(\mathrm{Id}_C \times \psi)^*(\widetilde{\Theta}) \simeq (j_{P_0,\underline{Q}} \times \mathrm{Id}_T)^*(\mathcal{L})$$

here we assume that T should be local. We have to show that already

$$(\mathrm{Id}_X \times \psi)^*(\widetilde{\Theta}) \simeq \mathcal{L}.$$

We have the two elements

$$(\mathrm{Id}_X \times \psi)^*(\widetilde{\Theta}), \mathcal{L} \in \mathcal{PIC}^0_{X/k}(A),$$

which are trivial in the special fibre and whose images under $j_{P_0,\underline{Q}}^*$ in $J(T)$ are equal. We consider the bundle $\mathcal{M} = (\mathrm{Id}_X \times \psi)^*(\widetilde{\Theta}) \otimes \mathcal{L}^{-1}$. It is trivial in t_0 and $j_{P_0,\underline{Q}}^*(\mathcal{M}) \sim_T \mathcal{O}_T$. In other words $j_{P_0,\underline{Q}}^*(\mathcal{M}) = p_2^*(\mathcal{N})$ where $\mathcal{N}$ is a line bundle on T. Hence we see that $\mathcal{M} \otimes p_2^*(\mathcal{N})^{-1}$ is trivial in t_0 and becomes trivial under the restriction $j_{P_0,\underline{Q}}^*$. We have to show that it trivial.

We consider the locus of triviality of $\mathcal{M}$ and we prove that it contains an open neighborhood of t_0. This is obviously enough because then it must be equal to T because T is connected.

Let A be the local ring at t_0,we just saw that it suffices to prove that $\mathcal{M}|C \times \mathrm{Spec}(A)$ is trivial. Let $\mathfrak{m}$ be the maximal ideal of A. Clearly it suffices to prove that

$$j_{P_0,\underline{Q}} : \mathcal{PIC}^0_{X/k}(A/\mathfrak{m}^N) \longrightarrow J(A/\mathfrak{m}^N) \text{ is injective for all } N,$$

because this implies that the ideal I, which defines the locus of triviality is contained in all $\mathfrak{m}^N$ and we know $\cap \mathfrak{m}^N = \{0\}$ by the Artin-Rees theorem.
For $N = 1$ both line bundles are trivial by assumption, we also have by definition $\mathcal{PIC}^0_{X/k}(A/\mathfrak{m}) = 0$. We get the two exact sequences (see 10.1.3)

$$\begin{array}{ccccccc} 0 & \to & H^1(X,\mathcal{O}_X)\otimes \mathfrak{m}^N/\mathfrak{m}^{N-1} & \to & H^1(X_N,\mathcal{O}^*_{X_N})(0) & \to & H^1(X_{N-1},\mathcal{O}^*_{X_{N-1}}) \to \\ & & \downarrow & & \downarrow & & \downarrow \\ 0 & \to & H^1(C,\mathcal{O}_C)\otimes \mathfrak{m}^N/\mathfrak{m}^{N-1} & \to & H^1(C_N,\mathcal{O}^*_{C_N})(0) & \to & H^1(C_{N-1},\mathcal{O}^*_{C_{N-1}}) \end{array}$$

The last arrow in the second row is surjective because we have $H^2(C,\mathcal{O}_C) = 0$.
Therefore, we get that $j_{P_0,\underline{Q}} : \mathcal{PIC}^0_{X/k}(A/\mathfrak{m}^N) \longrightarrow J(A/\mathfrak{m}^N)(0)$ is an isomorphism for all N, provided we can prove that $j^1_{P_0,\underline{Q}} : H^1(X,\mathcal{O}_X) \longrightarrow H^1(C,\mathcal{O}_C)$ is an isomorphism.

It is clear that $j^1_{P_0,\underline{Q}}$ is surjective because ϕ_Θ gives us a homomorphism from $H^1(C,\mathcal{O}_C) \longrightarrow H^1(X,\mathcal{O}_X)$ and we know the composition with $j^1_{P_0,\underline{Q}}$ is the identity. We have the morphisms

$$C^g \xrightarrow{q} C^g/\Sigma_g \xrightarrow{\pi} X,$$

which provide k-linear maps

$$H^1(X,\mathcal{O}_X) \xrightarrow{\pi^1} H^1(C^g/\Sigma_g,\mathcal{O}_{C^g/\Sigma_g}) \xrightarrow{q^1} H^1(C^g,\mathcal{O}_{C^g})^{\Sigma_g}.$$

We claim that these two maps are injective. To see this for the first morphism we recall that

$$\pi : C^g/\Sigma_g \longrightarrow X$$

is birational and projective. This implies that

$$\pi_*(\mathcal{O}_{C^g/\Sigma}) = \mathcal{O}_X$$

(See proof of Lemma 10.2.2) The edge homomorphism of the spectral sequence yields

$$0 \longrightarrow H^1(X,\mathcal{O}_X) \longrightarrow H^1(C^g/\Sigma_g,\mathcal{O}_{C^g/\Sigma_g}) \longrightarrow H^0(X,R^1\pi_*(\mathcal{O}_{C^g/\Sigma_g})) \longrightarrow$$

and the injectivity becomes obvious.

To prove the injectivity of q^1 we go back to the general principles of the computation of coherent cohomology. We obtained Θ from the line bundle $\mathcal{O}_{C^g/\Sigma_g}(D)$ on C^g/Σ_g and this bundle was obtained from the divisor D' on C^g (see 10.6.) Both divisors are ample and hence the sheaves $\mathcal{O}_{C^g/\Sigma_g}(rD)$,$\mathcal{O}_{C^g}(rD')$ are acyclic if $r >> 0$ (See Thm. 8.3.3). Hence we get two exact sequences

$$0 \longrightarrow \mathcal{O}_{C^g} \longrightarrow \mathcal{O}_{C^g}(rD') \longrightarrow \mathcal{L}'_{rD'} \longrightarrow 0$$

and

$$0 \longrightarrow \mathcal{O}_{C^g/\Sigma_g} \longrightarrow \mathcal{O}_{C^g/\Sigma_g}(rD) \longrightarrow \mathcal{L}_{rD} \longrightarrow 0$$

of sheaves, where the to sheaves at the right are defined as quotients.
Taking global sections we get the exact sequences

$$\begin{array}{ccccccc}
0 \longrightarrow H^0(C^g,\mathcal{O}_{C^g}) & \longrightarrow H^0(C^g,\mathcal{O}_{C^g}(rD')) & \longrightarrow H^0(C^g,\mathcal{L}'_{rD'}) & \xrightarrow{\delta} \\
\cup & \cup & \cup & \\
0 \longrightarrow H^0(C^g/\Sigma_g,\mathcal{O}_{C^g/\Sigma_g}) & \longrightarrow H^0(C^g/\Sigma_g,\mathcal{O}_{C^g/\Sigma_g}(rD)) & \longrightarrow H^0(C^g/\Sigma_g,\mathcal{L}_{rD}) & \xrightarrow{\delta_\Sigma}
\end{array}$$

$$\begin{array}{cc}
\xrightarrow{\delta} & H^1(C^g,\mathcal{O}_{C^g}) \longrightarrow 0 \\
& \uparrow q^1 \\
\xrightarrow{\delta_\Sigma} & H^1(C^g/\Sigma_g,\mathcal{O}_{C^g/\Sigma_g}) \longrightarrow 0
\end{array}$$

The modules in the top row are Σ_g modules, and the inclusions $\cup$ from the bottom line to the top line always go into the Σ_g invariants. We have a Σ_g invariant splitting of the sequences on the left end. To get this we choose a point $Q \in C(k)$ different from P_0 and get a Σ_g invariant linear form $\lambda_Q : H^0(C^g,\mathcal{O}_{C^g}(rD')) \longrightarrow k$, which is given by the evaluation at the point $(Q,Q,\dots,Q) \in C^g(k)$. Denoting the kernel of this linear form by adding a (0) at the left end we get shorter exact sequences

$$\begin{array}{cccc}
0 & \longrightarrow H^0(C^g,\mathcal{O}_{C^g}(rD'))(0) & \longrightarrow H^0(C^g,\mathcal{L}'_{rD'}) & \longrightarrow H^1(C^g,\mathcal{O}_{C^g}) \longrightarrow 0 \\
& \cup & \cup & \uparrow q^1 \\
0 & \longrightarrow H^0(C^g/\Sigma_g,\mathcal{O}_{C^g/\Sigma_g}(rD))(0) & \longrightarrow H^0(C^g/\Sigma_g,\mathcal{L}_{rD}) & \longrightarrow H^1(C^g/\Sigma_g,\mathcal{O}_{C^g/\Sigma_g}) \longrightarrow 0
\end{array}$$

If a class $\xi \in H^1(C^g/\Sigma_g,\mathcal{O}_{C^g/\Sigma_g})$ goes o zero under q^1 we represent it by a class $\eta \in H^0(C^g/\Sigma_g,\mathcal{L}_{rD})$. We send η to $\eta' \in H^0(C^g,\mathcal{L}'_{rD'})^{\Sigma_g}$ and since this element is going to zero in $H^1(C^g,\mathcal{O}_{C^g})$ it is in $H^0(C^g,\mathcal{L}'_{rD'})^{\Sigma_g} \cap H^0(C^g,\mathcal{O}_{C^g}(rD'))(0) = H^0(C^g,\mathcal{O}_{C^g}(rD'))(0)^{\Sigma_g}$. But from the definition of the symmetric product it follows that

$$H^0(C^g,\mathcal{O}_{C^g}(rD'))(0)^{\Sigma_g} = H^0(C^g/\Sigma_g,\mathcal{O}_{C^g/\Sigma_g}(rD))(0)$$

and hence we see that η itself is zero. □

The last theorem also gives us the local representability of the functor $\mathcal{PIC}^0_{J/k}$. We assume that we have a point $P \in C(k)$, the we can identify $i_P : J \xrightarrow{\sim} X$. Using this identification we get the divisor $i_P(\Theta)$ on J, we call it again Θ. We consider the diagram

$$J \times J \;\; \begin{array}{c} \xrightarrow{p_1} \\ \xrightarrow{m} \\ \xrightarrow{p_2} \end{array} \;\; J \tag{10.9}$$

and define the line bundle

$$\widetilde{\Theta} = m^*(\Theta) \otimes p_1^*(\Theta)^{-1} \otimes p_2^*(\Theta)^{-1} \tag{10.10}$$

on $J \times J$. It is a symmetric version of the bundle in 10.5. It is clear that $(J \times J,\widetilde{\Theta})$ provides a local representation of $\mathcal{PIC}^0_{J/k}$ over k despite of the fact that Θ may not be defined over k. Another choice of the point Q yields another $i_Q(\Theta)$, but it yields the same bundle $\widetilde{\Theta}$. We see from the definition of $\widetilde{\Theta}$ that we have canonical identifications

$$\widetilde{\Theta}|\{e\} \times J = \mathcal{O}_{\{e\}\times J}, \widetilde{\Theta}|J \times \{e\} = \mathcal{O}_{J\times\{e\}}.$$

The first identification is a trivialization $s \in H^0(\{e\} \times J,\widetilde{\Theta}|\{e\} \times J)$ and hence we get

The functor $\mathcal{PIC}^0_{J/k,e}$ is represented by $(J \times_k J,\widetilde{\Theta},s)$

10.2.4 The self duality of the Jacobian

This explains of course the self duality of J, the choice of the line bundle Θ (up to algebraic equivalence) provides a canonical isomorphism

$$\phi_\Theta : J \xrightarrow{\sim} \mathrm{Pic}^0_{J/k} = J^\vee. \tag{10.11}$$

Since J has the structure of a group scheme, we have a canonical choice of a rational point, namely the identity element $e \in J(k)$. This means that we have a canonical rigidification of $\mathcal{PIC}_{J/k}$.
In certain situations is better to forget the self duality of the Jacobian. Since we have a canonical point $e \in J(k)$ we can consider the functor $\mathcal{PIC}^0_{J/k,e}$. Our results imply that this functor is representable. We get a variant of Theorem 10.2.6
We can construct an abelian variety $J^\vee/k$ and a line bundle $\mathcal{N}$ (the Poincaré-bundle) on $J \times_k J^\vee$, which satisfies

$$\mathcal{N}|\{e\} \times J^\vee \xrightarrow{\sim} \mathcal{O}_{J^\vee},\mathcal{N}|J \times \{e\} \xrightarrow{\sim} \mathcal{O}_J, \tag{10.12}$$

and for which an isomorphism

$$s : \mathcal{N}_{e,e^\vee} \xrightarrow{\sim} k \tag{10.13}$$

is given.(This isomorphism also fixes the isomorphisms in 10.12.) Finally the triplet $(J \times_k J^\vee,\mathcal{N},s)$ represents the functor $\mathcal{PIC}^0_{J/k,e}$ i.e. for any T of finite type over k and for any line bundle $\mathcal{L}|J \times_k T$ and a given trivialization $s_T : \mathcal{L}|\{e\} \times T \xrightarrow{\sim} \mathcal{O}_T$ we find a unique $\psi : T \longrightarrow J^\vee$ and a unique isomorphism

$$\Psi : (\mathrm{Id} \times \psi)^*(\mathcal{N}) \xrightarrow{\sim} \mathcal{L} \text{ such that } \Psi(s) = s_T.$$

Mutatis mutandis the triplet $(J \times_k J^\vee,\mathcal{N},s)$ also represents $\mathcal{PIC}^0_{J^\vee,e^\vee}$.

All this is an obvious consequence of the Theorem 10.2.6 above. We have to extend the ground field such that the divisor Θ becomes available. By then we have proved representability of a functor over this extension, and then our general descend arguments work. The new formulation has the advantage that it actually considers J and $J^\vee$ as two different objects. Their identification via the choice of Θ is somewhat artificial. We will see this also in the next section, when we briefly discuss arbitrary abelian varieties.

10.2.5 General abelian varieties

We want to say a few words about arbitrary abelian varieties A/k. This subject is treated extensively in the book of D. Mumford ([Mu1]), we will be somewhat brief. We know that for any abelian variety A/k the local representability of $\mathcal{PIC}^0_{A/k}$ is equivalent to the representability of $\mathcal{PIC}^0_{A/k,e}$.
We have a simple general theorem

Theorem 10.2.7. *An abelian variety variety is a commutative group scheme, any morphism $f : A/k \longrightarrow G/k$ from an abelian variety to an arbitrary group scheme, which maps the identity element $e_A \in A(k)$ to the identity $e_G \in G(k)$ is a homomorphism.*

Consider any morphism $H : A \times_k A \longrightarrow G$, which satisfies $H(a,e_A) = H(e_A,b) = e_G$ for all $a,b \in A(\bar{k})$. Such a morphism must be constant. To see this look at an affine neighborhood $U \subset G$ of the identity element. Show that we can find an open neighborhood $V \subset A$ of e_A such that $H(A \times W) \subset U$. Then for any $w \in W(\bar{k})$ we get the map $H_w : A \times \{w\} \longrightarrow U$. But A is projective and connected and U is affine. Since any regular function on a connected projective variety must be constant, it follows that H_w must be constant. But then it is clear that the value of this constant is e_G. Since $A \times W$ is open in $A \times A$ it follows that H is constant. Now take $G = A$ and apply this argument to the commutator map $(a,b) \mapsto aba^{-1}b^{-1}$. For the case of a morphism f apply the argument to $(a,b) \mapsto f(a+b)f(a)^{-1}f(b)^{-1}$. □

Proposition 10.2.8. *Let A/k be an abelian variety and let $\mathcal{L}$ be an ample line bundle on it. Then the kernel of the homomorphism*

$$\phi_{\mathcal{L}} : A \longrightarrow \mathcal{PIC}^0_{A/k}$$

is a finite group scheme.

We can replace $\mathcal{L}$ be a very ample $\mathcal{L}^{\otimes n}$ then $\ker(\phi_{\mathcal{L}}) \subset \ker((\phi_{\mathcal{L}}^{\otimes n})$. Then any $x \in \ker(\phi_{\mathcal{L}})^{\otimes n})(\bar{k})$ defines an automorphism $\widetilde{T}_x^* : H^0(A,\mathcal{L}^{\otimes n}) \longrightarrow H^0(A,T_x^*(\mathcal{L}^{\otimes n})) \xrightarrow{\mu_x} H^0(A,(\mathcal{L}^{\otimes n}))$, where μ_x is induced by the choice of an isomorphism $\mathcal{L}^{\otimes n} \xrightarrow{\sim} T_x^*(\mathcal{L})^{\otimes n}$. This gives us a faithful representation

$$\rho : \ker(\phi_{\mathcal{L}}^{\otimes n}) \longrightarrow \mathrm{PGL}(H^0(A,T_x^*(\mathcal{L}^{\otimes n})) = \mathrm{GL}(H^0(A,T_x^*(\mathcal{L}^{\otimes n}))/\mathbb{G}_m.$$

The "projective linear" group $\mathrm{PGL} = \mathrm{GL}/\mathbb{G}_m$ is affine. If $\ker(\phi_{\mathcal{L}}^{\otimes n})$ is not finite then its connected component of the identity is an abelian variety. But a morphism of a connected projective scheme to an affine scheme is constant, the homomorphism ρ must be trivial on this connected component, hence we have a contradiction. □

Mumford uses this construction to prove that $\mathcal{PIC}^0_{A/k,e}$ is representable, he simply defines $A/\ker(\phi_{\mathcal{L}}) \xrightarrow{\sim} \mathrm{Pic}^0_{A/k,e} = A^\vee$ and constructs a universal bundle $\mathcal{N}$ on $A \times_k A^\vee$. This is not so easy, it is done in Chap. III, 13 of [Mu1].
We can also derive this from our results on Jacobians. We start from a

Lemma 10.2.9. : *If J/k is the jacobian of a curve C/k and if $A \subset J$ is an abelian subvariety, then we can find another subvariety $B/k \subset J/k$ such that the homomorphism $A \times B \longrightarrow J$ has finite kernel and is surjective.*

We start from the isomorphism $\phi_\Theta : J \longrightarrow J^\vee$. Now we use the general result that we can form the quotient $J/A = B$ (See [Ro], Thm. 2) . To prove this theorem Rosenlicht proceeds as follows. The subvariety A acts on the function field $k(J)$ by translations and it defines a fixed field L/k. This fixed field is big enough to rediscover A. (This step is not so easy.) The functions in L have divisors which are invariant under A and from this we can conclude that we can find an effective divisor Θ_A such that the connected component of the identity of $\ker(\phi_{\Theta_A})^{(0)} = A$ We consider the homomorphism $\phi_{\Theta_A} : J \longrightarrow J^\vee$, The image is an abelian subvariety $B^\vee \subset J^\vee$. But then $B = \phi_{\Theta_A}^{-1}(B^\vee) \subset J$ maps isomorphically back to $B^\vee$ under ϕ_Θ, this shows that $A \times B \longrightarrow J$ is surjective and that the intersection $A \cap B$ finite. □

A homomorphism $\phi : A \longrightarrow B$ between two abelian varieties is called an **isogeny** if it is surjective and its kernel is finite.
Finally we observe that for any abelian variety A/k we can construct a non trivial homomorphism $A \longrightarrow J_C$ where J_C is the Jacobian of a suitable curve C/k. To get such a homomorphisms we intersect $A \subset \mathbb{P}^n/k$ with a suitable number of generic hyperplanes and get a smooth curve $j : C \hookrightarrow A$. Then we have the restriction $j^* : \mathcal{PIC}^0_{A/k} \longrightarrow J_C^\vee = J_C$ and we can compose it with any $\phi_\mathcal{L} : A \longrightarrow \mathcal{PIC}^0_{A/k}$, $\mathcal{L}$ ample. The following is true

Proposition 10.2.10. *For any abelian variety A/k we can find a curve $j : C \hookrightarrow A$ such that $j^* \circ \phi_\mathcal{L} : A \longrightarrow J_C$ is non trivial.*

We cannot prove this result here, it depends on some finiteness properties of the Picard functor $\mathrm{Pic}^0_{X/k}$ for arbitrary projective schemes X/k. But we can explain some of the basic ideas of the proof. The first thing we need is that the line bundles on X, which are algebraically (or only numerically) equivalent to zero form a bounded family. (See [Kl], Thm. 6.3). This means that we can find a scheme $T \longrightarrow \mathrm{Spec}(k)$ of finite type and a line bundle $\mathcal{L}$ on $X \times T$ such that any bundle on $\mathrm{Pic}_{X/k}$ is obtained by evaluation at a point t_0.
We assume $\dim(X/k) \geq 2$. If now $X/k \hookrightarrow \mathbb{P}^n/k$ then we choose a $l/k \subset \mathbb{P}^n/k$, which is a $\mathbb{P}^{n-2}$ and a $\mathbb{P}^1/k \subset \mathbb{P}^n/k$ such that $l \cap \mathbb{P}^1 = \emptyset$. This defines a family of hyperplanes $H_{x,l}$, which are parameterized by the points $x \in \mathbb{P}^{\bar{k}}$, here $H_{x,l}$ is simply the hyperplane containing l and x. We can arrange our data in such a way $X \cap l$ is smooth and that for an open subset $U \subset \mathbb{P}^1$ the intersection $X_{x,l} = X \cap H_{x,l}$ is smooth. Now we blow up our $\mathbb{P}^n$ along the "line" l and get a diagram

$$\begin{array}{ccc} \widehat{X} & \longrightarrow & \widehat{\mathbb{P}}^n \\ \downarrow & & \downarrow \\ X & \longrightarrow & \mathbb{P}^n \end{array}$$

By definition of the blow up we have a morphism $\widehat{\mathbb{P}}^n \longrightarrow \mathbb{P}^1$ whose fibers are the hyperplanes $H_{l,x}$, and hence we get a morphism

$$\pi : \widehat{X} \longrightarrow \mathbb{P}^1$$

whose fibers are the hyperplane sections $H_{l,x} \cap X$. We have seen that $\mathrm{Pic}^0_{X/k} \longrightarrow \mathrm{Pic}_{\widehat{X}/k}$ is injective. (See argument in the proof of Lemma 10.2.2). Let η be the generic point in $\mathbb{P}^1$. Our considerations in the proof of the theorem of the cube can be applied to show that a bundle $\mathcal{L}$ in $\mathrm{Pic}^0_{X/k}$, which is trivial on $H_{l,\eta}$ is in fact trivial on $\widehat{X}$ and hence on X. Hence we see that for any $\mathcal{L} \in \mathrm{Pic}^0_{X/k}, \mathcal{L} \neq \mathcal{O}_X$ we can find a point $t \in U(\bar{k})$ such that

$\mathcal{L}|X_{l,t}$ is non trivial. Applying our considerations about the locus of triviality again we see that the bundles in our family, which are trivial on $X_{l,t}$ form a proper closed subscheme of T. Then it becomes clear that we can find a finite number of points $t_1,t_2,\dots,t_r$ such that $\mathrm{Pic}_{X/k} \longrightarrow \oplus_i \mathrm{Pic}_{X_{t_i}/\bar{k}}$ becomes injective.
We apply the same reasoning to the $X \cap H_{l,t_i}$ and eventually the resulting hyperplane sections will be curves. This implies our proposition above, but we also get

Theorem 10.2.11. *For any abelian variety A/k we can find a finite number of curves $j_\nu : C_\nu \hookrightarrow A$ such that $A \longrightarrow \oplus_\nu J_{C_\nu}$ has a finite kernel.*

This theorem has been proved in [La2], Chapter VIII, §2, Corollary 2, using Chow's theory of the K/k-trace and K/k image. It can be used to reduce the theory of the Picard functor $\mathrm{Pic}_{A/k}$ or more generally $\mathrm{Pic}_{X/k}$, to the theory of Jacobians.
We mentioned already that Mumford constructs in his book [Mu1] a dual abelian variety $A^\vee/k$ for any abelian variety A/k. On the product he constructs a Poincaré bundle $\mathcal{N}$ such that the pair $(A \times_k A^\vee, \mathcal{N})$ provides a local representation of $\mathcal{PIC}^0_{A/k}$ and the two homomorphisms

$$x \mapsto \mathcal{N}|\{x\} \times_k A^\vee, \text{ resp. } x \mapsto A \times_k \{x\}$$

are isomorphisms from $A \xrightarrow{\sim} A^{\vee\vee} = A$, (resp.) $A^\vee \xrightarrow{\mathrm{Id}} A^\vee$.
This result we have proved for the special case of Jacobians, We have the bundle $\widetilde{\Theta}$ on $J \times J$ and the isomorphism $\phi_\Theta : J \longrightarrow J^\vee$. Then we put

$$\mathcal{N} = (\mathrm{Id} \times \Theta^{-1})^*(\widetilde{\Theta})$$

and this is the Poincaré bundle. The case of arbitrary abelian varieties can be reduced to the case of Jacobians if we apply theorem 10.2.11 and investigates the behavior of the dual abelian variety under isogenies.
In the introduction to his book Mumford adopts the point of view that he stubbornly avoids to use the crutch of "reduction to the Jacobian". He considers the abelian varieties is the basic objects of interest and Jacobians are just special abelian varieties.
Our point of view is the opposite and more classical one. We consider the category of abelian varieties over an algebraically closed field k up to isogeny, this means that isogenies become isomorphisms. An abelian variety A/k is called simple if it does not contain any non trivial subvariety. Then we have seen that we can embed A into a suitable Jacobian J such that we have (up to isogeny)

$$J = A \oplus B.$$

From this we can derive easily that any Jacobian and hence any abelian variety A is isomorphic to a direct sum of simple abelian varieties and this means the category of abelian varieties (up to isogeny) over k is *semi simple.* For $k = \mathbb{C}$ it is the complete reducibility theorem of Poincaré.
Unfortunately we missed to state and prove this theorem in Vol. I. But in this case it is not so difficult to prove: An abelian variety over $\mathbb{C}$ is a triplet $(\Gamma, < \ , \ >, I : \Gamma_\mathbb{R} \longrightarrow \mathbb{R})$ (see Vol. I, 5.2.1.) , for which the resulting hermitian form is positive definite (see Vol. I , 5.2.21). An abelian subvariety is given by a sub lattice $\Gamma_1 \subset \Gamma$ such that $\Gamma_1 \otimes \mathbb{R}$ is stable under I. But then the positivity implies that the alternating form $< \ , \ >$ restricted to Γ_1 is rationally non degenerate and we can consider its complement Γ_2 with respect to $< \ , \ >$. Clearly $\Gamma_2 \otimes \mathbb{R}$ is stable under I and hence it defines a complement up to isogeny.

With this in mind we can say that understanding abelian varieties means to understand Jacobians and their decomposition into simple pieces. To get an understanding of this decomposition another object comes into play, namely the ring of correspondences of a curve. For a given smooth, projective, absolutely irreducible curve C/k define a ring structure on $A^1(C \times C)$, where the Chow ring is defined by any of the equivalence relations. If we have two codimension one cycles $T_1,T_2 \subset C \times C$ we consider the cycles $\widetilde{T_1} = T_1 \times C \subset C \times C \times C, \widetilde{T_2} = C \times T_2 \subset C \subset C \times C \times C$ and take their intersection product

$$\widetilde{T_1} \cdot \widetilde{T_2} \in A^2(C \times C \times C).$$

To this element we apply the projection $p_2^\bullet : A^2(C \times C \times C) \longrightarrow C \times C$, which is induced by the projection to the first and third factor. Then we put

$$T_1 \circ T_2 = p_2^\bullet(\widetilde{T_1} \cdot \widetilde{T_2}),$$

and this defines the ring structure on $A^1(C \times C)$ and this is the ring of correspondences. The ring has an identity element, it is given by the diagonal. Of course this construction also applies to varieties of higher dimension.
In the next section we study the ring of endomorphisms $\mathrm{End}(J_{C/k})$ and we will establish a relationship to $A^1(C \times C)$. It will turn out that we have to look for idempotent elements $p \in \mathrm{End}(J)$. Each such idempotent p defines a decomposition

$$J = A \oplus B,$$

where p is the identity on A and zero on B.
Hence we see that the construction and understanding of other abelian varieties besides the Jacobians is intimately linked to the understanding of the ring of correspondences of curves.

10.3 The ring of endomorphisms $\mathrm{End}(J)$ and the ℓ-adic modules $T_\ell(J)$

Some heuristics and outlooks

We resume briefly: Let k be a field, let k_s be a separable closure and let $\bar{k} \supset k_s$ be an algebraic closure. We start from a smooth projective curve C/k. We constructed the schemes $\mathrm{Pic}^g_{C/k}, \mathrm{Pic}^0_{C/k}$ as projective schemes over k. We will use the notation $\mathrm{Pic}^0_{C/k} = J/k$. Sometimes we drop the field k in the notation.
We also constructed the schemes $\mathrm{Pic}^0_{X/k}, \mathrm{Pic}^0_{J/k}$ under the assumption that $C(k) \neq \emptyset$. But as before it is clear that both schemes have a canonical descend datum hence they are well defined as schemes over k. The abelian variety $\mathrm{Pic}^0_{J/k}$ is called the dual of J/k and will be denoted by $J^\vee/k$. The theta divisor- which as a divisor is only defined over some separable extension- defines a isomorphism $\phi_\Theta : J \longrightarrow J^\vee$. This isomorphism is indeed defined over k because it does not depend on the choice of $P_0 \in C(k_s)$ and two such choices are algebraically equivalent. We apply proposition 10.2.1
A key tool for understanding the structure of J is its ring of endomorphisms. Of course an endomorphism is a morphism $\phi : J \longrightarrow J$, which respects the group structure. The endomorphism form a ring, where the multiplication is given by composition.

This endomorphism ring End(J/k) may become larger if we perform a base change with an extension L/k especially we may define $\mathrm{End}(J) := \mathrm{End}(J \otimes \bar{k}/\bar{k})$. On this ring we have an action of the Galois group, we recover the endomorphisms defined over k if we take the Galois-invariant endomorphisms.
Any endomorphism $\varphi : J \to J$ induces an endomorphism of the Picard functor

$$\varphi^* : \mathrm{Pic}_{J/k} \longrightarrow \mathrm{Pic}_{J/k}\,,$$

for a line bundle $\mathcal{L}$ on $J \times T$ (i.e. essentially an element in $\mathrm{Pic}_{J/k}(T)$) we define

$$\varphi^*(\mathcal{L}) = (\varphi \times \mathrm{Id}_T)^*(\mathcal{L}).$$

This is of course a homomorphism with respect to the group structure on $\mathrm{Pic}_{J/k}$. We want to study the properties of the function $\varphi \to \varphi^*$.

Let us assume that we have a point $P_0 \in C(k)$. We restrict $\widetilde{\varphi}^*$ to the sub functor $\mathrm{Pic}^0_{J/k}$, since this sub functor is locally represented by $(J \times_k J,\widetilde{\Theta})$ this restriction is given by an endomorphism ${}^t\varphi : J^\vee \longrightarrow J^\vee$.
We will show that $\varphi \longrightarrow^t \varphi$ is an additive homomorphism i.e. we have

$${}^t(\varphi + \psi) =^t \varphi +^t \psi.$$

In the case of Riemann surfaces the ring End(J) could be considered as a subring of the endomorphisms of the first homology group (Vol. I.5.3.2) and from this we get easily insight into the structure of End(J). For instance it is clear that End(J) is a finitely generated torsion free algebra over $\mathcal{Z}$. But these homology groups are not available in the algebraic context. Hence we have to look at this object as it is and get our insights from elsewhere.
The endomorphism φ^* also induces an endomorphism

$$\bar{\varphi}^* : NS(J) \longrightarrow NS(J).$$

We also should have in mind that $NS(J)$ is related to the second cohomology. In I.5.2.1 we explained that $NS(J)$ can be identified to a subgroup of $\mathrm{Hom}(\Lambda^2\Gamma,\mathbb{Z})$. This had the consequence that $\varphi \longrightarrow \bar{\varphi}^*$ was quadratic. Hence we expect that also in the algebraic context $\varphi \to \varphi^*$ is a quadratic function. This means that we expect hat

$$(\varphi + \psi)^* - \varphi^* - \psi^* = \langle\varphi,\psi\rangle$$

where $\langle\varphi,\psi\rangle \in \mathrm{End}(\mathrm{Pic}_{J/k})$ and $(\varphi,\psi) \mapsto \langle\varphi,\psi\rangle$ is biadditive.

The study of End(J)

Since the following considerations are more geometric in nature we assume that k is algebraically closed. We want a formula for the endomorphism ${}^t\varphi$ of $J^\vee$. The key is of course the representability of $\mathrm{Pic}^0_{J/k}$. We have the divisor Θ on J and the bundle $\widetilde{\Theta}$ on $J \times J^\vee$. Then the universality and the definition of ϕ^* yields the defining formula for ${}^t\varphi$:

$$(\varphi \times \mathrm{Id})^*(\widetilde{\Theta}) \sim_{J^\vee} (\mathrm{Id} \times^t\varphi)^*(\widetilde{\Theta}) \tag{10.14}$$

On the geometric points this specializes to

$$\varphi^*(T_x^*(\Theta)\otimes\Theta^{-1}) \xrightarrow{\sim} T^*_{{}^t\varphi(x)}(\Theta)\otimes\Theta^{-1}.$$

Since φ^* is a homomorphism this gives us the additivity

$$^t(\varphi+\psi) = {}^t\varphi + {}^t\psi \tag{10.15}$$

To understand the properties of $\varphi \longrightarrow \varphi^*$ we need

Theorem 10.3.1. *Let $\varphi,\psi,\eta \in \mathrm{End}(J)$ and let $\mathcal{L}$ be a line bundle on J. Then the bundle*

$$\begin{aligned}(\varphi+\psi+\eta)^*(\mathcal{L})\otimes(\varphi+\psi)^*(\mathcal{L})^{-1}\otimes(\varphi+\eta)^*(\mathcal{L})^{-1}\otimes(\psi+\eta)(\mathcal{L})^{-1}\otimes\\ \otimes\varphi^*(\mathcal{L})\otimes\psi^*(\mathcal{L})\otimes\eta^*(\mathcal{L})\otimes(0^*\mathcal{L})^{-1}\end{aligned} \tag{10.16}$$

is trivial (here $0 : J \to J$ is the zero homomorphism).

This is again a consequence of the theorem of the cube. We consider the threefold product $J \times J \times J$, and we consider the following 8 homomorphisms from $J\times J\times J$ to J

$$\begin{aligned}&m_{123} : J\times J\times J \to J \qquad \text{sum of all components}\\ &m_{23}\circ p_1, m_{13}\circ p_2, m_{12}\circ p_3\\ &p_{12}, p_{13}, p_{23}\\ &\text{and}\\ &0\end{aligned} \tag{10.17}$$

We consider the bundle

$$\begin{aligned}\mathcal{N}_0 = &m_{123}^*(\mathcal{L})\otimes(m_{23}\circ p_1)^*(\mathcal{L})^{-1}\otimes(m_{12}\otimes p_3)^*(\mathcal{L})^{-1}\otimes(m_{13}\circ p_2)^*(\mathcal{L})^{-1}\otimes\\ &\otimes p_{12}^*(\mathcal{L})\otimes p_{13}^*(\mathcal{L})\otimes p_{23}^*(\mathcal{L})\otimes(0^*\mathcal{L})^{-1}\end{aligned} \tag{10.18}$$

If we restrict this bundle to one of the subvarieties

$$e\times J\times J \quad , \quad J\times e\times J \quad , \quad J\times J\times e \quad ,$$

then in any case two of the 8 maps become equal and occur with opposite signs in the product. Hence the restriction becomes trivial. Then the theorem of the cube tells us that the bundle $\mathcal{N}_0$ is trivial.
To get our theorem we use φ,ψ,η to map

$$(\varphi,\psi,\eta) : J \longrightarrow J\times J\times J,$$

and the bundle in question is the pullback of $\mathcal{N}_0$, hence trivial. □

For any pair of endomorphisms $\varphi,\psi \in \mathrm{End}(J)$ we defined $(\varphi+\psi)^* - \varphi^* - \psi^* = \langle\varphi,\psi\rangle$ where now $\langle\varphi,\psi\rangle \in \mathrm{End}(\mathrm{Pic}_{J/k})$.

Perhaps this is a good place to summarize the properties of $\varphi \longrightarrow \varphi^*$.

Theorem 10.3.2. *i) If we have two endomorphism* φ,ψ *then*

$$(\varphi \circ \psi)^* = \psi^* \circ \varphi^*.$$

ii) The pairing

$$\begin{array}{ccc} \mathrm{End}(J) \times \mathrm{End}(J) & \longrightarrow & \mathrm{End}(\mathrm{Pic}(J)) \\ (\varphi,\psi) & \longrightarrow & \langle \varphi,\psi \rangle \end{array}$$

is biadditive in both variables.
iii)

$$^t(\varphi + \psi) =^t \varphi +^t \psi$$

iv) The endomorphism $\langle \varphi,\psi\rangle \in \mathrm{End}(\mathrm{Pic}_{J/k})$ *is trivial on* $\mathrm{Pic}^0_{J/k}$ *and hence it is an endomorphism of* $NS(J)$.

The assertion i) is obvious, iii) is the additivity of $\varphi \longrightarrow^t \varphi$. The assertion iv) is an obvious consequence of iii), It remains to prove ii). Let us replace φ by $\varphi_1 + \varphi_2$ and apply our formula above to the sum $\varphi_1 + \varphi_2 + \psi$. Then we see

$$\begin{gathered} (\varphi_1 + \varphi_2 + \psi)^*(\mathcal{L}) \otimes (\varphi_1 + \varphi_2)^*(\mathcal{L})^{-1} \otimes \psi^*(\mathcal{L})^{-1} = \\ (\varphi_1 + \psi)^*(\mathcal{L}) \otimes \varphi_1^*(\mathcal{L})^{-1} \otimes \psi(\mathcal{L})^{-1} \otimes (\varphi_2 + \psi)^*(\mathcal{L}) \otimes \varphi_2^*(\mathcal{L})^{-1} \otimes \psi(\mathcal{L})^{-1} \end{gathered} \tag{10.19}$$

the term on the left hand side is

$$\langle \varphi_1 + \varphi_2, \psi \rangle(\mathcal{L}),$$

and on the right hand side we get

$$\langle \varphi_1,\psi\rangle(\mathcal{L}) \otimes \langle \varphi_2,\psi\rangle(\mathcal{L}).$$

Since the expression $\langle \varphi,\psi\rangle$ is symmetric, the rest is clear. □

We have the isomorphism $\phi_\Theta : J \longrightarrow J^\vee$ and use it to identify these two abelian varieties. Hence we can interpret $^t\varphi$ also as an element in End(J). The map $\varphi \longrightarrow^t \varphi$ from End(J) into itself is called the **Rosati involution.** This notation is a little bit problematic, because the involution depends on the choice of the line bundle Θ,. In our situation this bundle is a "very canonical" choice of an ample bundle on J, this means for abelian varieties, which are given as the Jacobian of a curve, the Rosati involution is "canonical".

For arbitrary abelian varieties A/k we do not have a canonical choice of a class (modulo algebraic equivalence) of ample line bundles. We always find ample line bundles $\mathcal{L}$ on A, Such an ample bundle defines a homomorphism $\phi_\mathcal{L} : A \longrightarrow \mathcal{PIC}^0_{A/k,e} = A^\vee$. We showed that $\phi_\mathcal{L}$ has a finite kernel (See prop. 10.2.8). Again we have the homomorphism $\varphi \longrightarrow^t \varphi$ from End(A) to End($A^\vee$). Since $\phi_\mathcal{L}$ has a finite kernel we can find a a non zero integer n and a $\psi' : A^\vee \longrightarrow \mathcal{A}$ such that $\phi_\mathcal{L} \circ \psi' = n\,\mathrm{Id}$. Therefore, we can define an inverse $\varphi'_\mathcal{L} = (1/n)\psi'$ to $\phi_\mathcal{L}$ and get a Rosati involution

$$\varphi \mapsto (\phi_\mathcal{L})^{-1} \circ^t \varphi \circ \varphi'_\mathcal{L}$$

on $\mathrm{End}_\mathbb{Q}(A)$.

It can be shown-and we will show this in our special case of jacobians - that $\mathrm{End}(A)$ is a finitely generated and torsion free $\mathbb{Z}$-module and then $\mathrm{End}_{\mathbb{Q}}(A) = \mathrm{End}(A) \otimes \mathbb{Q}$ is a finite dimensional $\mathbb{Q}$- algebra.
The choice of an ample bundle $\mathcal{L}$ and the resulting morphism $\phi_{\mathcal{L}}$ is called a **polarization** of A. Such a polarization is called **principal polarization** if $\phi_{\mathcal{L}}$ is an isomorphism. Not all abelian varieties admit a principal polarization, but Jacobians of curves do.
Mutatis mutandis the results, which will be proved in the next section for the Jacobians, will be true for arbitrary abelian varieties.

The degree and the trace

For an endomorphism $\varphi : J \longrightarrow J$ we can define the kernel $\ker(\varphi) = \varphi^{-1}(0)$. This is a subgroup scheme. If φ is an isogeny then it is a finite group scheme over k and in this case we define the degree of φ as

$$\deg(\varphi) = \mathrm{Rank}(\varphi^{-1}(0))$$

If the kernel is not finite then we put $\deg(\varphi) = 0$. An endomorphism $\varphi : J \longrightarrow J$ with finite kernel is an isogeny of J. The degree is multiplicative, i.e. $\deg(\phi \circ \psi) = \deg(\phi)\deg(\psi)$. The following is quite clear:

If $\varphi : J \longrightarrow J$ is an isogeny then the morphism φ is finite and locally free of rank $\deg(\varphi)$. For any point $y \in J$ the fibre $\varphi^{-1}(y)$ is finte of rank $\deg(\varphi)$ over $k(y)$.

We apply our formula for the degree: Let $\mathcal{L}$ be an ample bundle on J. In section 8.4.2 we defined the g-fold intersection number $\mathcal{L}^g = \mathcal{L} \cdot \mathcal{L} \cdot \dots \cdot \mathcal{L}$. We have the formula

$$\deg(\varphi)\mathcal{L}^g = (\varphi^*(\mathcal{L}))^g \tag{10.20}$$

(To see this we can replace $\mathcal{L}$ by a power $\mathcal{L}^{\otimes n}$ so that it becomes very ample. Then we can write $\mathcal{L}^{\otimes n} = \mathcal{O}_J(D_i)$ where $D_1, \dots, D_g$ are divisors, which intersect transversally in $n^g\mathcal{L}^g$ points. Then the divisors $\varphi^{-1}(D_i)$ intersect in $\deg(\varphi)n^g\mathcal{L}^g$ points and this is the formula.)

We compute $(n\,\mathrm{Id})^*$. It is clear how to do this, we have

$$(n\,\mathrm{Id})^* = ((n-1)\,\mathrm{Id} + \mathrm{Id})^* = ((n-1)\,\mathrm{Id})^* + \mathrm{Id}^* + (n-1)\langle \mathrm{Id}, \mathrm{Id}\rangle$$

or since $\mathrm{Id}^* = \mathrm{Id}_{\mathrm{Pic}(J)} = \mathrm{Id}$

$$(n\,\mathrm{Id})^* - ((n-1)\,\mathrm{Id})^* = \mathrm{Id} + (n-1)\langle \mathrm{Id}, \mathrm{Id}\rangle.$$

Since $(\mathcal{O} \cdot \mathrm{Id})^* = 0$ we get

$$(n\,\mathrm{Id})^* = n\,\mathrm{Id} + \frac{n(n-1)}{2}\langle \mathrm{Id}, \mathrm{Id}\rangle. \tag{10.21}$$

We can also say something about $\langle \mathrm{Id}, \mathrm{Id}\rangle$. We have

$$\mathrm{Id} + (-\mathrm{Id}) = 0$$

as endomorphism on J. Then

$$0 = (\mathrm{Id} + (-\mathrm{Id}))^* = \mathrm{Id} + (-\mathrm{Id})^* - \langle \mathrm{Id}, \mathrm{Id}\rangle,$$

and hence we see

$$\langle \mathrm{Id}, \mathrm{Id}\rangle = \mathrm{Id}^* + (-\mathrm{Id})^*. \tag{10.22}$$

We want to evaluate this formula on a line bundle $\mathcal{L}$. We put $\mathcal{L}_- = (-\mathrm{Id})^*(\mathcal{L})$ and our formula yields

$$(n\,\mathrm{Id})^*(\mathcal{L}) \xrightarrow{\sim} \mathcal{L}^{\otimes n} \otimes \mathcal{L}^{\otimes \frac{n(n-1)}{2}} \otimes \mathcal{L}_-^{\otimes \frac{n(n-1)}{2}} \tag{10.23}$$

This allows us to compute the degree of $n\,\mathrm{Id}$. As we explained. we may take any ample line bundle $\mathcal{L}$ on J, we may even assume it to be very ample. We may replace $\mathcal{L}$ by $\mathcal{L} \otimes (-\mathrm{Id})^*(\mathcal{L})$, since $-\mathrm{Id}$ is an automorphism, it is clear that $(-\mathrm{Id})^*(\mathcal{L})$ is very ample and hence the tensor product is so too. Then we have the formula

$$\mathcal{L}^g \deg(n\,\mathrm{Id}) = (n\,\mathrm{Id})^*(\mathcal{L})^g$$

and $(n\,\mathrm{Id})^*(\mathcal{L}) = \mathcal{L}^n \otimes \mathcal{L}^{\frac{n^2-n}{2}} \otimes \mathcal{L}^{\frac{n^2-n}{2}}$ (because $(-\mathrm{Id})^*(\mathcal{L}) = \mathcal{L}$) and hence

$$(n\,\mathrm{Id})^*(\mathcal{L}) = \mathcal{L}^{\otimes n^2}.$$

And now $(\mathcal{L}^{\otimes n^2})^g = \mathcal{L}^g \cdot n^{2g}$ and it follows

$$\deg(n\,\mathrm{Id}) = n^{2g}. \tag{10.24}$$

Since we have seen that the group law $J \times J \to J$ induces the addition on the tangent space $T_{J,e}$ (see 7.5.6) we conclude that the multiplication by n on the tangent space. We conclude:

Theorem 10.3.3. *The kernel of the multiplication by n is a finite group scheme*

$$J[n] \longrightarrow \mathrm{Spec}(k)$$

of rank n^{2g}. If the characteristic p of k does not divide n, then this group scheme is étale. In this case

$$J[n](\overline{k}) \simeq (\mathbb{Z}/n\mathbb{Z})^{2g}.$$

We consider the function

$$n \longrightarrow \deg(\psi + n\,\mathrm{Id}),$$

we know how to express it in terms of intersection numbers. We choose an ample line bundle $\mathcal{L}$ on J and then we have

$$\deg(\psi + n\,\mathrm{Id}) \cdot \mathcal{L}^g = ((\psi + n\,\mathrm{Id})^*(\mathcal{L}))^g.$$

We can expand the right hand side and find

$$\deg(\psi + n\,\mathrm{Id}) \cdot \mathcal{L}^g = \cdots n^{2g-1}\mathcal{L}^{g-1} \cdot \langle \psi, \mathrm{Id}\rangle(\mathcal{L}) + n^{2g} \cdot \mathcal{L}^g.$$

This expression looks pretty much like a characteristic polynomial of an endomorphism and in the next section we will see that this is indeed the case. In view of this expectation we define

$$\mathrm{tr}(\psi) = \frac{\mathcal{L}^{g-1} \cdot \langle \psi, \mathrm{Id} \rangle (\mathcal{L})}{\mathcal{L}^g}.$$

In any case it is clear that $\deg(\psi + n\,\mathrm{Id})$ is a polynomial in n of degree $2g$ with rational coefficients and this polynomial takes integer values. If we have an ample bundle $\mathcal{L}$ with $\mathcal{L} = \mathcal{L}_-$ and an endomorphism $\phi = \sum_i^r n_i \varphi_i$ then it follows from 10.23

$$(n_1\varphi_1 + \cdots + n_r\varphi_r)^*(\mathcal{L}) = \prod_i^r \varphi_i^*(\mathcal{L})^{\otimes n_i^2} \bigotimes \otimes_{i<j} < \varphi_i, \varphi_j > (\mathcal{L})^{\otimes n_i n_j}$$

and if we now apply 10.20 and take the g-fold self intersection then we get that

$$\deg(n_1\varphi_1 + \cdots + n_r\varphi_r)$$

is a homogenous polynomial with integer coefficients in the n_i of degree $2g$ (10.25)

If we have two endomorphisms φ,ψ then we can look at the diagram

$$J \xrightarrow{\varphi \times \psi \circ \Delta} J \times J \begin{array}{c} \xrightarrow{p_1} \\ \xrightarrow{m} \\ \xrightarrow{p_2} \end{array} J$$

and for any line bundle $\mathcal{L}$ on J we have the formula

$$< \psi, \varphi > (\mathcal{L}) \xrightarrow{\sim} ((\varphi \times \psi) \circ \Delta))^*(m^*(\mathcal{L}) \otimes p_1^*(\mathcal{L})^{-1} \otimes p_2^*(\mathcal{L})^{-1})$$

by definition. If we apply this to Θ we get

$$< \psi, \varphi > (\Theta) \xrightarrow{\sim} ((\varphi \times \psi) \circ \Delta))^*(\widetilde{\Theta}) = \Delta^* \circ (\varphi \times \psi)^*(\widetilde{\Theta})$$

We have the defining relation for the transpose (see 10.14) and get

$$< \psi, \varphi > (\Theta) \xrightarrow{\sim} \Delta^* \circ ({}^t\psi\varphi \times \mathrm{Id})^*(\widetilde{\Theta}) \xrightarrow{\sim} \Delta^* \circ (\mathrm{Id} \times {}^t\varphi\psi)^*(\widetilde{\Theta}) \xrightarrow{\sim} <^t \psi\varphi, \mathrm{Id} > (\Theta) \xrightarrow{\sim} <^t \varphi\psi, \mathrm{Id} > (\Theta)$$

If we take $\psi = \mathrm{Id}$ then we get

$$\mathrm{tr}({}^t\varphi) = \mathrm{tr}(\varphi).$$

Now we are ready for the famous

Theorem 10.3.4. Positivity of the Rosati involution
The bilinear form

$$(\varphi, \psi) \mapsto \mathrm{tr}({}^t\varphi\psi)$$

is a symmetric positive definite form.

We have to prove the positivity. We have seen

$$< \psi, \psi > (\Theta) \xrightarrow{\sim} <^t \psi\psi, \mathrm{Id} > (\Theta)).$$

We multiply by Θ^{g-1} and observing the definition of the trace we get

$$\Theta^{g-1} < \psi, \psi > (\Theta) = \Theta^{g-1} <^t \psi\psi, \mathrm{Id} > (\Theta) = \mathrm{tr}({}^t\psi\psi)\Theta^g$$

Our formula for $<\mathrm{Id}\,,\mathrm{Id}>$ implies $<\psi,\psi>(\Theta) \xrightarrow{\sim} \psi^*(\Theta)\otimes\psi^*(-\mathrm{Id}(\Theta))$ and hence we find

$$\Theta^g\,\mathrm{tr}({}^t\psi\psi) = \Theta^{g-1}(\psi^*(\Theta)\otimes\psi^*(-\mathrm{Id}(\Theta))$$

Since Θ is ample we have $\Theta^g > 0$. We claim that also the intersection number on the right is strictly positive if $\psi \neq 0$.(Obviously this implies the theorem). To see this last point we consider the image $A = \varphi(J)$, this is an abelian subvariety of strictly positive dimension. We may assume that $A \not\subset \Theta$ or $A \not\subset (-\mathrm{Id})(\Theta)$ (otherwise we replace Θ by a suitable translate $\Theta + x$) Then $A\cap\Theta, A\cap(-\mathrm{Id})(\Theta)$ are of codimension 1 in A and hence $\varphi^{-1}(\Theta),\varphi^{-1}(\Theta_-)$ are subschemes of codimension 1 in J, from this we get a non zero divisor $\sum n_Y Y$ (see 9.4) where the coefficients at the components are strictly positive. The Θ restricted to any of the components is ample and therefore, $\Theta^{g-1}\cdot Y > 0$. □

The ℓ-adic modules

Now we can pick a prime ℓ, which is different from the characteristic of k, and we define

$$T_\ell(J) = \varprojlim_{\alpha} J[\ell^\alpha]$$

as before.

The group of geometric points is

$$T_\ell(J)(\bar{k}) \xrightarrow{\sim} \mathbb{Z}_\ell^{2g},$$

but we have to observe that the Galois group acts upon this module.

This is now the replacement for the cohomology groups, which were available in the transcendental case. But these ℓ-adic cohomology groups have the defect that they are $\mathbb{Z}_\ell$-modules and not $\mathbb{Z}$-modules. (It has been pointed out by J.-P. Serre that cohomology groups, which are free $\mathbb{Z}$-modules of rank $2g$ cannot exist.)

Of course it is clear that we get a homomorphism

$$\mathrm{End}(J) \longrightarrow \mathrm{End}(T_\ell(J)),$$

and hence we can define a trace $\mathrm{tr}_\ell(\varphi) = \mathrm{tr}(\varphi_\ell \mid T_\ell(J))$ and a determinant $\det_\ell(\varphi) = \det(\varphi_\ell \mid T_\ell)$. A priori these numbers are ℓ-adic numbers it is not clear how they depend on ℓ. But we have

Theorem 10.3.5. *The number* $\mathrm{tr}(\varphi_\ell \mid T_\ell)$ *and* $\det(\varphi_\ell \mid T_\ell)$ *are integers, which do not depend on* ℓ*. More precisely we have*

$$\det_\ell(\varphi) = \deg(\varphi), \mathrm{tr}_\ell(\varphi) = \mathrm{tr}(\varphi)$$

Looking at $\det(\varphi + n\,\mathrm{Id})$ it becomes clear that it suffices to prove the first assertion. Before we can prove this theorem we have to prove two more theorems:

Theorem 10.3.6. *The* $\mathbb{Z}$*-module* $\mathrm{End}(J)$ *is finitely generated and for any prime* ℓ *the natural map*

$$\mathrm{End}(J) \otimes \mathbb{Z}_\ell \longrightarrow \mathrm{End}(T_\ell(J))$$

is an inclusion.

Proof: Let $M \subset \mathrm{End}(J)$ be any finitely generated submodule, which is stable under the involution. The trace defines an integer valued pairing

$$\begin{array}{lcl} \mathrm{tr} & : & M \times M \longrightarrow \mathbb{Z} \\ \mathrm{tr} & : & \langle \varphi_1, \varphi_2 \rangle \longrightarrow \mathrm{tr}(\varphi_1 \, {}^t\varphi_2)\text{'} \end{array}$$

which is positive definite. We see that this bilinear pairing is non degenerate over $\mathbb{Q}$, i.e. if we take a basis $\varphi_1 \cdots \varphi_k$ of the $\mathbb{Z}$-module M, we can conclude that

$$\det(\mathrm{tr}(\varphi_i \, {}^t\varphi_j))$$

is a non zero integer.

Let us assume that $M \otimes \mathbb{Z}_\ell$ does not embed into $\mathrm{End}(T_\ell(J))$, this means that we can find $\alpha_1, \cdots, \alpha_k \in \mathbb{Z}_\ell$, which are not all congruent zero $\mod \ell$ such that the linear combination $\sum_{i=1}^k \alpha_i \varphi_i$ is zero. If we approximate the α_i by integers n_i such that $\alpha_i \equiv n_i \mod \ell^\alpha$, then the element

$$\psi = \sum n_i \varphi_i \in \mathrm{End}(J)$$

is zero on the group of ℓ^α division points. this implies that $\psi = \ell^\alpha \psi'$ where $\psi' \in \mathrm{End}(J)$. (We have the diagram

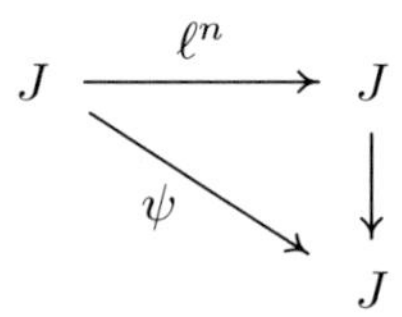

and since ψ is zero on the kernel of ℓ^α, we can see easily that we can find a $\psi' : J \longrightarrow J$ completing the diagram.)
We get a system of linear equations for the n_i:

$$\sum n_i \ \mathrm{tr}(\varphi_i \, {}^t\varphi_j) = \ell^\alpha \cdot \mathrm{tr}(\psi' \, {}^t\varphi_j)$$

where the $\mathrm{tr}(\psi' \, {}^t\varphi_j)$ are integers. We solve this system for the n_i using Cramer's rule and find

$$n_i = \ell^\alpha \ \frac{A_i}{\det(\mathrm{tr}(\varphi_i \, {}^t\varphi_j))}$$

where the A_i are integers. The maximal power of ℓ dividing the denominator does not depend on α. Hence we get a contradiction to the assumption that not all of the n_i are divisible by ℓ. For this we conclude that $\mathrm{End}(J)$ is a finitely generated $\mathbb{Z}$-module.

The homomorphism $\mathrm{End}(J)\otimes\mathbb{Z}_\ell \to \mathrm{End}(T_\ell(J))$ is called the ℓ-adic representation of the endomorphism ring.
We form the $\mathbb{Q}$-algebra $\mathrm{End}(J)\otimes_{\mathbb{Z}}\mathbb{Q} = \mathrm{End}(J)_{\mathbb{Q}}$, it is finite dimensional. If $\phi : J \longrightarrow J$ is an isogeny then we can find a non zero integer n such that $\ker(\phi) \subset \ker(n\,\mathrm{Id})$. This implies we can find another isogeny $\psi : J \longrightarrow J$ such that $\psi\circ\phi = n\,\mathrm{Id}$ and this tells that $\psi\otimes\frac{1}{n} \in \mathrm{End}(J)_{\mathbb{Q}}$ is the inverse of ϕ. This implies that the invertible elements in $\mathrm{End}(J)_{\mathbb{Q}}$ are the elements of the form $\psi\otimes\frac{1}{n}$ where ψ is an isogeny. The map $\phi \longrightarrow \deg(\phi)$ extends to a multiplicative function on $\mathrm{End}(J)_{\mathbb{Q}}$. An element $\phi\in\mathrm{End}(J)_{\mathbb{Q}}$ is invertible iff $\deg(\phi)\neq 0$. We need a second theorem.

Theorem 10.3.7. *The algebra* $\mathrm{End}(J)\otimes_{\mathbb{Z}}\mathbb{Q} = \mathrm{End}(J)_{\mathbb{Q}}$ *is a finite dimensional semi-simple* $\mathbb{Q}$*-algebra.*

A finite dimensional algebra A over a field k is called semi simple, if its radical is trivial. The radical is the maximal two sided ideal consisting of nilpotent elements.
To see this, we consider the radical $\mathfrak{a}$, which is the maximal two-sided ideal consisting of nilpotent elements. Clearly, $\mathfrak{a}$ is stable under the Rosati involution, and hence we see: If $\mathfrak{a}\neq 0$, then we find non zero $\varphi\in\mathfrak{a}$ with ${}^t\varphi = \pm\varphi$. But then $\varphi^2\neq 0$ since $\mathrm{tr}(\varphi^2) = \pm\,\mathrm{tr}(\varphi\,{}^t\varphi)$, and this is non zero by the positivity of the Rosati involution. But then φ cannot be nilpotent, we have a contradiction to $\mathfrak{a}\neq 0$. $\square$

Now we are ready for the proof of Thm. 10.3.5. The structure theory of semi-simple algebras advises us to consider the center Z_J of $\mathrm{End}(J)_{\mathbb{Q}}$. It is a finite dimensional commutative $\mathbb{Q}$ algebra and has no nilpotent elements $\neq 0$. We have seen in the Chapter on commutative algebra that there is a maximal set of orthogonal idempotent elements $e_1\cdots e_r\in Z_J$, such that

$$\begin{aligned} 1 &= \sum e_i \\ e_ie_j &= \begin{cases} e_i & if\ i=j\,. \\ 0 & else \end{cases} \end{aligned}$$

Then $Z_J = \oplus Z_Je_i$ is a direct sum of fields and

$$\mathrm{End}(J)_{\mathbb{Q}} = \bigoplus \mathrm{End}(J)_{\mathbb{Q}}e_i = \bigoplus_{i=1}^{r} A_i \tag{10.26}$$

where the A_i are central simple algebras with center Z_Je_i. The Rosati involution has to fix these e_i because of the positivity it can not send one of the idempotents into another one. Hence ${}^te_i = e_i$ for all i.
The e_i are not endomorphisms of J but, if we multiply them by an integer $d_i\neq 0$, then

$$E_i = d_i\; e_i$$

will be an endomorphism of J. Now we can consider the endomorphisms

$$\Phi_i = \sum_{j\neq i} E_j$$

and let $\widetilde{J}_i$ be the connected component of the kernel of Φ_i. Then we have of course that $\widetilde{J}_i$ is an abelian subvariety of J and it is clear that the map

$$\prod \widetilde{J}_i \longrightarrow J$$

is surjective with finite kernel, in other words, this map is an isogeny. The algebra $A_{i,\mathbb{Z}} = A_i \cap \mathrm{End}(J)$ acts trivially on the $\widetilde{J}_j$ with $j \neq i$ and it injects into $\mathrm{End}(\widetilde{J}_i)$ and is clearly of finite index in this ring of endomorphisms. It is clear that the Rosati involution is the identity on the e_i and hence it induces involutions on the A_i and on $\mathrm{End}(\widetilde{J}_i)$. Recall that we still want to prove the equality $\deg(\varphi) = \det(\varphi_\ell)$.
In the decomposition 10.26 we have $\mathrm{End}(\widetilde{J}_i)_{\mathbb{Q}} = A_i$. In other words, for any $\varphi \in \mathrm{End}(J)_{\mathbb{Q}}$ we find a non zero integer m such that

$$m\varphi = \varphi_1 + \cdots + \varphi_r$$

where $\varphi_i \in \mathrm{End}(\widetilde{J}_i)$. This means that we have a diagram

$$\begin{array}{ccc} \prod \widetilde{J}_i & \longrightarrow & J \\ \downarrow \underline{\Phi} & & \downarrow m\varphi \\ \prod \widetilde{J}_i & \longrightarrow & J \end{array}$$

where $\underline{\Phi} = (\cdots \varphi_i \cdots)$ and $\varphi_i \in \mathrm{End}(\widetilde{J}_i)$. This implies that $\deg(\Phi) = \prod \deg(\varphi_i)$. The same relation holds for the $\det(\Phi|T_\ell)) = \prod \det(\varphi_i|T_\ell(J_i)))$. Hence it is sufficient to prove our equality for the $\widetilde{J}_i$.
We have reduced the proof of theorem 10.3.6 to the case where $\mathrm{End}_{\mathbb{Q}}(J)$ is a central simple algebra over its center $F/\mathbb{Q}$. It is a well known theorem that such a central simple algebra is of the form $M_n(D)$, where D/F is a division algebra of dimension d_0^2 over F and $M_n(D)$ is the algebra of (n,n)-matrices with entries in D. Then $\dim_F \mathrm{End}_{\mathbb{Q}}(J) = n^2 d_0^2$. If we choose an embedding $\sigma : F \hookrightarrow \bar{\mathbb{Q}}$ then $\mathrm{End}_{\mathbb{Q}}(J) \otimes_{F,\sigma} \bar{\mathbb{Q}} = M_{nd_0}(\bar{\mathbb{Q}})$. We put $d = nd_0$. From this it follows that

$$\mathrm{End}_{\mathbb{Q}}(J) \times_{\mathbb{Q}} \bar{\mathbb{Q}} = \bigoplus_{\sigma : F \longrightarrow \bar{\mathbb{Q}}} M_d(\bar{\mathbb{Q}})$$

where σ runs over the set of embeddings of F into $\bar{\mathbb{Q}}$, where $d^2[F : \mathbb{Q}] = \dim_{\mathbb{Q}} \mathrm{End}_{\mathbb{Q}}(J)$. It is clear that the group of invertible elements $\mathrm{End}_{\mathbb{Q}}(J)^\times$ is the group of $\mathbb{Q}$ valued points of an algebraic group $G_J/\mathbb{Q}$ whose $\mathbb{Q}$ rational points are $M_n(D)^\times$, and for which

$$G_J \times_{\mathbb{Q}} \bar{\mathbb{Q}} = \prod_{\sigma : F \longrightarrow \bar{\mathbb{Q}}} \mathrm{GL}_d/\bar{\mathbb{Q}}.$$

Let π_σ be the projection from $G_J \times_{\mathbb{Q}} \bar{\mathbb{Q}}$ to its σ-component. The homomorphism $\deg : G_J(\mathbb{Q}) \longrightarrow \mathbb{Q}^\times$ is actually the evaluation of a group scheme homomorphism $\gamma_{\deg} : G_J/\mathbb{Q} \longrightarrow \mathbb{G}_m/\mathbb{Q}$, a so called rational character on $G_J/\mathbb{Q}$. (This is clear from 10.25). We have the rational characters

$$\det \circ \pi_\sigma = \det_\sigma : \prod_{\sigma : F \longrightarrow \bar{\mathbb{Q}}} \mathrm{GL}_d/\bar{\mathbb{Q}} \longrightarrow \mathbb{G}_m.$$

An arbitrary rational character $\gamma : G_J \times \bar{\mathbb{Q}} \longrightarrow \mathbb{G}_m$ is of the form

$$\gamma = \prod_{\sigma:F\longrightarrow\bar{\mathbb{Q}}} \det_\sigma^{n_\sigma}, \text{ where } n_\sigma \in \mathbb{Z}.$$

Such a rational character is defined over $\mathbb{Q}$ if and only if $n_\sigma = n_\tau$ for all σ,τ, i.e $n_\sigma = n$. Hence we can conclude that $\gamma_{\deg}$ must be a power of $\prod_{\sigma:F\longrightarrow\bar{\mathbb{Q}}} \det_\sigma$. The character $\det_\sigma$ has degree d hence the product over all σ has degree $d[F:\mathbb{Q}]$. Therefore we get that $d[F:\mathbb{Q}]$ divides $2g$ and with $r = 2g/(d[F:\mathbb{Q}])$ we get

$$\gamma_{\deg} = (\prod_{\sigma:F\longrightarrow\bar{\mathbb{Q}}} \det_\sigma)^r.$$

We apply the same consideration to

$$\det_\ell : (\mathrm{End}(J)\otimes\mathbb{Q}_\ell)^* \longrightarrow \mathbb{Q}_\ell^*.$$

This is not directly possible because $F\otimes\mathbb{Q}_\ell$ will not be a field in general. We need a little bit of number theory. We have a decomposition

$$F\otimes\mathbb{Q}_\ell = \bigoplus_{\mathfrak{l}|\ell} F_\mathfrak{l}$$

and hence we get a decomposition

$$G_J\otimes_\mathbb{Q}\mathbb{Q}_\ell = \prod_{\mathfrak{l}|\ell} G_{J,\ell}$$

where $G_{J,\ell}/\mathbb{Q}_\ell$ and its $\mathbb{Q}_\ell$ rational points are $M_n(D\otimes_F F_\mathfrak{l})$.
Now we can apply our above considerations to the individual factors, we choose an algebraic closure $\mathbb{Q}_\ell \hookrightarrow \bar{\mathbb{Q}}_\ell$ and find that

$$\det(\phi_\ell) = \prod_{\mathfrak{l}|\ell}(\prod_{\sigma:F_\mathfrak{l}\longrightarrow\bar{\mathbb{Q}}_\ell} \det_\sigma)^{r_\mathfrak{l}}$$

where we (only) know that $(\sum_{\mathfrak{l}|\ell}[F_\mathfrak{l}:\mathbb{Q}_\ell]r_\mathfrak{l})d = 2g$. Since we have $[F:\mathbb{Q}] = \sum_{\mathfrak{l}|\ell}[F_\mathfrak{l}:\mathbb{Q}_\ell]$ we have to show that $r_\mathfrak{l} = r$ for all $\mathfrak{l}|\ell$. We can find an embedding $\bar{\mathbb{Q}} \hookrightarrow \bar{\mathbb{Q}}_\ell$. The valuation $|\ |_\ell$ extends to a unique valuation also denoted by $|\ |_\ell$ on $\bar{\mathbb{Q}}_\ell$ then the set $\{\sigma : F \hookrightarrow \bar{\mathbb{Q}}\}$ gets divided into the subsets $\{\sigma : F_\mathfrak{l} \hookrightarrow \bar{\mathbb{Q}}_\ell\}$, these are the embeddings σ, which induce a given $\mathfrak{l}$ on F.
So far we have not yet used the definition of the function $\phi \longrightarrow \deg(\phi)$. If ϕ is not an isogeny then we have clearly $\deg(\phi) = \det_\ell(\phi) = 0$. Hence it suffices to consider the case that ϕ is an isogeny. In this case $\deg(\phi) = \mathrm{Rank}(\phi^{-1}(0))$. The group scheme $\phi^{-1}(0)$ has its ℓ- Sylow subgroup $\phi^{-1}(0)[\ell]$, this is simply the étale group scheme, which is annihilated by high powers of ℓ. The order of $\phi^{-1}(0)[\ell][\bar{k}]$ is the power of ℓ dividing $\deg(\phi)$. This is also the order of the kernel $\phi : J[\ell^N] \longrightarrow J[\ell^N]$ provided N is sufficiently large. But this is also the order of the kokernel of this homomorphism and hence the order of the kokernel of $\phi : T_\ell(J) \longrightarrow T_\ell(J)$. This order is now equal to the power of ℓ dividing the determinant $\det_\ell(\phi)$. In terms of ℓ-adic valuations this says

$$|\deg(\phi)|_\ell = |\det(\phi_\ell)|_\ell.$$

We choose an element $a \in F$, which is integral at all places $\mathfrak{l}|\ell$, which is a uniformizing element at one of these places say $\mathfrak{l}_0$ and a unit at all the others.
Now we have

$$\deg(\phi)(a) = \prod_{\sigma:F \longrightarrow \bar{\mathbb{Q}}_\ell} \sigma(a)^{rd}, \ \det(\phi_\ell) = \prod_{\mathfrak{l}|\ell}\Big(\prod_{\sigma:F_\mathfrak{l} \longrightarrow \bar{\mathbb{Q}}_\ell} \sigma(a)^{dr_\mathfrak{l}}\Big)$$

If we now take absolute values with respect to the ℓ-adic valuation then the factors corresponding to $\mathfrak{l} \neq \mathfrak{l}_0$ contribute by the factor 1. For the σ corresponding to $\mathfrak{l}_0$ we have $|\sigma(a)|_\ell = \ell^h$ with $h > 0$ and hence

$$|\deg(\phi)(a)|_\ell = \prod_{\sigma:F \longrightarrow \bar{\mathbb{Q}}_\ell} |\sigma(a)^{rd}|_\ell = \ell^{h[F_{\mathfrak{l}_0}:\mathbb{Q}]dr}$$

$$|\det(\phi_\ell)(a)|_\ell = \prod_{\sigma:F_{\mathfrak{l}_0} \longrightarrow \bar{\mathbb{Q}}_\ell} |\sigma(a)|_\ell^{dr_{\mathfrak{l}_0}} = \ell^{h[F_{\mathfrak{l}_0}:\mathbb{Q}]dr_{\mathfrak{l}_0}}$$

Now the equality of the ℓ-adic absolute values show that $r_{\mathfrak{l}_0} = r$ and hence we have proved theorem 10.3.5

The Weil Pairing

We consider a Jacobian J/k. for a moment we drop the assumption that k is algebraically closed. For any integer $n > 0$, which is not divisible by the characteristic p of k field we consider the endomorphism $n\,\mathrm{Id}$ we want to denote it by $[n]$.

$$J[n] = \ker(J \xrightarrow{[n]} J).$$

We have seen that $[n]$ is an étale morphism, the kernel is a finite $\acute{e}$tale group scheme and

$$J[n](\bar{k}) = J[n](k_s) \simeq (\mathbb{Z}/n\mathbb{Z})^{2g}.$$

On this group of n-torsion points we have an action of the Galois group $\mathrm{Gal}(\bar{k}/k)$.
We can also say that $J[n]$ is the Galois group of the covering $J \overset{[n]}{\to} J$, this means that any $a \in J[n](k_s)$ induces a translation

$$\begin{array}{ccc} J & \xrightarrow{T_a} & J \\ & {\scriptstyle [n]}\searrow \quad \swarrow{\scriptstyle [n]} & \\ & J & \end{array} \tag{10.27}$$

and for any affine open subset $V \subset J$, the open set $(n\,\mathrm{Id})^{-1}(V) = V' \subset J$ is affine, and

$$\mathcal{O}_J(V) = \mathcal{O}_J(V')^{J[n]},$$

i.e. the algebra downstairs is the algebra of invariants under $J[n]$.

Our aim is the construction of a non degenerate, Galois invariant, alternating pairing

$$w_0 : J[n] \times J[n] \to \mu_n = \text{ group of } n \text{ 'th roots of unity.}$$

(This corresponds to the alternating pairing on Γ in Vol. I, 5.2.1.)

We recall the construction of the bundle $\mathcal{N}$ (see 10.9.) If we pick a point $\xi \in J[n]$, then this gives us a line bundle

$$\mathcal{L}_\xi = \mathcal{N} \mid J \times \xi$$

on J. This line bundle is algebraically equivalent to zero and satisfies $\mathcal{L}_\xi^{\otimes n} = \mathcal{O}_J$.

We take the pull back of this line bundle under

$$[n] : J \longrightarrow J$$

and clearly we have $[n]^*(\mathcal{L}_\xi) = \mathcal{L}_\xi^{\otimes n} = \mathcal{O}_J$.

Hence we see that $[n]^*(\mathcal{L}_\xi)$ is trivial and we conclude that $H^0(J,[n]^*(\mathcal{L}_\xi))$ is a one dimensional k-vector space.

Since $[n]^*(\mathcal{L}_\xi)$ is a pull back of a bundle on J under $[n]$, we see that we have an action of $J[n]$ (the Galois group of the covering) on $[n]^*(\mathcal{L}_\xi)$ and hence an action of $J[n]$ on $H^0(J,[n]^*(\mathcal{L}_\xi))$, which then defines a homomorphism

$$\chi_\xi : J[n] \longrightarrow \mu_n \subset k^*.$$

One thing is clear: This homomorphism is trivial if and only if $\mathcal{L}_\xi$ is trivial in other words if $\xi = 0$.

We put

$$w_0(\xi,\eta) = \chi_\xi(\eta)$$

for $(\xi,\eta) \in J[n] \times J[n]$. It is linear in η by definition. But if $\xi = \xi_1 + \xi_2$, then we have

$$\mathcal{L}_\xi = \mathcal{L}_{\xi_1} \otimes \mathcal{L}_{\xi_2}$$

and this provides a non zero bilinear map

$$H^0(J,[n]^*(\mathcal{L}_{\xi_1})) \times H^0(J,[n]^*(\mathcal{L}_{\xi_2})) \to H^0(J,[n]^*(\mathcal{L}_\xi)),$$

which commutes with the action of $J[n]$. Hence we see that

$$\chi_{\xi_1+\xi_2} = \chi_{\xi_1} \cdot \chi_{\xi_2}$$

and this shows that w_0 is bilinear.

We show that $w_0(\xi,\xi) = \chi_\xi(\xi) = 1$, once we have done this, then it is clear that the map is alternating and non degenerate.
We assume that our element ξ is of order n. We can define the quotient $J/ < \xi >$, i.e. divide J by the cyclic subgroup generated by ξ. We get a diagram

$$\begin{array}{ccc} J & \xrightarrow{[n]} & J \\ & {\scriptstyle q_\xi}\searrow \quad \nearrow{\scriptstyle p_\xi} & \\ & J/<\xi> & \end{array} \qquad (10.28)$$

The Galois group of the covering q_ξ is the cyclic group $<\xi>$. Hence we see that $\chi_\xi(\xi) = 1$ holds if and only if the pull back of the line bundle $\mathcal{L}_\xi$ by p_ξ is trivial on $J/<\xi>$. This means that we have to prove that ${}^t p_\xi(\mathcal{L}_\xi) = \mathcal{O}_{J/<\xi>}$.
Our line bundle can be described in terms of divisors. Our bundle Θ is of the form $\mathcal{O}_J(D)$. Then $\mathcal{L}_\xi \xrightarrow{\sim} \mathcal{O}(T_{-x}(D) - D)$ Now $(n\,\mathrm{Id})^*(\mathcal{L}_\xi)$ is trivial, hence $(n\,\mathrm{Id})^{-1}(T_{-x}(D)) - D)$ is trivial, hence a divisor of a function g. Our aim is to prove that this g is already a function on $J/<\xi>$, this means that it is invariant under the translations by ξ.
We solve $n\eta = \xi, \eta \in J(k)$, let $E = (n\,\mathrm{Id})^{-1}(D)$ then

$$\mathrm{Div}(g) = (T_\eta^{-1}(E) - E).$$

The divisor is invariant under translations by elements in $J[n](k)$ and therefore, $T_\xi(E) = E$. Clearly

$$\mathrm{Div}(T_{\nu\eta}^*(g)) = T_{(\nu+1)\eta}^{-1}(E) - T_{\nu\eta}^{-1}(E)$$

and hence

$$\mathrm{Div}(\prod_{\nu=0}^{\nu=n-1}(T_{\nu\eta}^*(g))) = \sum_\nu (T_{(\nu+1)\eta}^{-1}(E) - T_{\nu\eta}^{-1}(E) = 0$$

This tells us that

$$h(x) = \prod_{\nu=0}^{\nu=n-1}(g(x+\nu\eta))$$

is constant. Then

$$1 = \frac{h(x+\eta)}{h(x)} = \frac{\prod_{\nu=0}^{\nu=n-1} g(x+(\nu+1)\eta)}{\prod_{\nu=0}^{\nu=n-1} g(x+\nu\eta)} = \frac{g(x+\xi)}{g(x)}$$

The Neron-Severi groups $NS(J)$,$NS(J\times J)$ *and* $\mathrm{End}(J)$

We resume the assumption that k is algebraically closed. Let C/k be a smooth, projective, absolutely irreducible curve, and let J/k be its Jacobian. We want to study the Neron-Severi $NS(J) = \mathrm{Pic}_{J/k} / \mathrm{Pic}^0_{J/k}$, and relate it to the endomorphism ring $\mathrm{End}(J)$. For any line bundle $\mathcal{L}$ on J we defined the homomorphism $\phi_\mathcal{L} : J \longrightarrow J^\vee$. On the other hand we have the principal polarization given by the line bundle Θ. For this line bundle $\phi_\Theta : J \longrightarrow J^\vee$ is an isomorphism. If we take the composition we get an element $(\phi_\Theta)^{-1} \circ \phi_\mathcal{L} \in \mathrm{End}(J)$, this yields a homomorphism

$$\Phi : NS(J) \longrightarrow \mathrm{End}(J).$$

We want to study this homomorphism.
We defined the involution $\phi \mapsto^t \varphi \in \mathrm{End}(J)$. This allows us to define $\mathrm{End}_{\mathrm{sym}}(J)$, which consists of those φ, which satisfy $\varphi =^t \varphi$.

We start the Neron-Severi group $NS(J \times J)$. The two projections $J \times J \longrightarrow J$ yield pull backs

$$NS(J) \rightrightarrows NS(J \times J),$$

and we have the two homomorphisms

$$NS(J \times J) \rightrightarrows NS(J)$$

given by restriction to $\{e\} \times J$ and $J \times \{e\}$.

This allows us to write

$$NS(J \times J) = N(J) \oplus NS'(J \times J) \oplus NS(J) \tag{10.29}$$

the two extremal summands are the images obtained from the restriction composed with the pullbacks, the summand in the middle is the kernel of the sum of the two restrictions. Let $\mathcal{L}$ be a line bundle on $J \times J$ whose Chern class (image in the Neron-Severi group) $c_1(\mathcal{L}) \in NS'(J \times J)$. Then we know that

$$\mathcal{L} \mid J \times e \quad \text{and} \quad \mathcal{L} \mid e \times J$$

lie in $\mathrm{Pic}^0_{J/k}$. After tensorization by bundles of the form $p_i^*(\mathcal{M}_i)$ we may assume that these restrictions of $\mathcal{L}$ are even trivial

Our theorem 10.2.6 implies that we find a unique $\psi : J \longrightarrow J$ such that

$$\mathcal{L} \simeq (\mathrm{Id} \times \psi)^*(\widetilde{\Theta}),$$

we have really an isomorphism, because $\mathcal{L}|\{e\} \times J \xrightarrow{\sim} \mathcal{O}_J$. The map $\psi \mapsto (\mathrm{Id} \times \psi)^*(\widetilde{\Theta})$ gives us a homomorphism

$$\Psi : \mathrm{End}(J) \longrightarrow NS'(J \times J) \tag{10.30}$$

which is surjective (as we just saw) and injective because ψ is unique. We formulate a theorem

Theorem 10.3.8. *The above homomorphism is an isomorphism. The homomorphism Φ yields an injective homomorphism*

$$\Phi : NS(J) \hookrightarrow \mathrm{End}_{sym}(J).$$

The image of Φ contains $2\,\mathrm{End}_{sym}(J)$.

(In the transcendental context in Vol. I , 5.2.3 we proved that Φ is actually isomorphism, this is also true in our situation here, but we do not give the proof (See [Mu1], §23)).

Now we consider the diagonal embedding

$$\Delta : J \longrightarrow J \times J,$$

and we put

$$\mathcal{L}_\psi = \Delta^* \circ (\mathrm{Id} \times \psi)^*(\widetilde{\Theta}),$$

this yields a homomorphism

$$\delta \circ \Psi : \mathrm{End}(J) \longrightarrow NS'(J \times J) \longrightarrow NS(J),$$

$$\psi \mapsto \mathcal{L}_\psi = \delta \circ \Psi(\psi)$$

We have the following formula

$$\Phi(\mathcal{L}_\psi) = \psi + {}^t\psi \tag{10.31}$$

To see this we start from the definition $\mathcal{L}_\psi = (\mathrm{Id} + \psi)^*(\Theta) \otimes \psi^*(\Theta)^{-1} \otimes \Theta^{-1}$. We compute $\phi_{\mathcal{L}_\psi}$ this sends the point $x \in J(k)$ to

$$T_x^*((\mathrm{Id} + \psi)^*(\Theta) \otimes \psi^*(\Theta)^{-1} \otimes \Theta^{-1}) \otimes \left((\mathrm{Id} + \psi)^*(\Theta) \otimes \psi^*(\Theta)^{-1} \otimes \Theta^{-1}\right)^{-1} =$$
$$(\mathrm{Id} + \psi)^*(T_{x+\psi(x)}^*)\Theta)) \otimes \psi^*(T_{\psi(x)}^*(\Theta)^{-1} \otimes T_x^*(\Theta^{-1}) \otimes \left((\mathrm{Id} + \psi)^*(\Theta)^{-1} \otimes \psi^*(\Theta) \otimes \Theta\right).$$

Rearranging the terms yields that this is equal to

$$(\mathrm{Id} + \psi)^*(T_{x+\psi(x)}^*(\Theta) \otimes \Theta^{-1}) \otimes \psi^*(T_{\psi(x)}^*(\Theta)^{-1} \otimes \Theta) \otimes (T_x^*(\Theta)^{-1} \otimes \Theta).$$

The argument in $(\mathrm{Id} + \psi)^*$ is in $\mathrm{Pic}^0_{J/k}$ and hence we can apply theorem 10.3.2 iii) and after expanding this term and looking at the cancellations we get

$$\psi^*(T_x^*(\Theta) \otimes \Theta^{-1}) \otimes T_{\psi(x)}^*(\Theta) \otimes \Theta^{-1}$$

and this is our formula above.

If on the other hand $\mathcal{L}$ is a bundle on J, then we get by the usual construction the bundle $\widetilde{\mathcal{L}} = m^*(\mathcal{L}) \otimes p_1^*(\mathcal{L})^{-1} \otimes p_2^*(\mathcal{L})^{-1}$ on $J \times J$. Again it is clear that $\widetilde{\mathcal{L}} \mid J \times e \simeq \mathcal{O}_J$ and $\widehat{\mathcal{L}} \mid e \times J \simeq \mathcal{O}_J$, and hence we find a $\psi : J \to J$ such that

$$\widetilde{\mathcal{L}} = (\mathrm{Id} \times \psi)^*(\widetilde{\Theta}).$$

Restriction to the diagonal yields

$$\Delta^*(\widetilde{\mathcal{L}}) = \mathcal{L}^{\otimes 2} = \mathcal{L}_\psi.$$

The shows that the image of the homomorphism $\delta \circ \Psi$ contains $2NS(J)$. This implies that

$$\Phi : NS(J) \longrightarrow \mathrm{End}_{\mathrm{sym}}(J) \subset \mathrm{End}(J)$$

because $\mathrm{End}(J)$ is torsion free and formula 10.31 implies that $2NS(J)$ is mapped into the symmetric tensors.

We also have seen that $\delta \circ \Psi \circ \Phi$ is the multiplication by two and hence we see that $NS(J)$ has at most 2-torsion and the kernel of Φ consists of 2-torsion elements.

We want to exclude 2-torsion. Here we assume that the characteristic of k is not equal to 2.

Let us consider a line bundle $\mathcal{L}$ on J, for which the $2c_1(\mathcal{L}) = 0$, i.e.

$$\mathcal{L}^{\otimes 2} = \mathcal{M} \in \mathrm{Pic}^0_{J/k}(k).$$

We can write $\mathcal{M} = \mathcal{M}_1^{\otimes 2}$ with $\mathcal{M}_1 \in \mathrm{Pic}^0_{J/k}(k)$ and hence we get

$$\left(\mathcal{L} \otimes \mathcal{M}_1^{-1}\right)^{\otimes 2} = \mathcal{O}_J$$

We consider the homomorphism $[2] : J \longrightarrow J$ this is an étale covering. The Galois group of this covering is $J[2](k)$. Clearly we have ${}^t[2](\mathcal{L}) = \mathcal{O}_J = ^t [2](\mathcal{O}_J)$. This means that $\mathcal{L}$ is a J-form of the trivial bundle (See 6.2.11), hence the isomorphism class of $\mathcal{L}$ determines a cohomology class in $H^1(J \xrightarrow{[2]} J,\mathbb{G}_m) = \mathrm{Hom}(J[2](k),\mu_2)$. But we have seen above that the elements of this cohomology group are in one two one correspondence to the elements in $J[2](k)$, hence we can conclude that $\mathcal{L}$ is isomorphic to one of the elements in $J[2](k)$. □

If the characteristic is 2, a similar argument should work if we replace the Galois-cohomology argument, by a computation in flat topology, i.e. we consider $J \xrightarrow{[2]} J$ as a covering in the flat topology. Then it is clear that the isomorphism classes of line bundles, which become trivial under the pullback $J \xrightarrow{[2]} J$ are in one to one correspondence with classes in $H^1(J \xrightarrow{[2]} J,J[2])$. If we exploit the structure theory for such finite group schemes it follows again that this cohomology group is equal to $J[2](k)$.

The ring of correspondences

We want to review section 5.3 of Volume I and reformulate our above results in terms of our given curve C/k. We assume that C/k has a rational point P_0. We had the morphisms $j_{P_0,\underline{Q}} : C \longrightarrow \mathrm{Pic}^g_{C/k}$. If we identify $\mathrm{Pic}^g_{C/k}$ to $\mathrm{Pic}^0_{C/k} = J$ we get a map $j : C \longrightarrow J$, we assume that $j(P_0) = e$. Hence we get a morphism

$$j \times \mathrm{Id}_J : C \times J \longrightarrow J \times J.$$

By construction we have $(j \times \mathrm{Id}_J)^* \widetilde{\Theta} = \mathcal{P}_0$, the isomorphism is determined by the rigidification. By restriction we get a homomorphism $NS(J \times J) \longrightarrow NS(C \times C)$. The two projections yield summands $p_1^*(NS(J)),p_2^*(NS(J)) \subset NS(J \times J)$ and $p_1^*(NS(C)),p_2^*(NS(C)) \subset NS(C \times C)$. As before this yields a decomposition

$$NS(C \times C) = p_1^*(NS(C)) \oplus NS'(C \times C) \oplus p_2^*(NS(C)).$$

The Neron-Severi group $NS(C \times C)$ is equal to the Chow group $A^1(C \times C)$ provided we define the Chow group using algebraic equivalence of cycles. We introduced a ring structure on $A^1(C \times C)$ (see section 10.2.5). Inside this ring we have the two classes obtained from the cycles $\{P_0\} \times C, C \times \{P_0\}$. It is clear from the definition of the product that the $\mathbb{Z}$- linear combinations of these two classes define an ideal $\mathcal{I}$. This is the ideal given by the first and the third summand in the above decomposition. If we divide by this ideal, we get the ring of correspondences

$$\mathbf{Cor}(C) = A^1(C \times C)/\mathcal{I} = NS'(C \times C)$$

Theorem 10.3.9. *The restriction*

$$r : NS'(J \times J) \longrightarrow NS'(C \times C)$$

is an isomorphism of rings.

An element in $NS'(C \times C)$ is uniquely (theorem of the cube) represented by a line bundle $\mathcal{L}$ whose restriction to $\{P_0\} \times C,C \times \{P_0\}$ is trivial. Hence we find a morphism $f : C \longrightarrow J$ such that $f(P_0) = e$ and $(\mathrm{Id} \times f)^*(\mathcal{P}_0) \xrightarrow{\sim} \mathcal{L}$. We extend this morphism to $J \times C \xrightarrow{\mathrm{Id}_J \times f} J \times J$ and we get

$$(j \times \mathrm{Id}_J)^*(\mathrm{Id}_J \times f)^*(\widetilde{\Theta}) = (\mathrm{Id} \times f)^*(\mathcal{P}_0) = (\mathrm{Id} \times f)^*(j \times \mathrm{Id}_J)^*(\widetilde{\Theta}) = \mathcal{L}.$$

By the same token the elements in $NS'(J \times C)$ are in one to one correspondence with the morphisms $f : C \longrightarrow J,f(P_0) = e$, hence we see that the restriction $NS'(J \times C) \longrightarrow NS'(C \times C)$ is an isomorphism. We apply the same argument to the restriction $NS'(J \times J) \longrightarrow NS'(J \times C)$: An element in $NS'(J \times C)$ is represented by a bundle $\mathcal{L}$ which is trivialized on $\{e\} \times C$ and $J \times \{P_0\}$ and hence given by an element $\widetilde{f} : J \longrightarrow J$ such that $\mathcal{L} = (\widetilde{f} \times \mathrm{Id}_C)^*(\mathrm{Id}_J \times j)^*(\widetilde{\Theta})$. But again these $\widetilde{f}$ are in one to one correspondence with elements in $NS'(J \times J)$ and we have shown that the restriction$NS'(J \times J) \longrightarrow NS'(C \times C)$ is an isomorphism.
We do not prove that this is actually a homomorphism of rings, to do this we need some Chow ring technology, especially the theory of cycle classes in a suitable cohomology theory. (See next section 7). This also applies to the followings assertions.
The set

$$\mathbf{Mor}((C,P_0),(J,e)) = \{\text{Morphisms } f : C \longrightarrow J | f(P_0) = e\} = \mathbf{Cor}(C)$$

and we have seen that we have a canonical bijection with $\mathrm{End}(J)$. We leave it as an exercise that this bijection is given by the rule

$$\phi \mapsto \phi \circ j,$$

i.e. that the elements in $\mathbf{Mor}((C,P_0),(J,e))$ are given by the composition of j with an endomorphism. If we have two elements $T_1,T_2 \in \mathbf{Cor}(C)$ and $T_i = \phi_i \circ j$ then we have

$$T_1 \circ T_2 = \phi_1 \circ \phi_2 \circ j.$$

We define the trace of an element $T = \phi \circ j$ simply by $\mathrm{tr}(T) = \mathrm{tr}(\phi)$ then we have the trace formula (See Vol. I ,Thm. 5.3.9)

$$\mathrm{tr}(T) = -T \cdot (\Delta - \{P_0\} \times C - C \times \{P_0\}).$$

A slight generalization of this formula yields for the intersection number of two cycles in $A_0^1(C)$

$$T_1 \cdot T_2 = -\,\mathrm{tr}(\phi_1 \circ^t \phi_2) = -T_1 \cdot^t T_2$$

and we see that the Hodge index theorem (See Thm. 9.7.1) translates into the positivity of the Rosati involution.
In a short summary we say that any curve C/k comes with a $\mathbb{Q}$-algebra $\mathbf{Cor}(C/k) \otimes \mathbb{Q}$. This algebra is semi simple, it has an involution ${}^t : \mathbf{Cor}(C/k) \otimes \mathbb{Q} \longrightarrow \mathbf{Cor}(C/k) \otimes \mathbb{Q}$, which on the level of cycles is given by interchanging the factors and it has a trace $\mathrm{tr} : \mathbf{Cor}(C/k) \otimes \mathbb{Q} \longrightarrow \mathbb{Q}$, which is $\mathbb{Q}$-linear. The bilinear form

$$(T_1,T_2) \mapsto \mathrm{tr}(T_1 \circ^t T_2) = -T_1 \cdot T_2$$

is positive definite.

This algebra depends on the ground field k (See p. 247) If we extend the ground field, we may find some new cycles. Since we have finite generation, we can find a finite extension $k_1 \subset k$ such that $A(C \times_k C \times_k k_1)$ is *saturated* , i.e. if we extend further the Chow ring does not change anymore. For the following we assume that already $A(C \times_k C)$ is saturated.
This algebra tells us among other things how the Jacobian J/k decomposes into simple abelian varieties (up to isogeny). Our considerations in 10.2.5 imply

Proposition 10.3.10. *An abelian variety A/k with saturated endomorphism ring* $\mathrm{End}(A/k)$ *is simple if and only if* $\mathrm{End}(A/k) \otimes \mathbb{Q}$ *is a field.*

We apply the considerations on p. 323 to $\mathbf{Cor}(C/k) \otimes \mathbb{Q}$. Our algebra $\mathbf{Cor}(C/k) \otimes \mathbb{Q}$ has a center Z_C which decomposes

$$Z_C = \oplus Z_C e_i,$$

and this yields a decomposition up to isogeny $J = \oplus J_i = \oplus e_i J$. Then $\mathrm{End}(e_i J) = M_{d_i}(D_i)$, where D_i is a division algebra over its center $Z_C e_i$ and $J_I = (J_i^{(0)})^{d_i}$ where $J_i^{(0)}$ is a simple abelian variety with $\mathrm{End}(J_i^{(0)}) = D_i$.
For a "generic" curve C/k we will have $\mathbf{Cor}(C/k) = \mathbb{Z}$ and the identity is the diagonal. This means that the ring of correspondences is not very interesting in this case.
In general it is difficult to compute $\mathbf{Cor}(C/k)$, even if the defining equations for the curve are explicitly given (see the theorems of Tate and Faltings further down).

Exercise 47. Consider the two elliptic curves over $\mathbb{Q}$, which are give by the equations $y^2 = x^3 - x, y^2 = x^3 - 1$. Show that their endomorphisms rings after extension of the scalars to $\mathbb{Q}(i), \mathbb{Q}(\rho), \rho$ third root of unity are $\mathbb{Z}[i]$, resp. $\mathbb{Z}[\rho]$.

These two curves are curves with complex multiplication.

Exercise 48. Let k be a field of characteristic zero, let C/k be a curve of genus 2, let us assume that $J_{C/k}$ is simple. Then $\mathbf{Cor}(C/k)$ is either $\mathbb{Q}$, or a quadratic extension $K/\mathbb{Q}$ or a totally imaginary extension $K/\mathbb{Q}$ of degree 4.

This is more an exercise in linear algebra that in algebraic geometry. We may assume that $k = \mathbb{C}$ and then we apply the methods of Vol. I, 5.3.
The following exercise might be considered as a little bit unfair.

Exercise 49. Let $k = \mathbb{F}_q$ a finite field and C/k a curve of genus one. Let $\mathbf{Cor}(C/k)$ be saturated. Then $\mathbf{Cor}(C/k) \otimes \mathbb{Q}$ is either an imaginary quadratic extension or it is the quaternion algebra over $\mathbb{Q}$, which is described in example 11 and which given by $\{a, b, -ab\} = \{p, -1, -p\}$.

This exercise requires some knowledge of class field theory.
The next exercise is also unfair

Exercise 50. Write down an elliptic curve over $\mathbb{Q}$, which has an endomorphism φ with $\varphi^2 = -163$.

10.4 Étale Cohomology

In the previous sections we gave some indications that the Tate-modules $T_\ell(J)$ should be considered a substitute for the first homology group of J. If our ground field is $\mathbb{C}$, and if

$$J(\mathbb{C}) = \mathbb{C}^g/\Gamma,$$

then $\Gamma = H_1(J(\mathbb{C}),\mathbb{Z})$ and

$$\Gamma \otimes \mathbb{Z}_\ell \simeq T_\ell(J).$$

The right hand side has a definition in purely algebraic terms and hence $H_1(J(\mathbb{C}),\mathbb{Z})\otimes\mathbb{Z}_\ell$ has a definition in purely algebraic terms.
Therefore, we can start from an abelian variety J/k over an arbitrary ground field k. We fix an algebraic closure $\bar{k}$, and let $k_s \subset \bar{k}$ the separable closure contained in it. We pick a prime ℓ, which is different from $\mathrm{char}(k) = p$, then we have seen that

$$J[\ell^m](\overline{k}) \simeq (\mathbb{Z}/\ell^m\mathbb{Z})^{2g},$$

and since the map $\ell^m : J \to J$ is separable, we even see that all the ℓ^m-division points are k_s-valued points.

Hence we get an action of the Galois group $\mathrm{Gal}(k_s/k) = \mathrm{Gal}(\overline{k}/k)$ on $J[\ell^m](\overline{k})$ and passing to the limit we get a continuous homomorphism

$$\rho : \mathrm{Gal}(\overline{k}/k) \to GL(T_\ell(J)).$$

This means that $T_\ell(J)$ has a much richer structure than just the structure of a free $\mathbb{Z}_\ell$-module. Observe that any abelian variety J/k is actually a base extension $J = J_0 \times_{k_0} k$ where $k_0 \subset k$ is a field, which is finitely generated over the prime field inside k. This field is far away from being algebraically closed and we get a very strong action of the Galois group $\mathrm{Gal}(\bar{k}_0/k_0)$ on the Tate modules $T_\ell(J_0)$, which contains a lot of information on the abelian variety J_0/k_0. (See further down.)

The cyclotomic character.

There is a much simpler construction of such Galois modules. We consider the multiplicative group scheme $\mathbb{G}_m$ (see section 7.5.7). Again we can pick a prime $\ell \neq p = \mathrm{char}(k)$, and consider the kernel $\mathbb{G}_m[\ell^n] = \ker(\mathbb{G}_m \xrightarrow{\ell^n} \mathbb{G}_m)$. This kernel is of course equal to μ_{ℓ^n}

$$\mu_{\ell^n}(\overline{k}) = \mu_{\ell^n}(\overline{k}_s) = \{\zeta \in \overline{k}^* \mid \zeta^{\ell^n} = 1\} \simeq \mathbb{Z}/\ell^n\mathbb{Z},$$

the last isomorphism is obtained by selecting a primitive ℓ^n-th root of unity. We can also can identify $T_\ell(\mathbb{G}_m) = \varprojlim_n \mu_{\ell^n}(\overline{k})$ to $\mathbb{Z}_\ell$ if we choose a consistent ℓ^n th root of unity for $n = 1,2,\ldots$
This gives us a representation of the Galois group

$$\alpha : \mathrm{Gal}(\overline{k}/k) \longrightarrow GL(T_\ell(\mathbb{G}_m)) = \mathbb{Z}_\ell^\times,$$

which does not depend on the choice of consistent roots of unity. For any $\sigma \in \mathrm{Gal}(\overline{k}/k)$ and any ℓ^n-th root of unity ζ we have by definition $\sigma(\zeta) = \zeta^{\alpha(\sigma)}$. This homomorphism is called the Tate character or the cyclotomic character. For any integer m we denote by $\mathbb{Z}_\ell(m)$ the module $\mathbb{Z}_\ell$, on which $\mathrm{Gal}(\overline{k}/k)$ acts by the character α^m.

10.4.1 Étale cohomology groups

A. Grothendieck introduced the étale cohomology group for schemes. We want to give some very general ideas how this can be done. The decisive step is to extend the notion of a topological space or the notion of a topology on a space.

Let us fix a field k and let us consider schemes $X/\operatorname{Spec}(k)$ of finite type. We define some new "topology" on X by saying that the open sets are morphisms

$$\widetilde{U} \longrightarrow U \subset X$$

where $U \subset X$ is Zariski open and $\widetilde{U} \to U$ is finite étale morphism. These new open sets form a category Et_X in an obvious sense. A morphism is a diagram

$$\begin{array}{ccc} \widetilde{U}_1 & \longrightarrow & \widetilde{U}_2 \\ \downarrow & & \downarrow \\ U_1 & \longrightarrow & U_2 \end{array}$$

where $U_1 \to U_2$ is an inclusion and the diagram is commutative (the vertical arrows being finite étale morphisms).

This category has fibered products. For an ordinary topology this fibered product of two objects (open subsets in the space) is given by the intersection of the two open subsets. In the general context this becomes more complicated. Another difference is this: For an ordinary topological space the set of morphisms between two subjects is either empty or consists of just one element. This is no longer true.

The étale topology is already extremely interesting for $X = \operatorname{Spec}(k)$. There are not so many Zariski open subsets, but an étale open set is simply a finite separable extension $\operatorname{Spec}(k') = X' \to X$. The fibered product is

$$X' \times_X X'' = \operatorname{Spec}(k' \otimes_k k''),$$

which then suddenly may consist of several points. If for instance k'/k is a finite normal extension, then

$$k' \otimes_k k' \simeq \bigoplus_{\sigma:\sigma\in \operatorname{Hom}_k(k',k')} k'.$$

If we have a morphism $X' \xrightarrow{f} X$, and if U is open in X, then $f^{-1}(U)$ is open in X' and if $\widetilde{U} \to U$ is an étale covering, then $\widetilde{U}' = U' \times_U \widetilde{U}$ is an étale covering of U'.

For an object $\widetilde{U} \to U$ is Et_X we can introduce the notion of a covering. This is a family of finite étale morphisms

$$\widetilde{U}_i \longrightarrow U_i \subset \widetilde{U} \qquad i \in I$$

such that $\bigcup U_i = \widetilde{U}$.

Now we can define sheaves. This are contravariant functors

$$\mathcal{F} : Et_X \longrightarrow \mathcal{C}$$

where $\mathcal{C}$ is any reasonable category (for instance sets, rings, abelian groups and so on), which satisfy the two sheaf conditions: For any covering $\{\widetilde{U}_i\}$ of $U \in Et_X$ we have

$$\mathcal{F}(U) \longrightarrow \prod_i \mathcal{F}(\widetilde{U}_i) \longrightarrow \prod_{(i,j)} \mathcal{F}(\widetilde{U}_i \times_U \widetilde{U}_j)$$

is an exact sequence. For any X and any abelian group we can define the constant $\underline{A}$, which is defined by the rule

$$\underline{A}(U) = \bigoplus_{\text{con. comp. of } U} A$$

Galois cohomology

This notion is already non trivial for the case $X = \operatorname{Spec}(k)$. What does it mean that $\mathcal{M}$ is a sheaf on X, say with values in the category of abelian groups?

This means that we have a functor $\mathcal{M}$, which attaches to any finite separable extension k'/k, an abelian group $\mathcal{M}(k')$ and for any morphism of k-algebras

$$\begin{array}{ccc} k' & \xrightarrow{\sigma} & k'' \\ & \nwarrow \quad \nearrow & \\ & k & \end{array}$$

we have a group homomorphism

$$\sigma' : \mathcal{M}(k') \longrightarrow \mathcal{M}(k'').$$

If our separable extension k'/k is not a field, then $k' = \oplus k'_i$ where these k_i are fields and $\mathcal{M}(k') = \oplus \mathcal{M}(k'_i)$.

We have to formulate what the sheaf conditions mean. A covering of $\operatorname{Spec}(k') \to \operatorname{Spec}(k)$ is a k-algebra homomorphism

$$\begin{array}{ccc} k' & \longrightarrow & L \\ & \nwarrow \quad \nearrow & \\ & k & \end{array}$$

i.e. $\operatorname{Spec}(L) \to \operatorname{Spec}(k')$, which sends the identity of k' to the identity of L (Note that k' is not necessarily a field.).

Now it is clear that it suffices to know the value of $\mathcal{M}$ on finite separable field extensions L/k.

If L_1/L is a finite separable and normal extension, then it is clear that the Galois group $\operatorname{Gal}(L_1/L)$ acts upon $\mathcal{M}(L_1)$. The sheaf condition then means: For any $L_1 \supset L \supset k$ as above we have

$$\mathcal{M}(L) \hookrightarrow \mathcal{M}(L_1),$$

and $\mathcal{M}(L)$ is exactly the module of fixed points under the action of $\operatorname{Gal}(L_1/L)$ on $\mathcal{M}(L_1)$.

We can reformulate this slightly. If we choose a separable closure k_s/k, then we can restrict the functor $\mathcal{M}$ to finite extensions

$$k \subset L \subset k_s.$$

If now $k \subset L_1 \subset L_2 \subset k_s$, then we have $\mathcal{M}(k) \subset \mathcal{M}(L_1) \subset \mathcal{M}(L_2)$ and we can form the limit

$$\mathcal{M} = \varinjlim_{K/k} \mathcal{M}(L) = \cup\, \mathcal{M}(L).$$

This is now a continuous module for the Galois group. The group $\mathrm{Gal}(k_s/k) = \varprojlim \mathrm{Gal}(L/k)$ acts on $\mathcal{M}$ and for any $m \in \mathcal{M}$ the stabilizer of m is an open subgroup in the Galois group. If in turn M is a module, on which we have a continuous action of $\mathrm{Gal}(k_s/k)$ (i.e. for any $m \in M$ we have a finite extension $k \subset L \subset k_s$ such that $\mathrm{Gal}(k_s/L)m = m$, then we can define $M(L) = M^{\mathrm{Gal}(k_s/L)}$ and $L \to M(L)$ defines an étale sheaf on $\mathrm{Spec}(k)$.)

This generalizes. If X,X_0 are two (irreducible) schemes and $\pi : X_0 \longrightarrow X$ a finite étale morphism, then we say that π is **normal** or **galois** if the fibered product decomposes into connected components, which are isomorphic to X_0, i.e.

$$X_0 \times_X X_0 = \bigcup_{\sigma: X_0 \longrightarrow X_0} X_0,$$

where the σ are of course automorphisms of $\pi : X_0 \longrightarrow X$. These σ form the Galois group $\mathrm{Gal}(X_0/X)$ and this Galois group is of course also the Galois group of the extension of the function fields of X,X_0. Whenever we have such a Galois extension $X_0 \longrightarrow X$ and a (finite) $\mathrm{Gal}(X_0/X)$- module $\mathcal{F}_0$, then this defines an étale sheaf on X, which is also denoted by $\mathcal{F}_0$ and whose restriction to X_0 is the constant sheaf $\underline{\mathcal{F}}_0$.

We return to the more general situation. We have our $X/\mathrm{Spec}(k)$. If we consider sheaves with values in (Ab) (or modules over some ring R), then we have the notion of an exact sequence of sheaves on Et_X: This is a sequence

$$0 \to \mathcal{F}' \to \mathcal{F} \to \mathcal{F}'' \to 0,$$

such that
a)for any $U \in Et_X$ the sequence

$$0 \to \mathcal{F}'(U) \to \mathcal{F}(U) \to \mathcal{F}''(U)$$

is exact and
b) the last arrow is " locally" exact: For any $s'' \in \mathcal{F}''(U)$ we can find a covering

$$\{\widetilde{U}_\alpha\}_\alpha \qquad \widetilde{U}_\alpha \to U_\alpha \subset U$$

such that for any α the restriction s''_α of s_α to $\widetilde{U}_\alpha$ is in the image of $\mathcal{F}(U_\alpha) \to \mathcal{F}''(U_\alpha)$.

Now it can be shown that we can define the derived functor to the functor

$$\mathcal{F} \longrightarrow \mathcal{F}(X) = H^0(X,\mathcal{F}),$$

and we call the derived group $H^i_{\text{ét}}(X,\mathcal{F})$. Our short exact sequence yields a long exact sequence in cohomology

$$
\begin{aligned}
0 \to & H^0_{\text{ét}}(X,\mathcal{F}') \to H^0_{\text{ét}}X,\mathcal{F}) \to H^0_{\text{ét}}(X,\mathcal{F}'') \xrightarrow{\delta} \\
& H^1_{\text{ét}}(X,\mathcal{F}') \to H^1_{\text{ét}}X,\mathcal{F}) \to H^1_{\text{ét}}(X,\mathcal{F}'') \xrightarrow{\delta}
\end{aligned}
\tag{10.32}
$$

(The only thing we need to know is the existence of enough injective objects.)

If for instance $X = \operatorname{Spec}(k)$, then we have see that an étale sheaf with values in (Ab) is simply a $\operatorname{Gal}(k_s/k)$-module M with continuous action. An exact sequence of such sheaves is nothing else than an exact sequence of such modules

$$0 \longrightarrow M' \longrightarrow M \longrightarrow M'' \longrightarrow 0.$$

But if we look at sections over $X = \operatorname{Spec}(k)$, then we get

$$0 \longrightarrow M'(k) \longrightarrow M(k) \longrightarrow M''(k),$$

and the last arrow needs not to be surjective. For $m'' \in M(k)$ we can find an extension L/k (normal over k) such that m'' lifts to an element m in $M(L)$ and the obstruction to lift m'' to an element in $M(k)$ lies in (See Vol. I, 5.3.)

$$H^1(\operatorname{Gal}(L/k),M'(L)).$$

Hence it is more or less clear that

$$H^1_{\text{ét}}X,M') = \varinjlim H^1(\operatorname{Gal}(L/k),M(L)) = H^1(\operatorname{Gal}(k_s/k),M).$$

In this case we are back in the situation of ordinary group cohomology. There is a slight difference to the situation in Volume I, since our group is the projective limit of finite groups. This requires some harmless continuity considerations. The groups $H^i_{\text{ét}}(X,M) = H^i(\operatorname{Gal}(k_s/k),M)$ are called the Galois cohomology groups.

The geometric étale cohomology groups.

From now on the exposition will become very informal, we will try to explain some of the fundamental ideas and formulate some basic results. But the proofs will be very sketchy and also the definitions will be somewhat imprecise. As a general references we can give SGA $4\frac{1}{2}$ and Milne's book [Mi].
Let us consider a scheme of finite type X/k. We choose a separable closure k_s and an algebraic closure $\overline{k} \supset k_s$. We form the scheme $\overline{X} = X \times_{\operatorname{Spec}(k)} \operatorname{Spec}(\overline{k}))$.
Let us assume that we have a sheaf $\mathcal{F}$ over X with values in (Ab) or modules over a ring R. Then it is clear that

$$H^0_{\text{ét}}(\overline{X},\mathcal{F}) = \mathcal{F}(\overline{X}) = \mathcal{F}(X \times_{\operatorname{Spec}(k)} \operatorname{Spec}(k_s))$$

is a module for the Galois group $\operatorname{Gal}(\overline{k}/k) = \operatorname{Gal}(k_s/k)$, and hence we see that all the cohomology groups

$$H^i_{\text{ét}}(\overline{X},\mathcal{F})$$

are modules for $\operatorname{Gal}(\overline{k}/k)$. These are the geometric cohomology groups.

We want to "compute" these cohomology groups in a couple of simple cases. Here "compute" means that we assume that the usual tools are available.

We consider a smooth, projective and absolutely irreducible curve C/k. On this curve we have the étale sheaf $\mathcal{O}^*_{C,\text{ét}}$: For any $\widetilde{U} \to U \subset C$ in Et_C we put

$$\mathcal{O}^*_{C,\text{ét}}(\widetilde{U}) = \mathcal{O}(\widetilde{U})^*$$

i.e. we consider the invertible regular functions on $\widetilde{U}$.
Let ℓ be a prime, which is different from the characteristic of k, let n be an integer > 0. For any $\widetilde{U}$ we can consider the homomorphism $\ell^n : \mathcal{O}(\widetilde{U})^* \longrightarrow \mathcal{O}(\widetilde{U})^*$ given by $x \mapsto x^{\ell^n}$. This defines a homomorphism of sheaves

$$\ell^n : \mathcal{O}^*_{C,\text{ét}} \longrightarrow \mathcal{O}^*_{C,\text{ét}}.$$

Here we see the whole point of the story. This map is certainly not surjective for the sheaf $\mathcal{O}^*_{C,\text{Zar}}$ – this is our sheaf restricted to the Zariski topology. But if we have an étale morphism $\widetilde{U} \to C$, $\widetilde{U}$affine and an element $f \in \mathcal{O}(\widetilde{U})^*$, then

$$\widetilde{U}' = \operatorname{Spec}(\mathcal{O}(\widetilde{U})\left[\sqrt[\ell^n]{f}\right]) \longrightarrow \widetilde{U} = \operatorname{Spec}(\mathcal{O}(\widetilde{U}))$$

is an étale covering, and $\sqrt[\ell^n]{f} \in \mathcal{O}(\widetilde{U}')^\times$ maps to f under ℓ^n. Hence we see that

$$\mathcal{O}^*_{C,\text{ét}} \xrightarrow{\ell^n} \mathcal{O}^*_{C,\text{ét}}$$

is surjective. The kernel is the sheaf μ_{ℓ^n} where

$$\mu_{\ell^n}(\widetilde{U}) = \{f \in \mathcal{O}_C(\widetilde{U})^* \mid f^{\ell^n} = 1\}.$$

This is the sheaf of ℓ^n-roots of unity. Hence we get an exact sequence

$$1 \to \mu_{\ell^n} \to \mathcal{O}^*_{C,\text{et}} \to \mathcal{O}^*_{C,\text{et}} \to 1$$

of sheaves for the étale topology.

Since we assume that the usual tools are available we can write down the long exact sequence in cohomology. But before we do that we want to make it plausible that

$$H^1_{\text{ét}}(C,\mathcal{O}^*_{C,\text{ét}}) = H^1_{\text{Zar}}(C,\mathcal{O}^*_C) = \operatorname{Pic}_{C/k}(k).$$

This is a consequence of Hilbert's Theorem 90 (see 7.5.23) This can be exploited to prove that any class $\xi \in H^1_{\text{ét}}C,\mathcal{O}^*_{C,\text{ét}})$ becomes trivial on a suitable small Zariski open subset, and this implies the assertion above.

We write our exact cohomology sequence, but we restrict the cohomology to $\overline{C} = C \times_k \overline{k}$. We get

$$\mathcal{O}_{\overline{C},\text{ét}}(\overline{C})^* = \overline{k}^*$$

and

$$\overline{k}^* \xrightarrow{\ell^n} \overline{k}^*$$

is surjective. Hence the sequence starts in degree one and we find

$$\begin{aligned} 0 \to H^1_{\text{ét}}(\overline{C},\mu_{\ell^n}) &\to H^1_{\text{Zar}}(\overline{C},\mathcal{O}^*_{\overline{C}}) \to H^1_{\text{Zar}}(\overline{C},\mathcal{O}^*_{\overline{C}}) \\ &\to H^2_{\text{ét}}(\overline{C},\mu_{\ell^n}) \to 0. \end{aligned} \tag{10.33}$$

(Is it entirely obvious why it stops in degree 2?)

Now we have by definition

$$H^1_{\mathrm{Zar}}(\overline{C},\mathcal{O}^*_{\overline{C}}) = \mathrm{Pic}_{C/k}(\bar{k})$$

and this sits in the exact sequence (recall $J = \mathrm{Pic}^0_{C/k}$)

$$0 \to J(\bar{k})(\bar{k}) \to \mathrm{Pic}_{C/k}(\bar{k}) \xrightarrow{deg} \mathbb{Z} \to 0,$$

and this implies

$$\begin{array}{rcl} H^1_{\text{ét}}(\overline{C},\mu_{\ell^n}) & \simeq & J_{\overline{C}}(\bar{k})[\ell^n] \approx (\mathbb{Z}/\ell^n\mathbb{Z})^{2g} \\ H^2_{\text{ét}}(\overline{C},\mu_{\ell^n}) & \simeq & \mathbb{Z}/\ell^n\mathbb{Z} \end{array}$$

and finally

$$H^0_{\text{ét}}(\overline{C},\mu_{\ell^n}) = \mu_{\ell^n}(\bar{k}).$$

Again we recall that these étale cohomology groups are not just abelian groups, but they come with an action of the Galois group $\mathrm{Gal}(\bar{k}/k)$, and of course it is as it must be: The exact sequence is an exact sequence of Galois modules. Therefore we see that the isomorphisms above are isomorphisms of Galois modules.

It may look a little bit strange that we took the coefficient sheaf μ_{ℓ^n}. Indeed we could also take the constant sheaf $\mathbb{Z}/\ell^n\mathbb{Z}$ on C, this is simply

$$\mathbb{Z}/\ell^n\mathbb{Z}(\widetilde{U}) = \bigoplus_{\text{components of } \widetilde{U}} \mathbb{Z}/\ell^n\mathbb{Z}.$$

This sheaf is certainly not isomorphic to μ_{ℓ^n}, but if we restrict to $\overline{C}$, these two sheaves become isomorphic. Hence we can construct an isomorphism

$$H^i_{\text{ét}}(\overline{C},\mathbb{Z}/\ell^n\mathbb{Z}) \simeq H^i_{\text{ét}}(\overline{C},\mu_{\ell^n}).$$

But this is not necessarily an isomorphism of Galois modules, this is already clear in degree zero: The cohomology group $H^0_{\mathrm{et}}(\overline{C},\mathbb{Z}/\ell^n\mathbb{Z}) = \mathbb{Z}/\ell^n\mathbb{Z}$ is the trivial Galois module whereas $H^0_{\text{ét}}(\overline{C},\mathbb{Z}/\ell^n\mathbb{Z}) = \mu_{\ell^n}(\bar{k})$ is non trivial, it is the Galois module of ℓ^n-th roots of unity.

This Galoismodule is also denoted by $\mathbb{Z}/\ell^n\mathbb{Z}(1)$, we can define

$$\mathbb{Z}/\ell^n\mathbb{Z}(-1) = \mathrm{Hom}(\mathbb{Z}/\ell^n\mathbb{Z}(1),\mathbb{Z}/\ell^n\mathbb{Z}),$$

and $\mathbb{Z}/\ell^n\mathbb{Z}(n)$ for any integer n.

It is at least very plausible that we have an isomorphism of Galois modules

$$H^i_{\text{ét}}(\overline{C},\mathbb{Z}/\ell^n\mathbb{Z}) \simeq H^i_{\text{ét}}(\overline{C},\mu_{\ell^n}) \otimes \mathbb{Z}/\ell^n\mathbb{Z}(-1)$$

and especially

$$H^1_{\text{ét}}(\overline{C},\mathbb{Z}/\ell^n\mathbb{Z}) \simeq J_{\overline{C}}(\bar{k})[\ell^n] \otimes \mathbb{Z}/\ell^n\mathbb{Z}(-1).$$

Now we may vary the n, we get a projective system, and we put

$$H^i_{\text{ét}}(\overline{C},\mathbb{Z}_\ell) =: \varprojlim H^i_{\text{et}}(\overline{C},\mathbb{Z}/\ell^n\mathbb{Z}),$$

and we showed

$$\begin{array}{rcl} H^0_{\text{ét}}(\overline{C},\mathbb{Z}_\ell) & = & \mathbb{Z}_\ell \\ H^1_{\text{ét}}(\overline{C},\mathbb{Z}_\ell) & \simeq & T_\ell(J)\otimes\mathbb{Z}_\ell(-1) \\ H_{\text{ét}}(\overline{C},\mathbb{Z}_\ell) & \simeq & \mathbb{Z}_\ell(-1) \end{array}$$

as modules under the Galois group. (It turns out that the reasonable approach is: We start from torsion sheaves and define cohomology with coefficients in $\mathbb{Z}_\ell$ by taking projective limits.)
The above definition of ℓ-adic cohomology groups works in full generality. We can take an arbitrary scheme X/k of finite type and we define

$$H^i_{\text{ét}}(\overline{X},\mathbb{Z}_\ell) = \varprojlim H^i_{\text{ét}}(\overline{X},\mathbb{Z}/\ell^n\mathbb{Z}),$$

and these modules are modules for the Galois group $\mathrm{Gal}(\overline{k}/k)$. After that we define

$$H^i_{\text{ét}}(\overline{X},\mathbb{Q}_\ell) = H^i_{\text{ét}}(\overline{X},\mathbb{Z}_\ell)\otimes\mathbb{Q}_\ell.$$

(Observe that the left hand sides could have an alternate meaning as cohomology with coefficients in the sheaf $\mathbb{Z}_\ell$,$\mathbb{Q}_\ell$, but these are not reasonable objects.)
(We may of course also consider cohomology groups

$$H^i_{\text{ét}}(X,\mathbb{Z}_\ell) = \varprojlim H^i_{\text{ét}}(\mathcal{X}.\mathbb{Z}/\ell^n\mathbb{Z})$$

but they are still much more complicated.)

We have a more general class of sheaves on a scheme X, this are the $\mathbb{Z}_\ell$ sheaves $\mathcal{F}$. They are projective systems of $\mathbb{Z}_\ell/(\ell^{n+1})$ module sheaves $\{\mathcal{F}_n\}_n$ on X together with morphisms $\phi_n : \mathcal{F}_n \longrightarrow \{\mathcal{F}_{n-1}\}$, which satisfy certain conditions. (See [De1] , [Rapport],2) Among these $\mathbb{Z}_\ell$ we have those, which could be called local systems (See Vol. I , 4.8). To get these we consider infinite a tower of finite Galois module morphisms over X

$$\longrightarrow X_{n+1} \longrightarrow X_n \longrightarrow \ldots \longrightarrow X_0 \longrightarrow X$$

and a compatible system of homomorphisms

$$\rho_n : \mathrm{Gal}(X_n/X) \longrightarrow \mathrm{GL}_r(\mathbb{Z}/\ell^{n+1}\mathbb{Z}).$$

For any n we get a sheaf $\mathcal{F}_n$ whose restriction to X_n is isomorphic to the trivial sheaf $(\mathbb{Z}/\ell^{n+1}\mathbb{Z})^r$ on X_n. The compatibility condition guarantees that these sheaves fit together to a $\mathbb{Z}_\ell$ sheaf $\mathcal{F} = \varprojlim \mathcal{F}_n$.

For any scheme X and any $\mathbb{Z}_\ell$ sheaf $\mathcal{F}$ we define

$$H^i_{\text{ét}}(X,\mathcal{F}) = \varprojlim H^i_{\text{ét}}(X,\mathcal{F}_n).$$

We list some basic results:

1) If we have a scheme X/k of finite type, then the cohomology groups

$$H^i_{\text{ét}}(\overline{X},\mathbb{Z}_\ell)$$

are finitely generated, and if X is irreducible, then $H^i(\overline{X},\mathbb{Z}_\ell) = 0$ for $i > 2\dim X$.

2) If X/k is an irreducible affine scheme, then we even have

$$H^i_{\text{ét}}(\overline{X},\mathbb{Z}_\ell) = 0 \quad \text{for} \quad i > \dim X$$

and

$$H^i_{\text{ét}}(\mathbb{A}^n_{\overline{k}},\mathbb{Z}_\ell) = \begin{cases} 0 & i > 0 \\ \mathbb{Z}_\ell & \text{for } i = 0. \end{cases}$$

If $X = \mathbb{P}^n/k$ then

$$H^\bullet_{\text{ét}}(\bar{X},\mathbb{Z}_\ell) = \bigoplus_{i=0}^{n} H^{2i}_{\text{ét}}(\bar{X},\mathbb{Z}_\ell)$$

and

$$H^{2i}_{\text{ét}}(\bar{X},\mathbb{Z}_\ell) = \mathbb{Z}(-i).$$

3) If we have a morphism $f : X \longrightarrow Y$ and a torsion sheaf $\mathcal{F}$ for the étale topology on X then we can define the higher direct image sheaves $R^q_{\text{ét}} f_*(\mathcal{F})$ and this is a torsion sheaf on Y. Of course we we have a spectral sequence

$$E_2^{p,q} = H^p_{\text{ét}}(Y,R^q_{\text{ét}} f_*(\mathcal{F})) \Rightarrow H^n_{\text{ét}}(X;\mathcal{F}).$$

In this context we have the following fundamental theorem

Theorem 10.4.1. *(* **Proper base change** *) If $\mathcal{F}$ is a finite sheaf for the étale topology on X and if $f : X \longrightarrow Y$ a proper morphism. Then for any point $y \in Y$ we have*

$$R^q_{\acute{e}t} f_*(\mathcal{F})_y = H^q_{\acute{e}t}(X_y,\mathcal{F}_y).$$

This is the analogue of theorem 4.4.17 in Vol. I.

4) If the field $k = \mathbb{C}$, then we can consider the étale topology on X but we also have the analytic topology on $X(\mathbb{C})$. A little bit of thinking yields that we can have a continuous map

$$X_{\text{ét}} \longleftarrow X_{\text{an}}(\mathbb{C}),$$

which induces a map in cohomology

$$H^i(X_{\text{ét}},\mathbb{Z}/\ell^n\mathbb{Z}) \longrightarrow H^1(X_{\text{an}}(\mathbb{C}),\mathbb{Z}/\ell^n\mathbb{Z}),$$

and the comparison theorem asserts that this map is an isomorphism.

5) If X/k is projective non singular and absolutely irreducible, then we have in addition

$$H^{2d}_{\text{ét}}(\overline{X},\mathbb{Z}_\ell) = \mathbb{Z}_\ell(-d),$$

and the cup product , which of course has to be defined, induces a non degenerate pairing

$$H^i_{\text{ét}}(\overline{X},\mathbb{Z}_\ell) \otimes \mathbb{Q}_\ell \times H^{2d-i}_{\text{ét}}(\overline{X},\mathbb{Z}_\ell) \otimes \mathbb{Q}_\ell \longrightarrow H^{2d}_{\text{ét}}(\overline{X},\mathbb{Z}_\ell) \otimes \mathbb{Q}_\ell \simeq \mathbb{Q}_\ell(-d).$$

6) We have a Künneth formula for products

$$H^m_{\text{ét}}(\overline{X} \times \overline{Y}, \mathbb{Q}_\ell) \simeq \bigoplus H^i_{\text{ét}}(\overline{X}, \mathbb{Q}_\ell) \otimes H^{m-i}_{\text{ét}}(\overline{X}, \mathbb{Q}_\ell).$$

7) If we have a subscheme $Y \subset X$, let us assume that it is absolutely irreducible of codimension r, then we can attach a cycle class

$$c(Y) \in H^{2r}_{\text{ét}}(\overline{X}, \mathbb{Q}_\ell)$$

to it. The subspace $\mathbb{Q}_\ell c(Y)$ is isomorphic to $\mathbb{Q}_\ell(-r)$ as a module under the Galois group $\mathrm{Gal}(\overline{k}/k)$.
We can extend this cycle map to arbitrary cycles in codimension r. First of all, if $Y \subset X$ is not irreducible, then $Y \times_k \overline{k} = \sum Y_i$ where the Y_i are permuted by the Galois group. Then

$$c(Y) = c(Y \times_k \overline{k}) = \sum c(Y_i),$$

and $\mathbb{Q}_\ell c(Y_i) = \mathbb{Q}_\ell(-r)$ under the Galois group $\mathrm{Gal}(\overline{k}/L_i)$ if $\mathrm{Gal}(\overline{\mathbb{Q}}/L_i)$ is the stabilizer of Y_i. Finally we extend this homomorphism to arbitrary cycles by linearity.

Of course we want: If we have two cycles Y_1, Y_2 in codimension r_1, r_2 respectively, and if they are intersecting in a reasonable manner, then

$$c(Y_1 \cdot Y_2) = c(Y_1) \cup c(Y_2).$$

Especially if D is a divisor, then we want

$$D^d = D \cdot - \cdot D = c_1(D)^d.$$

8) These formal properties imply a Lefschetz fixed point formula. If we have a morphism $f : X \to X$ of a smooth projective, absolutely irreducible variety into itself, then we may consider the two cycles

$$\Gamma_f \subset X \times X \ , \ \Delta \subset X \times X,$$

where Δ is the diagonal. Let us assume, (it is not really necessary) that Γ_f and Δ intersect transversally (See Def. 7.5.21). Then the intersection number

$$\overline{\Gamma}_f \cdot \overline{\Delta} = \#(\overline{\Gamma}_f \cap \overline{\Delta}) = \#\text{fixed points of } f.$$

Now we look at the cycle classes and apply the Künneth formula, then we get

$$c(\overline{\Gamma}_f) \cup c(\overline{\Delta}) = \sum_{i=0}^{2d} (-1)^i \operatorname{tr}(f^* \mid H^i_{\text{ét}}(\overline{X}, \mathbb{Q}_\ell)),$$

and we get (under our assumption) # fixed points of $f = \sum_{i=0}^{2d} (-1)^i \operatorname{tr}(f^* \mid H^i_{\text{ét}}(\overline{X}, \mathbb{Q}_\ell))$.

If the intersection is not proper, then we have to be more careful in counting the fixed points.

10.4.2 Schemes over finite fields

We apply these theorems to the special case of an absolutely irreducible projective, smooth variety over a finite field $\mathbb{F}_q$, let us write $X/\mathbb{F}_q$. We define the (geometric) Frobenius morphism $\Phi_q : X/\mathbb{F}_q \longrightarrow X/\mathbb{F}_q$, which on the geometric points acts by

$$\underline{a} = (a_0,a_1,\dots,a_n) \mapsto \Phi_q(\underline{a}) = (a_0^q,a_1^q,\dots,a_n^q).$$

(Remember that $X \subset \mathbb{P}^n$ and the defining equations have coefficients in $\mathbb{F}_q$.) The intersection of Γ_{Φ_q} and the diagonal is transversal, this gives us

$$\#\text{fixed points of } \Gamma_{\Phi_q} = \#X(\mathbb{F}_q) = \sum (-1)^i \operatorname{tr}(\Phi_q^* \mid H^i_{\text{ét}}(\overline{X},\mathbb{Q}_\ell)).$$

Of course we can apply this to all powers Φ_q^n, and then

$$t \cdot \frac{Z'_X(t)}{Z_X(t)} = \sum_{n=0}^{\infty} \#X(\mathbb{F}_{q^n})t^n =$$

$$\sum_{n=0}^{\infty}\Big(\sum_{i=0}^{2d}(-1)^i \operatorname{tr}(\Phi_q^n \mid H^i_{\text{ét}}(\overline{X},\mathbb{Q}_\ell))\Big)t^n \tag{10.34}$$

and this yields the formula

$$Z_X(t) = \prod_{i=0}^{2g} \det(\mathrm{Id} - t\Phi_q \mid H^i_{\text{ét}}(\overline{X},\mathbb{Q}_\ell))^{(-1)^{i+1}}$$

for the Z-function of X. This is Grothendieck's theorem. (See [De1],[Rapport])

Of course we know that

$$Z_X(t) = \frac{P_1 \cdots P_{2g-1}}{P_0 \cdots P_{2d}} \in \mathbb{Z}[[t]],$$

but for the individual factors we only know

$$P_i(t) = \det(\mathrm{Id} - t\Phi_q \mid H^i_{\text{ét}}(\overline{X},\mathbb{Q}_\ell)) \in \mathbb{Z}_\ell[t].$$

But then in 1973 P. Deligne proved the following theorem, which was anticipated or conjectured by A. Weil in [We3].

Theorem 10.4.2. (**Weil conjectures):**
1) The polynomials

$$P_i(t) = \det(\mathrm{Id} - t\Phi_q \mid H^i(\overline{X},\mathbb{Q}_\ell)) \in \mathbb{Z}[t]$$

and they are independent of ℓ.

2) If we write

$$P_i(t) = \prod_{\nu=1}^{\nu=b_i} (1 - \omega_{i\nu} t)$$

then he reciprocal roots $\omega_{i\nu}$ are algebraic integers and

$$|\omega_{i\nu}| = q^{i/2} \quad \textit{for all } \nu.$$

We can not prove this theorem here. But we almost prove it for curves $C/\mathbb{F}_q$ and their Jacobians $J/\mathbb{F}_q$, except that we replace $H^i(\overline{C},\mathbb{Q}_\ell)$ by the ad hoc defined groups $\mathbb{Q}_\ell(0)$, $T_\ell(J) \otimes \mathbb{Q}_\ell(-1) = T_\ell(J)(-1)$, $\mathbb{Q}_\ell(-1)$ and for the Jacobian we assume that $H^\bullet(\bar{J},\mathbb{Z}_\ell) = \Lambda^\bullet(T_\ell(J)\otimes\mathbb{Q}_\ell(-1)))$. The Frobenius defines an endomorphism $\Phi_q \in \mathrm{End}(J)$ (See Theorem 10.2.7). We assume for simplicity that the line bundle Θ is defined over $\mathbb{F}_q$. We claim that $\Phi_q^*(\Theta) = \Theta^q$. To see this we describe Θ by a 1-cocycle $g_{ij} \in \mathcal{O}(U_i \cap U_j)^\times$. This means that $\Phi_q^*(\Theta)$ is given by the cocycle $g_{ij}^q \in \mathcal{O}(U_i \cap U_j)^\times$ and this is the claim. This implies for the Weil-pairing $< \ , \ >: T_\ell(J)(-1) \times T_\ell(J)(-1) \longrightarrow \mathbb{Q}_\ell(-1)$ the rule

$$< \Phi_q(\xi),\Phi_q(\eta) >= q^2 < \xi,\eta > \tag{10.35}$$

and hence

$$^t\Phi_q\Phi_q = [q]. \tag{10.36}$$

We look at the eigenvalues ω_i of Φ_q on $T_\ell(J)(-1) \otimes \bar{\mathbb{Q}}_\ell$, for any of them we have a generalized eigenspace $X(\omega_i) \subset T_\ell(J)(-1)\otimes\bar{\mathbb{Q}}_\ell$. It has an orthogonal complement $X(\omega_i)^t$ and the quotient $T_\ell(J)(-1)\otimes\mathbb{Q}_\ell/X(\omega_i)^t$ is a generalized eigenspace with eigenvalue q/ω_i. This implies:

Proposition 10.4.3. *If we list the eigenvalues with multiplicities* $\{\omega_1,\omega_2,\dots,\omega_{2g}\}$ *then there exists a permutation* σ *of the indices such that* $\omega_{\sigma(i)} = q/\omega_i$. *Especially 10.35 implies that* Φ_q *and* $^t\Phi_q$ *have the same eigenvalues (with multiplicities) and* $\mathrm{tr}(\Phi_q) = \mathrm{tr}(^t\Phi_q)$

For any pair m,n of integers we consider the endomorphisms $(m\Phi_q + n\,\mathrm{Id})$, the positivity of the Rosati involution yields

$$\mathrm{tr}(m^t\Phi_q + n\,\mathrm{Id})(m\Phi_q + n\,\mathrm{Id}) = 2gqm^2 + nm(\mathrm{tr}(\Phi_q +^t \Phi_q) + 2gn^2 \geq 0 \tag{10.37}$$

This implies

$$16qg^2 \geq 4\,\mathrm{tr}(\Phi_q)^2$$

and hence we get

$$4qg^2 \geq \sum_i^{2g} \omega_i$$

We can base change our abelian variety to any finite extension $\mathbb{F}_q \longrightarrow \mathbb{F}_{q^n}$ and then our inequality says

$$4q^n g^2 \geq \sum_i^{2g} \omega_i^n \text{ for all integers } n > 0 \tag{10.38}$$

and this implies $|\omega_i| \leq \sqrt{q}$ for all indices i. But then our previous proposition implies that we must have equality.
Since we have $J(\mathbb{F}_q) = \{x \in J(\bar{\mathbb{F}}_q) | \Phi_q(x) = x\}$ and since $\mathrm{Id} - \Phi_q$ is separable we know

$$\# J(\mathbb{F}_q) = \deg(\mathrm{Id} - \Phi_q) = \prod_{i=1}^{2g}(1 - \omega_i) = \sum_i^{2g} \mathrm{tr}(\Phi_q | \Lambda^i((T_\ell(J) \otimes \mathbb{Q}_\ell(-1)) \tag{10.39}$$

and this also verifies the trace formula for Jacobians and hence also for abelian varieties. If we accept the trace formula and our computations above then we also proved the above theorem for curves $C/\mathbb{F}_q$. We get a formula for the number of $\mathbb{F}_q$ rational points:

$$\#C(\mathbb{F}_q) = q - \sum_{i=1}^{2g} \omega_i + 1 \tag{10.40}$$

We already proved such a formula in section 9.7.7, and we also proved $|\omega_i| = \sqrt{q}$. But there the ω_i were the zeroes of the Zeta-function and we did not have the interpretation as eigenvalues of the Frobenius.
We consider an arbitrary scheme $X/\mathbb{F}_q$ of finite type. Recall that we have the action of the Galois group $\mathrm{Gal}(\bar{\mathbb{F}}_q/\mathbb{F}_q)$ on the cohomology groups $H^i(\bar{X},\mathbb{Z}_\ell)$. The Galois group $\mathrm{Gal}(\bar{\mathbb{F}}_q/\mathbb{F}_q)$ is a profinite cyclic group, which is topologically generated by the Frobenius automorphism $\sigma_q : x \mapsto x^q$. Hence σ_q induces an automorphism also called σ_q on the cohomology groups $H^i(\bar{X},\mathbb{Z}_\ell)$.

Proposition 10.4.4. *For the action on the cohomology groups we have*

$$\sigma_q^{-1} = \Phi_q$$

To see this we consider the scheme $\bar{X} = X \times \mathrm{Spec}(\bar{\mathbb{F}}_q)$ this is a scheme (of infinite type) over $\mathrm{Spec}(\mathbb{F}_q)$. The automorphism σ_q acts on the first factor, the base extension $\Phi_q \times_{\mathbb{F}_q} \bar{\mathbb{F}}_q : \bar{X} \longrightarrow \bar{X}$, is a morphism over $\mathrm{Spec}(\bar{\mathbb{F}}_q)$. It is clear that the composition $\sigma_q \circ \Phi_q : \bar{X} \longrightarrow \bar{X}$ is the identity on the underlying topological space of $\bar{X}$ and it acts on the structure sheaf by raising every section $f \in \mathcal{O}_{\bar{X}}(U)$ into its q-th power, therefore, it is not the identity. But since it acts as the identity on the underlying topological space implies that the composition $\sigma_q \circ \Phi_q$ induces the identity on the cohomology.

The global case

We consider a smooth, projective and absolutely irreducible scheme $X/\mathrm{Spec}(K)$ over a number field K. It is clear that we can extend $X/\mathrm{Spec}(K)$ to a smooth projective scheme $\mathcal{X}/U$ where U is a non empty open subset of $\mathrm{Spec}(\mathcal{O}_K)$. For any $\mathfrak{p} \in U$ we have a finite residue field $k(\mathfrak{p}) = \mathcal{O}_\mathfrak{p}/\mathfrak{p}$. We can consider the reduction $\mathcal{X} \times_U k(\mathfrak{p})$, this is a smooth projective scheme over $k(\mathfrak{p})$.
We pick a prime ℓ and consider the etale cohomology groups $H^i(X \times_K \bar{K},\mathbb{Q}_\ell)$. This is a finite dimensional vector space with an action of the Galois group $\mathrm{Gal}(\bar{K}/K)$ on it. If we recall that

$$H^i_{\text{ét}}(X \times_K \bar{K},\mathbb{Q}_\ell) = \varprojlim H^i_{\text{ét}}(X \times_K \bar{K},\mathbb{Z}/\ell^m\mathbb{Z}) \otimes \mathbb{Q}_\ell,$$

then it is clear that we get a tower of extensions $K_{\ell,m}/K$ such that the Galois action on $H^i_{\text{ét}}(X \times_K \bar{K},\mathbb{Z}/\ell^m\mathbb{Z})$ factors over $\mathrm{Gal}(K_{\ell,m}/K)$. Now it follows from the base change theorem in ℓ-adic cohomology that the extension $\mathrm{Gal}(K_{\ell,m}/K)$ is unramified over all those primes $\mathfrak{p} \in U$, which in addition do not divide ℓ.
The choice of a prime $\mathfrak{P}_m$, which lies above $\mathfrak{p}$, gives us a Frobenius element $\sigma_{\mathfrak{P}_m} \in \mathrm{Gal}((K_{\ell,m}/K)$, its conjugacy class does not depend on the chosen prime (VII.2.4). Now it is more or less obvious that we get a conjugacy class $\{\Phi_\mathfrak{P}\}$ of elements in $\mathrm{Gal}(K_\ell/K) = \varprojlim \mathrm{Gal}(K_{\ell,m}/K)$. Hence it makes sense to speak of the characteristic polynomial

$$\det(\mathrm{Id} - \sigma_{\mathfrak{P}}^{-1}T | H^i_{\text{ét}}(X \times_K \bar{K}, \mathbb{Q}_\ell)).$$

Now we apply the proper base change theorem we see

$$\det(\mathrm{Id} - \sigma_{\mathfrak{P}}^{-1}T | H^i_{\text{ét}}(X \times_K \bar{K}, \mathbb{Q}_\ell)) = \det(\mathrm{Id} - \Phi\mathfrak{p}T | H^i_{\text{ét}}(\mathcal{X} \times_U \overline{k(\mathfrak{p})}, \mathbb{Q}_\ell)) \in \mathbb{Z}[T]$$

and if we apply Deligne's theorem then we see that this characteristic polynomial does not depend on ℓ.

Such a system of Galois modules (ℓ varying) is called a compatible system of Galois modules. This allows us to define the L-function of the smooth projective scheme $\mathcal{X}/U$: At any prime $\mathfrak{p} \in U$ we define $N(\mathfrak{p}) = \#k(\mathfrak{p})$ and we substitute $N(\mathfrak{p})^{-s}$ for T. Then

$$L(H^i_{\text{ét}}(\mathcal{X}), s) = \prod_{\mathfrak{p} \in U} \frac{1}{\det(\mathrm{Id} - \Phi_{\mathfrak{p}} N(\mathfrak{p})^{-s} | H^i_{\text{ét}}(\mathcal{X} \times_U \overline{k(\mathfrak{p})}, \mathbb{Q}_\ell))}.$$

Deligne's theorem implies that the denominator is a polynomial in $N(\mathfrak{p})^{-s}$ with coefficients in $\mathbb{Z}$ and this polynomial does not depend on ℓ. Hence we can view this denominator as a holomorphic function in the variable s. The same theorem provides estimates for the eigenvalues of $\Phi_{\mathfrak{p}}$ they are of absolute value $N(\mathfrak{p})^{\frac{i}{2}}$. Hence it is easy to see that this infinite product converges for $\mathrm{Re}\,(s) > \frac{i}{2} + 1$ and defines a holomorphic function in this halfspace.

We say that $L_{\mathfrak{p}}(H^i_{\text{ét}}(\mathcal{X}), s) = \det(\mathrm{Id} - \Phi_{\mathfrak{p}} N(\mathfrak{p})^{-s} | H^i_{\text{ét}}(\mathcal{X} \times_U \overline{k(\mathfrak{p})}, \mathbb{Q}_\ell))^{-1}$ is the local Euler factor of $\mathcal{X}/U$ at $\mathfrak{p}$. We observe that for $\mathfrak{p} \in U$ the inertia group $I_{\mathfrak{P}} \subset K_\ell$ (see 7.3.12) is trivial and hence the local local Euler factor can be defined without reference to the smooth model $\mathcal{X}/U$. For $\mathfrak{p} \notin U$ we may define

$$L_{\mathfrak{p}}(H^i_{\text{ét}}(X), s) = \det(\mathrm{Id} - \sigma_{\mathfrak{p}} N(\mathfrak{p})^{-s} | H^i_{\text{ét}}(\bar{X}, \mathbb{Q}_\ell)^{I_{\mathfrak{P}}})^{-1}.$$

Here we have to assume that these Euler factors are in $\mathbb{Z}[N(\mathfrak{p})^{-s}]$ and are independent of ℓ, this is not known in general. It is also possible to define Euler factors at the infinite places, they are products of Γ- factors depending on the Hodge numbers $h^{p,q}$ with $p+q = i$. (See [De2]) Putting all this together we can define a global L-function

$$L(H^i_{\text{ét}}(X), s) = L_\infty(H^i_{\text{ét}}(X), s) \prod_{\mathfrak{p}} L_{\mathfrak{p}}(H^i_{\text{ét}}(X), s) = L_\infty(H^i_{\text{ét}}(X), s) L_{\text{fin}}(H^i_{\text{ét}}(X), s).$$

We resume the discussion from the last pages of Volume I. These cohomology groups $H^i_{\text{ét}}(\bar{X}, \mathbb{Q}_\ell)$ together with their structure as modules for the Galois group belong to the most interesting and most fascinating objects of study in the field which is called arithmetic algebraic geometry. Here X should be a smooth projective scheme over $\mathbb{F}_q$ or over a number field K or even more generally a field which is finitely generated over the prime field in it. The bare cohomology groups, or for instance the Betti numbers alone, contain little information on X. If for instance, X is a curve then they merely give us the genus, but if we add the Galois-module structure then the amount of information grows considerably. For instance, if we have two such curves $X_1/K, X_2/K$ we may be able to rule out the existence of non trivial morphisms $f : X_1/K \longrightarrow X_2/K$ by looking at the Galois-module structure of $H^1_{\text{ét}}(\bar{X}_1, \mathbb{Z}_\ell)$ and $H^1_{\text{ét}}(\bar{X}_2, \mathbb{Z}_\ell)$.

This kind of question is related to the **Tate conjecture**. We look at the cohomology in even degree $2r$. In item 7) above we explained the cycle map $A^r(\bar{X}) \longrightarrow H^{2r}_{\text{ét}}(\bar{X},\mathbb{Q}_\ell)$ and said that the absolutely irreducible cycles yield a copy $\mathbb{Q}_\ell(-r) \subset H^{2r}_{\text{ét}}(\bar{X},\mathbb{Q}_\ell)$ where $\mathbb{Q}_\ell(-r)$. The Tate conjecture asserts

After a suitable finite extension K'/K the image of the cycle map $A^r(\bar{X}) \longrightarrow H^{2r}_{\acute{e}t}(\bar{X},\mathbb{Q}_\ell)$ is the maximal $\mathrm{Gal}(\bar{K}/K')$ *subspace of the form $\mathbb{Q}_\ell(-r)^m$, this subspace is a direct summand in $H^{2r}_{\acute{e}t}(\bar{X},\mathbb{Q}_\ell)$.*

This conjecture is proved in very few cases. If $r = 1$ then it amounts to the determination of the image of $NS(X) \longrightarrow H^2_{\text{ét}}(\bar{X},\mathbb{Q}_\ell)$. Even in this special case the conjecture is not known for surfaces over $\mathbb{F}_q$.
The Tate conjecture has a compagnon for smooth projective algebraic varieties over $\mathbb{C}$. If $X/\mathbb{C}$ is smooth projective then $X(\mathbb{C})$. In this case $X(\mathbb{C})$ has the structure of a Kähler manifold and we have the Hodge decomposition (See Vol. I, Thm. 4.11.15)

$$H^{2r}(X(\mathbb{C}),\mathbb{Q}) \otimes \mathbb{C} = H^{2r}(X(\mathbb{C}),\mathbb{C}) = \bigoplus_{p+q=2r} H^{p,q}(X(\mathbb{C}),\mathbb{C}).$$

Then the **Hodge conjecture** asserts

$$\mathrm{im}(A^r(X) \longrightarrow H^{2r}(X(\mathbb{C}),\mathbb{C}) = H^{2r}(X(\mathbb{C}),\mathbb{Q}) \cap H^{r,r}((X(\mathbb{C}),\mathbb{C}),$$

this conjecture is actually much older than the Tate conjecture. It is proved for $r = 1$ for any $X/\mathbb{C}$.
If $X/K = A/K$ is an abelian variety, then we can use the relationship between the Neron-Severi group and the ring of endomorphisms $\mathrm{End}(A)$, which we discussed in 10.3. We have the ℓ adic representation $j : \mathrm{End}(A) \otimes \mathbb{Z}_\ell \hookrightarrow \mathrm{End}(T_\ell(A))$. The action of the Galois group on $T_\ell(A)$ induces an action of the Galois group on $\mathrm{End}(T_\ell(A)$ clearly j sends $\mathrm{End}(A)$ into the Galois invariants. From this it follows easily that the Tate conjecture for the $H^2_{\text{ét}}(\bar{A},\mathbb{Q}_\ell)$ is a consequence of the following assertion

$$j : \mathrm{End}(A) \otimes \mathbb{Z}_\ell \xrightarrow{\sim} \mathrm{End}(T_\ell(A))^{\mathrm{Gal}(\bar{K}/K)}.$$

This last assertion has been proved by Tate in the case where K is a finite field (See [Ta]) and it is a celebrated theorem of Faltings ([Fa]) if K is a finitely generated extension of $\mathbb{Q}$. As already mentioned at the end of Volume I this implies the Mordell conjecture.

Another circle of problems concerns the deeper understanding of the L-functions $L(H^i_{\text{ét}}(\bar{X}),s)$ in the case where K is a number field. For the following we refer to [De2] .
It is conjectured that this $L-$ functions has a meromorphic continuation into the entire complex plane and satisfy a functional equation. Let d the dimension of X, we assume that X is absolutely irreducible. Then one expects a functional equation of the form

$$L(H^i_{\text{ét}}(X),s) = \epsilon(H^i_{\text{ét}}(X),s)L(H^{2d-i}_{\text{ét}}(X),2d-1-s).$$

where the so called ϵ-factors are product of local factors $\epsilon_{\mathfrak{p}}(H^i_{\text{ét}}(X),s) = N(\mathfrak{p})^{a_{\mathfrak{p}}s}$ for $\mathfrak{p} \notin U$ and a constant.
It is also expected that the finite part $L_{\text{fin}}(H^i_{\text{ét}}(X),s)$ has holomorphic continuation if i is odd and if $i = 2r$ then the only poles are at $s = r + 1$ and the order of the pole is the number m in the formulation of the Tate conjecture.

If $X/\mathbb{Q} = \mathbb{P}^n/\mathbb{Q}$ then it is clear from 3) above that $L_{\text{fin}}(H^{2i}_{\text{ét}}(X),s) = \zeta(i+s)$, where of course $\zeta(s)$ is the Riemann ζ function. In this case all the conjectures are true.

We can define a larger class of objects, to which we can attach L-functions. These are so called (pure) motives. If X/K is smooth projective and irreducible then we may consider correspondences $T \subset X \times_K X$, this are cycles of codimension $\dim(X)$. They induce endomorphisms $p_T^\bullet : H^\bullet_{\text{ét}}(\bar{X},\mathbb{Q}_\ell) \longrightarrow H^i_{\text{ét}}(\bar{X},\mathbb{Q}_\ell)$. If such an endomorphism of the cohomology turns out to be an idempotent, then we consider the pair $(X,p_T) = M$ and call it a motive. Whatever this object is, it has ℓ-adic cohomology groups namely

$$H^\bullet_{\text{ét}}(M) = \{c \in H^\bullet_{\text{ét}}(\bar{X},\mathbb{Q}_\ell) \mid p_T(c) = c\}.$$

Obviously we can attach an L-function to these motives M and we get a product decomposition

$$L(H^i_{\text{ét}}(X),s) = L(H^i_{\text{ét}}(M),s)L(H^i_{\text{ét}}(M'),s),$$

where of course $M' = (X,1-p_T)$.
In [De1] Deligne has formulated conjectures concerning special values of these L-functions attached to motives. At certain integer arguments, which he calls critical, these values divided by a suitable period are rational numbers. We do not give a precise statement here, we rather refer to his article. But it is certainly worth to mention that in the case $X = \mathbb{P}^0/\mathbb{Q}$, i.e. $X = \text{Spec}(\mathbb{Q})$ the conjecture says that for any integer $m > 0$

$$L_{\text{fin}}(X,2m) = \zeta(2m) = \pi^{2m} \times \text{rational number}$$

and this was known to L. Euler.

At the present time we have only one method to gain some insight into the mysteries of these L-function. This method is based on the fact that there exists a second class of L-functions namely the automorphic (or modular) L-functions. These L-functions are also products over all places of the algebraic number field

$$L(\pi,s,r) = \prod_v L(\pi_v,s,r),$$

where $\pi = \otimes\pi_v$ is an automorphic form on a reductive group G/K, and r is a representation of the Langlands dual group ${}^LG(\mathbb{C})$. (We refer to [Bo] for further explanation). For almost all finite primes $\mathfrak{p}$ the local component $\pi_\mathfrak{p}$ is determined by its Satake-parameter $\lambda(\pi_\mathfrak{p})$ and this Satake-parameter is a semi-simple conjugacy class in ${}^LG(\mathbb{C})$. This allows us to write down the local Euler factor of the automorphic L-function at these places

$$L(\pi_\mathfrak{p},s,r) = \det(\text{Id} - r(\lambda(\pi_\mathfrak{p}))N(\mathfrak{p})^{-s})^{-1}.$$

In this generality these L-functions have been introduced by Langlands, the origin of this concept goes back to Hecke. He attached an L-function to any holomorphic modular cusp form f of weight k for $\text{SL}(2,\mathbb{Z})$. If this cusp form is an eigenform for the Hecke operators T_p, i.e. $T_p(f) = a_p f$, then this L-function is given by the infinite product

$$L(f,s) = \frac{\Gamma(s)}{(2\pi)^s} \prod_p \frac{1}{1 - a_p p^{-s} + p^{k-1-2s}}.$$

(See Zagier's exposition in [1-2-3].) In this case the reductive group is $\mathrm{GL}_2/\mathbb{Q}$ and the Langlands dual group is $\mathrm{GL}_2(\mathbb{C})$. To our cusp form corresponds an automorphic form π and if we write $1-a_p p^{-s}+p^{k-1-2s} = (1-\alpha_p p^{-s})(1-\beta_p p^{-s})$, i.e. $\alpha_p+\beta_p = a_p, \alpha_p\beta_p = p^{k-1}$ then the Satake parameter at p is given by the matrix

$$\lambda(\pi_p) = \begin{pmatrix} \alpha_p & 0 \\ 0 & \beta_p \end{pmatrix}$$

The argument r is suppressed, since in this case r is the tautological representation of $\mathrm{GL}_2(\mathbb{C})$.
Now we can formulate a very challenging questions:

What is the relationship between automorphic and and motivic L-functions? Is every motivic L- function equal to a automorphic L-function? Which modular L-functions are motivic?

It is conjectured that there is a very strong connection between these two kinds of L-function. These conjectures are embedded in a much larger network of conjectures, which run under the name Langlands program or Langlands philosophy. The above questions are motivated by abelian class field theory and the Artin reciprocity law and the Shimura-Taniyama -Weil conjecture, which we discuss next.

The most spectacular result in this direction is the proof Shimura-Taniyama -Weil conjecture by Wiles-Taylor. This conjecture asserts that any elliptic curve $E/\mathbb{Q}$ is *modular* and this means that there exists a holomorphic cusp-eigenform f of weight 2 with rational eigenvalues such that we have an equality of L-functions

$$L(H^1_{\text{ét}}(\bar{E}),s)) = L(f,s).$$

The original proof of Wiles-Taylor made some assumptions on the reduction behavior of $E/\mathbb{Q}$, these assumptions have been removed in a paper by Christophe Breuil, Brian Conrad, Fred Diamond and Richard Taylor. It is common knowledge in the mathematical community that this result implies Fermat's last theorem.
I plan to write a third volume of this book, its title will be "Cohomology of arithmetic groups". In this book I will come back to the questions raised above. A preliminary version of this book exists on my home page http://www.math.uni-bonn.de/people/harder. There is a folder buch, which contains several .pdf files chap2.pdf-chap.6.pdf. It does not have a chapter I because the chapter I of Vol. III is essentially chapter I-IV of Volume I of this book. This text will undergo some changes but I will not remove it from my home-page before this Volume III goes into print.

The degenerating family of elliptic curves

We come back to the discussion in Vol. I Chap. V, 5.2.8. We start from the ring $R = \mathbb{Z}[\frac{1}{2}]$, we consider the power series ring $R[[q]]$ and the ring of Laurent series $R[[q]][\frac{1}{q}]$. Inside this ring we have the power series

$$\lambda(q) = 2 + 64q + 512q^2 + 2818q^3 + \dots,$$

which is explicitly written down in formula (5.164). We also consider the polynomial ring $R[t]$ and map it to $R[[q]]$ by sending $t \longrightarrow \lambda(q) - 2 = 64q + 512q^2 \cdots$ (See Vol. I p.262, there is a misprint, we have to insert a $+$ between the 2 and the 64). Again we consider the curve

$$\begin{array}{ccc} \mathcal{E} & \hookrightarrow & \mathrm{Spec}(R[t,t^{-1}]) \times_{\mathrm{Spec}(R)} \mathbb{P}^2 \\ & \searrow & \downarrow \\ & & \mathrm{Spec}(R[t,t^{-1}]) \end{array} \quad \hookrightarrow \quad \begin{array}{ccc} \bar{\mathcal{E}} & \hookrightarrow & \mathrm{Spec}(R[t]) \times_{\mathrm{Spec}(R]} \mathbb{P}^2 \\ & \searrow & \downarrow \\ & & \mathrm{Spec}(R[t]), \end{array}$$

which was defined by the equation

$$y^2 z = x^3 - (2+t)x^2 z + xz^2.$$

Its affine part lies in the affine plane $z \neq 0$.We can easily compute its j invariant

$$j = 256 \frac{(1+4t+t^2)^3}{t(t+4)}$$

We are interested to understand what happens locally at $t = 0$. We invert in addition $t+4$, then the restriction $\mathcal{E}' \longrightarrow \mathrm{Spec}(R[t,t^{-1},\frac{1}{t+4}])$ is an elliptic curve, but its extension $\bar{\mathcal{E}}' \longrightarrow \mathrm{Spec}(R[t,\frac{1}{t+4}])$ is not an elliptic curve anymore. The fiber $\bar{\mathcal{E}}'_0$ over $t = 0$ is given by the equation

$$y^2 z = x^3 - 2x^2 + x = xz(x-z)^2,$$

the point $(x,y,z) = (1,0,1)$ is singular, it is easy to see that is an ordinary double point. We can find a morphism $\psi : \mathbb{P}^1/R \longrightarrow \bar{\mathcal{E}}'_0/\mathrm{Spec}(R)$ and two "disjoint" points $P,Q \in \mathbb{P}^1(R)$ such that ψ induces an isomorphism

$$\psi : \mathbb{P}^1 \setminus \{P,Q\} \xrightarrow{\sim} \bar{\mathcal{E}}'_0 \setminus \{(1,0,1)\}$$

and maps P,Q to $(1,0,1)$. We say that the fiber $\bar{\mathcal{E}}'_0/\mathrm{Spec}(R)$ is a $\mathbb{P}^1$ with two points identified to a double point. At the point $t = 0$ the family of elliptic curve "degenerates" in a very specific way into a singular curve.

We may also do something else: We have the group scheme structure on $\mathcal{E}' \longrightarrow R[t,t^{-1},\frac{1}{t+4}]$, obviously this group scheme structure does not extend to a group scheme structure on $\bar{\mathcal{E}}' \longrightarrow R[t,\frac{1}{t+4}]$. But if we remove the singular point $(1,0,1) \in \mathcal{E}'_0$ from the fiber, then it is easy to see that the group scheme structure of $\mathcal{E}$ extends to a group scheme structure on

$$\begin{array}{ccc} \bar{\mathcal{E}} \setminus \{(1,0,1)\} & \hookrightarrow & \mathrm{Spec}(R[t]) \times_{\mathrm{Spec}(R]} \mathbb{P}^2 \\ & \searrow & \downarrow \\ & & \mathrm{Spec}(R[t]), \end{array}$$

and the fiber at $t = 0$ is $\mathbb{G}_m/\mathrm{Spec}(R)$.

Hence we learn that we have to pay a price: If we want to extend the projective scheme $\mathcal{E} \longrightarrow \mathrm{Spec}(R[t,t^{-1},\frac{1}{t+4}])$ to a scheme over $\mathrm{Spec}(R[t,\frac{1}{t+4}])$, then we can extend it to a projective scheme, but we get a singular fiber (with a very mild singularity), or we extend it to a group scheme and then it is not projective anymore.

We now assume that we have chosen a homomorphism $\mathbb{Z}[\frac{1}{2}] \longrightarrow k$ and we consider the base change of our curve

$$\mathcal{E} \times_{\mathrm{Spec}(\mathbb{Z}[\frac{1}{2}][t,\frac{1}{t},\frac{1}{t+4}])} \times \mathrm{Spec}(k[t,\frac{1}{t},\frac{1}{t+4}])$$

let us call it $\mathcal{E}$ again. If now n is coprime to the characteristic of our field k, then we know that the kernel $\mathcal{E}[n] = n^{-1}(0)$ is a finite etale covering of $\mathrm{Spec}(k[t,\frac{1}{t},\frac{1}{t+4}])$. We can find a normal finite connected covering $U_n \longrightarrow \mathrm{Spec}(k[t,\frac{1}{t},\frac{1}{t+4}])$ such that

$$\mathcal{E}[n] \times_{\mathrm{Spec}(k[t,\frac{1}{t},\frac{1}{t+4}])} U_n = \underline{(\mathbb{Z}/n\mathbb{Z})^2}_U$$

i.e. over this covering it becomes a constant sheaf. Then we get a homomorphism

$$\mathrm{Gal}(U_n/\mathrm{Spec}(k[t,\frac{1}{t},\frac{1}{t+4}]) \longrightarrow \mathrm{GL}_2(\mathbb{Z}/n\mathbb{Z}),$$

from which we can reconstruct $\mathcal{E}[n]$ (See above). Now we encounter the fundamental problem to obtain some information concerning the image of the Galois group. To get this information we return to the variable q.

We make a second base change to $k((q))$. We can also base change the scheme $\mathcal{E}[n]$ to $\mathrm{Spec}(k((q))$ and we get a separable extension $K/k((q))$, over which $\mathcal{E}[n] \times K$ becomes trivial. In Vol. I we wrote down sections in $\mathcal{E}(R[[q]])$ (formula 5.180)). This provides a homomorphism

$$(k[[q]][\frac{1}{q}])^\times / < q > \longrightarrow \mathcal{E}(k[[q]])$$

and a variant of a theorem of Tate asserts that this is even an isomorphism. Let us assume that n is odd. We can extend our field to $K = k((q))[q^{1/n},\zeta_n,]$ where ζ_n is a primitive n-th root of unity. We also extend our homomorphism to

$$K^\times \longrightarrow \mathcal{E}(K).$$

and this implies that the group of n-division points $\mathcal{E}[n](K)$ is the free $\mathbb{Z}/n\mathbb{Z}$-module generated by $q^{1/n},\zeta_n \mod < q >$.

We get an an exact sequence of Galois groups

$$1 \longrightarrow \mathrm{Gal}(K/k((q))[\zeta_n]) \longrightarrow \mathrm{Gal}(K/k((q))) \longrightarrow \mathrm{Gal}(k[\zeta_n]/k) \longrightarrow 1.$$

The Galois group acts on the generators by matrices

$$\sigma \mapsto \begin{pmatrix} 1 & u(\sigma) \\ 0 & \alpha(\sigma) \end{pmatrix}$$

where $\alpha : \mathrm{Gal}(k[\zeta_n]/k) \longrightarrow (\mathbb{Z}/n\mathbb{Z})^\times$ is the Tate character, which is defined by $\sigma(\zeta_n) = \zeta^{\alpha(\sigma)}$ and $\sigma(q^{1/n}) = q^{1/n}\zeta_n^{u(\sigma)}$.

It follows from elementary algebra that the homomorphism

$$\mathrm{Gal}(K/k((q))[\zeta_n]) \longrightarrow \{\begin{pmatrix} 1 & u(\sigma) \\ 0 & 1 \end{pmatrix} \mid u(\sigma) \in \mathbb{Z}/n\mathbb{Z}\} \xrightarrow{\sim} \mathbb{Z}/n\mathbb{Z}$$

is in fact an isomorphism.

Now we pick a prime ℓ, which should be different from the characteristic p of k and for convenience also different from 2. We choose $n = \ell^m$ where m runs over all integers. The projective limit

$$\varprojlim \mathcal{E}[\ell^m](\bar{k}((q))) = T_\ell(\mathcal{E}) \xrightarrow{\sim} \mathbb{Z}_\ell^2$$

is the so called Tate module of our curve. It is a module for the projective limit of Galois groups

$$\varprojlim \mathrm{Gal}(k((q))[q^{1/\ell^m}, \zeta_{\ell^m}]/k((q))) = \mathrm{Gal}(k((q))[q^{1/\ell^\infty}, \zeta_{\ell^\infty}]/k((q)))$$

we get a representation

$$\rho_{0,\mathcal{E}_\ell} : \mathrm{Gal}(k((q))[q^{1/\ell^\infty}, \zeta_{\ell^\infty}]/k((q))) \longrightarrow \mathrm{GL}_2(\mathbb{Z}_\ell) = Gl(\mathbb{Z}_\ell q^{1/\ell^\infty} \oplus \mathbb{Z}_\ell \zeta_{\ell^\infty}).$$

Our previous considerations imply:
The image of the subgroup $\mathrm{Gal}(k((q))[q^{1/\ell^\infty}, \zeta_{\ell^\infty}]/k((q)))[\zeta_{\ell^\infty}])$ *under the above representation is the group*

$$\{\begin{pmatrix} 1 & b \\ 0 & 1 \end{pmatrix} | b \in \mathbb{Z}_\ell\} = U(\mathbb{Z}_\ell).$$

Now we suggest to consider $\mathrm{Spec}(k[[q]])$ as an infinitesimally small disk around zero in $\mathrm{Spec}(k[t, \frac{1}{t}, \frac{1}{t+4}])$. As I explained above we have the projective system of finite, connected etale schemes $U_{\ell^m} \longrightarrow \mathrm{Spec}(k[t, \frac{1}{t}, \frac{1}{t+4}])$, which trivialize $\mathcal{E}[\ell^m]$ and we have the representation

$$\rho_{\mathcal{E}_\ell} : \varprojlim \mathrm{Gal}(U_{\ell^m} / \mathrm{Spec}(k[t, \frac{1}{t}, \frac{1}{t+4}]) \longrightarrow Gl(T_\ell(\mathcal{E})).$$

If we consider the restriction of these U_{ℓ^m} to $\mathrm{Spec}(k((q)))$ they decompose into connected components, if we pick one of these components, which we can identify to $\mathrm{Spec}(k((q))[q^{1/\ell^m}, \zeta_{\ell^m}])$, then we get an embedding

$$\varprojlim \mathrm{Gal}(\mathrm{Spec}(k((q))[q^{1/\ell^m}, \zeta_{\ell^m}]) / \mathrm{Spec}(k((q))) \hookrightarrow \varprojlim \mathrm{Gal}(U_{\ell^m} / \mathrm{Spec}(k[t, \frac{1}{t}, \frac{1}{t+4}])$$

and we conclude that the image of $\rho_{\mathcal{E}_\ell}$ contains the subgroup $N(\mathbb{Z}_\ell) = \{\begin{pmatrix} 1 & x \\ 0 & 1 \end{pmatrix} \mid x \in \mathbb{Z}_\ell\}$.
Now we need a fact classical result, which is called the irreducibility of the modular equation. Let us put $U = \mathrm{Spec}(k[t, \frac{1}{t}, \frac{1}{t+4}]))$. We choose $n = \ell$, i.e. $m = 1$

Theorem 10.4.5. *The image* $\rho_{\mathcal{E}_\ell}(\mathrm{Gal}(U_\ell/U)$ *in* $\mathrm{GL}_2(\mathbb{F}_\ell)$ *is not contained in*

$$B(\mathbb{F}_\ell) = \{\begin{pmatrix} t_1 & b \\ 0 & t_2 \end{pmatrix} \mid t_1, t_2 \in \mathbb{F}_\ell^\times, b \in \mathbb{F}_\ell\}$$

This is a consequence of Theorem 3 in [La], I. 5, in principle it is also in [Web] , Achter Abschnitt. For the case that the ground field k has positive characteristic it is Thm. 1 in [Ig].

The next thing we need is a simple lemma, which is easy to prove

Lemma 10.4.6. *If a subgroup H of* $\mathrm{GL}_2(\mathbb{Z}_\ell)$ *contains* $N(\mathbb{Z}_\ell)$ *and its reduction* $\mod \ell$ *is not contained in* $B(\mathbb{F}_\ell)$ *then this subgroup contains* $Sl_2(\mathbb{Z}_\ell)$.

A digression into representation theory: If we have any field k and a subgroup $H \subset \mathrm{GL}_n(k)$ then we may consider its *Zariski closure* $\bar{H}$: It is the smallest algebraic subgroup of GL_n/k, for which its group of k- rational points $\bar{H}(k)$ contains H. If for instance k is infinite then the Zariski closure of $Sl_n(k)$, (resp.)$\mathrm{GL}_n(k)$ is Sl_n/k(resp.) GL_n/k.
A representation $r : \mathrm{GL}_n/k \longrightarrow Gl(V)/k$ is r is a homomorphism between the two algebraic groups over k. Such a representation is called irreducible, if there is no non trivial subspace $W \subset V, W \neq (0)$, V which is invariant under GL_n/k (i.e. for any $k \longrightarrow A$ $W \otimes A \subset V \otimes A$ is invariant under $\mathrm{GL}_n(A)$.) The following is more or less clear by definition.

If $r : \mathrm{GL}_n/k \longrightarrow Gl(V)/k$ is irreducible and $H \subset \mathrm{GL}_n(k)$ is a subgroup. If the restriction from r to the algebraic group $\bar{H}/k$ is irreducible then V V is also irreducible under the abstract group H.

We apply this to the subgroup $\rho_{\mathcal{E}_\ell}(\varprojlim \mathrm{Gal}(U_{\ell^m}/U) \subset \mathrm{GL}_2(\mathbb{Q}_\ell)$. The Zariski closure of an open subgroup of $\mathrm{SL}_2(\mathbb{Q}_\ell)$ is $\mathrm{SL}_2/\mathbb{Q}_\ell$. Hence by our two observations above we conclude that the Zariski closure of the image of the Galois group contains $Sl_2/\mathbb{Q}_\ell$. We can say something else. If we pass to the geometric situation and consider the base extension $U \times_k \bar{k} \longrightarrow U$, then $U_{\ell^m} \times_k \bar{k} \longrightarrow U \times_k \bar{k}$ then $U_{\ell^m} \times_k \bar{k}$ will not be connected, the connected components are labelled by the primitive ℓ^m -roots of unity (the value of the Weil-pairing). Taking just one of the components, we get

$$\mathrm{Gal}((U_{\ell^m} \times_k \bar{k})^{(\nu)}/U \times_k \bar{k} = \mathrm{SL}_2(\mathbb{Z}/\ell^m\mathbb{Z}).$$

If we pass to the limit, then the set of connected components will be a profinite set, labelled by the choice of an ℓ^∞-th root of unity and we may say

$$\varprojlim(\mathrm{Gal}((U_{\ell^m} \times_k \bar{k})^{(\nu)}/U \times_k \bar{k}) = \mathrm{SL}_2(\mathbb{Z}_\ell) \tag{10.41}$$

The tensor product $T_{\mathcal{E}_\ell} \otimes \mathbb{Q}_\ell = T_\ell(\mathcal{E})_{\mathbb{Q}_\ell}$ is a $\mathrm{GL}_2(\mathbb{Q}_\ell)$ module, even better we can consider it as a module for $\mathrm{GL}_2/\mathbb{Q}_\ell$. The classical theory of representations tells us how r -fold tensor products

$$T_\ell(\mathcal{E})_{\mathbb{Q}_\ell} \otimes T_\ell(\mathcal{E})_{\mathbb{Q}_\ell} \otimes \cdots \otimes T_\ell(\mathcal{E})_{\mathbb{Q}_\ell} = T_\ell(\mathcal{E})_{\mathbb{Q}_\ell}^{\otimes r}$$

decompose into irreducibles. First of all we have that

$$\Lambda^2(T_\ell(\mathcal{E})_{\mathbb{Q}_\ell}) = \mathbb{Q}_\ell \otimes \det,$$

where this means that the group GL_2 acts via the determinant on $\mathbb{Q}_\ell$. Furthermore the symmetric tensors $\mathrm{Sym}^k(T_\ell(\mathcal{E})_{\mathbb{Q}_\ell})$ form an irreducible submodule in $(T_\ell(\mathcal{E})_{\mathbb{Q}_\ell})^{\otimes r}$, to which we can find a complement (complete reducibility of representations of GL_2). Finally we have

$$\mathrm{Sym}^r(T_\ell(\mathcal{E})_{\mathbb{Q}_\ell}) \otimes T_\ell(\mathcal{E})_{\mathbb{Q}_\ell} \xrightarrow{\sim} \mathrm{Sym}^{r+1}(T_\ell(\mathcal{E})_{\mathbb{Q}_\ell}) \oplus \mathrm{Sym}^{r-1}(T_\ell(\mathcal{E})_{\mathbb{Q}_\ell}) \otimes \det$$

where $\mathrm{Sym}^0(T_\ell(\mathcal{E})_{\mathbb{Q}_\ell}) = \mathbb{Q}_\ell$ is the module with trivial action.
It follows that

$$T_\ell(\mathcal{E})_{\mathbb{Q}_\ell}^{\otimes r} = \bigoplus_{\nu=0}^{\nu=[\frac{r}{2}]} (\mathrm{Sym}^{r-2\nu} \otimes \det{}^\nu))^{m(r,\nu)} \tag{10.42}$$

where the $m(r,\nu)$ are multiplicities.

The Weil conjectures for elliptic curves in the spirit of Weil I

In this last section we give another proof of the the Weil conjectures (see 9.7.7, 10.4.2) but only in the simplest case, namely for elliptic curves. The reason for doing this is, that this proof provides a toy model for the general strategy of Deligne ([Del]) to prove the Weil conjectures. The decisive point is that he puts the elliptic curves into a family such that we have large "monodromy"; then he applies the "Rankin method" to tensor powers of the sheaf of cohomology groups.

The family, which we will consider, is our elliptic curve $\mathcal{E} \xrightarrow{\pi} \mathrm{Spec}(k[t,\frac{1}{t},\frac{1}{t+4}])$ and we choose for k the finite field $\mathbb{F}_p$ with p elements, where $p \neq 2$. We have large "monodromy" because the image of the Galois action is as large as possible. (See 10.41).

$$T_\ell(\mathcal{E}) \otimes \mathbb{Q}_\ell(-1) = T_\ell(\mathcal{E})(-1) = R^1\pi_*(\mathbb{Q}_\ell)$$

and this is an $\ell-$ adic local system of rank 2 on $U = \mathrm{Spec}(k[t,\frac{1}{t},\frac{1}{t-4}])$. We have the Weil pairing

$$T_\ell(\mathcal{E})(-1) \times T_\ell(\mathcal{E})(-1) \longrightarrow \mathbb{Q}_\ell(-1) = R^2\pi_*(\mathbb{Q}_\ell).$$

The sheaf $R^2\pi_*(\mathbb{Q}_\ell)$ is geometrically trivial, i.e. if we base change to $U \times_{\mathbb{F}_q} \bar{\mathbb{F}}_q$ it becomes trivial. For any

$$\mathfrak{p} : \mathbb{F}_p[t,\frac{1}{t},\frac{1}{t-4}] \longrightarrow \mathbb{F}_q,$$

the (geometric) Frobenius $\Phi_\mathfrak{p}$ has two eigenvalues $\alpha_\mathfrak{p},\beta_\mathfrak{p}$, the sum $\alpha_\mathfrak{p} + \beta_\mathfrak{p} \in \mathbb{Z}$ and $\alpha_\mathfrak{p}\beta_\mathfrak{p} = N(\mathfrak{p}) = p^{f_\mathfrak{p}}$. This implies for our sheaf $\mathcal{F} = \mathrm{Sym}^{2k}(T_\ell(\mathcal{E})) \otimes \mathbb{Q}_\ell(-1)^{2k}$ that the trace of $\Phi_\mathfrak{p}$ on $\mathcal{F}_\mathfrak{p}$ is $(\alpha_\mathfrak{p}+\beta_\mathfrak{p})^{2k}$ and hence positive. We can define the cohomology groups with compact support $H^i_!(U \times_{\mathbb{F}_p} \bar{\mathbb{F}}_p,\mathcal{F}))$ (See [Del] [Arcata], 5).

Since our sheaf is a local system it vanishes for $i = 0$, it is complicated for $i = 1$ and for $i = 2$ it is the module of coinvariants as in the topological situation(See Vol. I, 4.8.5)

For this cohomology with compact supports we have a Grothendieck-Lefschetz fixed point formula. For this consider the function

$$Z_\mathcal{F}(t) = \frac{\det(\mathrm{Id} - \Phi_p t|\ H^1_!(U \times_{\mathbb{F}_p} \bar{\mathbb{F}}_p,\mathcal{F}))}{\det(\mathrm{Id} - \Phi_p t|\ H^2_c(U \times_{\mathbb{F}_p} \bar{\mathbb{F}}_p,\mathcal{F})},$$

and the fixed point formula says

$$t\frac{Z'_\mathcal{F}(t)}{Z_\mathcal{F}(t)} = \sum_n^\infty \sum_{\mathfrak{p}:f_\mathfrak{p}|n} \mathrm{tr}(\Phi_\mathfrak{p}|\mathcal{F}_\mathfrak{p})t^n = \sum_n^\infty (\sum_{\mathfrak{p}:f_\mathfrak{p}|n} (\alpha_\mathfrak{p}^{n/f_\mathfrak{p}} + \beta_\mathfrak{p}^{n/f_\mathfrak{p}})^{2k})t^n$$

We can compute the module of coinvariants for the action of $\mathrm{SL}_2(\mathbb{Z}_\ell)$ using the above formula 10.42. The only contribution comes from the term $\nu = k$. The coinvariants for the geometric fundamental group $\mathrm{SL}_2(\mathbb{Z}_\ell)$ form still a module for $\mathrm{Gal}(\bar{\mathbb{F}}_p/\mathbb{F}_p)$ and more precisely we have

$$\mathcal{F}_{\mathrm{SL}_2(\mathbb{Z}_\ell)} = \mathbb{Q}_\ell(-k)^{m(2k,k)}$$

i.e. the eigenvalues of the inverse Frobenius on these coinvariants are q^r. This implies that

$$H_c^2(U \times_{\mathbb{F}_p} \bar{\mathbb{F}}_p, \mathcal{F})) = H_c^2(U \times_{\mathbb{F}_p} \bar{\mathbb{F}}_p, \mathbb{Q}_\ell) \otimes \mathcal{F} = \mathbb{Q}_\ell(-k-1)^{m(2k,k)}.$$

Let $\beta_1, \dots \beta_t \in \bar{\mathbb{Q}}_\ell$ be the eigenvalues of Φ_q^{-1} on $H_c^1(U \times_{\mathbb{F}_p} \bar{\mathbb{F}}_p, \mathcal{F})$. We choose any embedding $\iota : \mathbb{Q}(\beta_1, \dots, \beta_t) \hookrightarrow \mathbb{C}$. Then we get

$$t\frac{Z'_{\mathcal{F}}(t)}{Z_{\mathcal{F}}(t)} = -\sum_i \frac{\iota(\beta_i)t}{1-\iota(\beta_i)t} + m(2k,k)\frac{p^{k+1}t}{1-p^{k+1}t} = \sum_n^{\infty} \Big(\sum_{\mathfrak{p}: f_{\mathfrak{p}} | n} (\alpha_{\mathfrak{p}}^{n/f_{\mathfrak{p}}} + \beta_{\mathfrak{p}}^{n/f_{\mathfrak{p}}})^{2k} \Big) t^n$$

The right hand side is a power series with positive coefficients, it has a radius of convergence r and since the coefficients are positive we must have a pole for $t = r$. This pole can only be one of the numbers $\iota(\beta_i^{-1})$ or p^{-k-1}. But since the poles are first order poles and since the value $r\frac{Z'_{\mathcal{F}}(r)}{Z_{\mathcal{F}}(r)} = +\infty$ the residue of the pole must be positive and hence we must have $r = p^{-k-1}$. But then any of the formulae for the radius of convergence implies that for any $\mathfrak{p} : \mathbb{F}_p[t, \frac{1}{t}, \frac{1}{t-4}] \longrightarrow \mathbb{F}_q$ $(q = p^{d_{\mathfrak{p}}})$ we there exists a constant $C_{\mathfrak{p}}$ such that

$$\sum_{\mathfrak{p}: f_{\mathfrak{p}} | n} (\alpha_{\mathfrak{p}}^{n/f_{\mathfrak{p}}} + \beta_{\mathfrak{p}}^{n/f_{\mathfrak{p}}})^{2k} \leq C_{\mathfrak{p}} p^{n(k+1)}.$$

If we take the Weil pairing into account, i.e. $\alpha_{\mathfrak{p}}\beta_{\mathfrak{p}} = N\mathfrak{p} = p^{f_{\mathfrak{p}}}$ we obtain

$$|\alpha_{\mathfrak{p}}^{2k}|, |\beta_{\mathfrak{p}}^{2k}| \leq N\mathfrak{p}^{k+1}$$

If we pass to the limit $k \longrightarrow \infty$ then we can conclude

$$|\alpha_{\mathfrak{p}}| = |\beta_{\mathfrak{p}}| = N(\mathfrak{p})^{1/2}.$$

and hence we gave another proof of the Riemann hypothesis for elliptic curves. This proof contains the beautiful central idea of Deligne's proof of the Weil conjecture in [De1]. Of course our argument also gives that $|\iota(\beta_i)| \leq p^{k+1}$, independently of the chosen embedding. This implies that the β_i must be algebraic numbers. If one of them were transcendent over $\mathbb{Q}$ we could choose a ι which sends it to any non zero number in $\mathbb{C}$. But the above estimate is by far not the best, the actual truth is that for a given i we have

$$|\iota(\beta_i)| = p^{k+1/2} \text{ or } = 1.$$

This we can not prove here, because we have not yet put $H_!^1(U \times_{\mathbb{F}_p} \bar{\mathbb{F}}_p, \mathcal{F}))$ into a family of cohomology groups.

I think its a good moment to stop here. I hope that I gave a motivating introduction into the fascinating field of algebraic geometry and an incentive for further study and research in this area of mathematics.

Bibliography

[Ab] Abrashkin, V. A. *pp-divisible groups over ZZ.* Izv. Akad. Nauk SSSR Ser. Mat. 41 (1977), no. 5, 987-1007, 1199.

[Arb] Arbarello, E. *Weierstrass points and moduli of curves* Compositio Mathematica 29 (1974), 325-342.

[ACGH] Arbarello, E. ,Cornalba M., Griffiths P. , Harris,J *The Geometry of Algebraic Curves Volume I* Grundlehren der mathematischen Wissenschaften 267, Springer-Verlag 1985, Berlin, Heidelberg, New York Volume II just appeared (see preface)

[At-McD] Atiyah, M. F.; Macdonald, I. G. *Introduction to commutative algebra* Addison-Wesley Publishing Co., Reading, Mass.-London-Don Mills, Ont. 1969 ix+128 pp 1995

[Art1] Artin, E. *Beweis des allgemeinen Reziprozitätsgesetzes* Hamburger Abh. 5 (1927), p. 353-363

[Art2] Artin, E. *Quadratische Körper im Gebiete der höheren Kongruenzen, I, II* Math. Z. 1 ,207-246, 1-2 153-206

[Bo] Borel, A. *Automorphic LL-functions. Automorphic forms, representations and LL-functions* (Proc. Sympos. Pure Math., Oregon State Univ., Corvallis, Ore., 1977), Part 2, pp. 27-61, Proc. Sympos. Pure Math., XXXIII, Amer. Math. Soc., Providence, R.I., 1979.

[Bo-Se] Borel, Armand; Serre, Jean-Pierre *Le théoréme de Riemann-Roch.* Bull. Soc. Math. France 86 1958 97–136.

[Br] Brieskorn, E. *Über die Auflösung gewisser Singularitäten von holomorphen Abbildungen.* Math. Ann. 166 1966 76–102.

[Ca-Se] Cartan,H.; Serre, J.-P. *Un théoréme de finitude concernant les variétés analytiques compactes.* C. R. Acad. Sci. Paris 237, (1953). 128–130

[Ca-Ei] Cartan, H.; Eilenberg, S. *Homological algebra.* With an appendix by David A. Buchsbaum. Princeton Landmarks in Mathematics. Princeton University Press, Princeton NJ, 1999. xvi+390 pp.

[Ch] Chow, W.-L. *On compact complex analytic varieties.* Amer. J. Math. 71, (1949). 893–914

[De-We] Dedekind, R. Weber H. *Theorie der algebraischen Funktionen einer Veränderlichen,* J. Reine Angew. Math 92 (1882) p.181-290.

[De] Deligne, Pierre *La conjecture de Weil. I*
Inst. Hautes Études Sci. Publ. Math. No. 43 (1974), 273–307.

[De1] Deligne, P. *Cohomologie Étale*
Séminaire de Géométrie Algébrique
du Bois-Marie SGA 4 1/2, Lecture Notes in Mathematics, 815. Springer, Berlin, 1980. x+175 pp

[De2] Deligne, P. *Valeurs de fonctions L et périodes d'intégrales.* With an appendix by N. Koblitz and A. Ogus. Proc. Sympos. Pure Math., XXXIII, Automorphic forms, representations and L-functions (Proc. Sympos. Pure Math., Oregon State Univ., Corvallis, Ore., 1977), Part 2, pp. 313–346, Amer. Math. Soc., Providence, R.I., 1979.

[Ei] Eisenbud, D. *Commutative algebra. With a view toward algebraic geometry.* Graduate Texts in Mathematics, 150. Springer-Verlag, New York, xvi+785 pp.

[Fa] Faltings, G. *Endlichkeitssätze für abelsche Varietäten über Zahlkörpern.* Invent. Math. 73 (1983), no. 3, 349–366.

[Fa2] Faltings, G. *Lectures on the arithmetic Riemann-Roch theorem.*
Notes taken by Shouwu Zhang. Annals of Mathematics Studies, 127.
Princeton University Press, Princeton, NJ, 1992. x+100 pp.

[Fa-Ch] Faltings, G.; Chai, C.-L. *Degeneration of abelian varieties.* With an appendix by David Mumford.
Ergebnisse der Mathematik und ihrer Grenzgebiete (3) 22. Springer-Verlag, Berlin, 1990. xii+316 pp.

[F-K] Freitag, E; Kiehl, R. *Étale cohomology and the Weil conjecture.*
Translated from the German by Betty S. Waterhouse and William C. Waterhouse.
With an historical introduction by J. A. Dieudonné.
Ergebnisse der Mathematik und ihrer Grenzgebiete (3) [Results in Mathematics and Related Areas (3)], 13. Springer-Verlag, Berlin, 1988. xviii+317 pp.

[Fo] Fontaine, J.-M. *Il n'y a pas de variété abélienne sur* $\mathbb{Z}$.
Invent. Math. 81 (1985), no. 3, 515-538.

[Fu] Fulton, W., *Intersection theory.*
Ergebnisse der Mathematik und ihrer Grenzgebiete. 3. Folge. A Springer-Verlag, Berlin, 1998. xiv+470 pp.

[Go] Godement, R. *Topologie algébrique et théorie des faisceaux.*
Publications de l'Institut de Mathèmatique de l'Université de Strasbourg, XIII.Actualités Scientifiques et Industrielles, No. 1252. Hermann, Paris, 1973. viii+283 pp.

[Gr-EGA I] Grothendieck, A. *Éléments de géométrie algébrique. I. Le langage des schémas.*
Inst. Hautes Études Sci. Publ. Math. No. 4 1960 228 pp.

[Gr-EGA II] Grothendieck, A. *Éléments de géométrie algébrique. II. Étude globale élémentaire de quelques classes de morphismes.*
Inst. Hautes Études Sci. Publ. Math. No. 8 1961 222 pp.

[Gr-EGA III] Grothendieck, A. *Éléments de géométrie algébrique. III. Étude cohomologique des faisceaux cohérents. I.* Inst. Hautes Études Sci. Publ. Math. No. 11 1961 167 pp.

[Gr-RH] Grothendieck, *A. Sur une note de Mattuck-Tate*
J. Reine Angew. Math. 200 , 1958 , p.208-215

[Gr-Ch] Grothendieck, A. *La théorie des classes de Chern.*
Bull. Soc. Math. France 86 1958 137-154.

[Gr-Ha] Griffiths, Ph.; Harris, J. *Principles of Algebraic Geometry.*
John Wiley & Sons Inc, 1994. 814 pp.

[Has] Hasse, H. *Ober die Kongruenzzetafunktionen,*
Sitz.-Ber. Preufl. Aksd. d. Wins., Phya-Math. KI. 1934. XVII.

[Ha] Hartshorne, R. *Algebraic geometry.*
Graduate Texts in Mathematics, No. 52. Springer-Verlag, New York-Heidelberg, 1977. xvi+496 pp.

[Hi] Hirzebruch, F. *Topological methods in algebraic geometry.*
Translated from the German and Appendix One by R. L. E. Schwarzenberger. Classics in Mathematics. Springer-Verlag, Berlin, 1995. xii+234 pp. ISBN: 3-540-58663-6 57-02 (01A75 14-02)

[Ho] Hodge, W. V. D. *The Theory and Applications of Harmonic Integrals.*
Cambridge University Press, Cambridge,England; Macmillan Company, New York, 1941. ix+281 pp.

[Hu] Husemöller, D. *Elliptic Curves*
Graduate Texts in Mathematics, 111 Springer-Verlag, New York, 1987. xvi+350 pp.

[Ig] Igusa J.-I. *Kroneckerian Model of Fields of Elliptic Modular Functions*
American Journal of Mathematics, Vol. 81, No. 3 (Jul., 1959), pp. 561-577.

[Ja-Sch] Jantzen, J.C.; Schwermer, J. *Algebra*
Springer Lehrbuch , Berlin, 2005. xii+464 pp.

[K-W] Kiehl, R.; Weissauer, R. *Weil conjectures, perverse sheaves and ℓ-adic Fourier transform.*
Ergebnisse der Mathematik und ihrer Grenzgebiete. 3. Folge.
Springer-Verlag, Berlin, 2001. xii+375 pp.

[Kl] Kleiman, S. *The Picard scheme. Fundamental algebraic geometry, 235–321,*
Math. Surveys Monogr., 123, Amer. Math. Soc., Providence, RI, 2005.

[La] Lang, S. *Elliptic Functions*
Addison-Wesley Publishing Company, 1993

[La1] Lang, S. *Algebra*
3rd edition Addison-Wesley Publishing Company, 1993

[La2] Lang, *Serge Abelian varieties.* Interscience Tracts in Pure and Applied Mathematics. No. 7 Interscience Publishers, Inc., New York; Interscience Publishers Ltd., London 1959 xii+256 pp.

[L-MB] Laumon, G.; Moret-Bailly, L. *Champs algébriques. [Algebraic stacks]* Ergebnisse der Mathematik und ihrer Grenzgebiete.
3. Folge. 39. Springer-Verlag, Berlin, 2000. xii+208 pp.

[Ma] Matsumura, H. *Commutative ring theory*
Cambridge Studies in Advanced Mathematics, 8. Cambridge University Press, Cambridge, 1989. xiv+320 pp.

[Ma-Ta] Mattuck, Arthur; Tate, John *On the inequality of Castelnuovo-Severi.*
Abh. Math. Sem. Univ. Hamburg 22, 1958, 295-299.

[Mi] Milne, James S. *Étale cohomology* Princeton Mathematical Series, 33. Princeton University Press, Princeton, N.J., 1980. xiii+323 pp.

[Mi1] Milne, J. S. *Jacobian varieties*
Arithmetic geometry (Storrs, Conn., 1984), 167–212, Springer, New York, 1986.

[Mu1] Mumford, D. *Abelian varieties.*
Tata Institute of Fundamental Research Studies in Mathematics, No. 5 Published for the Tata Institute of Fundamental Research, Bombay; Oxford University Press, London 1970 viii+242 pp.

[Mu2] Mumford, D. *Tata lectures on theta. I-II-III*
Progress in Mathematics, 28, 43, 97. Birkhäuser Boston, Inc., Boston, MA

[M-F-K] Mumford, D.; Fogarty, J.; Kirwan, F. *Geometric invariant theory.*
Third edition. Ergebnisse der Mathematik und ihrer Grenzgebiete (2) [Results in Mathematics and Related Areas (2)], 34. Springer-Verlag, Berlin, 1994. xiv+292 pp. ISBN: 3-540-56963-4

[Neu] Neukirch, J. *Algebraic number theory.*
Grundlehren der Mathematischen Wissenschaften 322. Springer-Verlag, Berlin, 1999

[Rie] Riemann, B. *Gesammelte mathematische Werke*
Springer-Verlag, Berlin,Heidelberg, New York, 1990

[Ro] Rosenlicht , M *Some Basic Theorems on Algebraic Groups* American Journal of Mathematics, Vol. 78, No. 2 pp. 401-443

[Sch] *Richard Dedekind, 1831-1981*
Eine Würdigung zu seinem 150. Geburtstag. Edited by Scharlau.
Friedr. Vieweg und Sohn, Braunschweig, 1981.

[Scm] Schmidt, F. K. *Analytische Zahlentheorie in Körpern der Charakteristik p* Math. Zeitschr: 88 (1931), S. 1-32.

[Se1] Serre, J.-P. *Géométrie algébrique et géométrie analytique.* Ann. Inst. Fourier, Grenoble 6 (1955–1956), 1–42.

[Se2] Serre, J.-P. *Un théorème de dualité* Comment. Math. Helv. 29, (1955). 9–26

[Se3] Serre, J.-P. *Algèbre locale. Multiplicités.* Cours au Collège de France, 1957–1958, rédigé par Pierre Gabriel. Seconde édition, 1965. Lecture Notes in Mathematics, 11 Springer-Verlag, Berlin-New York 1965 vii+188 pp.

[Sl] Slodowy, P. *Simple singularities and simple algebraic groups.* Lecture Notes in Mathematics, 815. Springer, Berlin, 1980. x+175 pp

[Ta] Tate, J. *Endomorphisms of abelian varieties over finite fields.* I inventiones. Math. 2 1966p. 134-144.

[Web] Weber, H.. *Lehrbuch der Algebra* Dritter Band. Chelsea Publishing Companyx+731 pp.

[We1] Weil, A. *On the projective embedding of Abelian varieties.* Algebraic geometry and topology. A symposium in honor of S. Lefschetz, pp. 177–181. Princeton University Press, Princeton, N. J., 1957.

[We2] Weil, A. *Introduction à l'étude des variétés kählériennes.* Publications de l'Institut de Mathématique de l'Université de Nancago, VI. Actualités Sci. Ind. no. 1267 Hermann, Paris 1958 175 pp.

[We3] Weil, A. *On the Riemann hypothesis in functionfields.* Proc. Nat. Acad. Sci. U. S. A. 27, (1941). 345-347.

[We4] Weil, A. *Numbers of solutions of equations in finite fields.* Bull. Amer. Math. Soc. 55, (1949). 497-508.

[Wi] Wiles, A *Modular elliptic curves and Fermat's Last Theorem* Annals of Mathematics, 142 (1995), 443-551

[1-2-3] Bruinier, J.-H; van der Geer, G.; Harder, G.; Zagier, D. *The 1-2-3 of modular forms.* Lectures from the Summer School on Modular Forms and their Applications held in Nordfjordeid, June 2004. Edited by Kristian Ranestad. Universitext. Springer-Verlag, Berlin, 2008. x+266 pp.

Index